Exploring Mathematics with *Mathematica*®

Exploring Mathematics with *Mathematica*®

Dialogs Concerning Computers and Mathematics

Theodore W. Gray
Jerry Glynn

ADDISON-WESLEY PUBLISHING COMPANY
The Advanced Book Program
Redwood City, California • Menlo Park, California • Reading, Massachusetts
New York • Don Mills, Ontario • Wokingham, United Kingdom • Amsterdam
Bonn • Sydney • Singapore • Tokyo • Madrid • San Juan

Mathematica is a registered trademark of Wolfram Research Inc. Derive is a registered trademark of Soft Warehouse, Inc. Macintosh is a registered trademark of Apple Computer, Inc. MacRecorder is a registered trademark and SoundEdit is a trademark of Farallon Computing, Inc. Maple is a trademark of the University of Waterloo. Milo is a trademark of Paracomp, Inc. NeXT is a trademark of NeXT Computer, Inc. PageMaker is a registered trademark of Aldus Corporation. PostScript is a registered trademark of Adobe Systems Incorporated. Windows is a trademark and MicroSoft and MS-DOS are registered trademarks of MicroSoft Corporation. UNIX is a registered trademark of AT&T. All other product names are trademarks of their producers.

Mathematica is not associated with Mathematica, Inc., Mathematica Policy Research, Inc., or MathTech, Inc.

Publisher: *Allan M. Wylde*
Marketing Manager: *Laura Likely*
Production Manager: *Jan V. Benes*
Production Assistant: *Karl Matsumoto*
Cover Design: *Andre Kuzniarek*

Library of Congress Cataloging-in-Publication Data

Gray, Theodore W.
Exploring Mathematics with Mathematica/Theodore W. Gray, Jerry Glynn
p. cm.
"The Advanced Book Program."
1. Mathematica (computer program) 2. Mathematics–Data processing.
I. Glynn, Jerry. II Title.
QA76.95.G73 1991
510'.285'536–dc20
ISBN 0-201-52809-6 – ISBN 0-201-52818-5 (pbk.)

Typeset in Palatino and Courier 11pt by the authors, using the *Mathematica* Macintosh Front End, designed and written by Theodore W. Gray. Typeset on a Compugraphic 9600 at Wadley Graphix, Champaign, Illinois. Color separations were done on a Scitex prepress system and output on a Dolev imagesetter at Flying Color Graphics, Pontiac, Illinois.

ABCDEFGHIJ-DO-9543210

Table of Contents

Part Four: Adventures in Mathematics

Part Five: For Teachers and Students

Appendices

Index

Foreword

Jerry and Theo explain why they wrote this book, and thank those who helped them.

Dialog

Jerry: This is a book of dialogs. In the pages that follow, you, the reader, can hear how Theo and I explore and learn about mathematics through *Mathematica*. In this preface, we will address you, the reader, directly: After this, you will eavesdrop as we chat with each other.

Theo: Since Jerry has had a lot of experience explaining math to people (including me), I thought he might be able to explain *Mathematica* to others as well.

Jerry: *Mathematica* is a huge system, something like the Indian Ocean, and we sometimes forget that we don't have to use or understand all of it at once. Those who have heard about or used *Mathematica* might be unsure about what it all means. We don't know what it all means either, and that's why the title of the book starts with "Exploring".

Theo: I've been working on *Mathematica* so long I sometimes forget that there are people who haven't used it at all. Jerry is very good at taking a big thing, like all of mathematics, and making it human-sized.

Jerry: Theo is always going off to write one function to do everything automatically. For him, *Mathematica* is the most natural thing in the world. I make him slow down and tell me how it works.

Theo: We decided to write down our conversations, because we thought they might be a way to help others understand *Mathematica* and mathematics.

Jerry: We've chosen a wide range of problems. Some of these problems are old and well understood, others we had never heard of before. We found them all very interesting.

Theo: We try to show you how these problems can be studied and experimented with in unique ways made possible by the computer. Many of the things we discover about these problems cannot be understood, let alone discovered, without having a computer at your side. I think that's very important.

Jerry: The teaching of math will be changed in serious ways by these new technologies. Students will uncover ideas that are foreign to both student and teacher. Teachers will need to take a more experimental approach. It's best to face up to this fact early, and try not to worry about it too much.

Theo: Along the way you'll learn a lot about the *Mathematica* language. We will show you functions we think are *really* important. We'll also show you ways to do things that are surprising and wonderful. What we have written is a printed book, and it is also an electronic book. People who have *Mathematica* can start where the book leaves off and begin their own explorations. The full text of the book with all the functions, formulas, plots, animations, and sounds is included along with each printed copy in the form of a CD-ROM disk, in *Mathematica*'s Notebook format.

Jerry: We wanted to have a picture in the Foreword, so here's one from the chapter about iteration and recursion, two subjects that always seem to end up making pretty pictures:

Theo and **Jerry:** Like all books, this one would not have been possible without the help, cooperation, cudgeling, suggestions, criticisms, interest, and encouragement of many people.

For constant supervision, we would like to thank David Eisenman (who edited *the whole thing*), Joyce Glynn, Eva Gray (who is translating the book into German), and John Gray (who tried to inject a little rigor).

For important suggestions about problems to work on, and for solutions to problems we started but couldn't finish, we would like to thank Doug Stein, Dana Scott (who will, sadly, not find nearly enough of his wonderful ideas here–maybe next time), Jim Beauchamp, Wally Dodge, and John Lux (who made flip-books for us before they were fashionable).

For being visiting professors and generally explaining things we didn't understand, Dan Grayson and Jerry Keiper were invaluable.

At Addison-Wesley we thank Allan Wylde, who first said "CD-ROM" (and whose initials are, suspiciously, A-W), Jan Benes, Karl Masumoto, and Laura Likely whose production, marketing and promotional efforts probably helped you to buy this book.

For reading the book in draft form and making suggestions, we thank Janet Coursey, Sandra Dawson, Richard Gaylord, Joyce Mast, Steve Omohundro, Horacio Porta, Shawn Sheridan, Ben Schwartz, Jerry Uhl, Doug West, and Stephen Wolfram.

We thank Paul Wellin and his class at Sonoma State University for using early drafts of the electronic edition in class, and for free psychoanalysis of the authors.

At Apple computer we thank Rob Wolff for encouraging the CD-ROM concept and helping us with its execution.

At NeXT Computer we thank Richard Crandall, and David Grady for encouragement.

At Wolfram Research, Inc we thank Joe Grohens, John Bonadies, Vicky Scoma, Renice Wernette, and everyone else, for help producing this book.

We thank all the authors of *Mathematica*.

And finally, we would like to thank Galileo Galilei for inspiring us to write a dialog.

This Book and Its Technology

The new world of electronic books.

Dialog

Theo: Most books are made out of paper, and don't really cause much trouble in the technology department. You can read *Exploring Mathematics with Mathematica* as if it were a normal paper book, but it is much more than that.

Jerry: What makes this book more than just a paper book?

Theo: It's an electronic book. Every word, every picture, and every function in it is included on a CD-ROM disk in every copy of the book. The electronic edition is in the form of *Mathematica* Notebook documents. These Notebooks can be read with *Mathematica*, if you have it (Version 2.0 is best). The Notebooks can also be read with the public domain program *MathReader*, which is included on the CD-ROM.

Depending on the equipment people have, different uses are possible for the CD:

> People who have an ordinary home audio CD player can listen to audio tracks on the disk. We have written several chapters about sound, and all the sounds from those chapters are included on the disk in the standard music format. (Don't play the first track, since it is the data track!)
>
> People who own a Macintosh or MS-DOS-compatible CD-ROM drive can read the whole book on-line, using the included copy of *MathReader*. With *Math-Reader* you can read the text, see the graphics, and, most importantly, run the animations in the book. You can also hear the sounds from the sound chapters through your computer.
>
> People who own a copy of *Mathematica* and a CD-ROM drive will get the most benefit from the electronic edition. They can use and modify the functions and programs we describe in the book to do their own explorations. Since everything is on-line, there is no need to type in long programs.

Jerry: What about people who have *Mathematica*, but not a CD-ROM drive?

Theo: They can order the floppy disk version of the electronic edition (from MathWare, 217-384-3196). Alternately, they can find someone who has a drive (perhaps a library), and copy the files onto their own disks (for personal use only, of course).

Short of buying a CD-ROM drive, the next best thing is to copy all or most of the files from the CD onto a harddisk. This is relatively easy to do, assuming you have enough space (several tens of megabytes) available on a portable harddisk, and you are able to connect both your harddisk and the CD-ROM drive to the same computer (trivial on a Macintosh or NeXT computer, potentially difficult on a MS-DOS computer).

NeXT computer owners can copy the files onto an optical disk or floppy disks, if they can find someone who owns a NeXT CD-ROM drive.

Another possibility is to copy files from the CD-ROM onto ordinary floppy disks. There is a folder on the CD that includes all the interesting functions and programs in a compact form that fits on one floppy disk. The files containing the chapters of text are quite large. It's possible to copy one or two chapters onto a floppy disk, but probably not practical to copy the whole thing.

Jerry: So, you're saying people should really think about buying a CD-ROM drive?

Theo: They should certainly think about it. The prices are quite reasonable, and most models can also be used as to play audio CDs as well.

People who don't have *Mathematica* might want to think about getting it, also, although it is by no means necessary.

Jerry: We would like people to use the functions to do what they want, rather than what we want. Some people may want to modify our functions to suit their own needs. With *Mathematica* this is quite easy.

■ Kinds of Computers You Can Use

Theo: This electronic book can be read with a variety of computers (the same CD-ROM disk will work with most computers). Let's list these computers and their requirements (in all cases you need a compatible CD-ROM drive and the appropriate software to connect it to your computer).

Macintosh
You can read the files and view most animations on a 1MB Mac Plus or larger model. To view the color animations you will need 4MB, and a color Mac model. To carry out your own mathematical explorations, you will need a

copy of *Mathematica* and at least 4MB, preferably 8MB. Some of the chapters describe functions that take quite a while to execute, and are not recommended for Mac Plus or SE users. (These are labeled with warnings.)

MS-DOS
You need MicroSoft Windows Version 3.0, and a compatible computer. You need a 386 with at least 2MB Extended memory together with Windows virtual memory. Any graphics adapter supported by Windows will work, with VGA, Super VGA, or better recommended.

NeXT
People who bought their NeXT computers as educational users already own a copy of *Mathematica* (double-click `/NextApps/Mathematica.app`). Other NeXT computers include a copy of the Notebook reader (double-click `/NextApps/MathReader.app`).

UNIX
There is a directory containing all the interesting functions and commands we use in the book in plain-text format. These files can be read with any UNIX text editor, or read into *Mathematica* directly. When a UNIX-based Notebook reader becomes available, it will be able to read the Notebooks containing the chapters of the book. (In the meantime, those Notebooks can be read with an ordinary UNIX text editor, but they are not formatted appropriately for this kind of editor.)

■ Different Versions of *Mathematica*

Theo: This book is based on Version 2.0 of *Mathematica*. Some of the things we talk about (like sound) do not exist in earlier versions, but many things do. If you intend to use this book with an earlier version of *Mathematica*, be aware that some chapters won't work, and that some functions in other chapters won't work properly.

I recommend upgrading to Version 2.0 if you intend to interact with the electronic edition of this book. All the chapter Notebooks are Version 2.0 Notebooks. Although earlier versions are able to open them, there are significant problems. When one of the chapters is opened in an earlier version of *Mathematica*, it takes a long time, and some of the formatting information is lost.

If you haven't upgraded your copy of *Mathematica*, it may be better to use the copy of *MathReader* on the CD-ROM to look at the files, instead of your copy of *Mathematica*.

■ Differences Between the Printed and Electronic Editions

Theo: We've tried very hard to make the printed and electronic editions of *Exploring Mathematics with Mathematica* as similar as possible. The main way of assuring this was to print the book directly from *Mathematica* itself. What you are reading (if you are reading it on paper) is a direct printout of an electronic Notebook. No page-layout or other publishing software was used other than *Mathematica*.

Jerry: I remember in the early days of our planning for this book you wanted to have the book written in *Mathematica*, and I thought we should use *PageMaker* instead. My earlier experience in writing *Exploring Math from Algebra to Calculus with Derive* convinced me that *PageMaker* was the way to go.

Theo: *PageMaker* is fine as a way of producing printed books, but that was not our real goal. We wanted to make an *electronic* book. *PageMaker* documents can be read on-line just like *Mathematica* Notebooks, but they are not designed for or suitable for the things we had in mind.

Jerry: Such as evaluating functions, making graphs, and looking at animations. Now that I think of it, making changes and revisions to my earlier book was not a pleasant experience. Going back and forth between Derive and *PageMaker* was a slow process. It would be nice to have an electronic edition of my Derive book, but there's no suitable format available for it.

Theo: My feeling was that if there was no suitable format available for publishing electronic mathematical books, then it was about time we made one. Since I was working on Version 2.0 of *Mathematica* at the same time we were writing this book, I took the opportunity to make sure that *Mathematica* was such a format.

Jerry: You're saying that *Mathematica* before Version 2.0 was not good enough?

Theo: Not quite. In order to make our book work, I had to improve several aspects of the Notebook front end. It's still not perfect; I'm sure our next book will be able to take advantage of many more features.

Jerry: Despite our best intentions, there are a few differences between the printed and electronic editions, particularly in the area of animations, color, and sound.

Theo: It's hard to put an animation in a printed book. We suggested flip-books, but our publisher discouraged this idea.

Jerry: And probably a good thing, too. We settled on printing the frames of each animation in a two-dimensional array. In the electronic edition, only the first frame of each animation is visible. Double-clicking on this frame starts the animation. In the printed edition, all the frames are visible on one page.

Theo: It was not always practical to include the same number of frames in the printed edition as in the electronic edition (some animations have so many frames that they won't fit on one page). For this reason, you may notice some discrepancies between the animation commands and the pictures that you see.

We used the function `Do` to make the animations. This will work in any Notebook version of *Mathematica*. People using *Mathematica* on systems that don't support Notebooks can usually use the function `Animate` in place of `Do` (they take the same arguments). We thought it was better to use `Do`, to avoid introducing a new function just for animations.

Jerry: Another difference is that we sometimes use different numbers of frames in the animations. It makes no sense to print 50 slightly different frames of an animation, so we take bigger steps and print fewer frames.

Theo: Another area of difference is color. In the electronic edition, color graphics are handled the same as other graphics, and are right where they belong. Unfortunately, due to the economics of present-day printing technology, in the printed edition all the color graphics are concentrated in one special section.

Sounds also present a problem for the printed edition. We've included all the sounds from the book in normal audio format on the CD-ROM. You can play the CD-ROM in a normal audio CD player. Each sound in the book has a note under it indicating which track to listen to on the CD.

■ How To Use *MathReader*

Theo: *MathReader* is a small, public-domain program. It lets you read Notebooks and look at their graphics and animations, but it does not allow you to create or edit Notebooks, or execute mathematical functions. To do these things you need a full copy of *Mathematica*.

Notebooks (like the chapters in this book) are usually divided into sections. When you first open a Notebook, you are shown the headings for the sections, but not their contents. To open up a section, double-click its angled bracket, as shown here:

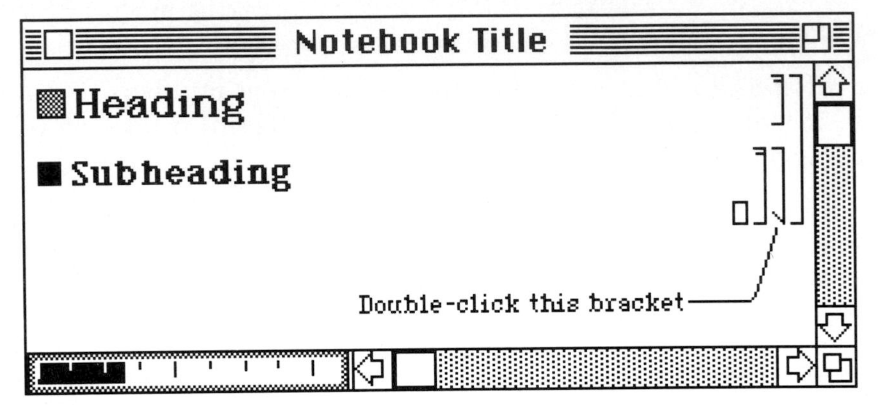

A Notebook consists of a sequence of "cells", each of which can contain text, graphics, or *Mathematica* input or output. Each cell is enclosed by a bracket (shown in blue on color systems) on the right-hand side of the window.

The cells in the Notebooks that make up the chapters of this book are arranged in sections. Each chapter contains one or more "days" of dialog, each of which has its own section.

The sections are indicated by additional brackets on the right-hand side of the window; the bracket for a section encloses the cells in that section.

A section of cells may be open or closed. When a section is open, all the cells in the section are displayed; when it is closed, only the first cell, which contains the section heading, is visible. The bracket of a closed section is angled at the bottom.

When you open a chapter of this book, you see the section-heading cells only, which form an outline of the contents of the chapter.

You can open a closed section by double-clicking the section's enclosing bracket (the one with an angle at the bottom). When you have a section of cells open, you can close it again by double-clicking the bracket that spans the group.

Jerry: What about graphics? How do people look at the animations we've created?

Theo: Graphics cells in Notebooks are like other cells, except they contain pictures instead of text. An animation is a sequence of graphics cells arranged into a section (you can recognize an animation by the extra section bracket it will have). To see the animation played out, double-click on the first frame (not on the cell bracket, on the picture itself).

You can control the speed and direction of an animation as it runs by clicking the palette of buttons that will appear at the bottom of the window. The first three buttons control the direction of the animation: Forwards, Back and Forth, and Backwards. The middle button is a pause button, and the last two buttons control the speed: Faster and Slower. (You can also type a number between 1 and 9 to set the speed.)

Chapter One
An Introduction

In which Jerry and Theo Introduce the Reader to Mathematica.

■ Dialog

Jerry: Many of us feel overwhelmed by *Mathematica*. Probably this is because our first attempt to learn about it is from the 992-page book, *Mathematica, a System for Doing Mathematics by Computer*. A lot of detail in that book is of no interest to most users, especially at the beginning. In this chapter we'll show you just what you need to get going. In later chapters we'll show you things so wonderful you won't mind having gone through this chapter.

Theo: Let's start with addition. I'll type in **5+7** in a *Mathematica* Notebook window:

```
5+7
12
```

Jerry: What is this "*Mathematica* Notebook window" stuff? Where are we typing? How did we get there?

Theo: If you are using *Mathematica* on a Macintosh or MS-DOS computer you should read the installation guide, copy the floppy disks onto your hard disk, and then double-click on the *Mathematica* icon.

If you are using a NeXT computer, read your manual to find out if *Mathematica* is already installed. If not, you can buy a copy from Wolfram Research, Inc.

If you are using a UNIX XWindows versions of the *Mathematica* Notebook Front End, you should probably talk to your system administrators.

In any case, after a little while a Notebook window will appear, ready for typing.

If you are using a version of *Mathematica* that does not support Notebooks (for example a plain UNIX, DOS, or VMS version) you will be typing into a terminal emulator window instead. Most of what you see in this book will work in this environment, although it may not look as pretty.

Jerry: So you're saying, all they have to do is double click on the icon, wait, and then start typing?

Theo: Right. Just start typing.

Jerry: After typing **5+7**, how do they get *Mathematica* to figure out **12**? They must type something special, or choose a menu command, or something....

Theo: Press Shift-Return (hold down the Shift key while pressing the Return key). That tells *Mathematica* to evaluate whatever you just finished typing. You can also press Enter instead of Shift-Return. I tend to use both. Many people try plain Return at first. This begins a new line, allowing you to type in multi-line expressions, but does not evaluate anything.

Jerry: OK, I get the idea. Type something, then press Shift-Return. So, from now on whenever we see an expression in boldface followed by a result in plain text, we can assume that someone pressed Shift-Return or Enter to get from one to the other.

Theo: Right. *Mathematica* uses **boldface** for all expressions you type in, and `plain text` for all its answers.

Jerry: Next step. Let's try some more commands. How about raising a number to a power?

Theo: *Mathematica* uses the "caret" symbol, **^**, for raising to a power, like this:

```
2^37
137438953472
```

Jerry: Am I supposed to believe that 2 to the 37th power is this 12 digit number? I would prefer 2 to the 3rd power, since I know that answer is supposed to be 8:

```
2^3
8
```

Theo: Satisfied? *Mathematica* also knows about a whole bunch of other functions, such as trigonometric functions. One of the weird things about *Mathematica* that tends to annoy people for a while is that you have to use square brackets and capital letters. For example:

```
Sin[1.2]
0.932039
```

Jerry: In other words, you're saying I can't type `sin 1.2` with no parentheses, or `sin(1.2)`, or `Sin(1.2)`, or `sin[1.2]`. I must type `Sin[1.2]`, exactly as you did. That seems like a real imposition.

Theo: Yes, you have to type `Sin[1.2]`, exactly. There are good reasons for both requirements, and we'll see why in later chapters. If you use one of the variations you suggested above, *Mathematica* will warn you that you are probably making a mistake. All of your variations are legal *Mathematica* input, but they don't mean what you want. (For example, `sin(1.2)` means the variable named `sin` multiplied by `1.2`.)

Jerry: OK, I'll live with the funny brackets for now. Can I use round parentheses for anything?

Theo: Yes, you can use round parentheses for changing the order in which things are evaluated, just as in regular mathematical notation. Consider the following two examples:

```
2^3 + 4
12
```

```
2^(3 + 4)
128
```

Jerry: I see. That looks fairly familiar. How do I know when to use capital letters and when not?

Theo: There's a simple rule: All built-in *Mathematica* functions, commands, and constants, like `Sin`, `Cos`, `Tan`, `Log`, `Print`, `Expand`, and `Catalan`, always start with a capital letter. Some functions have names which consist of two words, like `NestList`. In this case the first letter of each word is capitalized. One of the strangest consequences of this rule is that what most people call *e* (the base of the natural logarithms) and *i* (the square root of -1) are called `E` and `I` in *Mathematica*. The ratio between the circumference and the diameter of a circle is called `Pi`.

Jerry: How about some action? I'm willing to put up with weirdness if something interesting begins to happen. Enough talk. How about a picture?

Theo: We can use the **`Plot`** command to make a plot. Let's make a plot of the **`Sin`** function **`Sin[x]`**, with **`x`** running from **`0`** to **`2 Pi`**.

```
Plot[Sin[x], {x, 0, 2 Pi}];
```

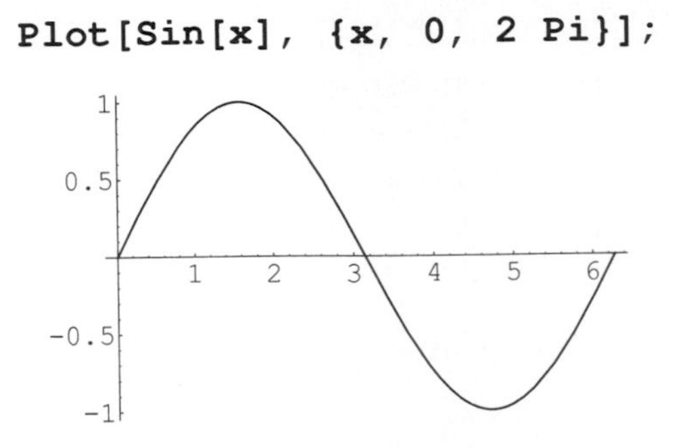

Jerry: Ah... now that's worth while. But wait a minute, you didn't tell me anything about curly brackets. You've got two curly brackets in there, and I want an explanation!

Theo: Curly brackets are your friends. Aside from square brackets, curly brackets are the most important things in *Mathematica*. You use curly brackets for making lists, and it just so happens that the second argument to the **`Plot`** command is a list that tells it what variable to use, and over what domain to plot.

Jerry: So, the first argument to the **`Plot`** command is the function you want to plot, and the second argument is a list of three things. First the variable you plot with, then the lower limit, and then the upper limit. How about the step size? How does it know how many points to plot?

Theo: It's quite clever about that. Instead of picking a fixed number of points to plot, it analyses how "curvy" the plot is and then adds as many points as it needs to make the line smooth.

Jerry: That's impressive. Does this graph consist of points, or is it points connected by lines?

Theo: It's points connected by lines. That way when you print the plot on a high resolution printer it still looks smooth.

Jerry: Can you show me the points that are calculated and then connected to make the plot?

Theo: Sure, we can draw a vertical line at each calculated point:

```
test = Plot[Sin[x], {x, 0, 2 Pi},
            DisplayFunction -> Identity];
Show[
    Graphics[{
        Thickness[0.001],
        Map[
            Line[{{#[[1]], 0}, #}]&,
            Nest[First, test, 4]
        ]
    }],
    Axes->Automatic
];
```

And I propose that we put off–for about 11 chapters–explaining the code that I just used to answer your question.

Jerry: Fine, but that graph is lovely. I see that there are lots of points where the function is curvy, and not many where it's almost straight. That's very clever. But now I want to plot just points, no lines. How do I do that?

Theo: First you have to make up a list of points. In *Mathematica* we always use curly brackets for lists of things, including points. We could type in a list of points by hand:

```
{{1, 0}, {2, 1}, {3, 2}, {4, 1}, {5, 1}}
{{1, 0}, {2, 1}, {3, 2}, {4, 1}, {5, 1}}
```

These two lines are identical, except that the first one is bold (signifying input) while the second one is not (signifying output). They are the same because the input contained nothing that *Mathematica* could transform.

Jerry: I see two curly brackets at the beginning, and two at the end. This tells me it's a list of lists?

Theo: A point in two dimensions is a list of two numbers. So, if you want to have a list of points, that's a list of pairs, where each pair is a list of two numbers.

Jerry: Does that mean yes, it's a list of lists?

Theo: I suppose so.

You wanted to plot a list of points, and you're in luck because there is a *Mathematica* function called **ListPlot**, which, not surprisingly, is used for plotting lists of numbers:

```
ListPlot[{{1, 0}, {2, 1}, {3, 2}, {4, 1}, {5, 1}}];
```

2
1.5
1
0.5
2 3 4 5

Jerry: This is going to get pretty tiring if we have to type in the lists of points by hand. Is there some way we could have *Mathematica* automatically make us a list of points?

Theo: The **Table** command is just what you want. It works sort of like the **Plot** command:

```
Table[Sin[x], {x, 0, 6}]
{0, Sin[1], Sin[2], Sin[3], Sin[4], Sin[5], Sin[6]}
```

The only thing different about this command is that we replaced the word **Plot** with the word **Table**. The two arguments are the same. The first argument is the function we want tabulated, and the second argument is a list giving the variable, its starting value, and its ending value. **Table** assumes a step size of one, which is different from the **Plot** command. It's important to notice that the arguments to the **Table** command are essentially the same as the arguments to the **Plot** command. You only have to learn once.

Jerry: Why is the first element of the list **0** instead of **Sin[0]**?

Theo: **Sin[0]** is exactly **0**, but there is no such exact value for **Sin[1]** and the rest. *Mathematica* doesn't make approximations unless you ask it to, which you can do using the **N** function:

```
N[Table[Sin[x], {x, 0, 6}]]
{0, 0.841471, 0.909297, 0.14112, -0.756802, -0.958924,
  -0.279415}
```

Jerry: What if I don't like the step size of one? How do I make it smaller?

Theo: You can add a fourth element to the list, like this:

```
Table[Sin[x], {x, 0, 6, 0.5}]
{0, 0.479426, 0.841471, 0.997495, 0.909297, 0.598472,
  0.14112, -0.350783, -0.756802, -0.97753, -0.958924,
  -0.70554, -0.279415}
```

Jerry: Why did we get approximate answers even though we didn't use the **N** function?

Theo: In the first **Table** command, we started at exactly zero, and our step size was exactly one, by default. In the second list we had a step size of 0.5. Now, when we type in 0.5, *Mathematica* does not assume that we mean exactly 1/2; it treats 0.5 as "about" 1/2, because 0.5 might really mean 0.50001, rounded off. *Mathematica* is perfectly happy to convert **Sin[0.5]** into an approximate decimal number, since it considers 0.5 only approximate to start with.

Notice that the first element is exactly zero (you can tell because it doesn't have a decimal point). That's because the starting point was exact. The second number is approximate, because it includes the step size, which is approximate.

If we make this same list using 1/2 as our step size we get this:

```
Table[Sin[x], {x, 0, 6, 1/2}]
```

$$\{0, \mathrm{Sin}[\tfrac{1}{2}], \mathrm{Sin}[1], \mathrm{Sin}[\tfrac{3}{2}], \mathrm{Sin}[2], \mathrm{Sin}[\tfrac{5}{2}], \mathrm{Sin}[3], \mathrm{Sin}[\tfrac{7}{2}], \mathrm{Sin}[4], \mathrm{Sin}[\tfrac{9}{2}], \mathrm{Sin}[5], \mathrm{Sin}[\tfrac{11}{2}], \mathrm{Sin}[6]\}$$

Jerry: Does this mean that **Sin[6]** is different from **Sin[6.]**?

```
Sin[6]
Sin[6]
Sin[6.]
-0.279415
```

I guess so.

But anyway, we were trying to make lists of points so we could plot them. I notice that the lists we've made with **Table** so far are not lists of points. They're just lists of single numbers. Can we use **Table** to make a list of lists?

Theo: Sure. **Table** will make a list of anything you put as its first argument:

```
Table[{x, Sin[x]}, {x, 0, 6, 0.5}]
{{0, 0}, {0.5, 0.479426}, {1., 0.841471},
  {1.5, 0.997495}, {2., 0.909297}, {2.5, 0.598472},
  {3., 0.14112}, {3.5, -0.350783}, {4., -0.756802},
  {4.5, -0.97753}, {5., -0.958924}, {5.5, -0.70554},
  {6., -0.279415}}
```

We can use **ListPlot** to plot this list of points. We'll combine the functions, using the **Table** command as the argument to the **ListPlot** command:

```
ListPlot[Table[{x, Sin[x]}, {x, 0, 6, 0.5}]];
```

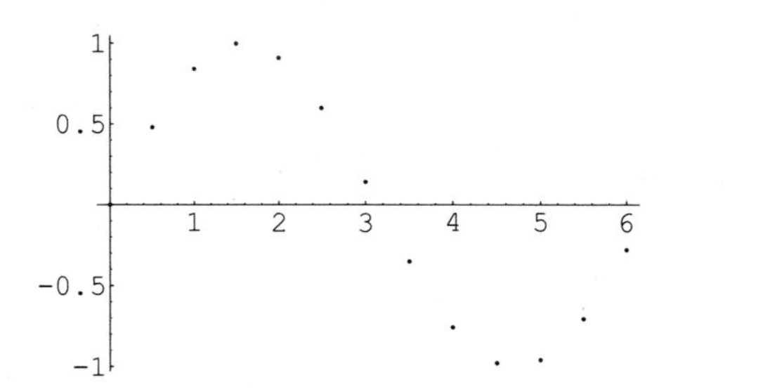

Jerry: Do I have to print out the list of pairs before I make the graph?

Theo: No.

Jerry: Is the list being generating behind our back, and just not being shown to us?

Theo: Yes. The result of the **Table** command is the list, which is passed directly on to the **ListPlot** command. The **ListPlot** command uses the numbers from the list, and then throws them away.

Jerry: Let's reduce the step size to get something more substantial looking:

```
ListPlot[Table[{x, Sin[x]}, {x, 0, 6, 0.1}]];
```

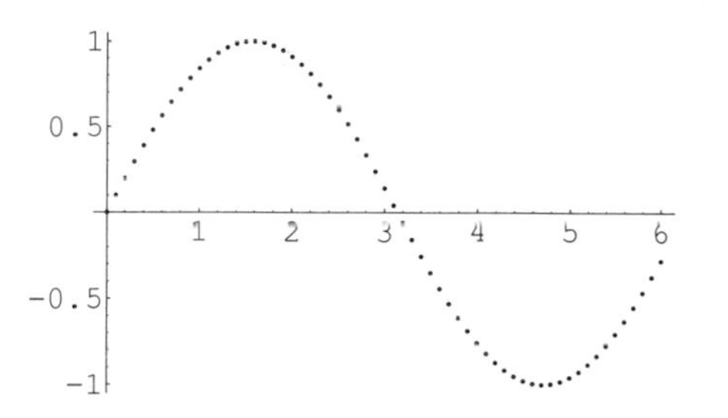

Jerry: Very pretty. I like **`Table`**, **`Plot`**, and **`ListPlot`**. What's another command like these?

Theo: Well, the **`Do`** command is sort of like **`Table`**, except that instead of making up a list of elements, it just generates each element and then discards it.

Jerry: That sounds sort of silly. You mean if I say:

```
Do[Sin[x], {x, 0, 6}]
```

Then I will get absolutely no output?

It appears to do nothing.

Theo: Right. The **`Do`** command is used for looping. For example, if we want to print out some numbers, we can use the **`Print`** command (which prints its arguments on the screen) together with the **`Do`** command.

```
Do[Print[x], {x, 0, 6}]
0
1
2
3
4
5
6
```

Jerry: Is that a list?

Theo: No, it's not at all a list. In fact, it's not even anything you can manipulate in *Mathematica*. It's just the side effects of a bunch of **`Print`** commands.

Jerry: What's this about side effects? What's the difference between a side effect and a result?

Theo: A function like **Sin[x]** returns a value when you evaluate it. You can use this value in further calculations, like in **Sqrt[Sin[x]]**. **Print[x]**, on the other hand, produces some text on the screen as a side effect, but does not give anything back as a value. So, **Sqrt[Print[x]]** is not sensible.

Jerry: What do you mean, not sensible? What happens if we try it anyway?

```
Sqrt[Print[x]]
```

```
x

Sqrt[Null]
```

Theo: The "x" is the side effect of the **Print** command. **Null** is a symbol used to mean "no result".

Jerry: How about another example in which **Do** makes sense?

Theo: The most fun thing to do with **Do** is to make plots with it. Let's try this:

```
Do[
    ParametricPlot[
        {Sin[n t], Sin[(n + 1) t]},
        {t, 0, 2Pi}
    ],
    {n, 1, 3}
]
```

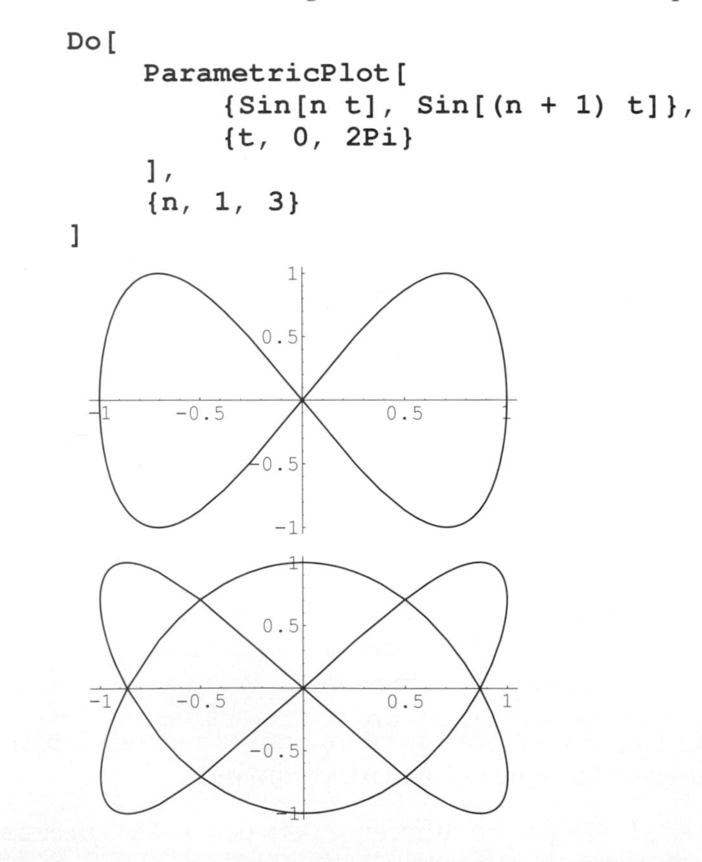

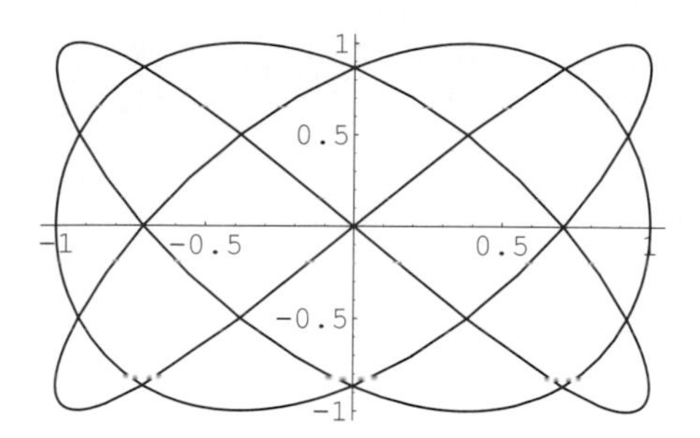

Jerry: What is this **ParametricPlot**? You didn't warn me that you were going to use a new function!

Theo: Don't worry, **ParametricPlot** is just like **Plot**, except that its first argument is a list of two functions, instead of just a single function. The two functions give the x and y coordinates of the curve as a function of the parameter specified by the second argument.

Jerry: I notice that the second argument is again a list of three things -- a variable, and lower and higher limits. The step size is automatic, as in **Plot**?

Theo: Right. **ParametricPlot** uses the same automatic mechanism as **Plot**.

Jerry: Are there many commands that use the same three-things-in-a-curly-bracket structure?

Theo: We've seen several so far: **Plot**, **ParametricPlot**, **Table**, and **Do**. Others include **Sum**, **Product**, **Series**, **Plot3D**, **ContourPlot**, **DensityPlot**, and **Integrate**. They all use lists of three or four elements very much like we've seen already. It's really nice to be able to use the same basic idea in many different ways. By the way, the three-things-in-a-curly-bracket structure is known as an "iterator" when used this way.

Jerry: Could we have a summary of the grammar of *Mathematica*? I'd like to get all the important rules together in one place, now that we've seen a few of them in action.

Theo: Sure, here are the basic rules: First, simple arithmetic expressions can be written pretty much as you would anywhere. A Space character can be used for multiplication, or you can use * instead (spaces around other operators, like + or / don't, of course, mean multiplication). Round parentheses are used to change the order of evaluation. Here are a few examples:

```
3+4
```

7

```
3*4
12

3 4
12

3 2 + 4
10

3 (2 + 4)
18

3 / (2 + 4)
1
-
2
```

Jerry: I notice that we got a rational number instead of a decimal. I suppose that if we type a decimal point in one of the numbers we will get a decimal result:

```
3. / (2 + 4)
0.5
```

Theo: Right. Putting a decimal point in a number makes it into an approximate number instead of an "exact" integer. You can use letters or words as variables in your expressions. It's usually a good idea to use lower case letters for your variables, because that way you can be sure you won't accidentally use a name that's already used by *Mathematica* (since all built-in *Mathematica* names start with capital letters). Here are some examples of using variables:

```
x (3 + 4)
7 x

x + 3 x^2 + 7 x
            2
8 x + 3 x

(x + y) / (x - y)
x + y
-----
x - y
```

Jerry: You say that by using lower case letters we can avoid conflicts with built-in functions. Does that mean that *Mathematica* cares about the case of the letters in variable names?

Theo: Absolutely! The variable **X** is completely different from the variable **x**. There are many situations where having case as a way of distinguishing variable names is very useful. For example I often use the variables **i**, **j**, and **k** as array index variables. I couldn't do that if the built-in variable **I** (the square root of -1) were considered the same as my variable **i**.

Jerry: I suppose that's why I have to use **Sin** instead of **sin** for the sine function. Now, what about square brackets? Why can't I use **Sin(x)** instead of **Sin[x]**?

Theo: Good question! There is, in fact, a good reason. Ordinary mathematical notation is inconsistent here. Round parentheses are used to mean two completely different things in traditional notation: first, order of evaluation; second, function arguments. Consider the expression **k(b + c)**. Does this mean **k** times the quantity **b + c**, or does it mean the function **k** with the argument **b + c**? Unless you know from somewhere else that **k** is a function, or that **k** is a variable, you can't tell. It's a mistake to use the same symbols to mean these two completely different things, and *Mathematica* corrects this mistake by using round parentheses only for order of evaluation, and square brackets only for function arguments.

Jerry: That's a nice point. I never thought of that before. It shows how easily we adapt to nonsense. Aside from that, are you saying that mathematicians have been sloppy for centuries? That's a pretty strong statement!

Theo: Yes. Although I'm all in favor of interesting, quirky languages for writing novels and poetry (English comes to mind), it's really a bad idea to use an ambiguous language for something like mathematics. One of the great contributions of computer science to the world has been a powerful set of tools for thinking about what makes a language "good".

An alternative would be to insist on using a * for all multiplication. Then **k(b + c)** would always mean the function **k**, and if you wanted it to mean multiplication you would have to use **k*(b + c)**. We decided it was better to remove an inconsistency than to force people to use an extra symbol. Another option would have been to have *Mathematica* "know" what was a variable and what was a function. This turns out to have serious consequences, and it's really not a good idea.

Jerry: Well, I didn't expect a lecture!

Theo: Sorry. Let's get back to the matter at hand. For functions, you use square brackets. Let's use the **Sin** function together with some round parentheses, to see how they fit:

```
Sin[1.2(3 + 4)] (4 + 5)
7.69139
```

Jerry: This means, Find the sine of 1.2 times 7 and multiply that answer by 9.

Theo: Yes.

Now, the last bit of syntax that everyone needs to know is the curly bracket. Curly brackets are used for lists. For example, we can make the following lists:

```
{1, 2, 3, 4}
{1, 2, 3, 4}
{a, b, c, d}
{a, b, c, d}
{Sin[1.5], Sin[2.5], Sin[3.5]}
{0.997495, 0.598472, -0.350783}
```

There are also many commands that use lists in special ways. For example the **`Plot`** command uses a list as its second argument, to specify the variable and domain to be plotted. You saw examples of these uses of lists earlier in this chapter.

Jerry: I think it's time for a reward, after all this grammar. Could we use everything we've learned to make something really nice?

Theo: The combination of **`Do`** and **`Plot`** has real potential. Let's use it to make an animation. An animation is a sequence of pictures that you flip through quickly. If the pictures are related to each other in some sensible way, you get the illusion of motion.

Jerry: How can we flip through pictures in a book? Perhaps we need to make a flip-book.

Theo: There are various technologies for looking at animations made with *Mathematica*. The easiest is to look at them directly inside the Notebook where they were generated. There's a menu command in *Mathematica* to animate any sequence of plots. For printing animations in the paper book we've decided to put all the frames of the animation into a matrix on a single page. See the section This Book and Its Technology for more information about animations.

Jerry: So to make an animation using **`Do`**, we need to plot a function that depends in some way on the variable being changed by the **`Do`** command.

Theo: Right. Let's plot **`Sin`** functions whose frequencies get higher and higher. We can do this by plotting **`Sin[n x]`**, where **`x`** is the variable we plot with respect to, and **`n`** is the **`Do`** loop parameter:

```
Do[Plot[Sin[n x], {x, 0, 2Pi}], {n, 1, 3, 1/4}]
```

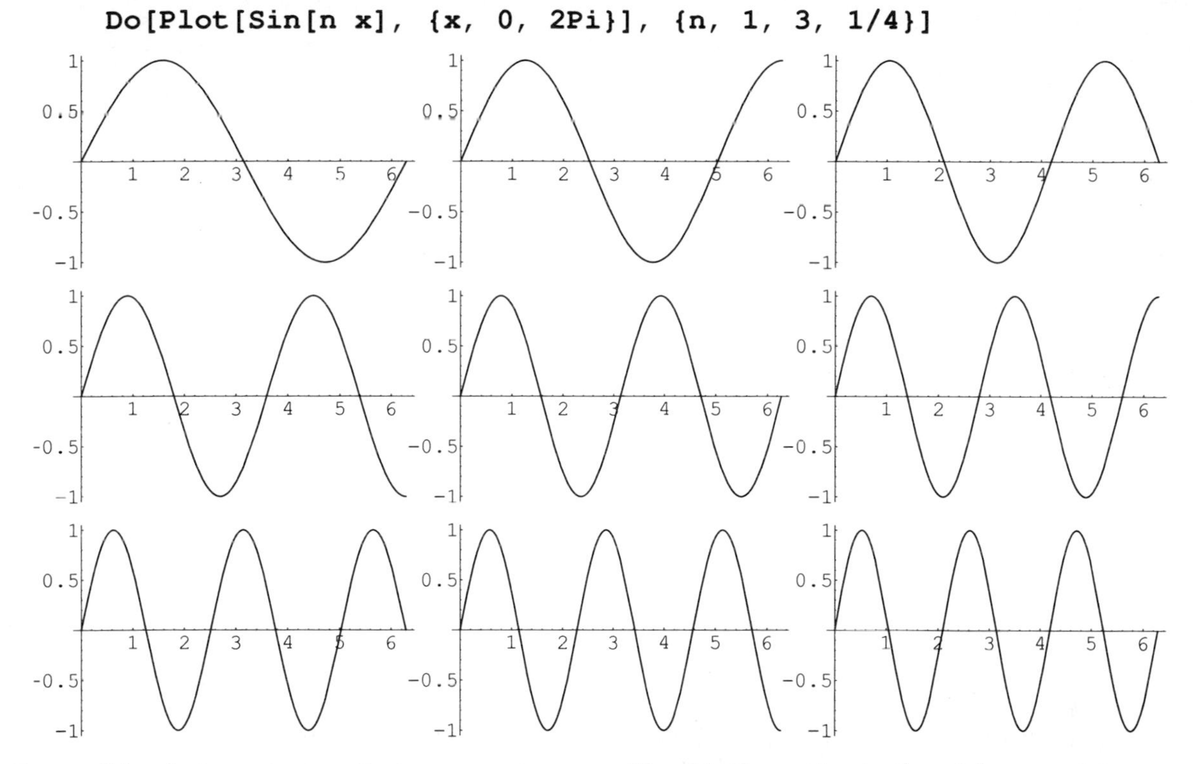

Jerry: It's obvious to me that one reason you like *Mathematica* is the richness of functions that are built-in. A problem for a new or infrequent user is deciding which, from this cornucopia of functions, to use. [Since Theo is one of the authors of *Mathematica*, I've devised a process to extract from him the short list we all need]:

> *Theo, if you were stranded on a desert island, and could only take eight functions with you, what would they be?*

Theo: Assuming you are giving me simple things like addition, numerical functions like **Sin**, and lists for free, here's my list:

```
Table
Plot
Play
Do
Factor
Expand
Integrate
D
```

There are all sorts of things you can do with just these functions. Here are some amusing ones:

```
Table[Expand[(a + b)^n], {n, 1, 5}]
```

$$\{a + b,\ a^2 + 2\,a\,b + b^2,\ a^3 + 3\,a^2\,b + 3\,a\,b^2 + b^3,\ a^4 + 4\,a^3\,b + 6\,a^2\,b^2 + 4\,a\,b^3 + b^4,\ a^5 + 5\,a^4\,b + 10\,a^3\,b^2 + 10\,a^2\,b^3 + 5\,a\,b^4 + b^5\}$$

```
Plot[x Sin[x], {x, -10, 10}];
```

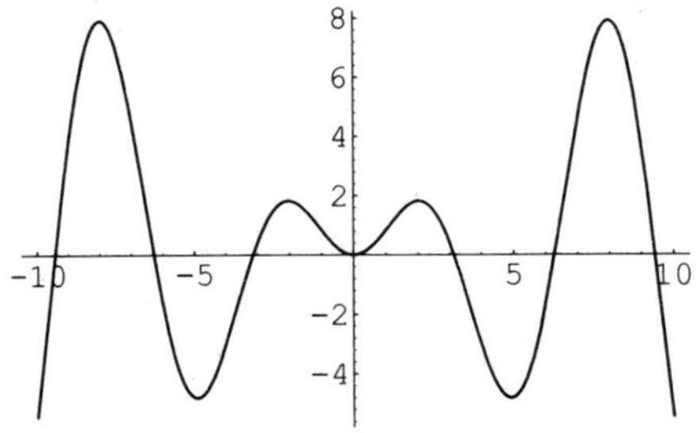

Play is like **Plot**, except that it makes a sound instead of a plot. The first argument is interpreted as an amplitude waveform. The variable range is interpreted as being in seconds (the following command, for example, plays a 5 second sound). The picture you see is an "icon" that tries to represent the waveform, sort of. It doesn't mean very much. To hear the sound (in the electronic edition), double click the small speaker-like icon at the top of the cell bracket.

```
Play[Sin[1000 x Sin[5 x]], {x, 0, 5}];
```

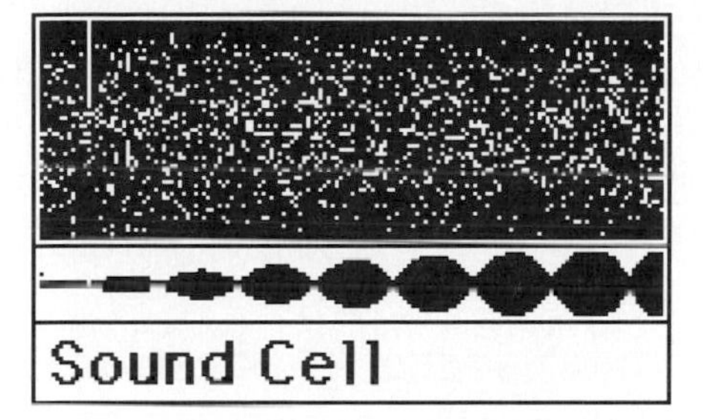

(Play track 2 of the CD to hear this sound.)

```
Play[Sin[1000 x Sin[Pi/6 x] (1 + Sin[120 x]/10)],
     {x, 0, 6}];
```

(Play track 3 of the CD to hear this sound.)

Do is often used in combination with plotting commands to make animations, and in combination with **Print** to make interesting printouts, like this one:

```
Do[
    Print["Remarkably, ", n, " squared is ", n^2, "."],
    {n, 1, 10}
]
Remarkably, 1 squared is 1.
Remarkably, 2 squared is 4.
Remarkably, 3 squared is 9.
Remarkably, 4 squared is 16.
Remarkably, 5 squared is 25.
Remarkably, 6 squared is 36.
Remarkably, 7 squared is 49.
Remarkably, 8 squared is 64.
Remarkably, 9 squared is 81.
Remarkably, 10 squared is 100.
```

`Factor` and **`Expand`** factor and expand polynomials, like this:

```
Factor[x^24 - 1]
```

```
                            2             2             2
(-1 + x) (1 + x) (1 + x ) (1 - x + x ) (1 + x + x )

          4        2    4        4    8
   (1 + x ) (1 - x  + x ) (1 - x  + x )
```

```
Expand[(1+x)^10]
```

```
                  2        3        4        5        6
1 + 10 x + 45 x  + 120 x  + 210 x  + 252 x  + 210 x  +

        7       8       9    10
   120 x  + 45 x  + 10 x  + x
```

`Integrate` and **`D`** integrate and differentiate expressions with respect to the variable named in their second argument:

```
Integrate[1/(1-x^3), x]
```

```
        1 + 2 x
ArcTan[-------]                                   2
       Sqrt[3]     Log[-1 + x]   Log[1 + x + x ]
--------------- - ----------- + ---------------
    Sqrt[3]            3                6
```

```
D[Sin[Tan[x]], x]
```

```
                      2
Cos[Tan[x]] Sec[x]
```

Jerry: Your life on the desert island would be quite bleak with only these 8 functions. Give us a rough estimate of the number of functions you consider central to *Mathematica*.

Theo: 35, but I won't tell you which ones.

Jerry: Roughly how many command in *Mathematica* would you never use?

Theo: 35, but I won't tell you which ones.

Jerry: So you would use the other 800 or so commands every once in a while?

Theo: That's about the way it is.

Chapter Two
Consistency, and Why You Want It

In which Jerry and Theo discuss Consistency in all its forms and variations.

■ Dialog

Theo: *Mathematica* is a model of consistency, most of the time.

Jerry: Do examples spring to mind?

Theo: Sure. There are many kinds of consistency, and *Mathematica* exhibits many of them. For example, consistency in the naming of commands. Many programs and languages use ridiculous abbreviations for commands. Several computer algebra systems use the word "int" to mean integration. Others use it to mean "integer part". Both are silly, since the better thing to do in both cases is to use the word "integrate" for integration, and thereby avoid any possible confusion.

Jerry: Initially, I found onerous the requirement in *Mathematica* that I type out a command, but the advantages of reduced ambiguity eventually overcame my hostility.

Theo: It's true that people sometimes complain about having to type long names. They don't have to, since they can use the automatic command completion feature available in *Mathematica*.

To use this feature, you type the first few letters in a command and then press Command-K (or choose the Complete Selection menu command). *Mathematica* completes the command for you, or gives you a list of possible completions if there are more than one. You can also use the Escape key, which is the same as Command-K except that it does *not* give you a menu of possible completions, it just fills in as much as is common to all possible completions.

Jerry: Other than to avoid confusing two different commands, what good is using complete words for command names?

Theo: It makes *Mathematica* programs much easier to read. Most of the commands are, after all, plain English words. By combining these command names with nice, long variable names of your own, it's possible to make very beautiful programs.

Jerry: We have a dichotomy in our society between "word people" and "math/science/computer people", and an awareness of your second reason could close that gap a little bit.

Theo: It's all language. *Mathematica* is just a language for which there is an interpreter other than another person. (This is one of the failings of the English language--it can be understood only by humans.) A third reason for using full English words is that it's much easier to guess the name of a command that you haven't used yet, without having to look it up. For example, it's hard to know if "factor" should be abbreviated as "fac" or "fact" (I've seen both in other systems). But if the command is just called "factor" there's not much to argue about.

Jerry: It would be nice if a person unfamiliar with *Mathematica* could think, "I would like to factor 268" and be able to go to *Mathematica* for the first time and type `Factor[268]` and have it work.

Theo: It would also be nice if people who know only French could come to America and speak to people more or less without trying. Both are unrealistic. In the case you mention, the `Factor` command actually works only on polynomials, not integers. There's a fairly good reason for this, and as a result there has to be a second command, called `FactorInteger`, that factors integers but not polynomials. A good analogy might be that whereas a French speaker has little chance of coming to America and being able to communicate, there is enough internal consistency in the English language that a native speaker does not have to memorize everything (plurals, for example, are usually formed by adding an "s" to the end).

Jerry: Are there exceptions in *Mathematica*, as there are in English?

Theo: Sure. There are four commands, `Factor`, `FactorInteger`, `GCD`, and `PolynomialGCD` which are not consistent with each other. In each pair of commands one works on integers and the other works on polynomials. But in the case of `Factor` and `FactorInteger` it's the one that works on integers that has the special name, whereas for `GCD` it's `PolynomialGCD` that has the special name. This is a mistake, but it's too late to fix it.

(Note that inconsistencies like this are just as much a mistake in English as they are in *Mathematica*. It's an interesting fact that when things like this come up in *Mathematica,* the *authors* are scolded; but when the same kinds of things show up in English, it's usually the innocent *students* who are scolded – for not knowing about them. My grades in English classes would probably have been higher if my English teachers understood this. They somehow think that English doesn't have mistakes, only "exceptions", which is not a productive way of thinking about it.)

Jerry: I notice that all *Mathematica* commands start with capital letters, and many have some more capitals inside.

Theo: Yes, all built-in *Mathematica* functions, commands, and constants start with capital letters. Commands which are made up of several words (like **FactorInteger** or **IdentityMatrix**) have internal capitalization. Functions that are named after a person have both the name of the person and the commonly used symbol, both capitalized, like **BesselJ**, **BesselK**, and **LegendreP**. For functions and variable names you define yourself, you can do what you want, although a good rule is to use internal capitalization the same as for built-in functions, but always start the name with a lower case letter. Since *Mathematica* is a case-sensitive language (and proud of it) this prevents name conflicts with built-in functions.

Jerry: What do you mean, "case-sensitive"?

Theo: Case-sensitive means that the name **a** does not have the same meaning as the name **A**, and that the name **medium** is not the same as the name **Medium**. This is important because a lot of people might want to use a variable called **medium**, and many of them might not know that **Medium** is a built in *Mathematica* function.

Jerry: What about the curly brackets and the commas that show up constantly in *Mathematica* code?

Theo: In the last chapter we talked about curly brackets *vs.* square brackets *vs.* round brackets, and I think we pretty much dealt with that issue.

There are lots of interesting questions of semantic consistency.

Let's talk about positional arguments. If you have two functions, call them **f** and **g**, and they both take two arguments, then the following two expressions are syntactically consistent:

```
f[a, b]
g[a, b]
```

These functions may not be semantically consistent, because the two functions may assign different meanings to their two arguments. For example, take integration and differentiation. Both integration and differentiation need two arguments: the expression to work on, and the variable in the expression to work on. We could make the two functions semantically inconsistent like this:

```
Integrate[expression, x]
Differentiate[x, expression]
```

No reasonable language would do this, but it's surprising how many languages get things very much like this wrong. In *Mathematica* the function for differentiation is called **D** and it works like the *Mathematica* function for integration, which is called **Integrate**, as in **D[expression, x]**. Another function that's consistent with **Integrate** and **D** is **Solve[expression, x]**. (You might think that having **D** as the function name for differentiation is inconsistent, since it is an abbreviation. The excuse is that "d" is a very common abbreviation for differentiation in the mathematics literature, as in df/dx. That's probably not enough of a reason, but there you have it.)

Jerry: Are there are any more examples of functions that are consistent with this scheme?

Theo: I can't think of any right now, but if you widen the definition of the second argument to include variables together with a range specification (e.g. "x running from 1 to 10) then all the plotting commands are consistent. For example:

```
Plot[expression, {x, xmin, xmax}]
ContourPlot[expression, {x, xmin, xmax}]
DensityPlot[expression, {x, xmin, xmax}]
Plot3D[expression, {x, xmin, xmax}, {y, ymin, ymax}]
```

Also functions like **Do**, **Table**, and **Animation** are the same:

```
Do[expression, {x, xmin, xmax}]
Table[expression, {x, xmin, xmax}]
Animation[expression, {x, xmin, xmax}]
```

Jerry: Sometimes it's confusing trying to remember the order in which arguments should go. Having things consistent between functions helps, but what if you don't know how any of them work?

Theo: That's a fundamental problem in using the position of an argument to determine its meaning. Once you have more than 3 or 4 positions in a function, it becomes very hard to remember the order in which the arguments are supposed to go. *Mathematica* deals with this problem using a scheme called "named arguments" or "options" for functions that have many different potential arguments. A good example is the **Plot** command.

Jerry: You mean that by naming the arguments we make them have the same meaning regardless of what position they are in? How does this naming work?

Theo: Named arguments are specified this way: **`Name -> Value`**. For example, the choice of whether to have a plot surrounded by a frame is determined by the **`Frame`** argument. To make a plot **`Frame`**, you include an argument of the form **`Frame -> True`**. Let's try **`Plot`** commands with and without the **`Frame`** argument:

```
Plot[Sin[x], {x, 0, 10}];
```

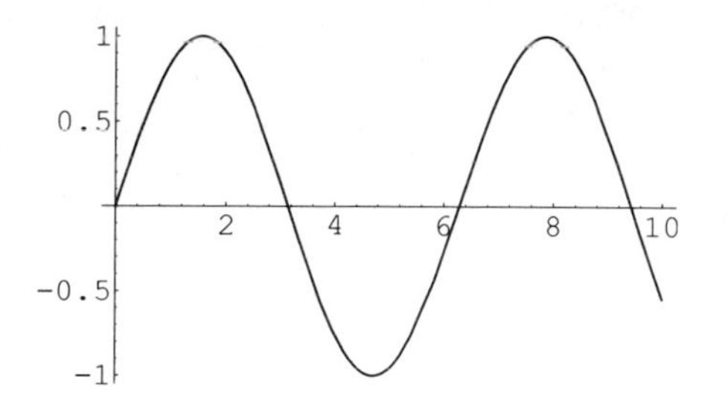


```
Plot[Sin[x], {x, 0, 10}, Frame -> True];
```

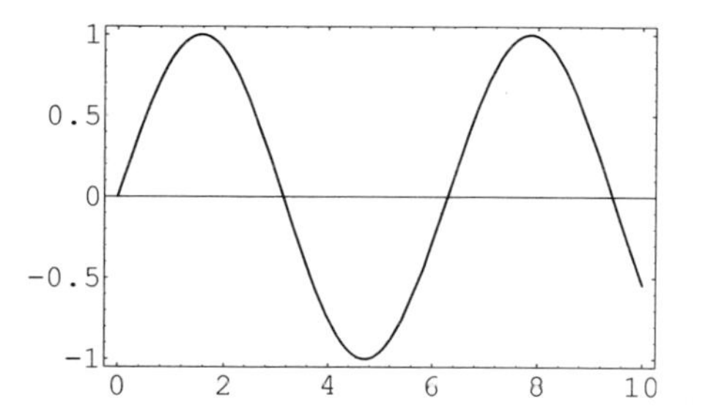

Jerry: It doesn't matter where you put that argument, so long as it's after the first two arguments (which are the expression to be plotted and the variable plus range to plot over). How does **`Plot`** know that if there is no **`Frame`** argument it's not supposed to put a frame in?

Theo: *Mathematica* has a default for each of its named arguments. You can find out the whole list of default settings by asking for **`Options[Plot]`**:

```
Options[Plot]
```

```
                      1
{AspectRatio -> -----------, Axes -> Automatic, AxesLabel -> None,
                GoldenRatio
  AxesOrigin -> Automatic, AxesStyle -> Automatic,
  Background -> Automatic, ColorOutput -> Automatic, Compiled -> True,
  DefaultColor -> Automatic, Epilog -> {}, Frame -> True,
  FrameLabel -> None, FrameStyle -> Automatic, FrameTicks -> Automatic,
  GridLines -> None, MaxBend -> 10., PlotDivision -> 20.,
  PlotLabel -> None, PlotPoints -> 25, PlotRange -> Automatic,
  PlotRegion -> Automatic, PlotStyle -> Automatic, Prolog -> {},
  RotateLabel -> True, Ticks -> Automatic, DefaultFont :> $DefaultFont,
  DisplayFunction :> $DisplayFunction}
```

Jerry: These default arguments can be changed, I assume, by changing the settings in **`Options[Plot]`**?

Theo: Sure. If you want to make **`Plot`** always make a frame you can use the **`SetOptions`** command like this (**`SetOptions`** returns the new list of options, so you can see what it did):

```
SetOptions[Plot, Frame -> True]

                      1
{AspectRatio -> -----------, Axes -> Automatic, AxesLabel -> None,
                GoldenRatio
  AxesOrigin -> Automatic, AxesStyle -> Automatic,
  Background -> Automatic, ColorOutput -> Automatic, Compiled -> True,
  DefaultColor -> Automatic, Epilog -> {}, Frame -> True,
  FrameLabel -> None, FrameStyle -> Automatic, FrameTicks -> Automatic,
  GridLines -> None, MaxBend -> 10., PlotDivision -> 20.,
  PlotLabel -> None, PlotPoints -> 25, PlotRange -> Automatic,
  PlotRegion -> Automatic, PlotStyle -> Automatic, Prolog -> {},
  RotateLabel -> True, Ticks -> Automatic, DefaultFont :> $DefaultFont,
  DisplayFunction :> $DisplayFunction}
```

Theo: Now if we make a plot, we will get a frame, even if we don't include the option in the **`Plot`** command itself:

```
Plot[Sin[x], {x, 0, 10}];
```

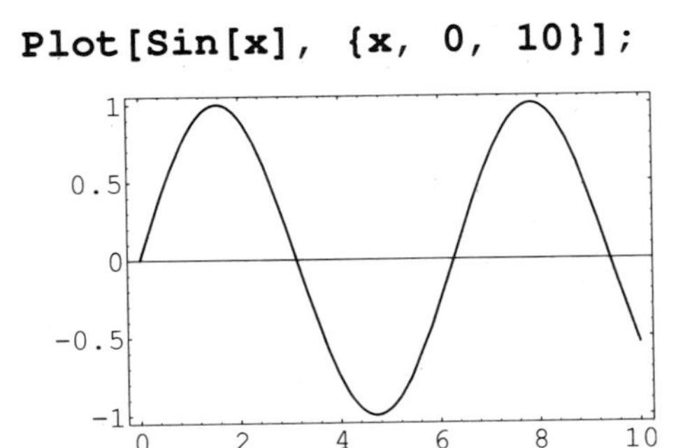

This change stays in effect until you quit and restart *Mathematica*, or until you use **`SetOptions`** to change it again. If you want to make the change permanent, you can put the **`SetOptions`** command in a special file called `init.m`, which is loaded automatically each time *Mathematica* is started. See your *Mathematica* installation guide for information about where this file is located on your system.

Jerry: What would the *Mathematica* code look like if the named arguments were replaced with positional arguments? That way I wouldn't have to type all these silly names and arrows.

Theo: We could imagine just putting the values of all the options in the same order they show up when you type **`Options[Plot]`**. If we did that, the **`Plot`** command would look like this:

```
Plot[expression, {x, xmin, xmax},
    GoldenRatio^(-1), True, None, Automatic,
    Automatic, Automatic, Automatic, False,
    Automatic, {}, False, None, Automatic, Automatic,
    None, 10., 20., None, 25, Automatic, Automatic,
    Automatic, {}, True, Automatic,
    $DefaultFont, $DisplayFunction]
```

If you wanted to know what all these arguments mean, you would have to look in the manual or at some sort of reference card or help feature.

Jerry: This is pretty terrifying. Would we really have to include *all* the arguments, if we were doing things this way?

Theo: No, you could leave out as many from the end as you wanted to. Since the way meaning is assigned to the arguments would be through their position, you couldn't leave out any in the middle. If you wanted to change the setting of the tenth argument, you would have to include the first nine as well, just to make the tenth argument *be* the *tenth* argument.

Jerry: Well, I can see that there is a lot to be said for named arguments. If they are such a great thing, why not have all arguments be named arguments?

Theo: Let's think about what that would be like for the `Integrate` function:

```
Integrate[Function -> expression, Variable -> x]
```

Instead of:

```
Integrate[expression, x]
```

The version with named arguments is overly wordy. People seem to be comfortable dealing with functions that have 2 or 3 or sometimes even 4 positional arguments, but after that it's a good idea to switch to named arguments. The named arguments are also ideal for settings that are rarely changed from their defaults. Most people don't even have to know that a given argument exists, unless they want to change it from the default.

Jerry: So, named arguments are a way of making commands more powerful, without the beginner having to know about the power until they need it. That seems like an important quality. We all want a system that works quite automatically when we begin using the system. The only problem with such a system is that people often dislike the first results and begin to say "Can't it put the graph where I want it?" or "We don't expect to see complex numbers in the answer at our level", or other statements of dissatisfaction. People want something that's easy to use, but then they get unhappy when it doesn't do what they want. Having these named arguments gives them the flexibility to change things later, without having to know about these options at first. If only the options weren't so confusing....

Theo: The problem is that flexibility means a lot of choices to make, and having a lot of choices is always confusing and, oddly, frustrating. Writer's block (sitting in front of a blank piece of paper not knowing what to write) is often a question of too much choice. Which of the 300,000 possible English words should I write down next? It can be very frustrating, but few writers would seriously suggest that the number of words in the language should be reduced in order to make their job easier.

Jerry: I agree that flexibility means having to choose a lot, and that making choices is sometimes difficult. I don't see why it has to be so confusing, though.

Theo: It doesn't have to be, it's just that making it less confusing is extremely difficult. I don't mean difficult in the "oh, that would be really hard" sense, I mean difficult in the "if I could do that they would give me the next seven Nobel prizes" sense. There are several steps in designing a system that is both powerful and easy to use (to quote a popular cliché). The first step is to make a powerful, flexible system. Without that you've got nothing to work on. The next step is to think very very carefully before choosing the default settings for everything. This is the stage *Mathematica* is in right now. It's very powerful, and things like the **`Plot`** command have carefully-thought-out default settings.

Jerry: That sounds like a good start. Many people first using a system get the message that the system's creators claim that its features are the only way to go, and are put off by this. Knowing that it's hard to make a powerful and easy-to-use system and that the authors of the system fall short of some ideal might make it easier for new users.

Theo: The next step is to make a system that translates relatively vague desires on the part of the user into specific modifications of the default settings. There are various projects in different areas of the AI (Artificial Intelligence) community that are trying to design systems in which the computer guides the user intelligently through a large number of complicated choices. Progress is being made, although irregularly.

Jerry: Is there anything like that in *Mathematica* right now?

Theo: No, unless you count something like **`Integrate`**, which makes many mathematical choices within its limited domain. The next step after this is to design a system that anticipates what choices the user is going to want to make, from the context and from what that user has wanted in the past. People are the only systems we know of that are any good at doing all these things in a complicated subject area.

Jerry: I have noticed that using *Mathematica* is a lot easier using you as a front end.

Theo: That's a good way of putting it. Think of the system of *Mathematica*+Theo as being roughly like what the *Mathematica* of the future may be.

Chapter Three
Many **Sin** Functions Make a Square Function

In which Theo and Jerry learn that combining many round things results in a square thing.

Dialog

Theo: Many people think that iterated and recursive functions are strange and mysterious things, and that they have nothing to do with the real world.

Jerry: That's a shame, since iteration is easy to understand and often reflects the real world. You can solve a lot of interesting problems using iteration. Iteration just means applying a function to some starting value and getting a result, then applying the same function again to that result, etc. (For example, we've applied the "Eisenman Editing Function" to this chapter at least six times.)

Theo: Let's try some simple iteration, using the *Mathematica* **Nest** function. **Nest** is iteration, pure and simple. If you have a function **f**, and you want to apply it to **x**, say, **3** times, you can do it like this:

```
Nest[f, x, 3]
f[f[f[x]]]
```

Or 5 times:

```
Nest[f, x, 5]
f[f[f[f[f[x]]]]]
```

Or zero times, which is just the same as the starting point:

```
Nest[f, x, 0]
x
```

Jerry: Many functions that we use every day react in uninteresting ways to iteration. They either get very large or head for zero. For example, if the function is **x^2** and we nest it 4 times we will get **(((x^2)^2)^2)^2** which is **x^16**. If **x** starts bigger than one its value quickly gets large, while if it starts between zero and one it heads for zero. The situations we end up studying are those that manage to "live" between these two extremes.

Theo: The problem with a function like **x^2** is that it just gets bigger and bigger, or smaller and smaller. We need a function that gives small values even for large **x** values, so that we don't get this runaway effect. **Sin[x]** might be a good bet, since its value is always between **-1** and **1**, no matter how large **x** is.

```
Nest[Sin, x, 3]
Sin[Sin[Sin[x]]]
```

Theo: Let's try it with a number as the starting point, instead of **x**:

```
Nest[Sin, 1.2, 3]
0.71933
```

Jerry: So, **Sin[Sin[Sin[1.2]]] = 0.71933**. We could try making a plot of this as a function of the starting point:

```
Plot[Nest[Sin, x, 3], {x, -2Pi, 2Pi}];
```

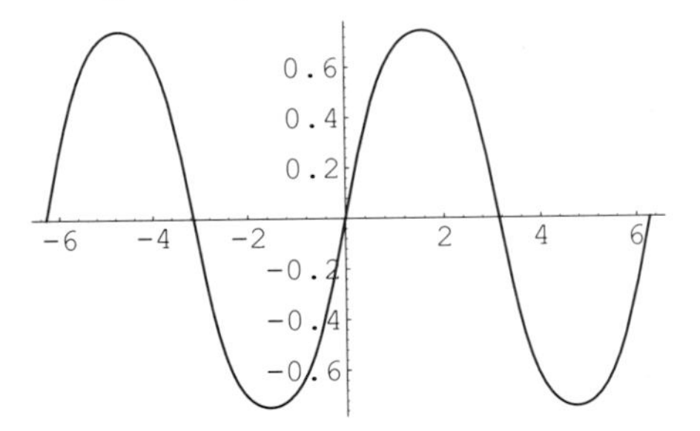

Theo: What do you suppose happens if we nest the **Sin** functions more deeply:

```
Plot[Nest[Sin, x, 100], {x, -2Pi, 2Pi}];
```

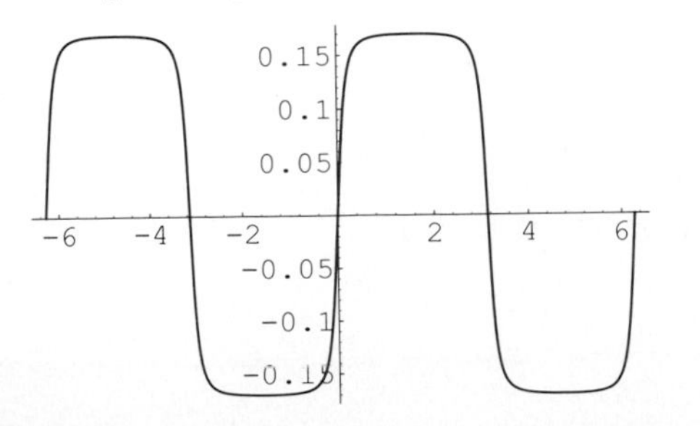

Jerry: It's starting to look more square than round. Hey! Notice how the y range has gotten much smaller. The range of the plot is only from **-0.15** to **0.15**. Let's try something really extreme:

```
Plot[Nest[Sin, x, 2000], {x, -2Pi, 2Pi}];
```

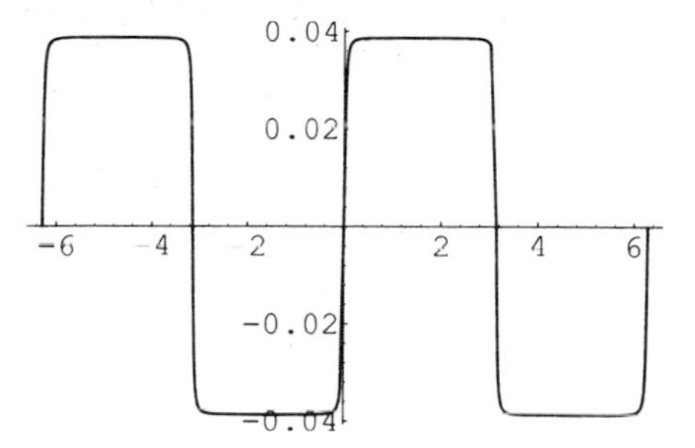

Jerry: Wow, that's really small and square! Let's try to understand what's happening. If we animate this, we can see how the function moves from the familiar sine wave to this peculiar nearly-square wave.

Theo: To make an animation, we have to make a sequence of plots in which we nest progressively more and more **Sin** functions. We can use the **Do** function to make all the plots at the same time. We have to specify the **y** range manually, to prevent *Mathematica* from automatically scaling each plot differently. We'll also label each plot with the number of **Sin** functions it includes.

```
Do[
    Plot[
        Nest[Sin, x, n], {x, -2Pi, 2Pi},
        PlotRange -> {-1, 1},
        PlotLabel -> n
    ],
    {n, 1, 50}
]
```

(The printed version includes fewer frames, for space reasons.)

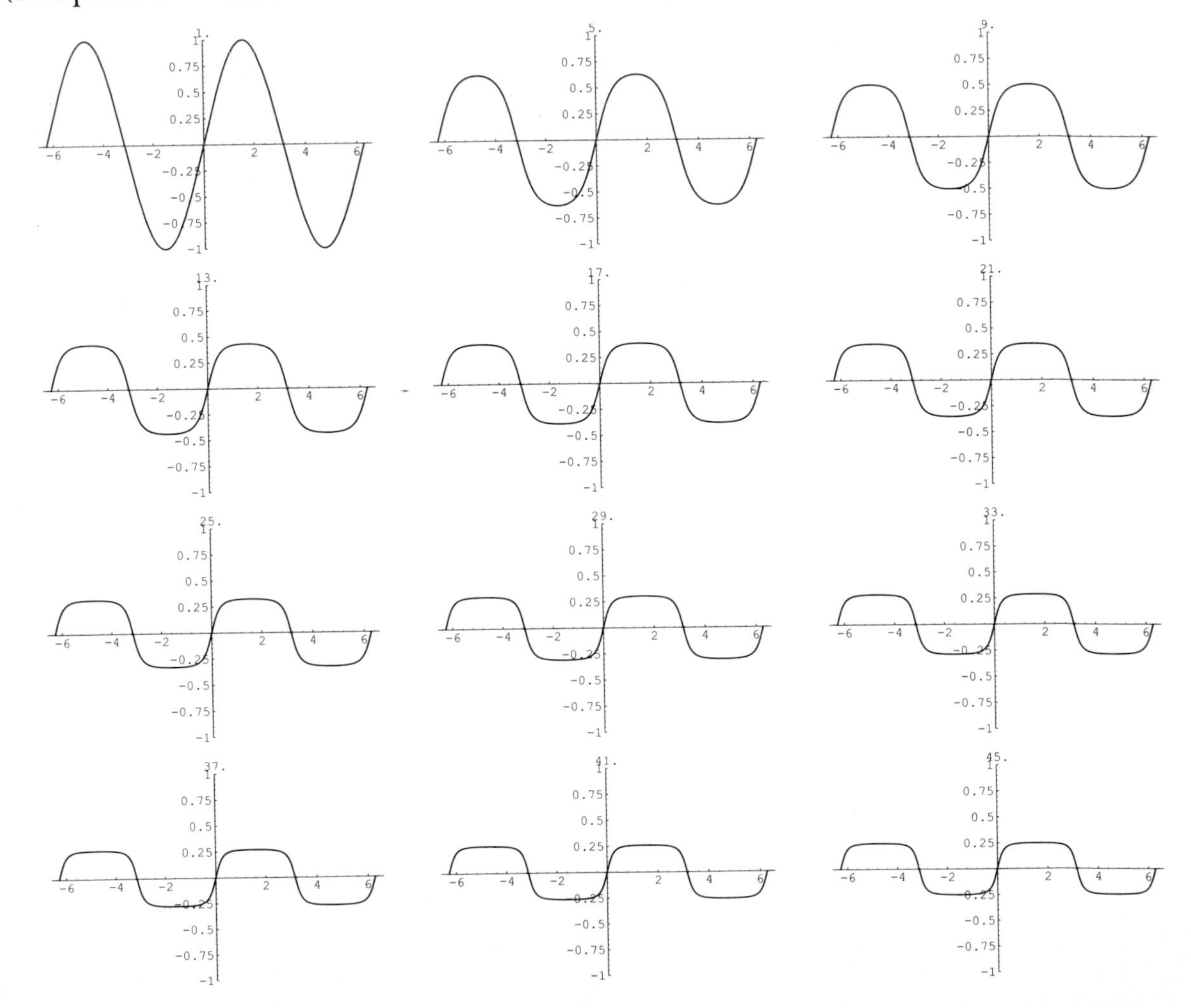

Jerry: This is an outstanding use of animation. I can see clearly what is happening.

Theo: The whole plot slowly collapses toward zero, but different parts collapse at different rates, making the plot get more and more square. I find this very surprising.

Everyone seems to agree that if you were to nest an infinite number of **Sin** functions you would get zero everywhere, but most mathematicians I've told about the getting square part don't believe me until I show them the pictures.

Jerry: We puzzled about the question of why it gets square for a long time, until Theo went off to Carnegie-Mellon University, and happened to talk to Dana Scott, a computer scientist there.

Theo: Professor Scott suggested that we should look at the function **f[x_] := Sin[x]/Sin[1]** (like many mathematicians, he seemed reluctant just to explain the answer). We tried making plots and animations like the one above using this new function, and indeed we did figure out a very interesting general principle that explains not only why the **Sin** function does what it does, but also shows that a whole class of other functions will do the same thing. Our visiting professor, Jerry Keiper, will explain this in his section below.

So far we have always chosen a fixed number of nested **Sin** functions and made a plot of the starting point (from **-2Pi** to **2Pi**) vs. the value of the nested **Sin** functions. What if we choose a fixed starting point and make a plot of the *number* of nested **Sin** functions vs. the value of the nested **Sin** functions at this fixed starting point? For example, with **1.2** as our starting point, we can use the **Table** function to make a table of **1.2, Sin[1.2], Sin[Sin[1.2]]**, etc:

```
Table[Nest[Sin, 1.2, n], {n, 0, 20}]
```

```
{1.2, 0.932039, 0.802837, 0.71933, 0.658881, 0.612232,
  0.574696, 0.543579, 0.517203, 0.494451, 0.474548,
  0.456937, 0.441201, 0.427026, 0.414166, 0.402426,
  0.391652, 0.381716, 0.372513, 0.363957, 0.355975}
```

Jerry: This says that we start with a value of 1.2 and take the **Sin** of it, which produced **0.932039**. Then take the **Sin** of **0.932039** and that produces **0.802837**, and so on.

Theo: Yes. We can make a plot of these numbers using **`ListPlot`**. Note that **`List-Plot`** labels the x-axis so that the first point is at x=1, even though we think of it as the zero-th iteration.

```
ListPlot[Table[Nest[Sin, 1.2, n], {n, 0, 15}]];
```

1.2

2.5 5 7.5 10 12.5 15

0.8

0.6

0.4

Jerry: I love *Mathematica*, BUT there are a few things I don't want to adjust to. In this plot, the axes cross at (0, 1), a choice that I don't like.

Theo: *Mathematica* always tries to make the axes cross in reasonable places (for example at integers), while at the same time not wasting too much space by having an axis far away from any of the points in the plot. We can force the axes to cross where we want, like this:

```
ListPlot[Table[Nest[Sin, 1.2, n], {n, 0, 15}],
    AxesOrigin -> {0, 0}];
```

1.2

1

0.8

0.6

0.4

2.5 5 7.5 1012.515

Jerry: I prefer this second plot, but the big *gap* seems mysterious and the lack of a label for 0 on the horizontal is distracting.

Theo: I agree that the lack of a label for zero is bad. However, the gap is actually a good thing, and is described and endorsed in the excellent book *The Visual Display of Quantitative Information* by Edward Tufte (Graphics Press, 1983). The idea is to draw the axes only over the interval of values occupied by the actual data points. That way you can see the domain and range just by looking at the axes.

Jerry: It does take some getting used to; your explanation will motivate me to adjust.

Theo: I prefer the first plot; I don't think there's any reason to have that inch-high blank space. I don't mind at all having the axes cross at one instead of zero.

Jerry: Back to the plot. Let me get this straight: The first point on the left is **`1.2`**, the second point over is **`Sin[1.2]`**, the third point is **`Sin[Sin[1.2]]`**, and the fourth point is **`Sin[Sin[Sin[1.2]]]`**, right?

Theo: Exactly. The farther we go to the right, the more deeply we have nested the **`Sin`** functions.

Jerry: This idea of making a table of more and more nested functions sounds worthwhile. There must be a lot of problems you could approach with this technique.

Theo: Probably so; in fact *Mathematica* has a built-in function that does exactly what the **`Table`** command we used above does. It's called **`NestList`**, and here's an example:

```
NestList[Sin, x, 5]
```

```
{x, Sin[x], Sin[Sin[x]], Sin[Sin[Sin[x]]],
  Sin[Sin[Sin[Sin[x]]]], Sin[Sin[Sin[Sin[Sin[x]]]]]}
```

The following command would, in more compact form, produce exactly the same plot as above:

```
ListPlot[NestList[Sin, 1.2, 15]];
```

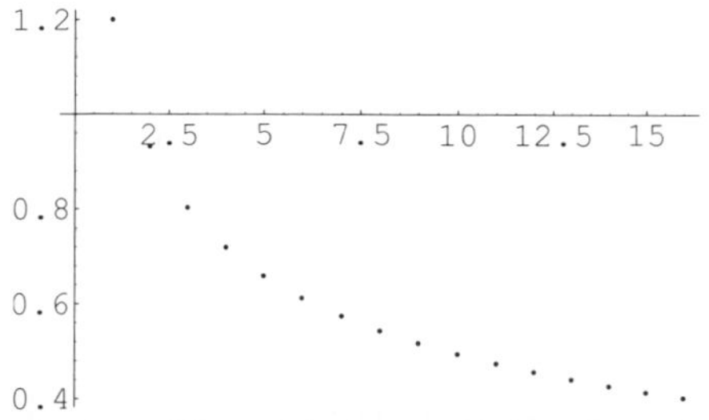

Jerry: Now we've made two different kinds of plot. First, we made a plot in which we vary the starting point, given a fixed number of nested **`Sin`** functions. Then we made a plot in which we vary the number of nested **`Sin`** functions, given a fixed starting point. Could we combine both plots together, to show how they fit?

Theo: Sure. We could make a three-dimensional plot, with one dimension being the starting point and the second dimension being the degree of nesting, and the third (or vertical) dimension being the value of the nested **Sin** functions.

```
Plot3D[Nest[Sin, x, Round[n]],
        {x, -2Pi, 2Pi}, {n, 1, 15},
        AxesLabel -> {"x", "n", ""},
        PlotPoints -> {30, 15},
        Shading -> False];
```

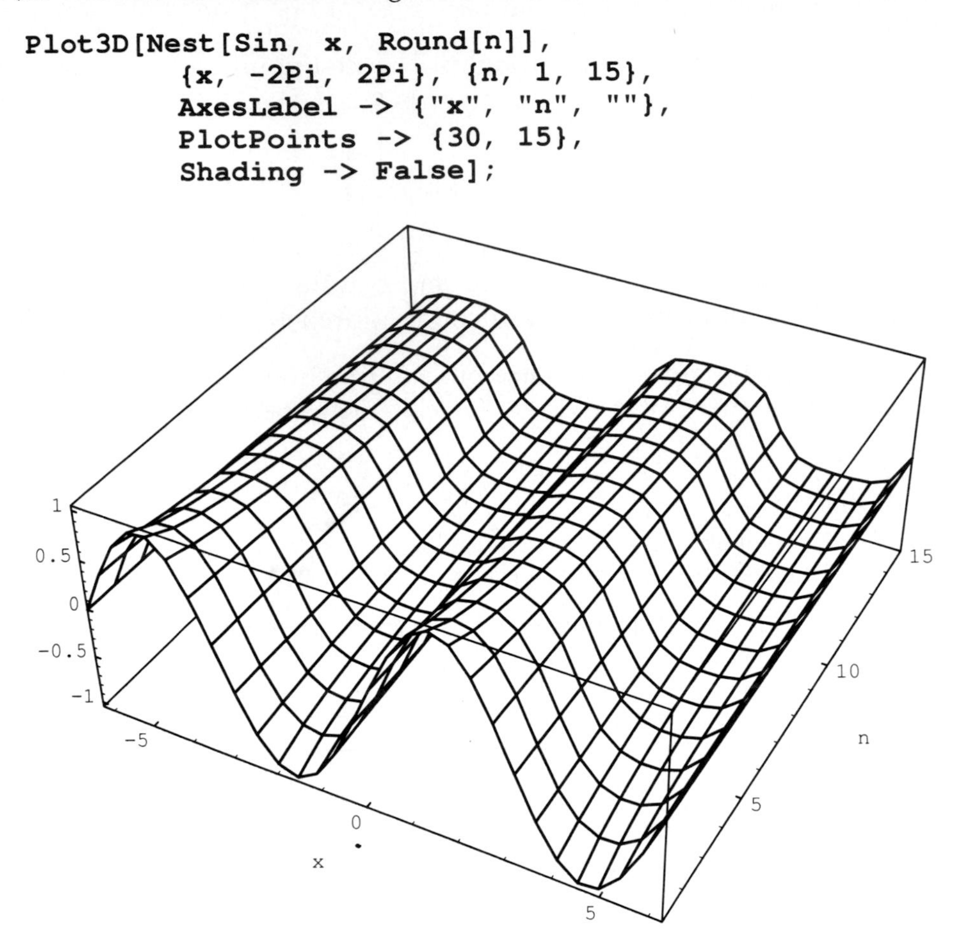

Theo: We have to use **Round[n]**, because **Nest** expects to get an integer for the number of times to nest, whereas **Plot3D** assigns floating point values to **n** while making the plot.

Jerry: The front of the plot looks like **Sin[x]**, which makes sense, because the front **n** is 1, which means nest **Sin** only once. The third grid line from the front must be a plot of **Sin[Sin[Sin[x]]]**. The front-to-back grid lines correspond to the **ListPlot** plots we made before. Perhaps this would be clearer if we looked at the 3D plot from the front and then from the side. How will we tell *Mathematica* to show us the plot from these angles?

Theo: We can add a **ViewPoint** option to the plotting command. The **ViewPoint** option gives the (x, y, z) coordinates from which you want to look at the plot. For example, to look at the plot head-on, we want to look at it from a place on the y axis, at some reasonable negative y coordinate. The y axis corresponds to having both x and z equal to zero, so let's try (0, -5, 0):

```
Plot3D[Nest[Sin, x, Round[n]],
        {x, -2Pi, 2Pi}, {n, 1, 15},
        AxesLabel -> {"x", "n", ""},
        PlotPoints -> {30, 15},
        Shading -> False,
        ViewPoint -> {0, -5, 0}];
```

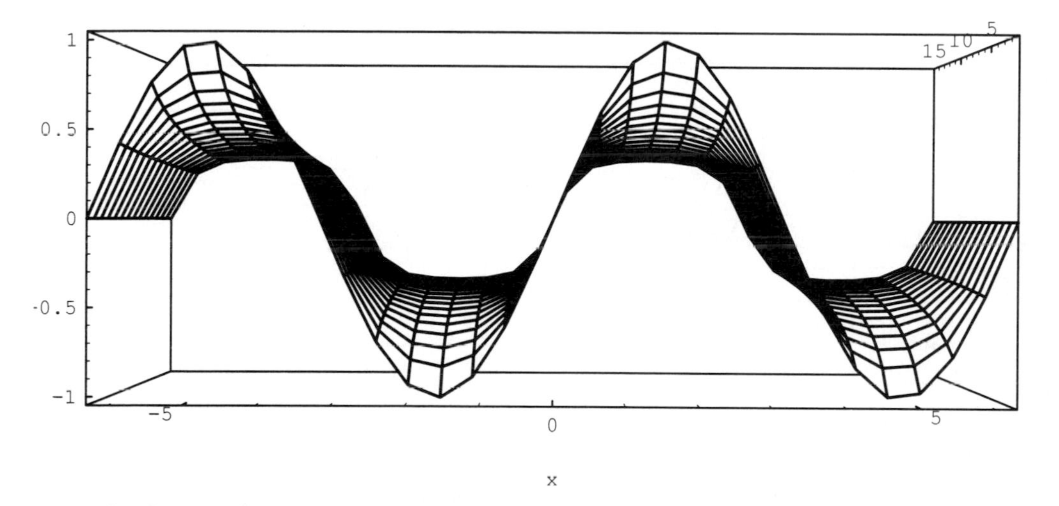

Jerry: It looks good to me. Now tell me, did you make up this viewpoint just by thinking about it, as you imply in your description?

Theo: No, I used an interactive viewpoint selector to get a rough idea of what sort of coordinates I wanted, then typed in the exact numbers you see above. The rationalization came after the fact.

Jerry: This looks like the animation we made before, only all on the same plot. How about from the side?

```
Plot3D[Nest[Sin, x, Round[n]],
        {x, -2Pi, 2Pi}, {n, 1, 15},
        AxesLabel -> {"x", "n", ""},
        PlotPoints -> {30, 15},
        Shading -> False,
        ViewPoint -> {5, 0, 0}];
```

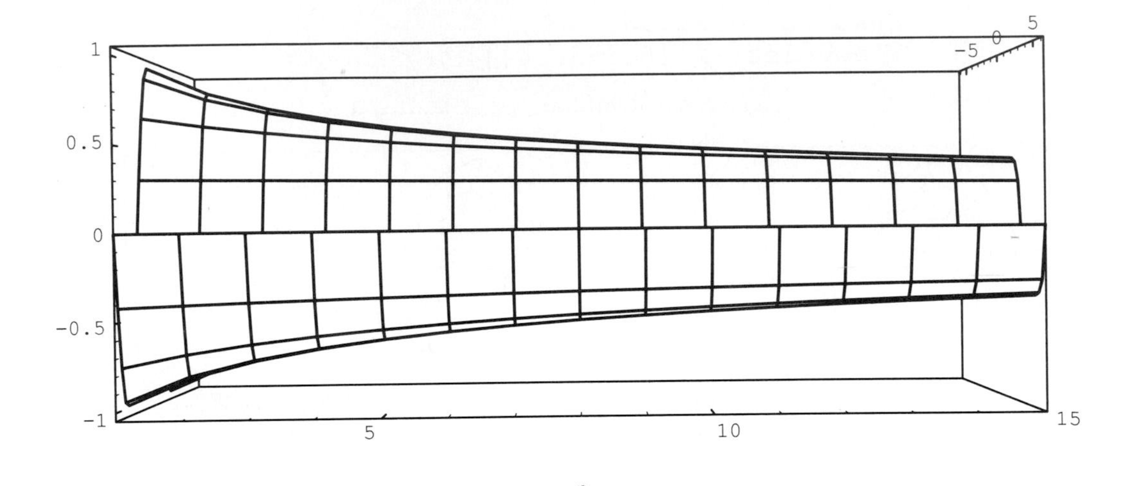

Jerry: We can see that this gets smaller to the right, just like the plot we made before, using **ListPlot**.

Let me pull together for myself what's happened so far. We've looked at the idea of iteration and recursion and the commands **Nest** and **NestList**, and looked at their effects on the **Sin** function. We've looked at this numerically in table form, and in 2D plots (which we've also animated), and in 3D plots from multiple viewpoints. I'm most impressed when I consider that exactly the same development could go on for a different function like a cosine by just changing the word **Sin** to **Cos** everywhere, like using global replace in a word processor.

Theo: This *is* a word processor. *Mathematica* has a global replace feature.

Jerry: Oh, great!

Theo: We could probably get interesting effects using almost any function to replace the **Sin** function -- particularly, functions like **Cos**, **BesselJ**, **AiryAi**, or other functions that tend to oscillate instead of just getting constantly bigger. Functions like **5 - x^2** that go up and down should be interesting, too.

■ Visiting Professor: Jerry Keiper

Jerry Keiper: To see that as we nest **Sin** deeper and deeper the resulting waveform approaches a square wave, let us examine the behavior of two sequences. Given x_0 and y_0 we wish to consider the sequences x_n and and y_n where x satisfies the recurrence relation $x_{n+1} = \sin(x_n)$ and y satisfies the recurrence relation $y_{n+1} = \sin(y_n)$. Now it is clear that $-1 \le x_1 \le 1$ and similarly for y. We will consider the case where $0 < x_1 < y_1$; other cases are similar. It is clear from the definition of the sine function that the sequences x_n and y_n are monotonically decreasing and that for all $n > 0$ we have $0 < x_n < y_n$.

We will use the following theorem from real analysis which we give without proof: Every bounded monotonic sequence approaches a finite limit.

We first note that both sequences converge to zero: Clearly by the theorem they both converge and the only fixed point for the iteration is 0.

Next we note that for n sufficiently large x_n/y_n is monotonically increasing:

$$\frac{x_{n+1}/y_{n+1}}{x/y}$$

$$= \frac{\sin(x_n)}{x_n}\frac{y_n}{\sin(y_n)}$$

$$= \frac{1 - \frac{x_n^2}{6} + \frac{x_n^4}{120} + O(x_n^6)}{1 - \frac{y_n^2}{6} + \frac{y_n^4}{120} + O(y_n^6)}$$

$$= 1 + y_n^2 - x_n^2 + O(x_n^4) + O(y_n^4)$$

Now since $y_n > x_n$ and since both sequences tend to 0 we have that for n sufficiently large x_n/y_n is monotonically increasing.

Now again by the theorem, since x_n/y_n is bounded above by 1, we have that x_n/y_n converges to a finite limit.

Next we examine the sequence $(1 - x_n/y_n) / x_n$. We note that if we can show that this sequence converges to a finite limit then x_n/y_n must converge to 1 since x_n converges to 0.

We first rewrite $(1 - x_n/y_n) / x_n$ as $(y_n - x_n) / (x_n y_n)$. Now we look at the ratio of consecutive terms of this sequence:

$$\frac{(y_{n+1} - x_{n+1}) / (x_{n+1} y_{n+1})}{(y_n - x_n) / (x_n y_n)}$$

$$= 1 - \frac{x_n y_n}{6} + O(x_n y_n^3) + O(x_n^3 y_n)$$

Thus for n sufficiently large we have that the ratio of consecutive terms is less than 1, i.e. the sequence is monotonically decreasing. But clearly $(y_n - x_n) / (x_n y_n)$ is bounded below by 0. Hence, by the theorem we have that the sequence $(1 - x_n/y_n) / x_n$ converges to a finite value and x_n/y_n converges to 1.

Now, the two points x_0 and y_0 were arbitrary (up to the condition that $1 < x_1 < y_1$) so although the limiting waveform of nested **`Sin`** is identically 0, if we keep scaling the function up (as **`Plot`** automatically does) we get, in the limit, a square wave.

Chapter Four
Finding Gold in Continued Fractions

In which Jerry and Theo find Gold at the end of the NestList.

■ Dialog

Jerry: Let's try using **NestList** with some other sorts of functions. As I recall, **NestList[f, x, n]** gives the list **{x, f[x], f[f[x]], f[f[f[x]]], ...}** up to **n**.

How about using **1/x** as the function?

Theo: OK. First we define a function, **f**, which is **1/x**:

```
f[x_] := 1/x;
```

Then we can use it:

```
NestList[f, x, 10]
```

$\{x, \frac{1}{x}, x, \frac{1}{x}, x, \frac{1}{x}, x, \frac{1}{x}, x, \frac{1}{x}, x\}$

This is sort of boring. It just alternates, since **1/(1/x)** is just **x**. How about **1/(1+x)**:

```
f[x_] := 1/(1+x)
NestList[f, x, 6]
```

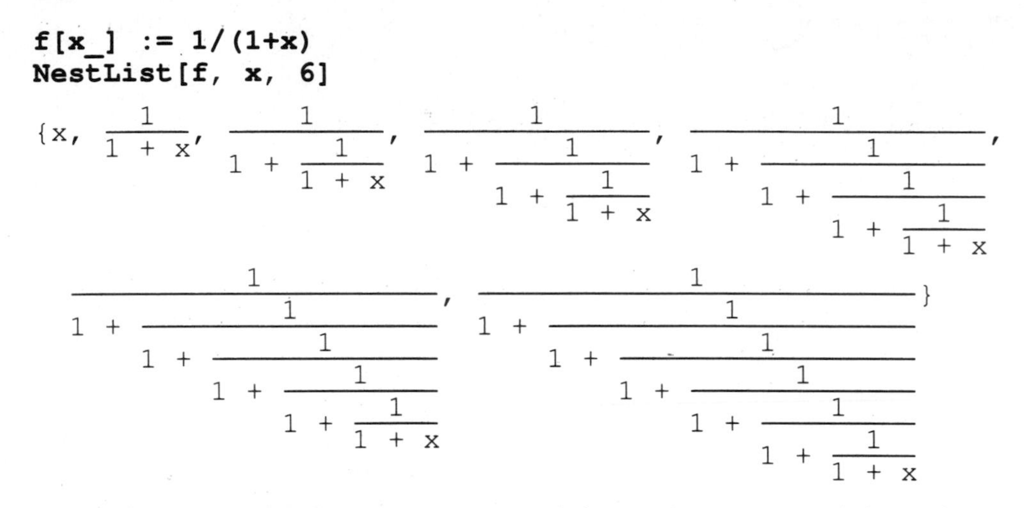

Jerry: That's amazing! With just two commands we get continued fractions! Let's see what happens if we use **1** as the starting point, instead of **x**:

```
f[x_] := 1/(1+x)
NestList[f, 1, 20]
```

$$\{1, \frac{1}{2}, \frac{2}{3}, \frac{3}{5}, \frac{5}{8}, \frac{8}{13}, \frac{13}{21}, \frac{21}{34}, \frac{34}{55}, \frac{55}{89}, \frac{89}{144}, \frac{144}{233}, \frac{233}{377}, \frac{377}{610}, \frac{610}{987}, \frac{987}{1597}, \frac{1597}{2584}, \frac{2584}{4181}, \frac{4181}{6765}, \frac{6765}{10946}, \frac{10946}{17711}\}$$

Theo: I notice right away that both the numerator and the denominator of these fractions follow the Fibonacci sequence.

Jerry: You mean, if we look just at the numerators, we see that the sequence is 1, 1, 2, 3, 5, 8, 13, 21, ..., and if we look just at the denominators the sequence is 1, 2, 3, 5, 8, 13, 21, ..., which is clearly the same sequence. By Fibonacci sequence you mean the sequence you get by saying that each subsequent number in the sequence is equal to the sum of the previous two. The first two numbers you choose to be 1 and 1. So the third number is 1+1 = 2, the fourth number is 1+2 = 3, the fifth number is 2+3 = 5, and so on.

Theo: The Fibonacci connection is an interesting thing to notice about these fractions. If we wanted to calculate what one of them was, we would not actually have to work out the whole continued fraction, we could just use the Fibonacci numbers, which are easy to calculate.

Jerry: It also means that the value of each fraction is equal to the ratio of two successive Fibonacci numbers. I seem to remember something about that sequence of ratios.

Theo: Let's look at these numbers in decimal form. Usually it's much easier to tell how big a number really is by looking at its decimal form instead of its fractional form:

```
f[x_] := 1/(1+x)
N[NestList[f, 1, 20]]
```

```
{1., 0.5, 0.666667, 0.6, 0.625, 0.615385, 0.619048,
  0.617647, 0.618182, 0.617978, 0.618056, 0.618026,
  0.618037, 0.618033, 0.618034, 0.618034, 0.618034,
  0.618034, 0.618034, 0.618034, 0.618034}
```

Jerry: Well, it certainly does look like these numbers settle down to about 0.618 pretty quickly, and don't change much after that. This number looks familiar, isn't it something like the Golden Ratio?

Theo: Let's find out: We'll ask *Mathematica* for the value of the Golden Ratio, which just happens to be a built-in function:

```
N[GoldenRatio]
```

```
1.61803
```

Jerry: Our number looks like it's 1 less than the Golden Ratio, or the reciprocal of the Golden Ratio, which happens to be the same number:

```
1 / N[GoldenRatio]
```

```
0.618034
```

Jerry: It's very convenient that *Mathematica* has the Golden Ratio built in. What other nice numbers are built in?

Theo: Looking through the *Mathematica* book, I found the following mathematical constants. I'm sure there are others:

```
N[{Pi, E, EulerGamma, Catalan}]
```

```
{3.14159, 2.71828, 0.577216, 0.915966}
```

Theo: Let's try printing both the Golden Ratio and the numbers from the **`NestList`** with a few more decimal places:

```
N[GoldenRatio, 20]
```

```
1.6180339887498948482
```

```
f[x_] := 1/(1+x)
N[NestList[f, 1, 20], 20]
```

```
{1., 0.5, 0.66666666666666666667, 0.6, 0.625,
  0.61538461538461538462, 0.61904761904761904762,
  0.61764705882352941176, 0.61818181818181818182,
  0.61797752808988764045, 0.61805555555555555556,
  0.61802575107296137339, 0.61803713527851458886,
  0.61803278688524590164, 0.61803444782168186424,
  0.61803381340012523482, 0.61803405572755417957,
  0.61803396316670652954, 0.61803399852180339985,
  0.61803398501735793897, 0.61803399017559708656}
```

Jerry: It looks like the last number here is the same as the Golden Ratio up to about 7 decimal places. Let's just check that for a really large number of iterations:

```
f[x_] := 1/(1+x)
N[Nest[f, 1, 200], 50]
```

```
0.61803398874989484820458683436563811772030917980576
```

```
N[GoldenRatio, 50]
```

```
1.6180339887498948482045868343656381177203091798058
```

Jerry: Now it's the same to about 48 decimal places. I wonder how fast it converges to that many decimal places. Could we make an animation?

Theo: Animation? Of numbers converging? Well I suppose we could try. We need to make a sequence of plots where each plot has a number from the **NestList**. Since we want a separate plot for each number, it's probably easier to use an individual **Nest** command for each plot, rather than starting with the **NestList** that contains all the numbers at once. We can use the **Text** graphics function to place the number in the plot.

Jerry: Tell me more about the **Text** function.

Theo: **Text** is a graphics function, just like **Line** or **Polygon**. Here's an example:

```
Show[Graphics[Text["Hello", {0, 0}]]];
```

```
Hello
```

Theo: The second argument to **Text**, **{0, 0}**, gives the coordinates of the text to be displayed. (Note that in the example above the coordinates actually don't make any difference, since *Mathematica* will automatically choose the scale of the plot to center the text. If you have other graphical elements in the plot, however, this pair of numbers will determine where the text shows up relative to these other elements.) By default the text is centered about the coordinates you give. If you want the text to be right or left (or top or bottom) aligned with the coordinates instead, you can add an optional third argument to text, like this: **`Text["Hello", {0, 0}, {-1, 0}]`**. This means left aligned. The *Mathematica* book describes in more detail how to use the third argument to specify alignment. Left aligned is what we want for our numbers.

Now we can make our animation. Since these numbers are going to be much wider than they are tall, we want the overall aspect ratio (ratio of height to width) of the plot to be very small. We can use the **`AspectRatio`** option to do this: I'm going to guess that our long, wide numbers have an aspect ratio of about 0.05.

Jerry: It is weird to think of a number as a plot, but I'm getting used to it.

Theo: We use a **`Do`** loop to make all the plots automatically. In this case we run **`n`** from 5 to 100, and use **`Show`** to make a plot for each value of **`n`**.

Jerry: Why are you starting at 5?

Theo: Well, as we saw earlier, for values of **`n`** less than 5 the decimal result is much shorter than 50 digits, and it makes the animation look sort of jumpy. The right edges of the numbers still jump around a bit because the number of digits in the result of an **`N`** command is not necessarily the same as the number of digits you ask for. *Mathematica* starts calculating with the number of digits you ask for, but if there is a loss of accuracy during the calculation, it prints fewer digits in the result.

```
f[x_] := 1/(1+x)
Do[
    Show[Graphics[
        Text[N[Nest[f, 1, n], 50], {0, 0}, {-1, 0}]],
    AspectRatio -> 0.05,
    PlotRange -> {{0, 1}, {-1, 1}}
    ],
    {n, 5, 100}
]
```

(The printed version includes fewer steps, for space reasons.)

```
0.61538461538461538461538461538461538461538461538462
0.61797752808988764044943820224719101123595505617978
0.61803278688524590163934426229508196721311475409836
0.61803396316670652953838794546759148529060033484812
0.61803398820532505147084481976480441078968489374324
0.618033988738303006852732437963934058996629636795
0.61803398874964810153097189343288748385352407282646
0.6180339887498895958965978196611962223644305342 8
0.61803398874989473640271811083789570455902134247698
0.61803398874989484582474584327826103106370428462961
0.61803398874989484815392897678666292899490820535427
0.61803398874989484820350851924118133676198756188312
0.618033988749894848204563881095144601405717319774 8
0.61803398874989484820458634577689963189281388520305
0.61803398874989484820458682396542280014262219373988
0.61803398874989484820458683414425667739651612931195
0.618033988749894848204586834360925740079722792662 5
0.61803398874989484820458683436553780893654203456614
0.61803398874989484820458683436563598252383974068658
0.61803398874989484820458683436563807227001268644385
0.61803398874989484820458683436563811675284343091515
0.61803398874989484820458683436563811769971547530896
0.61803398874989484820458683436563811771987081734654
0.61803398874989484820458683436563811772029984871889
```

Jerry: This certainly converges pretty convincingly to the Golden Ratio.

Theo: It just goes to show that at the end of every **`NestList`** lies a pot of gold.

In this section we learned the value of arbitrary precision arithmetic. It's one thing to find that the continued fractions we examined seem to converge to about the Golden Ratio. It's quite another thing to watch the digits all the way out to 50 places and notice the first 40 places stop changing as we increase the number of nestings to 100.

Jerry: We also learned that plots need not have lines and dots in them. Some plots consist of nothing but numbers.

Theo: Yes. It's quite interesting to think about numbers and algebraic formulas as graphics. I'm sure there are a lot of interesting animations one could make using the technique we used in this chapter.

■ Visiting Professor: Jerry Keiper

If the iteration $x_{n+1} = 1/(1 + x_n)$ converges (as it certainly appears to do) it is clear that it must converge to a solution to the equation

```
q == 1 / (1 + q)
```

$$q == \frac{1}{1 + q}$$

i.e.

```
Solve[q == 1 / (1 + q), q]
```

$$\{\{q \to \frac{-1 + \mathrm{Sqrt}[5]}{2}\}, \{q \to \frac{-1 - \mathrm{Sqrt}[5]}{2}\}\}$$

Likewise, running the recursion backwards, i.e. $x_n = 1 / x_{n+1} - 1$, leads to the same equation for the fixed point and the same solution. Clearly if the original recursion approaches a limit, the reverse recursion must instead move *away* from that limit. Apparently one of the above solutions is the limit for the original recursion and the other is the limit for the reverse recursion.

Let us see how we can decide which of the solutions is the correct limit, or in the language of mathematics which solution is stable and which is unstable. Instead of examining the iteration $x_{n+1} = 1/(1 + x_n)$ let us think of the iteration in terms of a small perturbation of the actual limit, i.e. let

$$x_n = q + e_n$$

where e_n is the perturbation. Then we have

$$q + e_{n+1} = 1 / (1 + q + e_n)$$

and using the fact that $q^2 + q - 1 = 0$ we get

$$e_{n+1} = -e_n\, q \,/\, (1 + q + e_n).$$

Now when the "error" e_n is very small the "error" at the next step will be magnified by the factor $q \,/\, (1 + q + e_n)$ or approximately just $q \,/\, (1 + q)$. For our problem the two different values of q give the values 0.381966 and 2.61803 for the ratio $q \,/\, (1 + q)$. Thus the value

```
-1 + Sqrt[5]
------------
     2
```

is the stable solution.

Chapter Five
I^I^I

In which Theo and Jerry discover the number three in a most unexpected place.

■ Dialog

Jerry: Another function we could try with **Nest** and **NestList** might be **I^x**, where the "imaginary" number, **I**, is the square root of -1. Let's use **I** as the starting value for **x**:

```
f[x_] := I^x
NestList[f, I, 6]
```

$$\{I,\ I^{I},\ I^{I^{I}},\ I^{I^{I^{I}}},\ I^{I^{I^{I^{I}}}},\ I^{I^{I^{I^{I^{I}}}}},\ I^{I^{I^{I^{I^{I^{I}}}}}}\}$$

Theo: The leaning towers of **I**? I think we might see something interesting if we looked at the decimal values of these numbers.

```
f[x_] := I^x
N[NestList[f, I, 6]]
```

```
{1. I, 0.20788, 0.947159 + 0.320764 I,
  0.0500922 + 0.602117 I, 0.387166 + 0.0305271 I,
  0.782276 + 0.544607 I, 0.142562 + 0.400467 I}
```

I guess not. Except perhaps for the third number, which, as Arkady Borkovsky pointed out to us, happens to be:

```
N[E^(-Pi/2)]
```

```
0.20788
```

It's interesting that you can get three of the most common mathematical constants into a single equation:

```
  I    -(Pi/2)
I   == E
```

Otherwise known as:

```
 Pi I
E     == -1
```

Jerry: Moving right along, perhaps we should try plotting the numbers in the complex plane. Maybe then we can see a pattern.

Theo: OK, we can use the **ListPlot** function. Since **ListPlot** expects to get pairs of real numbers, we need to write our own little variation of **ListPlot**, called **complexListPlot**, which takes a list of complex numbers and splits it into a list of pairs of numbers, namely the real and imaginary parts. We can do it like this:

```
complexListPlot[list_, options___] :=
    ListPlot[Transpose[{Re[list], Im[list]}],
        options,
        AspectRatio -> Automatic,
        PlotRange -> All,
        PlotStyle -> PointSize[0.0075],
        AxesLabel -> {"Re", "Im"}]
```

Jerry: This function, **complexListPlot**, looks very complicated, but I recognize some of the elements in it. I've seen **Transpose** before, and I remember the **AspectRatio**, **PlotRange**, and **PlotStyle** options, too. Maybe you should explain in more detail how the **complexListPlot** function works.

Theo: There are really two things going on in the **complexListPlot** function. First, a list of complex numbers is being transformed into a list of pairs of numbers. Second, any extra arguments given to the **complexListPlot** function are being passed on unchanged to the **ListPlot** function, by way of the **options** argument. You can read about the triple-underbar syntax in the *Mathematica* book, so we won't worry about it here. We also add in a few default values for **AspectRatio, PlotRange,** and **AxesLabel**, just to make things look nicer.

Here's how the list of complex numbers is being reformulated. Let's start with a list of complex numbers, call it **myList**:

```
myList =
    {1.00 I, 0.207, 0.947 + 0.320 I, 0.050 + 0.602 I};
```

We can use the **Re** function to get the real parts of the numbers in this list. **Re** is "Listable" which means that, if you apply **Re** to a list of numbers, it goes inside and applies itself to each element of the list, like this:

```
Re[myList]
{0, 0.207, 0.947, 0.05}
```

Likewise we can use the **Im** function to get the imaginary parts:

```
Im[myList]
{1., 0, 0.32, 0.602}
```

If we do both at the same time, in a list, we get this:

```
{Re[myList], Im[myList]}
{{0, 0.207, 0.947, 0.05}, {1., 0, 0.32, 0.602}}
```

Jerry: This is a matrix or a list of lists with two rows. We can show it in a rectangular form, like this:

```
MatrixForm[{Re[myList], Im[myList]}]
0       0.207  0.947  0.05
1.      0      0.32   0.602
```

Theo: Right. This has all the information we want, but instead of being a list of pairs of numbers, it's a list of all the real parts, and then a list of all the imaginary parts. We want to take the first elements of each of the two lists and make them into a pair, and so on for the other elements. **Transpose** does exactly this, as you can see:

```
Transpose[{Re[myList], Im[myList]}]
{{0, 1.}, {0.207, 0}, {0.947, 0.32}, {0.05, 0.602}}
```

Jerry: What a versatile function! We can see how this relates to the matrix-like nature of these lists by looking at the rectangular form again:

```
MatrixForm[Transpose[{Re[myList], Im[myList]}]]
0       1.
0.207   0
0.947   0.32
0.05    0.602
```

Let's look at that plot now:

```
f[x_] := I^x
complexListPlot[N[NestList[f, I, 6]]];
```

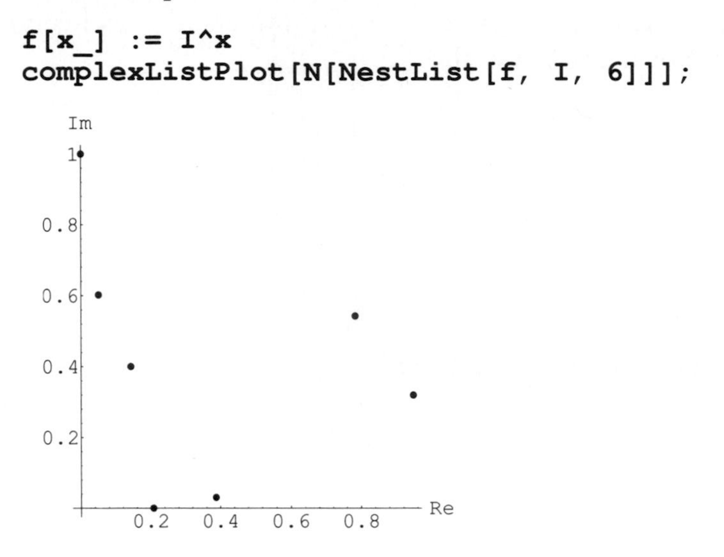

Theo: That doesn't look very inspiring yet. Let's try nesting deeper:

```
f[x_] := I^x
complexListPlot[N[NestList[f, I, 50]]];
```

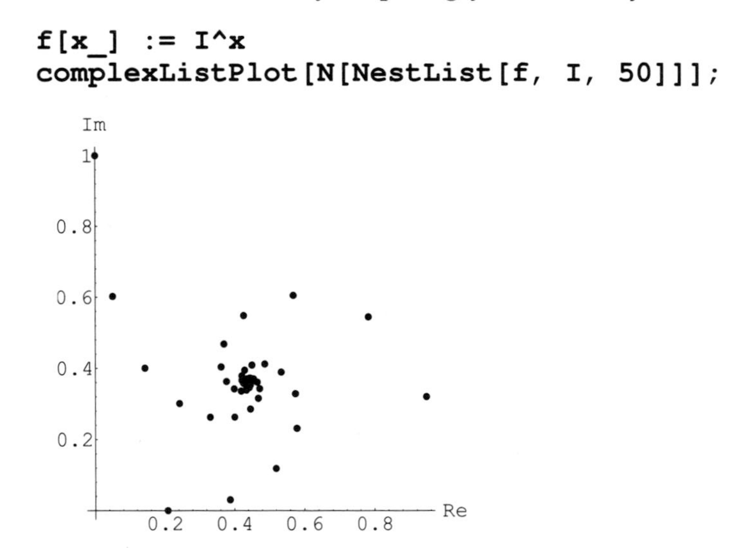

Jerry: It's obviously a triple spiral. I can sort of understand a spiral, but why a triple spiral? Why not a double spiral, or just a single spiral, or even a quadruple spiral? What's so special about the number three? In what order do you suppose the points occurred in the **NestList**?

Theo: By adding the option **PlotJoined->True** we can have *Mathematica* draw a line connecting all the points in the order they occurred in the **NestList**.

```
f[x_] := I^x
complexListPlot[
    N[NestList[f, I, 50]],
    PlotJoined -> True];
```

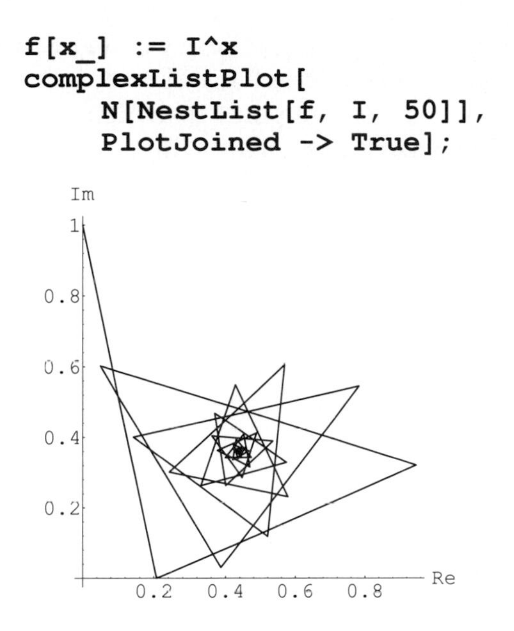

So the list bounces around from one branch of the spiral to the other. I suppose that makes sense, since it would be surpising if the list sort of filled in one arm of the spiral, and then went back out to fill in the next arm.

Jerry: Three certainly does seem a bit arbitrary. I wonder if we can change the function a little, and get different sorts of spirals. Let's try **`(1 + I)^x`**, starting at **`x = 1 + I`**:

```
f[x_] := (1+I)^x
complexListPlot[N[NestList[f, 1+I, 50]],
                AxesOrigin->{0, 0}];
```

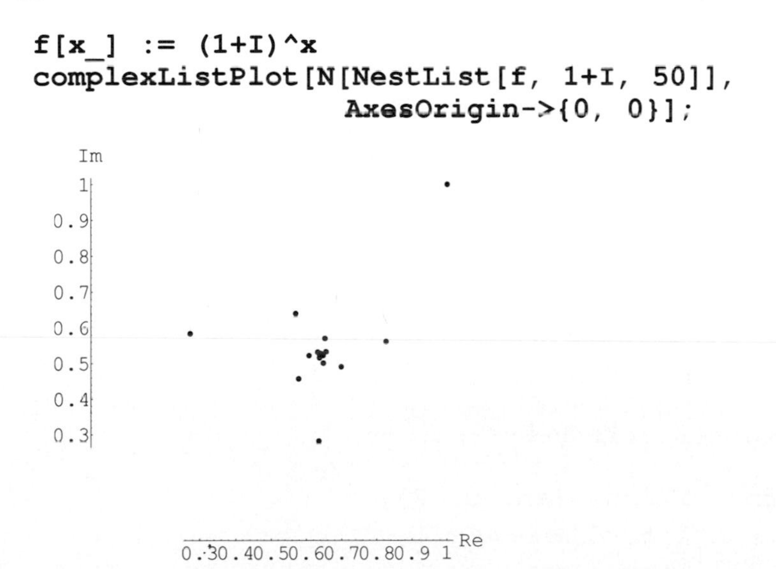

Theo: It's still a triple spiral, but it seems to be converging to a different point.

I wonder if we can see something interesting by comparing where the **NestList** converges to, depending on what power we use. We've tried **I** and **1+I**, but what about all the other possible complex numbers? To see the point to which the sequence converges we can use **Nest** instead of **NestList**. This will give us the last number in the list we would get from **NestList**:

```
f[x_] := (1+I)^x
N[Nest[f, 1+I, 100]]
```

```
0.641026 + 0.523628 I
```

Now we have a function that will tell us the point to which any of these spirals converges. This seems like a nice place to stop and let our visiting professors talk.

Visiting Professors: Jerry Keiper and Dan Grayson

Jerry Keiper: Consider the iteration $x_{n+1} = z\text{^}x_n$. In our problem we have $z = I$, but the more general problem is just as easy to work with. We want to see how the sequence of x_n approaches its limit. As in most iteration problems where the generated sequence approaches a limit, a useful approach is to look at how the successive perturbations from the limit affect each other. To this end let a be the limit of the sequence x_n and let

$$x_n = a + e_n$$

The iteration then becomes

$$a + e_{n+1} = z\text{^}a\ z\text{^}e_n$$

or

$$e_{n+1} = a\ (z\text{^}e_n - 1)$$

since a satisfies the equation $a = z\text{^}a$.

Thus the ratio of consecutive e_n is given by

```
a Series[(z^en - 1)/en, {en, 0, 2}]
```

```
                      2                 3   2
              a Log[z]  en      a Log[z]  en           3
a Log[z] + -------------- + ---------------- + O[en]
                  2                 6
```

Eventually e_n gets small and e_{n+1}/e_n approaches the limit

$$a\ log(z) = log(a).$$

Now in the case where $z = I$ we have $log(a) = -0.566417 + 0.688453\ I$. So the magnitudes of the perturbations decrease by a factor of 0.891514 with each iteration and the perturbations rotate through an angle of about 129.445 degrees. This rotation is close to 120 degrees so the graph appears as a triple spiral.

We can choose z to give us almost any kind of desired behavior:

```
f[z_] := Block[
    {a, loga},
    loga = Log[a] /. FindRoot[a == z^a, {a, {.2+.3I, .3+.4I}}];
    {Abs[loga], Arg[loga] 180/N[Pi]}
]
```

By adjusting the value of z manually (or using **FindRoot** on the function **f**) we get that $z = 0.335664 + 1.33566\ I$ gives a triple spiral with no rotation at all.

Likewise, $z = -0.29931 + 1.29931\ I$ gives a "spiral" with only rotation and no shrinking at all. Actually with this value of z we do not get convergence so our perturbations do not approach 0 and the analysis above does not apply. Nevertheless the plot is rather interesting.

Dan: Let's actually show how to find the limit point of the original sequence with **FindRoot**.

```
FindRoot[I^z-z, {z,1}]
{z -> 0.438283 + 0.360592 I}
```

Let's set **a** to that value.

```
a = z /. %
0.438283 + 0.360592 I
```

And let's check **I^a**:

```
I^a
0.438283 + 0.360592 I
```

It's the same as **a**, as desired.

If we define a function:

```
f[z_] := I^z
```

then `f[a]==a`. Consider a number `z` very near to `a`, then `f[z]-f[a]` is about equal to `f'[a](z-a)`, more or less by the definition of the derivative. Since `f[a]==a` we see that `f[z]-a` is about equal to `f'[a](z-a)`. So replacing `z` by `f[z]` has approximately the same effect as multiplying `z-a` by `f'[a]`. So the question is, what happens to a complex number when it gets multiplied by `f'[a]`? Its length gets multiplied by the absolute value of `f'[a]`, and its angle (or argument) gets increased by the argument of `f'[a]`. Let's examine these two numbers:

```
Abs[N[f'[a]]]
```

```
0.891514
```

Here we get a number less than one, so our iteration decreases the size of `z-a`. This means that further iteration will get us closer and closer to `a`, so that `a` is a *stable* fixed point of the function `f`.

```
N[ Arg[ f'[a] ] / Degree ]
```

```
129.445
```

We divided by `Degree` here so that the answer would be a number of degrees rather than a number of radians. The answer tells us that iteration adds about 129 degrees to the angle of the line from a to z, which, as Jerry Keiper said, is nearly 1/3 of a full rotation.

■ Functions That We Defined

Theo: In this section we defined the **complexListPlot** function:

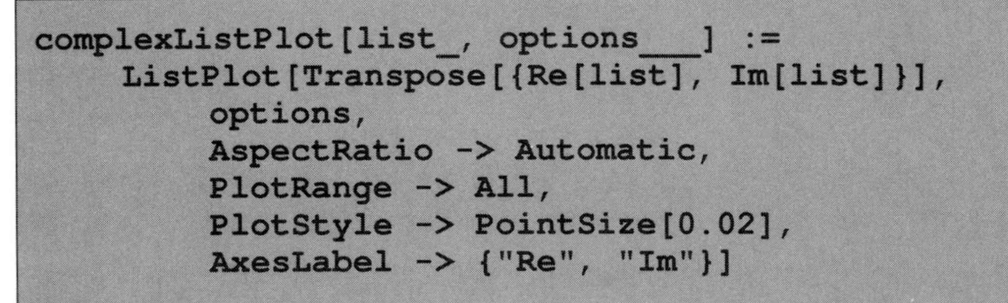

```
complexListPlot[list_, options___] :=
    ListPlot[Transpose[{Re[list], Im[list]}],
        options,
        AspectRatio -> Automatic,
        PlotRange -> All,
        PlotStyle -> PointSize[0.02],
        AxesLabel -> {"Re", "Im"}]
```

This function takes a list of complex numbers as its first argument and plots them in the complex plane.

Chapter Six
If I Have Seen Far It Is Because I Climbed a Tree

In which Jerry and Theo find that while NestLists don't grow on trees, trees do grow on NestLists.

■ Dialog–First Day

Theo: I think we could use **NestList** to make some interesting graphics. Let's try to make a branching tree.

Jerry: OK. I know how to make lines in *Mathematica*, using the **Line** function. Perhaps we could use lines like this to make the branches of a tree:

```
Show[Graphics[Line[{{0, 0}, {1, 1}}]],
    AspectRatio -> Automatic];
```

This just means a line from (0, 0) to (1, 1). The **Graphics** function means two dimensional graphics, and **Show** makes it actually draw the line. **AspectRatio ->**

Automatic makes the line come out at a 45 degree angle, like it should -- it would also make a circle look like a circle, not an ellipse.

Theo: Of course, we don't have to restrict ourselves to just one line. We can, for example, show a table of 20 random lines:

```
Show[Graphics[
    Table[
        Line[{{Random[], Random[]},
              {Random[], Random[]}}],
        {20}
    ]
]];
```

Or we can show a line with 20 random bends in it:

```
Show[Graphics[
    Line[Table[{Random[], Random[]}, {20}]]
]];
```

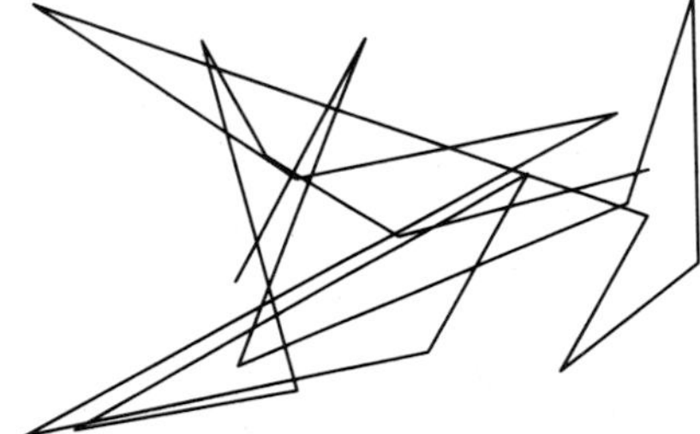

To make a tree, we're going to have to be more selective about our lines. The basic thing we need to be able to do is to take a single line and make it split into two or more branches. For now, let's say that whenever a branch splits, it splits exactly in half, and each of the two smaller branches goes off symmetrically from the original branch. To give you a picture of what I mean, I'll draw one branch manually:

That's sort of the idea. Now, if we could split each of the upper branches the same way, and then split each of those branches, we would start to have a tree.

Jerry: This certainly sounds like a job for **NestList**. I think the function we want to nest is one that takes a single line and makes it into two lines.

Theo: Yes, and to do that we're going to have to decide two more things: How much we want the two branches to be rotated from the original branch, and how much shorter (or longer) than the original we want the branches to be. Let's choose some variables for those two things, and give them trial values:

```
rotation = 0.3;
shrinkage = 0.8;
```

We're also going to need a two dimensional rotation matrix, to calculate the location of the new branches, which will be rotated a certain amount from the original branch. We can get such a rotation matrix from the standard Rotations package, which we can load like this:

```
Needs["Geometry`Rotations`"]
```

This contains, among other things, a function **Rotate2D**, which explains itself like this:

```
?Rotate2D

Rotate2D[vec,theta,(pt:{0,0})] rotates the vector vec by
   angle theta about point pt.
```

Now we can write the function, like this:

```
twinLine[Line[{start_, end_}]] :=
     {
          Line[{
               end,
               end + shrinkage Rotate2D[
                         end - start,
                         rotation,
                         {0, 0}
                    ]
          }],
          Line[{
               end,
               end + shrinkage Rotate2D[
                         end - start,
                         -rotation,
                         {0, 0}
                    ]
          }]
     }
```

I should probably explain a little bit about how vector manipulations work in *Mathematica*. You can add two vectors, say **start** and **end**, by saying **start+end**. (Where **start** and **end** are each a list of two numbers, that is, vectors.) You can multiply a vector by a scalar by saying, for example, **shrinkage*end** (where **shrinkage** is a number and **end** is a vector). If you multiply two vectors, as in **start*end**, *Mathematica* multiplies each of the corresponding elements. If you want the dot-product, use **end.start**, and if you want the cross-product, use the **Cross** function (defined in the standard package `Vectors.m`).

Jerry: Let's try to understand the **twinLine** function. First, the pattern on the left hand side of the **:=** says that this is a function that can be applied only to a line with two endpoints (which are going to be called **start** and **end**). The result of the function is a list of two **Line** functions, each of which starts at the **end** point of the original line. To calculate the end point of the first new line, we rotate the original line about its starting point, and then add that to the end point of the old line, after first shrinking its length by the specified amount. The end point of the second new line is just the same, except rotated in the negative direction.

Theo: Right. Let's try it out on a line from **{0, 0}** to **{1, 1}**. We'll show both the original line and the twinned lines:

```
Show[
    Graphics[{
        Line[{{0, 0}, {1, 1}}],
        twinLine[Line[{{0, 0}, {1, 1}}]]
    }],
    AspectRatio -> Automatic
];
```

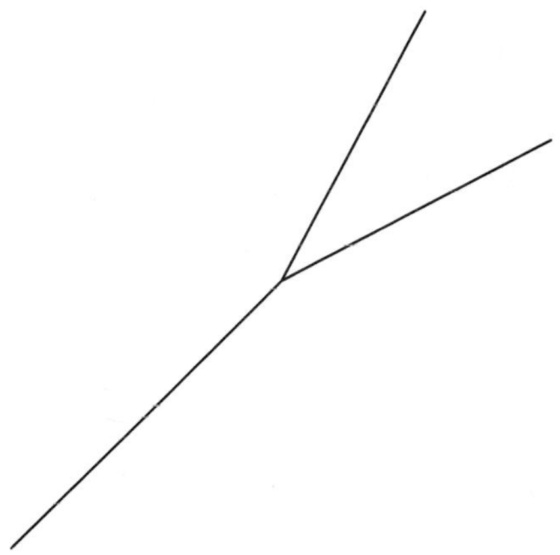

This is pretty good, but it's not quite enough. The function we really need is one that works on a whole list of lines, splits them all into twins, and then returns the whole list of twins. Let's call this function **doTwins**, and define it using the **twinLine** function from above:

```
doTwins[lines_] :=
    Map[twinLine, lines]
```

Let's try it out on two simple lines:

```
doTwins[{Line[{{0, 0}, {1, 1}}],
         Line[{{0, 0}, {-1, 1}}]}]
{{Line[{{1, 1}, {2.00069, 1.52785}}],
   Line[{{1, 1}, {1.52785, 2.00069}}]},
  {Line[{{-1, 1}, {-1.52785, 2.00069}}],
   Line[{{-1, 1}, {-2.00069, 1.52785}}]}}
```

Well, this is not quite right. Notice it did not really return a list of **Line** functions. It returned a list of two lists of **Line** functions. This is not good, since we are going to want to re-apply **doTwins** to this list, and **doTwins** doesn't know what to do with lists of lists of **Line** functions. Let's modify **doTwins** so it flattens out the list before returning it:

```
doTwins[lines_] :=
    Flatten[Map[twinLine, lines]]
```

Again:

```
doTwins[{Line[{{0, 0}, {1, 1}}],
         Line[{{0, 0}, {-1, 1}}]}]
```

```
{Line[{{1, 1}, {2.00069, 1.52785}}],
  Line[{{1, 1}, {1.52785, 2.00069}}],
  Line[{{-1, 1}, {-1.52785, 2.00069}}],
  Line[{{-1, 1}, {-2.00069, 1.52785}}]}
```

Jerry: So, **Flatten** takes nested lists and makes them into a single list?

Theo: Right. It's almost as useful as **Transpose**. It can also take additional options to restrict its flattening action to certain levels in a complex nested list. You can read more about it in the *Mathematica* book.

Now we can start nesting these **doTwins**:

```
Nest[doTwins, {Line[{{0,0}, {0,1}}]}, 2]
```

```
{Line[{{0.236416, 1.76427}, {0.597787, 2.29248}}],
  Line[{{0.236416, 1.76427}, {0.236416, 2.40427}}],
  Line[{{-0.236416, 1.76427}, {-0.236416, 2.40427}}],
  Line[{{-0.236416, 1.76427}, {-0.597787, 2.29248}}]}
```

Let's look at that:

```
Show[Graphics[
    Nest[doTwins, {Line[{{0,0}, {0,1}}]}, 2]
]];
```

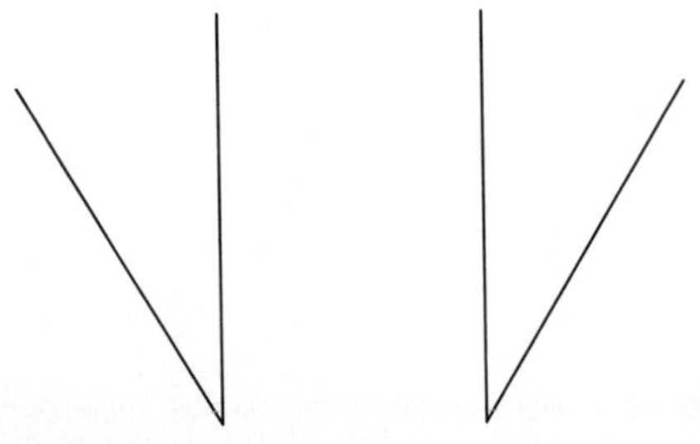

These are just the last branches, without the ones we started with. If we use **NestList** instead of **Nest**, we will get a list of lists which includes not only the final branches but also all the intermediate branches, like this:

```
NestList[doTwins, {Line[{{0,0}, {0,1}}]}, 2]
{{Line[{{0, 0}, {0, 1}}]},
  {Line[{{0, 1}, {0.236416, 1.76427}}],
   Line[{{0, 1}, { 0.236416, 1.76427}}]},
  {Line[{{0.236416, 1.76427}, {0.597787, 2.29248}}],
   Line[{{0.236416, 1.76427}, {0.236416, 2.40427}}],
   Line[{{-0.236416, 1.76427}, {-0.236416, 2.40427}}],
   Line[{{-0.236416, 1.76427}, {-0.597787, 2.29248}}]}}
```

We can show these lines. I've added a few options to the **Show** command, just to make things look nicer. I've also nested one level deeper:

```
Show[
    Graphics[
        NestList[doTwins, {Line[{{0,0}, {0,1}}]}, 3]
    ],
    AspectRatio -> Automatic,
    PlotRange -> All
];
```

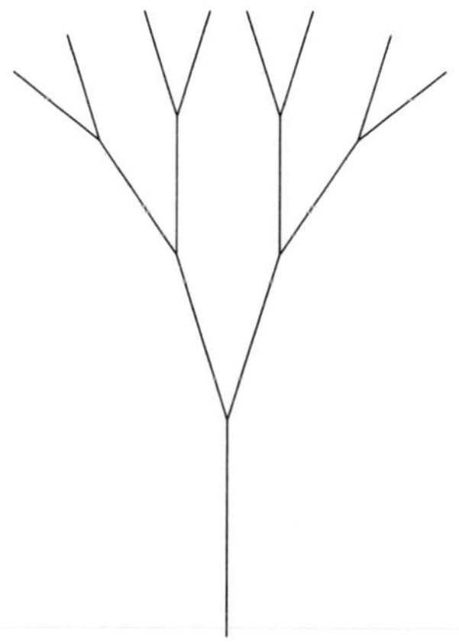

Or, starting with a different initial line, we can get a bent tree:

```
Show[
    Graphics[
        NestList[doTwins, {Line[{{0,0}, {1,1}}]}, 3]
    ],
    AspectRatio -> Automatic,
    PlotRange -> All
];
```

Jerry: Very pretty! I bet we could make a lot of different shapes by changing the angle and shrinkage variables, which we've been leaving constant so far.

Theo: I'm sure we could. To make this more convenient, let's make a function, **`tree`**, which generates any tree just by specifying the number of times we want to nest, the bend angle, and the shrinkage:

```
tree[nestingDepth_, rotationP_,
                    shrinkageP_, options___] :=
    Block[{},
        rotation = rotationP;
        shrinkage = shrinkageP;
        Show[
            Graphics[
                NestList[
                    doTwins,
                    {Line[{{0,0}, {0,1}}]},
                    nestingDepth
                ]
            ],
            options,
            AspectRatio -> Automatic,
            PlotRange -> All
        ]
    ]
```

Jerry: What are the variables called **rotationP** and **shrinkageP** for?

Theo: We already have the function **twinLine**, which refers to the global variables **rotation** and **shrinkage**. We need to reset the values of those global variables to the values passed as arguments to the **tree** function. We have to use arguments with different names to avoid confusion. In the Functions That We Defined section below we'll use a more sensible method to get around this naming problem. Let's try **tree** out on some different values:

```
tree[3, 0.2, 0.8];
```

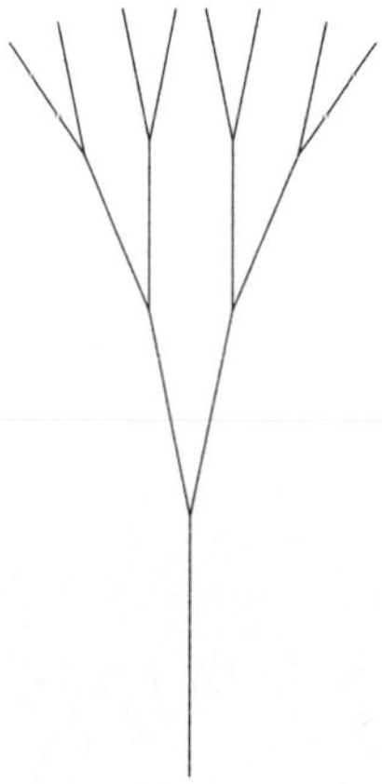

```
tree[7, 0.3, 0.8];
```

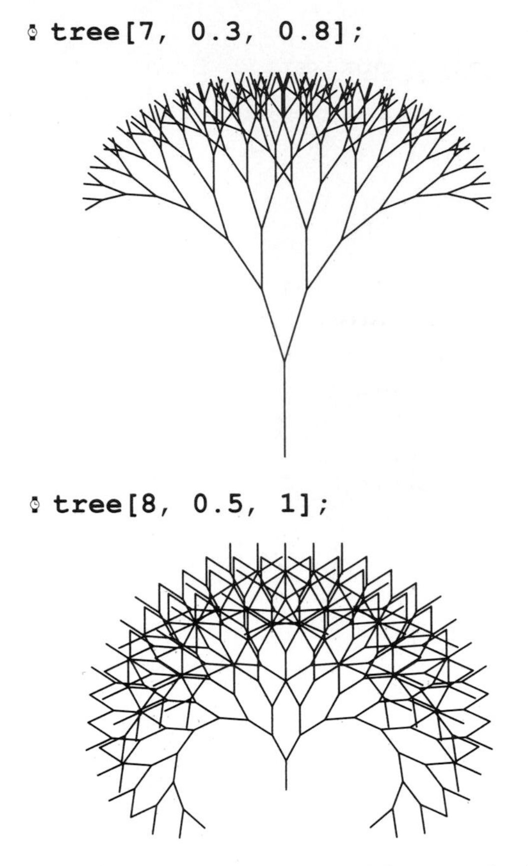

```
tree[8, 0.5, 1];
```

Jerry: You know, the only problem with these admittedly very pretty trees is that in real trees the trunk is generally thicker than the branches. The branches get thinner as well as shorter the further out you go.

Theo: You're right about that. Not to fear, we can fix our mathematical trees to form a more perfect model of Nature. We need two more parameters, the ratio by which the branches get thinner at each split (this is just like the **shrinkage** parameter, only for thickness instead of length), and the initial thickness of the trunk. We'll call the first parameter **thicknessShrinkage** and the second one **original-Thickness**.

```
thicknessShrinkage = 0.8;
originalThickness = 0.05;
```

Amazingly, all we have to do to make our **tree** function include this new feature is to add the following rule for the **twinLine** function, and make a minor change to **tree** itself:

```
twinLine[Thickness[th_]] :=
                Thickness[th thicknessShrinkage];
```

Jerry: Let me understand this. The definition you made now does not replace the (much longer) definition from near the beginning of this chapter. Instead it just adds to it?

Theo: Right. *Mathematica* is a pattern-based language. We now have two rules for the twinLine function, one for the pattern `twinLine[Line[{start_, end_}]]` and one for the pattern `twinLine[Thickness[th_]]`. When we apply the `twinLine` function to an expression, one or the other of these rules will be used. In other words, the `twinLine` function does two completely different things depending on whether you apply it to a `Line` or to a `Thickness`. If we use `doTwins` to map `twinLines` onto a list which includes both `Line` and `Thickness` functions, each will be transformed in its own special way.

Jerry: What's `Thickness` anyway?

Theo: `Thickness` is a graphics function which specifies how thick the lines following it are to be drawn. To finish adding this new feature, we need to add one more argument to the `tree` function, and include an initial `Thickness` as well as an initial `Line`:

```
Clear[tree];
tree[nestingDepth_, rotationP_,
          shrinkageP_, thicknessShrinkageP_,
          originalThicknessP_, options___] :=
    Block[{},
        rotation = rotationP;
        shrinkage = shrinkageP;
        thicknessShrinkage = thicknessShrinkageP;
        originalThickness = originalThicknessP;
        Show[
            Graphics[
                NestList[
                    doTwins,
                    {
                        Thickness[originalThickness],
                        Line[{{0,0}, {0,1}}]
                    },
                    nestingDepth
                ]
            ],
            options,
            AspectRatio -> Automatic,
            PlotRange -> All
        ]
    ]
```

Jerry: Let's try out a few of these:

```
tree[6, 0.3, 0.8, 0.7, 0.05];
```

```
tree[8, 0.5, 1, 0.75, 0.05];
```

I'm ready to go climbing! How much more of Nature can you include in our tree?

Theo: I think we can show a tree growing by making an animation where the thickness, shrinkage and the bending angle grow slowly bigger.

The **`PlotRange`** option (which gets passed on to the **`Show`** inside of **`tree`**) just makes sure the automatic scaling usually used on graphics doesn't mess up our movie. Note that I spent quite a while picking out the exact formulas to use for the various parameters to **`tree`**:

```
Do[
     tree[7, r, 0.5+r, 0.75, 0.005 + 0.07 r,
          PlotRange->{{-6, 6}, {0, 8}}
     ],
     {r, 0, 0.5, 0.02}
]
```

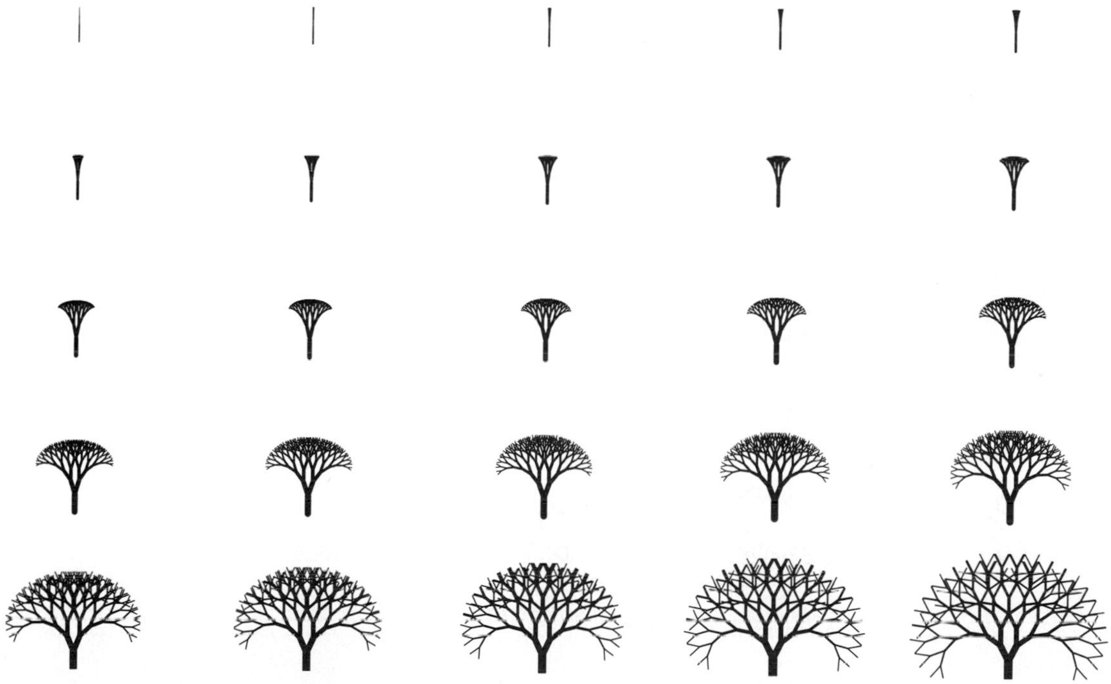

Jerry: So you really can use **NestList** to grow trees.

Theo: Isn't mathematics wonderful?

Dialog–Second Day

Jerry: Growing perfectly symmetrical trees is interesting, but just a little bit Type A. Can we try to make our trees looser, more relaxed-looking?

Theo: Yes, I think we should. As any bonsai expert will tell you, symmetry is unacceptable in trees. We need to introduce some randomness into our trees.

In the Functions That We Defined section below you can see the definition of a new, improved function, called **`BeautifulTree`**. I'm not going to go through this function in detail, but if you want more information you can look down where it's defined. This new function works the same way the older **`tree`** function worked, but it automatically introduces randomness. It multiplies the rotation angle you specify by a random number between 0.5 and 1.5 each time before using it.

Jerry: So, the angle we specify is sort of like a suggestion, and it uses some angle like that, but not exactly the one we wanted? That sounds natural.

Theo: Also, **`BeautifulTree`** uses named options instead of multiple arguments to specify all its options. This is less confusing, since each value is explicitly named, as you can see in this example:

```
BeautifulTree[
      Generations -> 10,
      BranchRotation -> 0.65,
      BranchShortening -> 0.75,
      BranchThinning -> 0.7,
      OriginalThickness -> 0.07
      ];
```

Jerry: We discussed this trade-off between arguments and options in Chapter 2. This looks like a good use of options, even though it is a bit wordy.

Theo: It is sort of long, but you don't have to include all the options in every use of **BeautifulTree**. Any option can be omitted, in which case **BeautifulTree** will use a default value. The values above happen to be the same as the defaults, so we can get the same tree using this much simpler command:

```
BeautifulTree[];
```

Jerry: Ahem. That's clearly not the same tree. Oh, I forgot about the randomness. I guess we will never get exactly the same two trees, even if we use identical commands. That seems pleasingly un-computerlike.

Theo: It's not the same tree, but it was grown using the same rules. The following tree is clearly of a different species, because it is grown using different rules:

```
BeautifulTree[BranchRotation -> 0.3];
```

Jerry: While were at it, I haven't seen many black trees lately. Can't we make this tree a bit more colorful?

Theo: Naturally. In fact, I've already done that. **`BeautifulTree`** can be given an option named **`BranchColors`** that determines how the branches are colored. **`BranchColors`** should be a list of colors (given in the form of **`RGBColor`** graphics primitives). The first color in the list is used for the trunk of the tree. The second color is used for the two branches connected to the trunk. The third color is used for the four branches connected to these two, and so on. If there are more generations of branches than colors in the list, the colors are reused as many times as needed. (For example, if there is only one color in the list, it will be used for all the branches.)

Jerry: If we were to make the first few generations of branches brown, and the last few green, the tree might look like it had brown branches and green leaves. How do we specify "brown" in an **`RGBColor`** object?

Theo: The easiest way is to use the automatic "Color Selector" command in the Notebook Front End. This command (found in the Prepare Input submenu of the Action menu, or with Shift-Command-R) puts up a standard color selector dialog box that allows you, interactively, to pick a color from a color wheel. The color is then pasted into your current text insertion point. Using this feature, I made up the following list of colors:

```
BeautifulTree[BranchColors ->
      {
      RGBColor[0.562, 0.236, 0.071],
      RGBColor[0.547, 0.229, 0.069],
      RGBColor[0.500, 0.210, 0.063],
      RGBColor[0.469, 0.196, 0.059],
      RGBColor[0.033, 0.281, 0.035],
      RGBColor[0.046, 0.395, 0.050],
      RGBColor[0.055, 0.469, 0.059],
      RGBColor[0.070, 0.602, 0.076],
      RGBColor[0.085, 0.727, 0.092],
      RGBColor[0.109, 0.937, 0.118],
      RGBColor[0.013, 0.750, 0.028]
      }
];
```

See picture on color plates page i.

I like this tree. The colors help give a three-dimensional effect, because you can see the leaves in front of the branches. Let's try making a Red Maple:

```
BeautifulTree[BranchColors ->
        {
        RGBColor[0.562, 0.236, 0.071],
        RGBColor[0.547, 0.229, 0.069],
        RGBColor[0.500, 0.210, 0.063],
        RGBColor[0.469, 0.196, 0.059],
        RGBColor[0.033, 0.281, 0.035],
        RGBColor[0.562, 0.154, 0.084],
        RGBColor[0.750, 0.206, 0.112],
        RGBColor[0.875, 0.240, 0.131],
        RGBColor[0.902, 0.248, 0.135],
        RGBColor[0.902, 0.160, 0.125],
        RGBColor[0.977, 0.048, 0.039]
        }
  ];
```

See picture on color plates page i.

Jerry: Well, don't give up your day job yet. I think the only people who would say that looks like a Red Maple are people who've never seen a Red Maple. But that's what I think of most of this computer-imitating-life-with-fractals business.

Theo: So, it's not perfect, but it is red. Maybe the problem is that we didn't use Maple™ to make it.

Functions That We Defined

Theo: In this chapter we defined several functions that all work together. We can build a single function that includes all the functions in this chapter as sub-functions. Here's the **tree** function and both functions it relies on, in one:

```
Needs["Geometry`Rotations`"]
```

```
tree[nestingDepth_, rotation_,
            shrinkage_, thicknessShrinkage_,
            originalThickness_, options___] :=
Block[{twinLine, doTwins},

    twinLine[Line[{start_, end_}]] :=
        {
            Line[{
                end,
                end + shrinkage Rotate2D[
                        end - start,
                        rotation,
                        {0, 0}
                    ]
            }],
            Line[{
                end,
                end + shrinkage Rotate2D[
                        end - start,
                        -rotation,
                        {0, 0}
                    ]
            }]
        };
    twinLine[Thickness[th_]] :=
        Thickness[th thicknessShrinkage];

    doTwins[lines_] :=
        Flatten[Map[twinLine, lines]];
```

```
        Show[
            Graphics[
                NestList[
                    doTwins,
                    {
                        Thickness[originalThickness],
                        Line[{{0,0}, {0,1}}]
                    },
                    nestingDepth
                ]
            ],
            options,
            AspectRatio -> Automatic,
            PlotRange -> All
        ]
    ]
```

We don't use any global variables anymore, and **twinLine** and **doTwins** are "local" functions, which exist only inside the **Block** command. Global variables are usually a bad idea if you can avoid them.

On the second day we defined a new and improved function, **BeautifulTree**, that includes several important improvements. First, the rotation angle specified is multiplied by a random number between 0.5 and 1.5 (by default) before being used, causing a pleasing randomness in the shape.

Second, the color of successive generations of branches can be specified. The colors are specified as a list of **RGBColor** objects that give the colors of successive generations of branches. The list is used cyclically if there are more generations than colors in the list.

Third, the long string of arguments to the old **tree** function has been replaced with a set of named options. It is not necessary to specify all the options in every use of **BeautifulTree**; all options have default values that will be used if they are missing.

In addition, the "local" functions used in the old **tree** function have been moved out to be global functions again. Local variables are used to supply them with the necessary values. In this particular case, both approaches are valid and the choice of which to use is largely a matter of taste. I wanted to demonstrate both techniques.

The utility function **FilterOptions** is reprinted from Roman Maeder's book [1].

```
Clear[rotateLine];
rotateLine[Line[{start_, end_}], angle_] :=
        Line[{
            end,
            end + branchShortening *
                Rotate2D[
                    end - start,
                    angle Random[Real, {0.5, 1.5}]
                ]
        }];
```

```
Clear[twinLine];
twinLine[Line[points_]] :=
    {
        rotateLine[Line[points], branchRotation],
        rotateLine[Line[points], -branchRotation]
    };
```

```
twinLine[Thickness[th_]] :=
    Thickness[th branchThinning];
```

```
twinLine[RGBColor[r_, g_, b_]] :=
    (
    branchColors = RotateLeft[branchColors];
    First[branchColors]
    );
```

```
Clear[doTwins];
doTwins[lines_] :=
    Flatten[Map[twinLine, lines]];
```

```
Options[BeautifulTree] =
     {
          Generations -> 10,
          BranchRotation -> 0.65,
          BranchShortening -> 0.75,
          BranchThinning -> 0.7,
          OriginalThickness -> 0.07,
          BranchColors ->
               {
                    RGBColor[0, 0, 0]
               }
     };
```

```
BeautifulTree[options___] := Block[
     {generations, branchRotation,
          branchShortening, branchThinning,
          originalThickness, branchColors},

     {generations, branchRotation,
          branchShortening, branchThinning,
          originalThickness, branchColors} =
     {Generations, BranchRotation,
          BranchShortening, BranchThinning,
          OriginalThickness, BranchColors} /.
               {options} /. Options[BeautifulTree];

     Show[
          Graphics[
               NestList[
                    doTwins,
                    {
                         First[branchColors],
                         Thickness[originalThickness],
                         Line[{{0, 0}, {0, 1}}]
                    },
                    generations
               ]
          ],
          FilterOptions[Show, options],
          AspectRatio -> Automatic,
          PlotRange -> All
]];
```

```
FilterOptions[ command_Symbol, opts___ ] :=
    Block[{keywords = First /@ Options[command]},
        Sequence @@ Select[ {opts},
                        MemberQ[keywords, First[#]]& ]
    ]
```

■ References

1) Roman Maeder, *Programming In Mathematica*, Addison-Wesley, 1989.

Chapter Seven
Bifurcations Forever

In which Theo and Jerry use NestList to go from order to chaos **[1]**.

Dialog–First Day

Jerry: There are connections between recursion, iteration, and chaos. For example, the famous Mandelbrot set is generated by applying a simple function iteratively to numbers in the complex plane. Those numbers for which the value of the function keeps getting larger, the more deeply nested you go, are outside of the Mandelbrot set. All the numbers for which the value stays small are inside the set.

Theo: Another example of iteration and chaos, and one that's particularly easy to study, is the famous "iterated quadratic mapping". The idea is to take a function, namely the quadratic function `lambda x (1 - x)`, and iteratively apply it to some starting value. Depending on the value of the `lambda` parameter, different things can happen. We can define this function in *Mathematica*:

```
quad[x_] := lambda x (1 - x);
```

Jerry: Let's try looking at this function using `NestList`:

```
lambda = 2.9;
NestList[quad, 0.4, 30]
```

```
{0.4, 0.696, 0.613594, 0.68758, 0.62296, 0.681154,
  0.629831, 0.676117, 0.63505, 0.672109, 0.639098,
  0.66889, 0.642281, 0.666293, 0.644805, 0.664191,
  0.64682, 0.662487, 0.648434, 0.661106, 0.64973,
  0.659984, 0.650775, 0.659074, 0.651617, 0.658336,
  0.652296, 0.657737, 0.652845, 0.657251, 0.653289}
```

The sequence seems to settle down to a single number. Let's try it for a different value of **lambda**:

```
lambda = 3.1;
NestList[quad, 0.4, 30]
```

```
{0.4, 0.744, 0.590438, 0.749645, 0.5818, 0.754257,
  0.574595, 0.75775, 0.569051, 0.760219, 0.565087,
  0.761868, 0.562419, 0.762922, 0.560703, 0.763577,
  0.559634, 0.763976, 0.558982, 0.764215, 0.55859,
  0.764358, 0.558355, 0.764443, 0.558216, 0.764494,
  0.558133, 0.764524, 0.558085, 0.764541, 0.558056}
```

It looks like it settles down to alternating between two numbers. Let's try to think of a way of visualizing these numbers, because I have the feeling that we're going to have trouble understanding what's going on just by looking at strings of numbers.

Theo: OK. Let's start with **lambda = 2.9** first. What does the function look like?

```
lambda = 2.9;
Plot[quad[x], {x, 0, 1}];
```

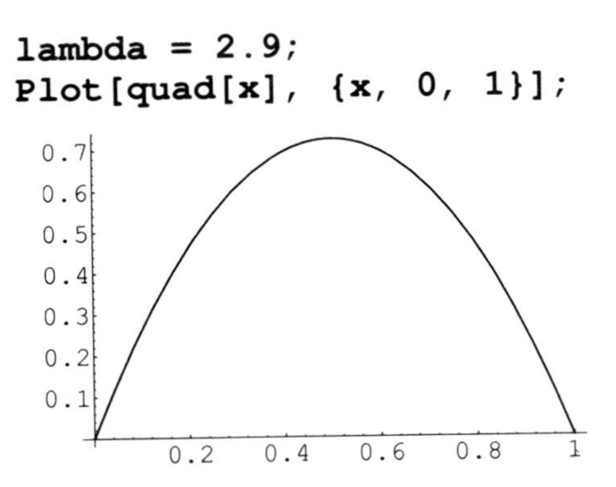

When we apply this function to a starting value, say **0.2**, we can read off the value of the function by drawing a vertical line from the **x** axis to the curve, at **x = 0.2**, and then a horizontal line from that point to the **y** axis, like this:

```
lambda = 2.9;
Plot[quad[x], {x, 0, 1},
    Epilog -> {
        Line[{{0.2,    0},         {0.2, quad[0.2]}}],
        Line[{{0.2, quad[0.2]}, {  0, quad[0.2]}}]
        }
];
```

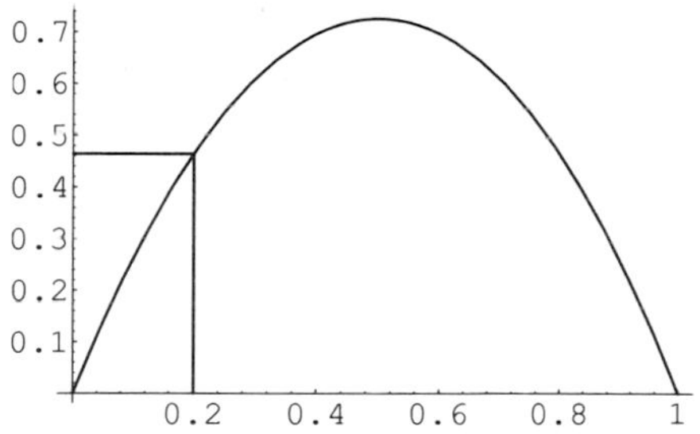

Jerry: That's an interesting option. What does **`Epilog`** mean?

Theo: The **`Epilog`** option to the **`Plot`** command lets you add any arbitrary two dimensional graphics primitives (like **`Line`**, **`Polygon`**, **`Circle`**, etc.) to a plot. **`Epilog`** adds the graphics after all the elements of the plot have been drawn. There is another option, **`Prolog`**, which adds its graphics before the rest of the plot is drawn. This lets you control who overwrites whom.

Jerry: Back to the plot: This shows us that when you start with **`x = 0.2`**, you get out about **`y = 0.5`**. Now we want to feed that **`0.5`** back in as the next **`x`** value.

Theo: Right. In other words, we want to translate the **`y`** value into an **`x`** value, and plug it back in. To translate a **`y`** value into an **`x`** value graphically, go to the line where **`x = y`**. Let's add that line to our plot, and draw the horizontal line we now have to the **`x = y`** line instead of to the **`y`** axis:

```
lambda = 2.9;
Plot[quad[x], {x, 0, 1},
    Epilog -> {
        Line[{{0, 0}, {1, 1}}],
        Line[{{0.2,           0}, {0.2,          quad[0.2]}}],
        Line[{{0.2, quad[0.2]}, {quad[0.2], quad[0.2]}}]
        }
];
```

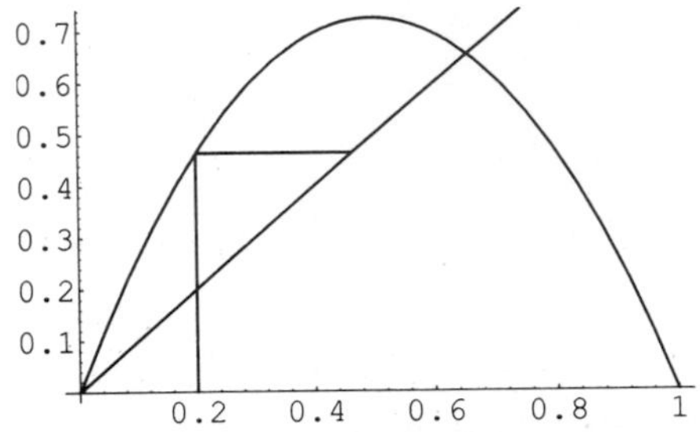

We see that where the horizontal line meets the diagonal line we can read off the next **x** value to use. We just have to draw another vertical line up to the **quad** curve, to get the next **y** value.

Jerry: I see a **NestList** coming!

Theo: Exactly. All we need to do now is to make an automatic way to repeat this process many times. We start with the list of numbers from using **NestList** on this **quad** function:

```
lambda = 2.9;
NestList[quad, 0.2, 5]
{0.2, 0.464, 0.721242, 0.583051, 0.704997, 0.603131}
```

For each number we want to draw two lines, one vertical up or down from the **x=y** line, and one horizontal over to the **x=y** line again. We can write a function which makes these two lines out of a single number, like this:

```
makeLines[x_] :=
    {
    Line[{{x,        x},  {x,        quad[x]}}],
    Line[{{x, quad[x]}, {quad[x], quad[x]}}]
    }
```

All we have to do now is to map this function onto our list of numbers. We use the **Map** function, which applies the function given by its first argument to all the elements of the list given by its second argument.

```
lambda = 2.9;
Map[makeLines, NestList[quad, 0.2, 3]]
{{Line[{{0.2, 0.2}, {0.2, 0.464}}],
   Line[{{0.2, 0.464}, {0.464, 0.464}}]},
  {Line[{{0.464, 0.464}, {0.464, 0.721242}}],
   Line[{{0.464, 0.721242}, {0.721242, 0.721242}}]},
  {Line[{{0.721242, 0.721242}, {0.721242, 0.583051}}],
   Line[{{0.721242, 0.583051}, {0.583051, 0.583051}}]},
  {Line[{{0.583051, 0.583051}, {0.583051, 0.704997}}],
   Line[{{0.583051, 0.704997}, {0.704997, 0.704997}}]}}
```

Let's use this list in the **Epilog** option to our **Plot** command, and let's also use a much smaller starting point for **x**:

```
lambda = 2.9;
Plot[quad[x], {x, 0, 1},
    Epilog -> {
        Line[{{0, 0}, {1, 1}}],
        Map[makeLines, NestList[quad, 0.01, 30]]
    }
];
```

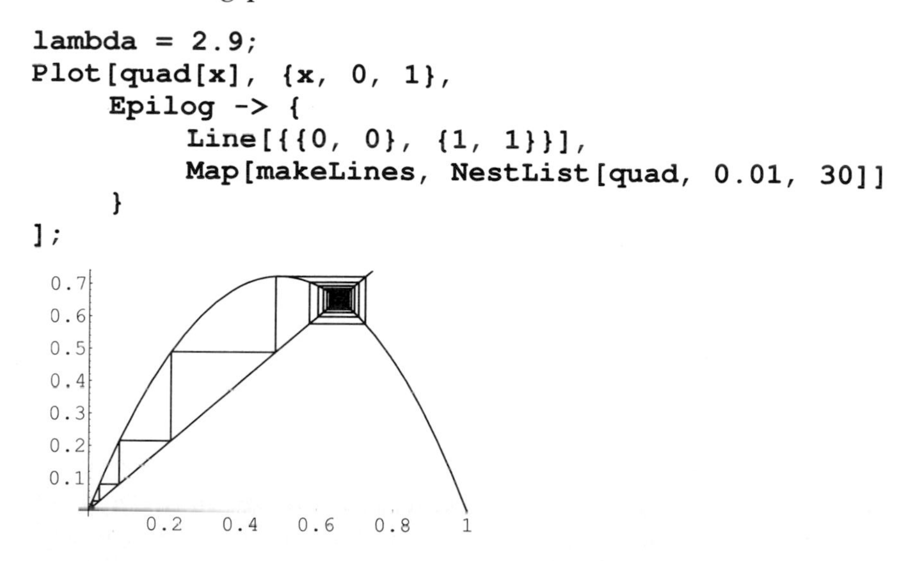

Jerry: Wow, it really spirals into a single value, doesn't it. That black rectangle means it moved in more and more slowly. Let's see what this would look like for the other value we tried, 3.1, which seemed to settle on alternating between two values:

```
lambda = 3.1;
Plot[quad[x], {x, 0, 1},
     Epilog -> {
          Line[{{0, 0}, {1, 1}}],
          Map[makeLines, NestList[quad, 0.01, 30]]
     }
];
```

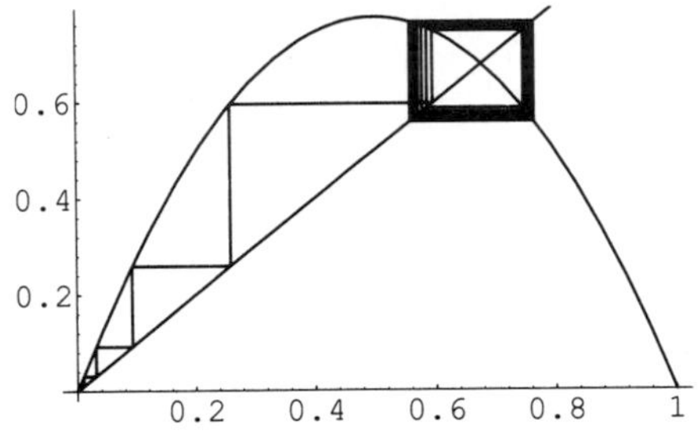

```
0.6
0.4
0.2
     0.2  0.4  0.6  0.8  1
```

Theo: That one leaves a hollow rectangle, because it never settles down to a single value, but just keeps alternating between two different values.

Jerry: This plot really has too many lines. All we really need is a line from one point in the list to the next. That way we could see more directly where the series is heading.

Theo: OK. I think we can use a clever function, **`Partition`**, to do this very quickly. We could calculate each point by looking at the value in the list and applying the **`quad`** function to it to get the corresponding y value. But the number we get by doing that is the next number in the list. So we can save time by not recalculating the value of the function, but instead using the next element of the list. What we need is a way to take a list of numbers and make it into a list of pairs of numbers, where the second number in each pair is the next number in the original list.

Let's talk about the **`Partition`** function. We'll soon see how it can do what we want. The simplest form of the **`Partition`** function breaks a list into sublists with **`n`** elements, like this:

```
Partition[{a, b, c, d, e, f, g, h, i}, 3]
{{a, b, c}, {d, e, f}, {g, h, i}}
```

or

```
Partition[{a, b, c, d, e, f, g, h, i}, 2]
{{a, b}, {c, d}, {e, f}, {g, h}}
```

Jerry: That looks like it. We have a list of pairs of numbers, and the second number in each pair is always the next number in the list.

Theo: That's true, but we would be missing half the points. In this example, the pair **{b, c}** is also a valid point on our curve, but it's nowhere to be found in this list of pairs.

Jerry: So, we want to break the list into pairs, starting at each element anew.

Theo: Right, and **Partition** can take an optional third argument that says how far to move along the list before making up the next **n** element sublist of elements. The default is to move exactly **n** elements along, which means that none of the elements in the original list ends up in more than one of the sublists. For example (study this closely to see what's happening):

```
Partition[{a, b, c, d, e, f, g, h, i}, 3, 2]
{{a, b, c}, {c, d, e}, {e, f, g}, {g, h, i}}
```

Notice that the last element of each sublist is duplicated as the first element of the next sublist. This is because **Partition** moved only two elements down the list (from **a** to **c** in the first case) before picking up the next 3 elements. This is the command we want:

```
Partition[{a, b, c, d, e, f, g, h, i}, 2, 1]
{{a, b}, {b, c}, {c, d}, {d, e}, {e, f}, {f, g}, {g, h},
  {h, i}}
```

Jerry: I see; **b** is not only the **y** value of the first point, it's also the **x** value of the second point, and **c** is the **y** value of the second point and the **x** value of the third point.

Theo: Exactly.

Well, after all that, it's sort of anticlimactic to see how we actually use **Partition**:

```
lambda = 3.1;
Plot[quad[x], {x, 0, 1},
    Epilog ->
        Line[
            Partition[NestList[quad, 0.01, 30], 2, 1]
        ]
];
```

0.6
0.4
0.2
0.2 0.4 0.6 0.8 1

We're most interested in what happens once the sequence settles into its stable, long term behavior (if any). Let's start at an arbitrary point, say 0.4, nest 400 times, and then look at the next 100 iterations. To do this, we use the **Nest** function to nest 400 times, and use the result as the starting point for the **NestList** command that does the next 100 iterations. This is faster than using **NestList** to generate all 500 iterations, because we don't actually store the first 400 values anywhere.

```
lambda = 3.1;
Plot[quad[x], {x, 0, 1},
    Epilog -> Line[
        Partition[
            NestList[quad, Nest[quad, 0.4, 400], 100],
            2, 1
            ]
        ]
];
```

0.6
0.4
0.2
0.2 0.4 0.6 0.8 1

We see that, after the first 400 nestings, the sequence has settled into a completely stable oscillation between two values. The oscillation is so stable that we see only a single line between the two values.

Jerry: I think now we have a way to start looking at lots of different values of **lambda**, to see what we can see.

Theo: I agree, and I think before we go on we should make up a function that encapsulates the things we've been doing. Here is such a function (it lets you choose a value of lambda, the starting **x** value, the number of cycles to skip before starting to draw, and the number of cycles to draw):

```
lambdaPlot[lam_, startX_, skipCycles_, showCycles_] :=
    Module[{quad},

        quad[x_] := lam x (1 - x);

        Plot[quad[x], {x, 0, 1},
            PlotLabel->SequenceForm["Lambda = ", lam],
            PlotRange->{0, 1},
            AspectRatio->Automatic,
            Epilog->
                Line[Partition[
                    NestList[
                        quad,
                        Nest[
                            quad,
                            startX,
                            skipCycles
                        ],
                        showCycles
                    ],
                    2, 1
                ]]
        ]
    ]
```

We can use it to produce the same plot as above. (The new plot looks taller because of the **AspectRatio -> Automatic** option included in **lambdaPlot**.)

```
lambdaPlot[3.1, 0.4, 400, 100];
```

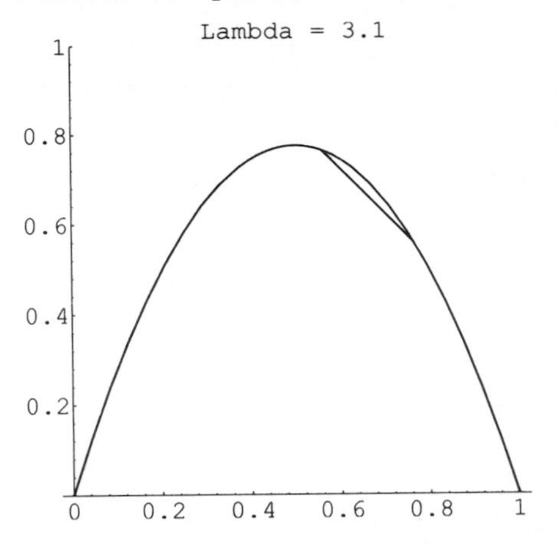

Jerry: Let's make plots for a whole range of values of **lambda**:

```
Do[lambdaPlot[l, 0.4, 400, 100], {l, 3.0, 4.0, 0.1}]
```

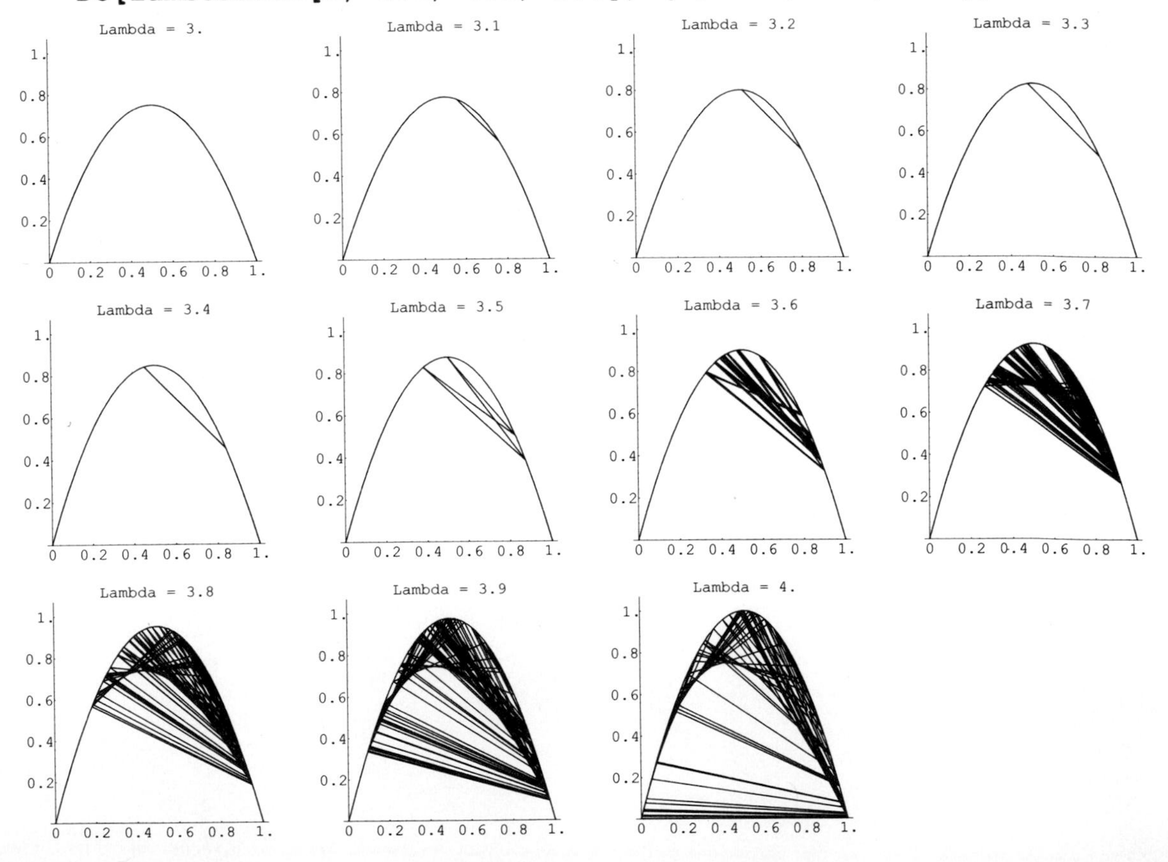

Theo: That certainly got exciting in a hurry! I'm particularly interested in the value 3.5, where it looks as if there is a stable cycle of four different values. Let's zero in on the range from 3.4 to 3.6. But before we do that, let's make a new version of **`lambdaPlot`** that will show just the area of interest, as big as possible. We don't really care about the exact slope of these lines, so we can let *Mathematica* scale the plot to fit the available area.

```
lambdaPlotB[lam_, startX_, skipCycles_, showCycles_] :=
    Module[{localQuad},

        localQuad[x_] := lam x (1 - x);

        Show[
            Graphics[{
                Text[
                    SequenceForm["Lambda = ", lam],
                    Scaled[{1, 1}],
                    {1, 1}
                ],
                Line[
                    Partition[
                        NestList[
                            localQuad,
                            Nest[
                                localQuad,
                                startX,
                                skipCycles
                            ],
                            showCycles
                        ],
                        2, 1
                    ]
                ]
            }],
            AspectRatio->Automatic
        ]
    ]
```

Here's the range from 3.4 to 3.6:

```
Do[
    lambdaPlotB[1, 0.4, 500, 100],
    {l, 3.4, 3.6, 0.01}
]
```

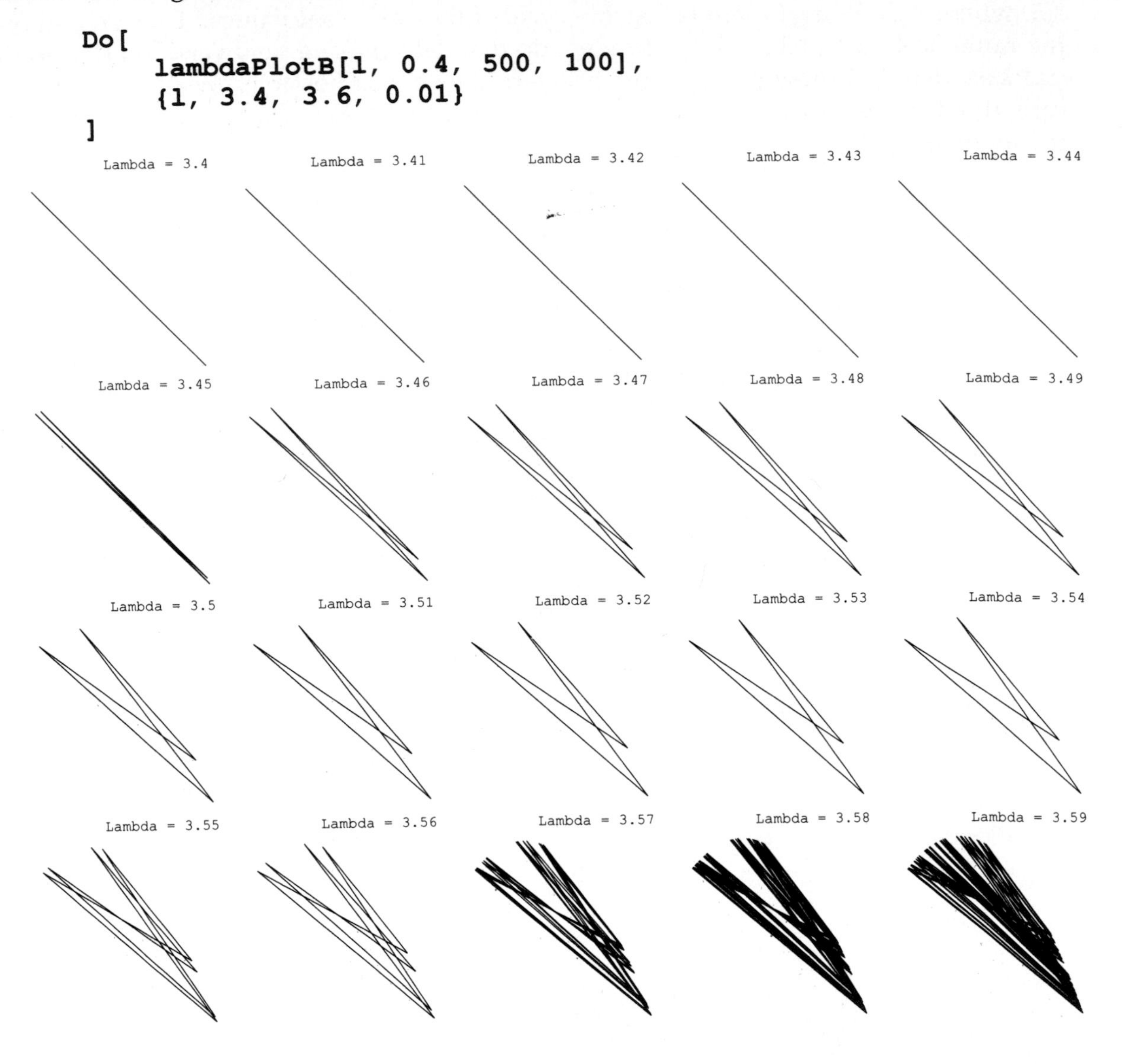

Dialog–Second Day

Jerry: These are interesting animations, but I still don't have a good overall feeling for how these changing cycles fit together. Could we make a plot in which the cycles for many different values of **lambda** are combined on one plot?

Theo: Well, I guess it was inevitable. There is a standard plot that seems to be included in every book that has anything thing to do with this subject, and a good many that don't. Here is about the simplest way of making this plot I've seen anywhere (note that we have to clear the value of **lambda** before we can use it as a variable in the **Table** command):

```
Clear[lambda];
ListPlot[
     Flatten[Table[
          Transpose[{
               Table[lambda, {129}],
               NestList[quad, Nest[quad, 0.5, 500], 128]
          }],
          {lambda, 0, 4, 0.01}
     ], 1],
     PlotStyle -> PointSize[0.001]];
```

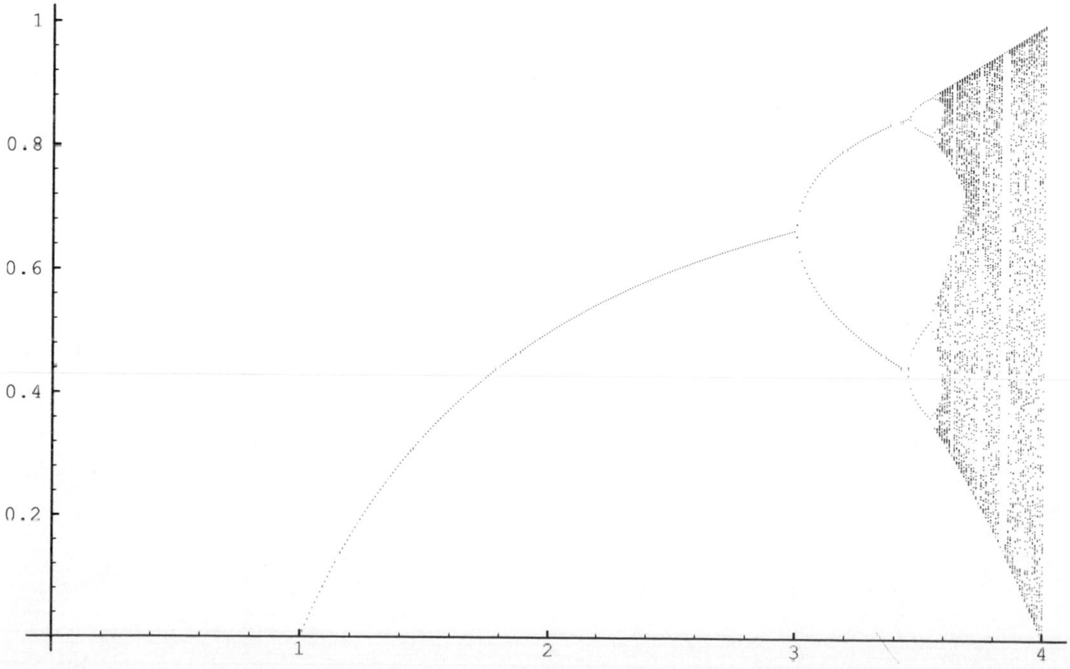

The horizontal axis is **lambda**, and the vertical axis represents the values in the **NestList**.

Jerry: So, each dot on this plot represents one of the end points of the lines we drew above. In this picture we can see how the number of values in the cycle goes from one to two to four, and so on. We can also see where the values don't settle down into any simple cycle.

I can understand why people like this plot.

Theo: I suppose we have to narrow in on the interesting area, just as everyone else does:

```
Clear[lambda];
ListPlot[
    Flatten[Table[
        Transpose[{
            Table[lambda, {257}],
            NestList[quad, Nest[quad, 0.5, 1000], 256]
        }],
        {lambda, 3, 4, 0.0025}
    ], 1],
    PlotStyle -> PointSize[0.001]];
```

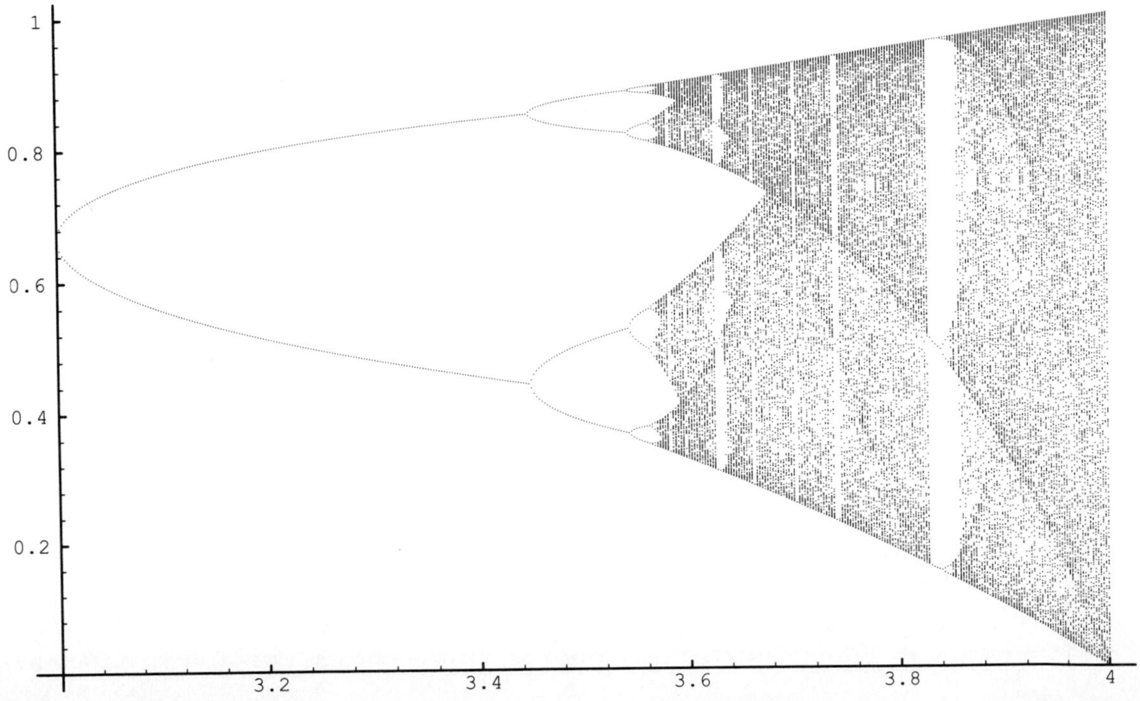

1
0.8
0.6
0.4
0.2
3.2 3.4 3.6 3.8 4

Jerry: In this expanded view I notice that there is an area around `lambda = 3.85` in which the period seems to be 3. Can we zoom in on this area?

Theo: This is *so* predictable! Of course we can zoom in on the classic period-3 area:

```
Clear[lambda];
ListPlot[
    Flatten[Table[
        Transpose[{
            Table[lambda, {257}],
            NestList[quad, Nest[quad, 0.5, 1000], 256]
        }],
        {lambda, 3.82, 3.86, 0.0001}
    ], 1],
    PlotStyle -> PointSize[0.001]];
```

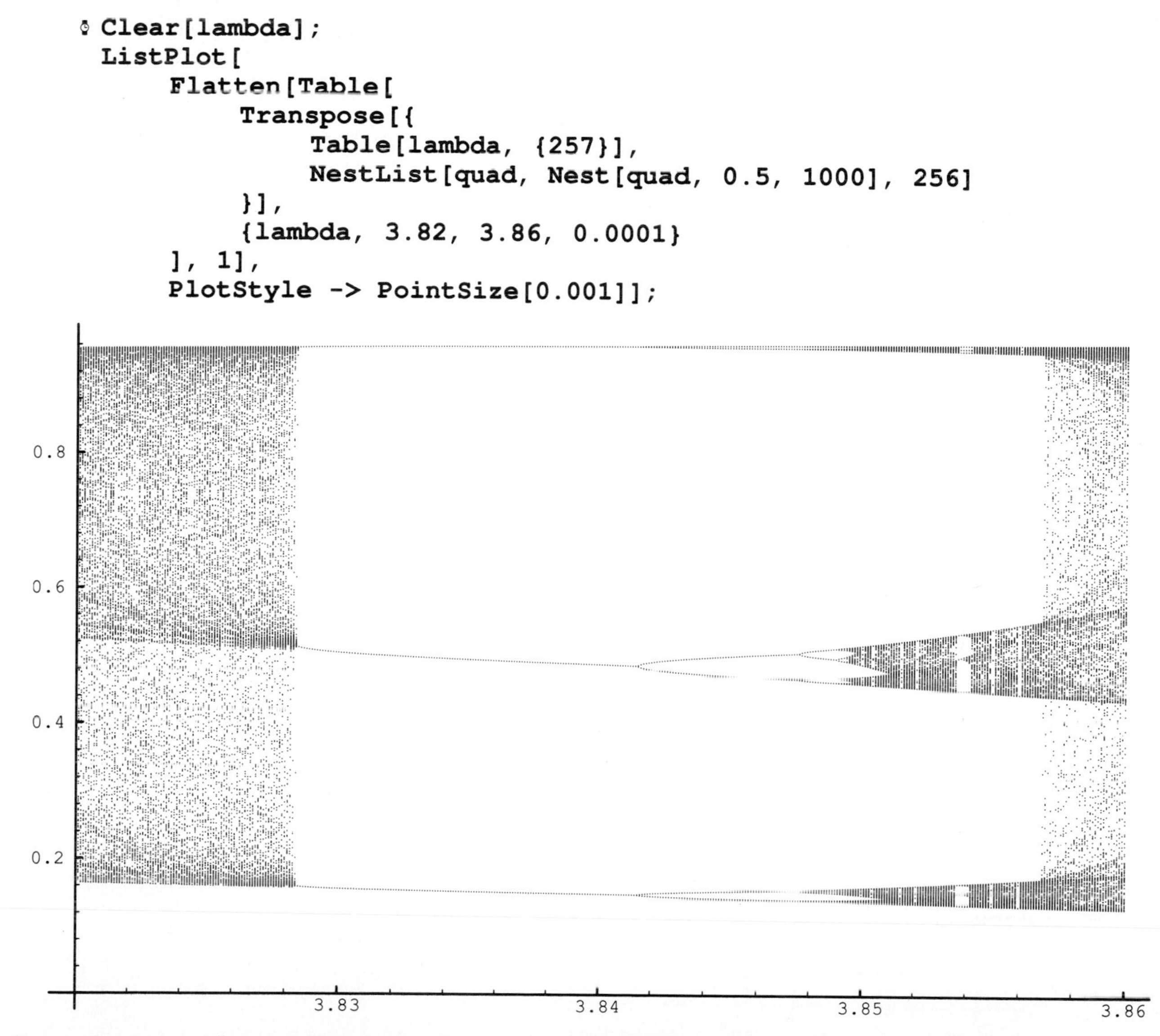

Jerry: I agree with you that these pictures have become commonplace but, now that we have them under our control, I have many more questions and interests than I did before. When the pictures sat in books, such as Chaos by Gleick **[3]** or in any of many fractal books **[4-6]**, I reacted in a more passive way than I do now that they are in electronic form. I'm sure I'm not the only person who will react this way -- I'm sure readers of this book will want to zoom in on their own favorite areas.

Theo: Yes, being able to make these pictures on demand is very nice. One slight problem is that there is a huge amount of information in each one of these plots (about 100,000 points), so they take a while to generate (an hour or so on a Mac II). Later on in this section we'll develop a way of making these plots *much* faster and in a way that occupies much less memory.

Jerry: One big advantage to having control over these pictures is that I can see clearly where the bifurcations occur, which I can't in books that leave out the scale. I also wish *they* would publish a simple list of the lambda values at which bifurcations begin. Can you fill this gaping void?

Theo: We can try. We're going to have to get a bit more sophisticated if we want to accurately calculate values for these bifurcation points. We may need a third day to do this.

Dialog–Third Day

Jerry: Could **`FixedPoint`**, a function I've been worrying about recently, help us calculate **`lambda`**s? **`FixedPoint`** is a very interesting function, somewhat like **`Nest`**. You give it a function and a starting point. It applies the function to the starting point and gets a new value. It then applies the function again to this new value, and so on. It stops when (and only when) it applies the function to the current value and gets back the same value.

As some people may know, the limit of **`Cos[Cos[Cos[...Cos[x]]]]`** is a number around 0.7 or so. We can use **`FixedPoint`** to find this number:

```
FixedPoint[Cos, 0.3]
0.739085
```

Theo: This result didn't take much time, presumably because the sequence converged quickly to the final value. Let's try **`Sin`** instead:

```
FixedPoint[Sin, 0.3]
$Interrupted
```

Jerry: I know that this is supposed to converge to zero, but this took so long we had to interrupt it. It might never have finished. Fortunately the Command-Period feature worked in this case (hold down the Command key and type the period key to interrupt a calculation). Can you predict when we can successfully interrupt *Mathematica*, and when we can't? I've had trouble with this in the past.

Theo: It's hard to know when it will work and when not. It should *always* work, but unfortunately there are cases in which it does not. It gets better with each new version, so I wouldn't want to say where there are problems; I would probably be wrong.

Jerry: Can we use **NestList** to see why the **Sin** was taking so long?

Theo: Sure. **NestList** is quite similar to **FixedPoint**, except that it returns a list of all the values it got along the way, and instead of stopping when the values stop changing you have to tell it a fixed number of iterations to do.

```
NestList[Sin, 0.3, 20]
```

```
{0.3, 0.29552, 0.291238, 0.287138, 0.283208, 0.279438,
  0.275815, 0.272331, 0.268978, 0.265746, 0.262629,
  0.25962, 0.256714, 0.253903, 0.251184, 0.248551, 0.246,
  0.243526, 0.241126, 0.238796, 0.236533}
```

Jerry: Well, these values are getting smaller, but not very quickly. Can we see just the tail end of a much longer list like this?

Theo: The best way to do this is to use **Nest** to calculate the starting point we give to **NestList**. **Nest** takes the same arguments as **NestList**, but returns only the last value, not all the ones in between.

```
NestList[Sin, Nest[Sin, 0.3, 3000], 20]
```

```
{0.0314345, 0.0314293, 0.0314241, 0.031419, 0.0314138,
  0.0314086, 0.0314035, 0.0313983, 0.0313931, 0.031388,
  0.0313828, 0.0313777, 0.0313725, 0.0313674, 0.0313622,
  0.0313571, 0.031352, 0.0313468, 0.0313417, 0.0313366,
  0.0313314}
```

After 3000 cycles it's still got a ways to go. Let's try a few more:

```
NestList[Sin, Nest[Sin, 0.3, 30000], 20]
```

```
{0.00999377, 0.0099936, 0.00999344, 0.00999327,
  0.0099931, 0.00999294, 0.00999277, 0.0099926,
  0.00999244, 0.00999227, 0.00999211, 0.00999194,
  0.00999177, 0.00999161, 0.00999144, 0.00999127,
  0.00999111, 0.00999094, 0.00999078, 0.00999061,
  0.00999044}
```

The values are closer to zero, but clearly taking their own good time getting there. Calculating this list took much less time than we allowed **FixedPoint** before interrupting it, indicating that **FixedPoint** did many more than 30000 iterations without reaching zero.

Jerry: A common approximation is that **Sin[x]** is about equal to **x** if **x** is small. This probably has something to do with why it is converging so slowly, since each new value is approximately (but not exactly) the same as the previous one.

Theo: So what you're saying is, the only thing that gets us any closer to zero is the error in this common approximation.

Jerry: Can we see a picture that would make this clearer? Let's try plotting **y = x** together with **y = Sin[x]**:

```
Plot[{x, Sin[x]}, {x, 0, Pi},
    AspectRatio -> Automatic];
```

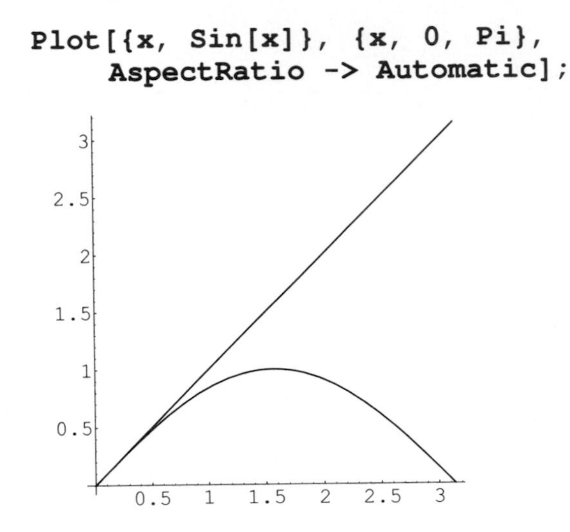

Let's look in closer to zero:

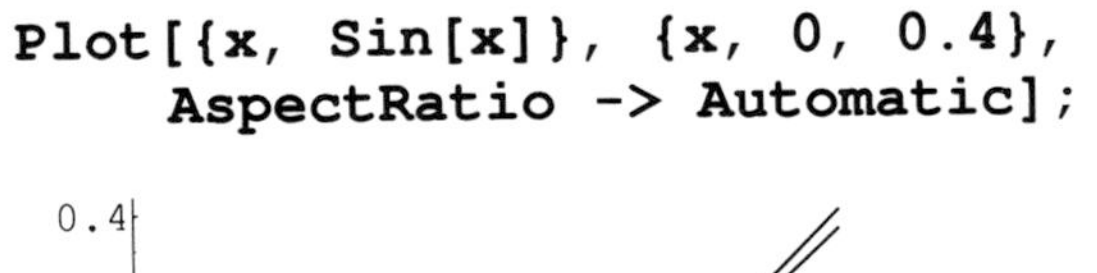

```
Plot[{x, Sin[x]}, {x, 0, 0.4},
    AspectRatio -> Automatic];
```

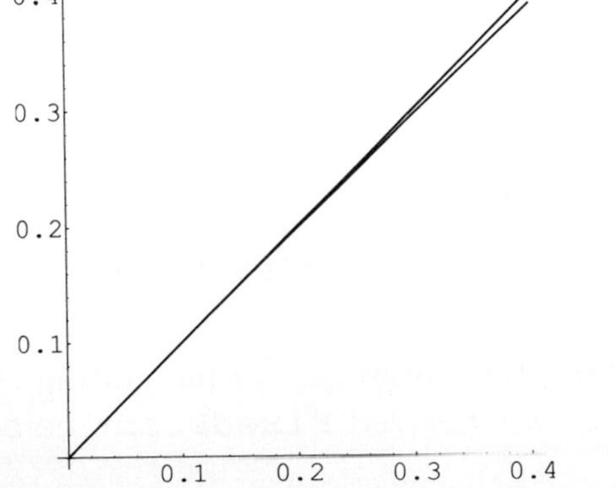

We can just barely see the difference between the two curves.

Theo: Another problem we haven't discussed is numerical round-off error. Even if such a numerical sequence does converge at a reasonable rate, the final value may jump around by a few units in the last significant digit. Generally one has to tell **FixedPoint** to ignore the last few places when comparing successive values.

Jerry: Well, what makes us think that the answer we get is going to be the same as the real answer. On the other hand, what makes us think that the answer we get with any number of digits is the "right" answer? Clearly if we looked only at the first digit, and stopped when that stopped changing, we would not have a useful answer. So, how many digits is enough? Two? Three? Sixteen?

Theo: That's the problem with floating point numbers: it's all approximations and hand-waving. Of course, most of the time people just say, "Sixteen digits is good enough for me, stop bothering me with all this talk".

Jerry: What if we try to use **FixedPoint** on the iterated quadratic map we explored earlier in this chapter? Can it detect the alternating between 2 values that we see in the following example?

```
lambda = 3.1
NestList[quad, 0.3, 40]
{0.3, 0.651, 0.704317, 0.645589, 0.709292, 0.639211,
  0.714923, 0.631805, 0.721145, 0.623394, 0.727799,
  0.614133, 0.734618, 0.604358, 0.741239, 0.594592,
  0.747262, 0.58547, 0.752354, 0.577584, 0.75634,
  0.571298, 0.759241, 0.566661, 0.761225, 0.563461,
  0.762515, 0.561366, 0.763326, 0.560044, 0.763824,
  0.559231, 0.764124, 0.558739, 0.764304, 0.558444,
  0.764411, 0.558269, 0.764475, 0.558165, 0.764512}
```

Theo: No, **FixedPoint** will just keep calculating forever if you give it such a sequence. We can make a more intelligent version that does detect cycles, using the clever technique described by Martin Gardner in one of his Scientific American articles **[2]**. Here is the function:

```
CycleFixedPoint[f_, start_] := Block[
    {nth, twonth, counter, result, value},

    nth = f[start];
    twonth = f[f[start]];
    counter = 1;

    While[nth =!= twonth,
        nth = f[nth];
        twonth = f[f[twonth]];
        ++counter;
    ];

    result = {nth};
    value = f[nth];
    While[value =!= nth,
        AppendTo[result, value];
        value = f[value]
    ];

    result
]
```

The problem is how to detect that there is a cycle in the numbers, without knowing how long the cycle is, or even when it starts (since the numbers may take a while to stabilize). The **CycleFixedPoint** function does this in a very clever way. It keeps two current values of the sequence, let's call them column A and column B. In column A we always advance from one number to the next by applying the function once. In column B we advance by applying the function twice. In other words, column B is moving at double speed.

Each time we advance columns A and B, we compare the current value in both columns. If they are the same, we stop. If it took us n steps before we stopped, we know that the nth and the 2nth values are the same, which means that the cycle must repeat an integer number of times between the nth and the 2nth step.

Now we can start at the nth value and iterate until we find a value that is the same as the nth value. This will be at most the 2nth value, but it might come sooner if the cycle repeats more than once between the nth and 2nth value.

This algorithm has the advantage that it can detect cycles of any length, without having to store up long lists of values. In the very end it builds a list of the elements of the cycle, but only after it knows exactly how long that list is going to be.

Jerry: That seems clever, but does it really work? Let's try it on our **quad** function:

```
lambda = 3.1
CycleFixedPoint[quad, 0.3]
{0.764567, 0.558014}
```

Same answer as we got from **NestList** above. It's pretty impressive that **CycleFixedPoint** was able to recognize when the numbers went from changing arbitrarily (while the sequence was converging) to alternating regularly, and was then able to extract just the minimal description of this cycling. Can it do the same for a longer cycle, such as this one?

```
lambda = 3.5
CycleFixedPoint[quad, 0.3]
{0.38282, 0.826941, 0.500884, 0.874997}
```

Ah, but what about a cycle of 3, as we should find at **lambda = 3.835**?

```
lambda = 3.835
CycleFixedPoint[quad, 0.3]
{0.494514, 0.958635, 0.152074}
```

Let's give it a really hard one. I expect a cycle of 12 at **lambda = 3.848**:

```
lambda = 3.848
CycleFixedPoint[quad, 0.3]
{0.1407, 0.465238, 0.95735, 0.157118, 0.509597, 0.961646,
  0.141927, 0.468624, 0.958212, 0.154081, 0.501548,
  0.961991}
```

Theo: We could use this new function to make the same sorts of plots as we made above using **Nest** and **NestList**. One complication is that if you choose a value of **lambda** that results in a chaotic sequence of values, **CycleFixedPoint** will never finish. To overcome this problem, I've written a version (see the Functions That We Defined section) that can take an optional third option that specifies a maximum number iterations before giving up (the built-in **FixedPoint** function also understands a third argument like this).

Jerry: Let's try it:

```
lambda = 3.75
CycleFixedPoint[quad, 0.3, 500]
0.724134
```

Why did it return a value, since we know that it didn't converge to any stable single value or cycle?

Theo: It returns the last value it had before giving up, in case you're interested. You can tell that it didn't converge, because if it had it would have returned one or more numbers inside list brackets. For example, this is what it looks like when `CycleFixedPoint` detects a cycle of one (otherwise known as a single value):

```
lambda = 2.5
CycleFixedPoint[quad, 0.3, 500]
{0.6}
```

Jerry: So, what can we do with `CycleFixedPoint` that we couldn't do with `Nest` and `NestList`?

Theo: Well, for making the kinds of plots we made earlier in this chapter, `Nest` and `NestList` are ideal. The strength of `CycleFixedPoint` is that it can tell us how long the cycle actually is, instead of relying on the fact that successive cycles will overlap when drawn on a plot (giving the illusion of a smaller number of points being plotted than actually are).

The plots we made gave us only a very crude estimate of where the transitions between different length cycles happen. It might be nice to know, for example, at exactly what value of `lambda` the length of the cycle goes from 2 to 4.

Jerry: We could find out by trying out different values of `lambda`, and seeing how long the cycle is for each value. If we found one value that had a cycle of 2, and another value that had a cycle of 4, we would know that the threshold was somewhere in between. We could pick a value between the two we tried, see how long the cycle was, and repeat until we had narrowed down the range as much as we wanted.

Theo: We could do this manually, or we could write a program to do it automatically. Better yet, we could be incredibly clever and realize that such a program already exists, and is called `Plot`.

Jerry: `Plot`! What are you talking about? `Plot` is for making plots, not recursively narrowing down the location of bifurcation points. I think you've finally gone off the deep end.

Theo: You must never underestimate the versatility of *Mathematica*. The key word here is recursively. I've mentioned several times that `Plot` uses a recursive subdivision algorithm to get nice, smooth curves. It starts with a regular grid of points (20 by default), and calculates the value of the function at each point. Then, it looks at the angle formed by the pair of line segments determined by each triple of neighboring points. If that angle is too far away from 180 degrees, it adds a new point between the existing ones. This process is repeated until all the angles are nearly straight, or it has subdivided a certain maximum number of times (20 by default).

Jerry: That's very interesting, but what does it have to do with our problem? What are we going to plot?

Theo: We are going to make a plot of the length of the cycle as a function of **lambda**. This should give us a step function with plateaus at 1, 2, 4, 8, etc, and another set of plateaus at 3, 6, 12, etc. Whenever you plot a discontinuous function like this, **Plot** works very hard to refine the location of the discontinuity, since this point causes sharp angles no matter how close together you space the calculated points.

Jerry: This is starting to sound like a good idea after all.

Theo: Let's try a simple example, avoiding all the dangerous areas. (Note that we have to clear the value of **lambda** before using it as the variable; otherwise **Plot** complains.)

```
Clear[lambda];
Plot[
     Length[CycleFixedPoint[quad, 0.5, 2000]],
     {lambda, 3.3, 3.5}
];
```

4
3
2
1
3.35 3.4 3.45 3.5

Jerry: I notice two strange things about this plot. First, there are places where the value is zero, which is hard to understand since the smallest cycle you can have is a constant value, which should have a length of one.

Theo: The zero points represent places where **CycleFixedPoint** failed to converge and thus returned a single number instead of a list, as we saw above. The length of any single number is zero. We can just remember that zero means failed to converge.

Jerry: The other thing I notice is that in the transition between 2 and 4 the curve jumps up and down several times. This can't be what it actually does: it must be some sort of error in **CycleFixedPoint**. Can we really trust it?

Theo: Not really. Since we are working with floating point numbers, we have to be prepared to deal with a certain amount of "noise" caused by numerical round off errors. When the limit cycle bifurcates from a cycle of 2 into a cycle of 4, it starts out with 4 values that come in two pairs of two very close values. Here are two specific examples showing what I mean:

```
lambda = 3.43
CycleFixedPoint[quad, 0.3, 5000]
```

```
{0.44458, 0.846965}
```

```
lambda = 3.46
CycleFixedPoint[quad, 0.3, 5000]
```

```
{0.467486, 0.861342, 0.413234, 0.838952}
```

Jerry: So, you're saying that this cycle of 4 values is similar to a cycle of 2 values, since the first value is close to the third value, and the second value is close to the fourth value. Clearly, if we are very close to the point where the 2-cycle turns into a 4-cycle, it must be extremely difficult to tell the difference between one and the other.

Theo: Exactly. It may take a very long time for the sequence to become stable enough to tell the difference. Worse, if there is any round-off error (as there always is) we can be tricked into thinking that a 2-cycle is really a longer cycle, because the round off may not cancel itself out for several cycles. It is possible to use very sophisticated analyses of numerical error to understand and precisely control these kinds of problems, but I think we can get away with a simpler approach.

`CycleFixedPoint` can be given an option (**`CycleTest`**) that specifies the test to be applied to two values in the sequence to determine if they are "the same" or not. This test can be any function of two variables that returns **`True`** or **`False`** for any two values. (The default is to use the built in function **`SameQ`** which, when applied to floating point numbers, compares all significant figures in both numbers and returns **`True`** only if they are identical.)

We can use a test that compares only, say, 12 significant digits. Since we are using machine-precision floating-point numbers, which are typically accurate to about 16 to 18 digits, this new test will ignore about 4 digits worth of potential noise. Here is such a test function:

```
FuzzySameQ[x_, y_] := Abs[x - y] < 10^-12
```

Jerry: That looks surprisingly sensible and straight forward. I was expecting something quite peculiar. This inequality is clearly true only if the difference between **`x`** and **`y`** is less than 10^{-12} which, for numbers whose value is near 1 means they are the same to about 12 decimal places.

Theo: There are two different test we can specify: First, the test to be used to determine when the sequence has converged (called the **SameTest**), and second the test to be used to determine how long the cycle is (called the **CycleTest**). Let's try using full equality (the default) for the **SameTest**, and our new fuzzy test for the **CycleTest**:

```
lambda = 3.46
CycleFixedPoint[quad, 0.3, 500,
                        CycleTest -> FuzzySameQ]
{0.467486, 0.861342, 0.413234, 0.838952}
```

Jerry: Let's use it very near the border between 2-cycles and 4-cycles, and see if we can get a different answer with and without this new comparison function.

```
lambda = 3.446
CycleFixedPoint[quad, 0.3, 5000]
{0.440778, 0.849414, 0.440778, 0.849414}
```

```
lambda = 3.446
CycleFixedPoint[quad, 0.3, 5000,
                        CycleTest -> FuzzySameQ]
{0.440778, 0.849414}
```

Using the looser test, we got a shorter cycle.

Theo: Let's define a function to put some of these things together in one place:

```
lambdaCycle[lambda_, maxCycles_] :=
    CycleFixedPoint[
        lambda # (1 - #)&,
        0.5,
        maxCycles,
        CycleTest -> FuzzySameQ,
        IterationCount -> True
    ]
```

This function takes two arguments: a value for **lambda**, and a maximum number of cycles we want to try before giving up. It returns a list of two things: first, a list of the cycle of values that the function converges to; second, the number of iterations required for convergence, which is something I think we might find interesting. Here's an example:

```
lambdaCycle[3.4, 2000]
{{0.451963, 0.842154}, 202}
```

Jerry: So, this says that at **`lambda = 3.4`**, the function converged to a cycle of two values after 202 iterations.

Theo: Right. Using this function, we can make a considerably more elaborate version of the plot we made above. Before we do, though, I'd like to introduce one more change. As anyone experimenting interactively with these functions will have discovered, they tend to run for quite a while. I have therefore written a small program in the C language that we can run automatically, from inside *Mathematica*, to do these calculations much faster.

Jerry: That sounds like something we haven't done before, and therefore strange, confusing, and potentially dangerous.

Theo: No, no, it's not bad at all. All you have to know is that there is a function called **`lambdaCycleC`** that does exactly the same thing as **`lambdaCycle`**, except that it does it about 500 times faster (I'm not kidding, it's that much faster). People who want to know how it's implemented, or who want to use it on their own computers, can read the Functions That We Defined section. (You can't use **`lambdaCycleC`** unless you are running the *Mathematica* kernel on a multi-tasking computer, such as a UNIX machine, and you have properly installed and compiled the required C program, as explained in the Functions That We Defined section. You can always use **`lambdaCycle`** instead, which will always work.)

Here is what **`lambdaCycleC`** does:

```
lambdaCycleC[3.4, 2000]
{{0.451963, 0.842154}, 232}
```

Jerry: That looks less threatening, in fact it looks exactly like **`lambdaCycle`**, just as you said it would. I guess we need this little bit of C to make things go faster, but we're not really changing the way we work. We can still work in *Mathematica*, writing *Mathematica* functions, without having to worry about the fact that when we use **`lambdaCycleC`** some sort of UNIX mumbo-jumbo goes on.

Theo: Exactly. This is a very good example of why *Mathematica* is such a powerful system. People often complain that *Mathematica* is slow when doing simple, iterative calculations. That's absolutely true, as proven by the fact that **`lambdaCycleC`** is about 500 times faster than **`lambdaCycle`**. However, *Mathematica* is a very nice environment to work in, particularly when it comes to manipulating, sorting, and graphing the results you get from programs like **`lambdaCycleC`**. It's quite easy to write 20-line C program to calculate iterated functions, but it's an incredible bother to write C programs to plot numbers, extract unique elements, and so on.

With *Mathematica*, it's about as painless as possible to have the best of both worlds. We have picked out the one real time-sink and moved it into C. This required only a very small amount of C program code, and a truly minimal amount of *Mathematica* code to support it. Now, we can use **lambdaCycleC** exactly as if it were any other *Mathematica* function, without having to worry about how it works.

Jerry: OK, let's use it to make a plot.

Theo: For the time being, let's ignore the iteration count returned by **lambda-CycleC**, and just plot the length of the first element of the list it returns. We'll increase the default number of points that **Plot** picks, and also allow it to subdivide a larger number of times. These two changes will allow us to see the bifurcations more closely (they will also increase the running time, of course).

```
Clear[lambda];
Plot[
    Length[First[lambdaCycleC[lambda, 20000]]],
    {lambda, 2, 4},
    PlotRange -> All,
    PlotPoints -> 100,
    PlotDivision -> 50
];
```

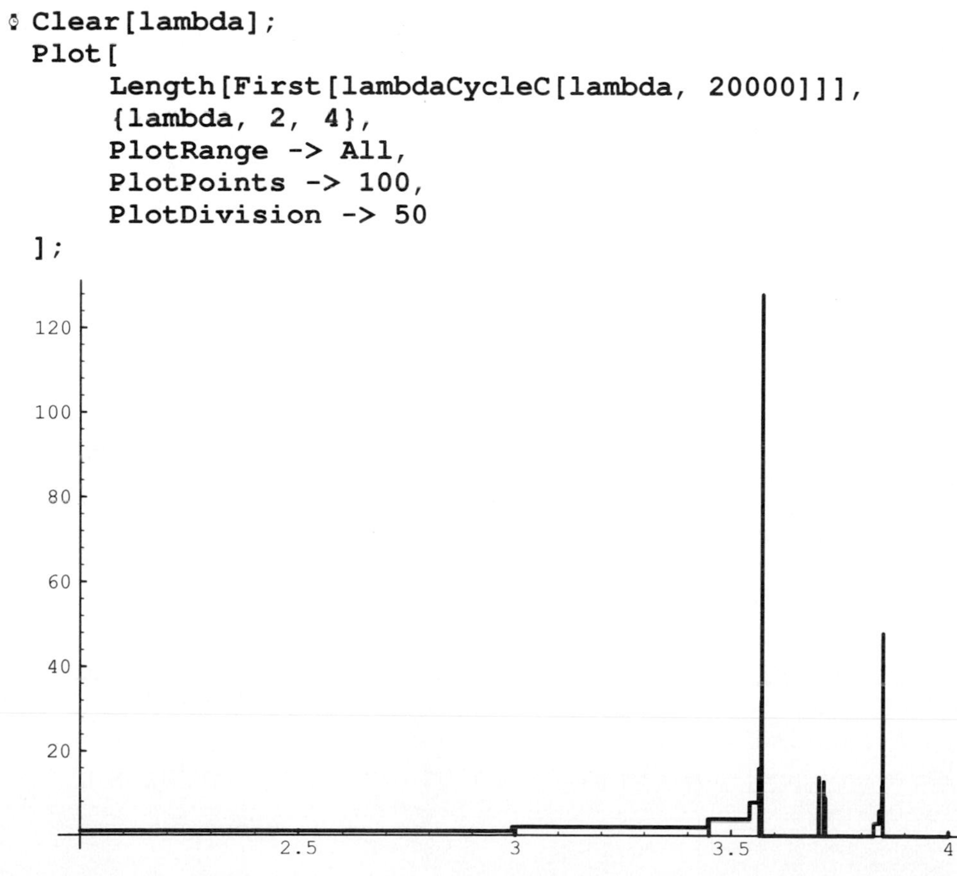

We can clearly see the bifurcations from 1, 2, 4, 8, 16, 32, 64, to 128 values. We can also see a smaller set from 3, 6, 12, 24, to 48 values, as we did earlier with the dot-plots.

Jerry: How are we going to use these plots to determine the **`lambda`** values at which the bifurcations occur? Although **`Plot`** may have narrowed in on the values very closely, as you claimed it would, we can't see that from the plot.

Theo: Yes, that's true. We are going to have to use a little trick to get **`Plot`** to tell us all the values it calculated, so we can look at them later.

`Plot` doesn't care what we give it as its first argument, so long as it returns a single number in the end. We can put an extra statement in to save all the values being generated by **`Plot`** in a list (which we initialize to be an empty list).

Notice the semi-colons (`;`) inside the following **`Plot`** command. The "first argument" to **`Plot`** is actually three separate statements, separated by semi-colons. The first one calculates the current cycle and number of iterations.

Jerry: What does it mean when you assign the result of **`lambdaCycleC`** to a list containing two variable names?

Theo: `lambdaCycleC` returns a list of two values. The assignment assigns the first element of the list to the first variable (**`values`**), and the second element of the list to the second variable (**`steps`**). This is a very useful technique for functions that return more than one value as their result.

The second statement appends a list consisting of the current value of **`lambda`**, the length of the cycle, and the number of steps to convergence to our growing list of results. The third statement gives the value to be plotted. **`Plot`** sees only the value of the last statement.

Let's concentrate on the area where the first set of bifurcations happens, between 2.5 and 3.6:

```
Clear[lambda];
results = {};
Plot[
    {values, steps} = lambdaCycleC[lambda, 40000];
    AppendTo[results, {lambda, Length[values], steps}];
    Print[Last[results]];
    Length[values],

    {lambda, 2.5, 3.6},
    PlotRange -> All,
    AxesOrigin -> {2.5, 0},
    PlotPoints -> 200,
    PlotDivision -> 100
];
```

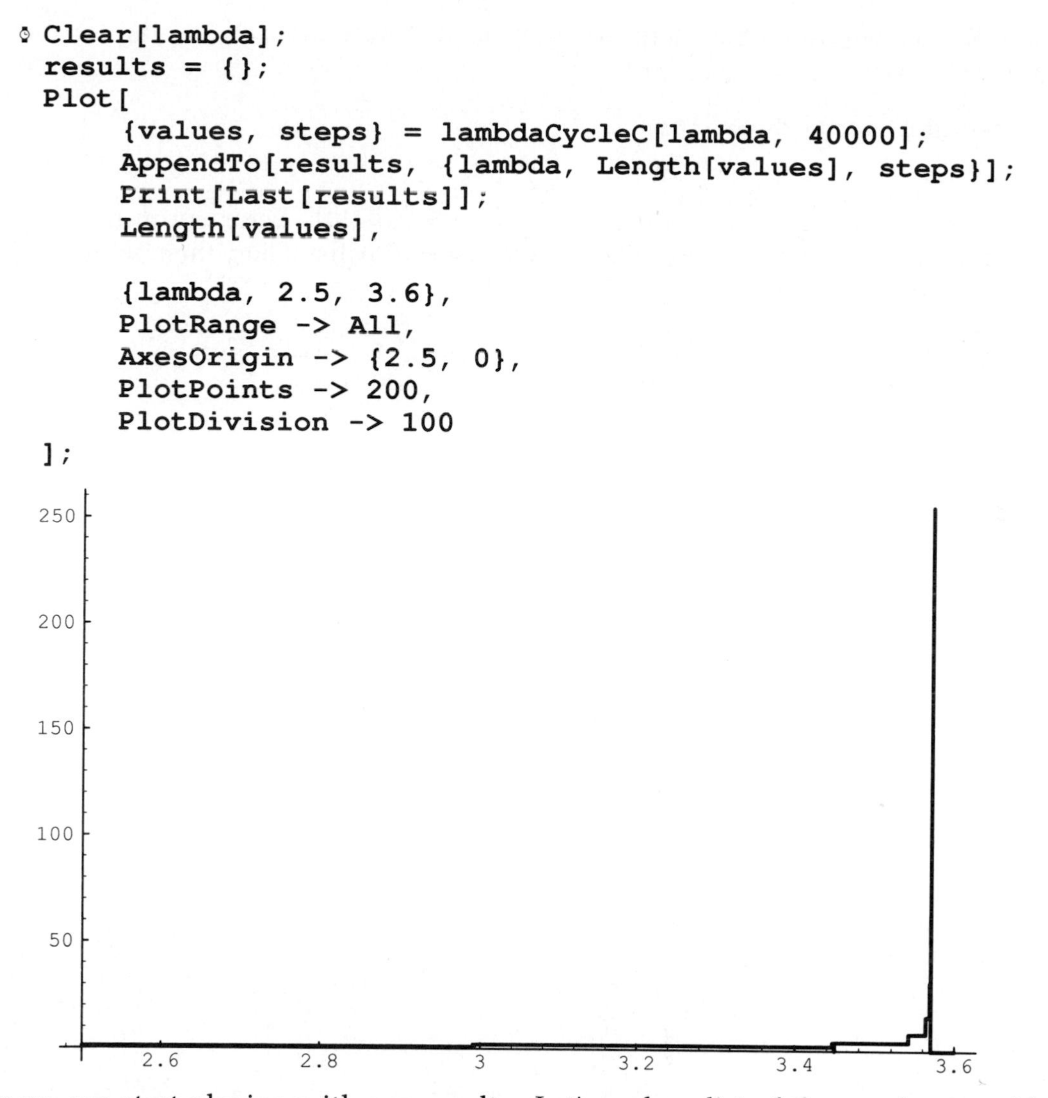

Now we can start playing with our results. Let's make a list of the results sorted by **lambda** values (**Plot** does not necessarily generate the results in order):

```
sortedResults = Sort[results];
```

Jerry: Wait a minute! I've heard a lot of talk about numbers and results, but I haven't *seen* anything yet. I want to print out these sorted results and look at them.

Theo: OK, but there's a *lot* of them. We can see just how many by asking for the length of the list:

```
Length[sortedResults]
382
```

I think it would be a good idea to rearrange this list slightly before printing it out. If we combine it into triples, merge those triples into flat lists, and then print them out in table form, we will get a three-column table:

```
Map[Flatten, Partition[sortedResults, 3]]//TableForm
```

2.5	1	50	2.50553	1	52	2.51106	1	52
2.51658	1	53	2.52211	1	53	2.52764	1	55
2.53317	1	55	2.53869	1	56	2.54422	1	57
2.54975	1	58	2.55528	1	58	2.5608	1	61
2.56633	1	62	2.57186	1	62	2.57739	1	64
2.58291	1	64	2.58844	1	66	2.59397	1	66
2.5995	1	68	2.60503	1	70	2.61055	1	71
2.61608	1	70	2.62161	1	72	2.62714	1	73
2.63266	1	76	2.63819	1	76	2.64372	1	78
2.64925	1	78	2.65477	1	82	2.6603	1	83
2.66583	1	85	2.67136	1	86	2.67688	1	86
2.68241	1	90	2.68794	1	92	2.69347	1	92
2.69899	1	94	2.70452	1	100	2.71005	1	98
2.71558	1	102	2.72111	1	104	2.72663	1	106
2.73216	1	110	2.73769	1	110	2.74322	1	114
2.74874	1	118	2.75427	1	123	2.7598	1	126
2.76533	1	130	2.77085	1	130	2.77638	1	134
2.78191	1	136	2.78744	1	144	2.79296	1	146
2.79849	1	152	2.80402	1	152	2.80955	1	160
2.81508	1	164	2.8206	1	168	2.82613	1	174
2.83166	1	184	2.83719	1	192	2.84271	1	200
2.84824	1	202	2.85377	1	214	2.8593	1	223
2.86482	1	228	2.87035	1	240	2.87588	1	254
2.88141	1	268	2.88693	1	276	2.89246	1	296
2.89799	1	306	2.90352	1	322	2.90905	1	342
2.91457	1	372	2.9201	1	396	2.92563	1	428
2.93116	1	464	2.93668	1	500	2.94221	1	538
2.94774	1	604	2.95327	1	664	2.95879	1	748
2.96432	1	858	2.96985	1	1014	2.97538	1	1228
2.9809	1	1636	2.98643	1	2268	2.98781	1	2424
2.9892	1	2780	2.98989	1	3014	2.99058	1	3138
2.99066	1	3164	2.99075	1	3198	2.99079	1	3172
2.99082	1	3238	2.99084	1	3234	2.99086	1	3292
2.99088	2	3184	2.99092	2	3194	2.99094	2	3220
2.99097	2	3220	2.99099	1	3318	2.99101	1	3358
2.9911	1	3288	2.99127	1	3398	2.99144	1	3512
2.99161	1	3494	2.9917	1	3626	2.99179	1	3660
2.99181	1	3676	2.99183	2	3546	2.99185	2	3574
2.99187	1	3622	2.9919	1	3610	2.99192	2	3608
2.99196	2	3620	2.99198	2	3634	2.992	2	3616
2.99202	1	3716	2.99205	1	3694	2.99213	1	3776

2.99218	1	3802	2.9922	1	3766	2.99222	1	3782
2.99224	1	3762	2.99226	2	3718	2.99228	1	3790
2.99231	2	3774	2.99233	1	3890	2.99235	1	3894
2.99239	1	3932	2.99248	1	3944	2.99265	1	4084
2.99274	1	4018	2.99282	1	4176	2.99285	1	4102
2.99287	2	4084	2.99289	1	4190	2.99291	1	4150
2.99293	2	4118	2.99295	1	4154	2.99297	2	4078
2.993	2	4142	2.99334	2	4304	2.99472	2	5452
2.99749	2	11152	3.00302	2	4574	3.00854	2	1686
3.01407	2	1026	3.0196	2	738	3.02513	2	580
3.03065	2	470	3.03618	2	398	3.04171	2	340
3.04724	2	300	3.05276	2	262	3.05829	2	238
3.06382	2	214	3.06935	2	192	3.07487	2	176
3.0804	2	162	3.08593	2	150	3.09146	2	138
3.09698	2	128	3.10251	2	118	3.10804	2	112
3.11357	2	102	3.1191	2	96	3.12462	2	90
3.13015	2	86	3.13568	2	78	3.14121	2	74
3.14673	2	68	3.15226	2	64	3.15779	2	62
3.16332	2	56	3.16884	2	54	3.17437	2	50
3.1799	2	46	3.18543	2	44	3.19095	2	40
3.19648	2	36	3.20201	2	34	3.20754	2	32
3.21307	2	28	3.21859	2	26	3.22412	2	22
3.22965	2	18	3.23518	2	12	3.2407	2	16
3.24623	2	20	3.25176	2	24	3.25729	2	28
3.26281	2	32	3.26834	2	32	3.27387	2	36
3.2794	2	42	3.28492	2	42	3.29045	2	48
3.29598	2	50	3.30151	2	56	3.30704	2	58
3.31256	2	62	3.31809	2	68	3.32362	2	72
3.32915	2	76	3.33467	2	84	3.3402	2	88
3.34573	2	96	3.35126	2	104	3.35678	2	112
3.36231	2	120	3.36784	2	132	3.37337	2	142
3.37889	2	158	3.38442	2	168	3.38995	2	190
3.39548	2	212	3.40101	2	240	3.40653	2	276
3.41206	2	320	3.41759	2	376	3.42312	2	456
3.42864	2	580	3.43417	2	776	3.4397	2	1248
3.44523	2	2748	3.44661	2	4048	3.44669	2	4136
3.44672	4	4136	3.44674	4	4196	3.44676	4	4188
3.44678	2	4332	3.4468	2	4356	3.44682	4	4204
3.44685	4	4296	3.44687	4	4368	3.44695	4	4468
3.4473	4	5108	3.44747	4	5584	3.44764	4	5968
3.44773	4	6336	3.44777	4	6628	3.44782	4	6724
3.44786	4	6932	3.44788	4	6804	3.4479	2	7172
3.44793	4	7212	3.44795	4	7252	3.44797	4	7280
3.44799	2	7608	3.44801	4	7396	3.44803	4	7736
3.44805	4	7744	3.44808	4	7952	3.44816	4	8456
3.44834	4	9440	3.44868	4	13448	3.44885	4	16916
3.44903	4	23248	3.44911	4	26748	3.44916	4	30588
3.4492	4	35540	3.44922	4	37728	3.44924	4	39624
3.44926	0	40000	3.44929	0	40000	3.44937	0	40000
3.44946	0	40000	3.4495	0	40000	3.44954	0	40000
3.44957	0	40000	3.44959	0	40000	3.44961	4	39440
3.44963	4	34176	3.44972	4	21412	3.45006	4	9160
3.45075	4	4280	3.45628	4	816	3.46181	4	428
3.46734	4	276	3.47286	4	196	3.47839	4	140
3.48392	4	104	3.48945	4	76	3.49497	4	48

3.48392	4	104	3.48945	4	76	3.49497	4	48
3.5005	4	36	3.50603	4	64	3.51156	4	96
3.51709	4	136	3.52261	4	192	3.52814	4	280
3.53367	4	460	3.53643	4	624	3.5392	4	1008
3.53989	4	1184	3.54058	4	1392	3.54092	4	1520
3.54127	4	1712	3.54144	4	1848	3.54161	4	1992
3.54164	4	2016	3.54166	4	2024	3.54168	8	1992
3.5417	8	1992	3.54172	4	2064	3.54174	4	2088
3.54179	4	2128	3.54183	4	2152	3.54185	4	2176
3.54187	8	2144	3.54196	8	2224	3.54472	8	3552
3.55025	8	272	3.55578	8	104	3.55854	8	240
3.55992	8	376	3.56131	8	592	3.56269	8	1120
3.56303	8	1456	3.56321	8	1648	3.56338	8	1888
3.56347	8	2128	3.56349	16	2128	3.56351	16	2160
3.56353	8	2272	3.56355	8	2288	3.56357	8	2368
3.5636	8	2416	3.56362	8	2552	3.56364	16	2512
3.56366	8	2640	3.56368	8	2704	3.5637	8	2848
3.56372	16	2768	3.56407	16	5648	3.56545	16	720
3.56683	16	160	3.56753	16	480	3.56787	16	768
3.56804	16	960	3.56822	16	1376	3.5683	16	1632
3.56835	16	1904	3.56837	16	2016	3.56839	16	2112
3.56841	32	2176	3.56843	32	2368	3.56848	32	2784
3.56856	32	3936	3.56891	32	2208	3.56925	32	192
3.56943	32	896	3.56951	32	1504	3.56955	32	2048
3.5696	32	3360	3.56962	64	4288	3.56964	32	6208
3.56966	64	9536	3.56968	64	25088	3.56973	64	3264
3.56977	64	1088	3.56979	64	512	3.56981	64	768
3.56986	128	3584	3.5699	128	5504	3.56992	128	2304
3.56994	256	3840	3.56997	0	40000	3.56999	0	40000
3.57003	0	40000	3.57012	0	40000	3.57029	0	40000
3.57098	0	40000	3.57236	0	40000	3.57789	0	40000
3.58342	0	40000	3.58894	0	40000	3.59447	0	40000

Jerry: I could stare at this list for hours.

In fact, I did, and there are many interesting things to see in it (after I got used to the fact that I have to read across each row first, instead of reading down the columns).

I was amazed that, as lambda changes, the number of iterations moves very consistently up or down by small steps, except in a few places where it explodes. For example, the fact that the number of iterations stays constant or increases in every step from **`lambda = 2.5`** all the way to **`lambda = 2.98`** is amazing.

Theo: It might be useful to make a list of pairs, **`lambda`** and length of cycle, and another list of pairs, **`lambda`** and number of cycles to convergence:

```
cycleLengths = Map[
    {#[[1]], #[[2]]}&, sortedResults];

iterationCounts = Map[
    {#[[1]], #[[3]]}&, sortedResults];
```

We can plot each of these using **ListPlot**:

```
ListPlot[cycleLengths,
        PlotRange -> {0, 64},
        AxesOrigin -> {2.5, 0}];
```

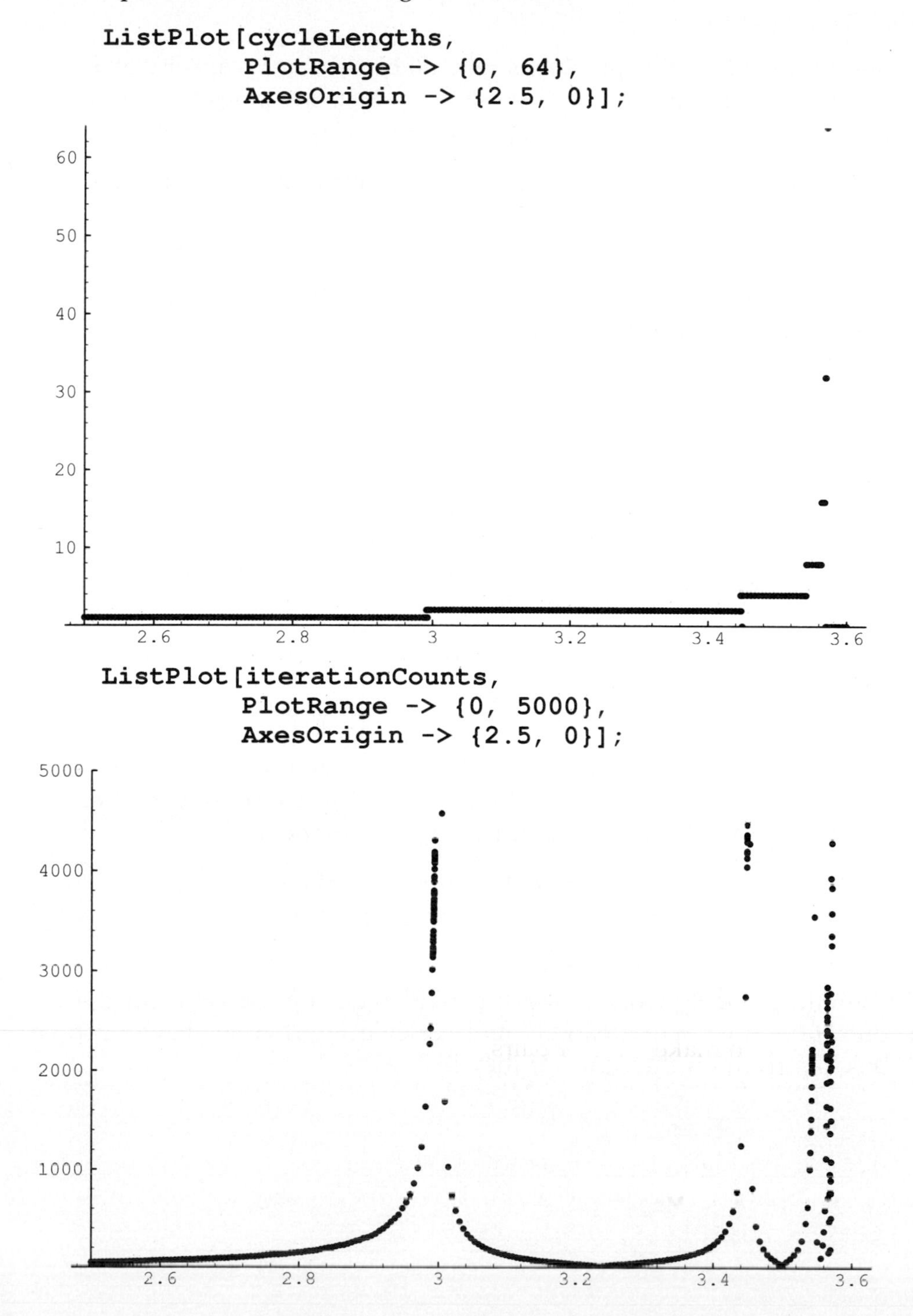


```
ListPlot[iterationCounts,
        PlotRange -> {0, 5000},
        AxesOrigin -> {2.5, 0}];
```

Jerry: This is a striking plot. It takes *much* longer to converge right at the points when the cycle is changing length.

Theo: It certainly does. In fact, this plot looks remarkably like heat capacity at a phase transition. I would be willing to bet that those spikes we see are poles that can be described by critical exponents, just like heat capacities.

Jerry: Can you say more about the heat-capacity-of-a-phase-transition connection to this graph? That sounds interesting.

Theo: Heat capacity means how much heat you have to add to something to raise its temperature by a given amount. Imagine a block of ice melting in a container. If the ice starts out well below zero, you have to add some heat to bring it up to the melting point. The more heat you add, the higher the temperature gets, indicating that the block of ice has a finite heat capacity.

Once the block starts melting, things change. As you add heat, all the heat goes into melting the ice, not into raising the temperature. You go from something which is mostly ice at zero degrees to something which is mostly water at zero degrees. As long as there is both ice and water in the container, the temperature will not change. Since you added a finite amount of heat without changing the temperature, the ratio of heat added to temperature change is infinite.

Jerry: The tall spikes in the graph above are proportional to the heat capacity at the phase transition, you say?

Theo: Right. There is a deep and beautiful theory of phase transitions that predicts the shape of the spikes in the many quantities (like heat capacity) that describe phase transitions. This theory has been applied both to physical phase transitions, and to phase transitions in various mathematical formulas and models.

Jerry: In the meantime, we can probably use those spikes. Maybe, to identify where the switches in cycle length happen, we could look for the `lambda` value that took the longest to converge.

Theo: Yes, this would probably work. Unfortunately, to get a good value for the Feigenbaum constant you have to do a LOT more calculating than we have. People reading this book can try it on their own, if they like.

■ Dialog–Fourth Day

Jerry: You said earlier when we were looking at the plots with lots of dots (Second Day) that we would be able to make them more efficiently later on. Can we use **`lambdaCycle`** to do this?

Theo: Yes, it's exactly what we need to make those plots quickly and accurately. I've made a function (see the Functions That We Defined section) that automatically makes orbit diagrams. It takes four arguments: lower lambda value; upper lambda value; number of subdivisions; and **`True`** or **`False`**, depending on whether you want to use the C program version (which is much faster, but will not run unless the correct programs are installed--see below).

It doesn't actually use **`lambdaCycle`**; it uses a new function called **`sloppy-LambdaCycle`**, which is exactly the same except that it uses a much looser test for convergence (only 6 decimal places instead of 14), because the resolution of our plots is low enough that we couldn't see the difference anyway.

We can use this new function to make the plot between 3 and 4:

```
orbitDiagram[3, 4, 300, True];
```

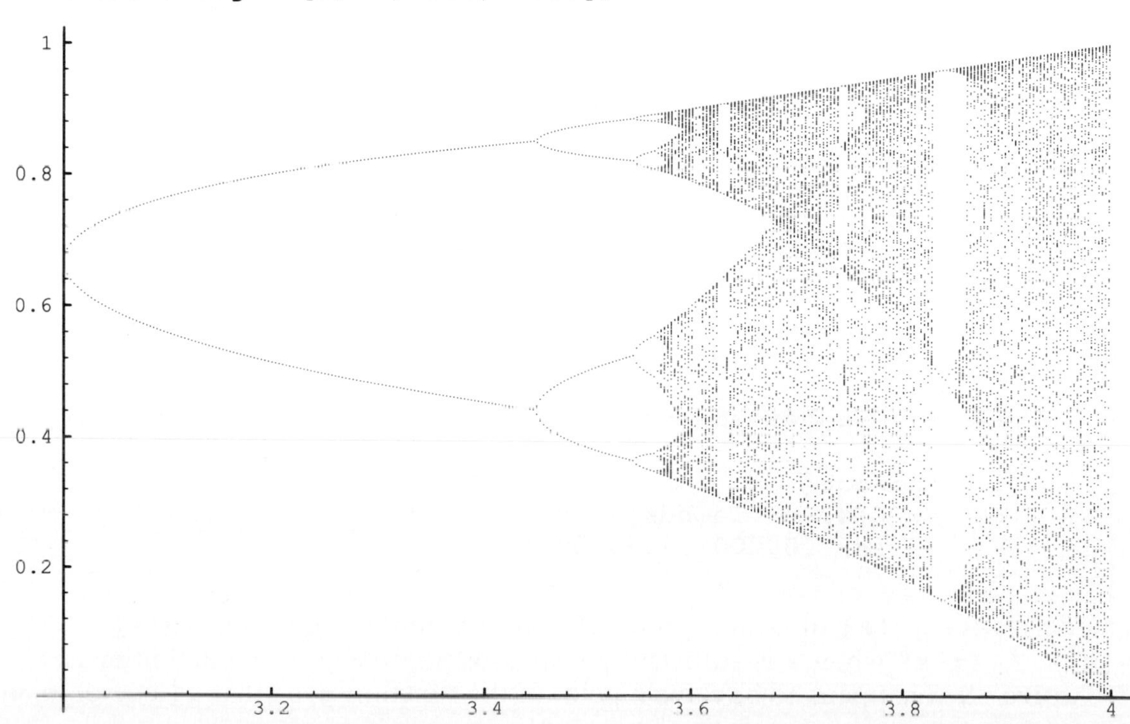

Visiting Professor: Dan Grayson

Dan: I'll show you how to use *Mathematica* to accurately compute the locations of the period-doubling bifurcations in the orbit diagram for the mapping that sends **x** to **lambda x (1 - x)**.

First we define the mapping function **f** and its derivative **fp**.

```
f[x_, lambda_] := lambda x (1 - x)
```

```
fp[x_, lambda_] = D[f[x, lambda], x];
```

We prefer to iterate the function for each requested numerical evaluation. This is better than iterating the function symbolically and storing the resulting polynomial. Here is the function that gives **f** iterated **n** times.

```
f[x_?NumberQ, n_, lambda_] :=
                Nest[f[#, lambda]&, x, n]
```

We implement here a recursive formula for the derivative of the n-th iterate of **f** based on the chain rule. The function **ff** takes the pair of values consisting of the value and derivative of some iterate of **f**, and computes the value and derivative of the next iterate.

```
ff[{val_, der_}, lambda_] :=
                {f[val, lambda], fp[val, lambda] der}
```

The function **fprime** computes the derivative of the **n**th iterate of **f** at the point **x**, by starting with the value and derivative for the 0th iterate, **x** and **1**.

```
fprime[x_?NumberQ, n_, lambda_] :=
          Nest[ff[#, lambda]&, {x, 1}, n][[2]]
```

The function **findx** takes a given value of **lambda** and locates a fixed point for the **n**th iterate of **f**. This amounts to finding a point in an orbit of length **n** for **f**.

```
findx[n_, lambda_?NumberQ] :=
     x /.
     FindRoot[
         f[x, n, lambda] - x,
         {x, .000001, .999999}
     ]
```

The derivative **g'[x]** of a function **g[x]** is nearly equal to the ratio **(g[z]-g[x]) / (z-x)** when **z** is sufficiently near to **x**, according to the definition of derivative. If **x** happens to be a fixed point of the function **g**, so that **g[x]==x**, then we can say that **g'[x]** is nearly equal to the ratio **(g[z]-x)/(z-x)**, or equivalently. Thus, if the absolute value of **g'[x]** is less than 1, then **g[z]-x** is smaller than **z-x**, which means that **g[z]** is closer to **x** than **z** is; in this case we say that **x** is an attracting fixed point of **g**, because the sequence of points **z, g[z], g[g[z]], g[g[g[z]]],...** will approach **x**. Conversely, if the absolute value of **g'[z]** is bigger than 1, this means that **g[z]** is farther from **x** than **z** is; in this case we say that **x** is a repelling fixed point, because the sequence of iterates will depart from the vicinity of **x**.

The points that get plotted in the orbit diagram are the attracting fixed points. When **lambda** increases in such a way that a period-doubling occurs, what has happened is that some attracting fixed points become repelling fixed points, which means that the derivative of the iterate of **f** passes through the value 1. The next function makes use of this observation to locate the points where period doubling occurs.

The function **find** accepts a range of values for **lambda** and looks for a value of **lambda** at which the derivative of the **n**th iterate of **f**, evaluated at a fixed point, has value 1 or -1. The parameter **s** specifies whether the value is 1 or -1. The value returned for **lambda** is the place where period-doubling occurs.

```
find[n_, s_, {lambda0_, lambda1_}] :=
    lambda /.
    FindRoot[
        fprime[ findx[n, lambda], n, lambda ] - s,
        {lambda, {lambda0, lambda1}}
    ]
```

Now we run our routine. (The search ranges below were determined by a bit of experimentation ahead of time, and don't represent some sort of dishonest prescience.)

```
{
find[2^0, -1, {2.9,3.1}],
find[2^1, -1, {3.4,3.5}],
find[2^2, -1, {3.5,3.6}],
find[2^3, -1, {3.55,3.58}],
find[2^4, -1, {3.56,3.59}],
find[2^5, -1, {3.569,3.57}],
find[2^6, -1, {3.5698,3.5699}]
}
{3., 3.44949, 3.54409, 3.56441, 3.56876, 3.56969, 3.56989}
```

The recipe for computing Feigenbaum's constant from these numbers is to take neighboring ratios of neighboring differences of these numbers. So first we take differences:

```
Drop[%, 1] - Drop[%, -1]
{0.449489, 0.0945995, 0.0203179, 0.00435208, 0.000932212,
  0.000199648}
```

and then we take ratios:

```
Drop[%, -1] / Drop[%, 1]
{4.7515, 4.65596, 4.66856, 4.66856, 4.66927}
```

These numbers appear to be converging to a limiting value of 4.669 or so. The actual limiting value, which we've computed to only three or four decimal places, is called *Feigenbaum's constant*. Other authors have computed this constant, too, and found that it is 4.66920, to five decimal places, so we have come pretty close!

Functions That We Defined

Theo: In this section we defined the **quad** function, which we iterated many times to discover chaos:

```
quad[x_] := lambda x (1 - x);
```

We also defined two different plotting functions, **lambdaPlot** and **lambdaPlotB**, which both plot the successive values of the iterated **quad** function.

```
lambdaPlot[lam_,startX_,skipCycles_,showCycles_] :=
    Block[{localQuad},

        localQuad[x_] := lam x (1 - x);

        Plot[localQuad[x], {x, 0, 1},
            PlotLabel->SequenceForm["Lambda = ", lam],
            PlotRange->{0, 1},
            AspectRatio->Automatic,
            Epilog->
                Line[
                    Partition[
                        NestList[
                            localQuad,
                            Nest[
                                localQuad,
                                startX,
                                skipCycles
                            ],
                            showCycles
                        ],
                        2, 1
                    ]
                ]
        ]
    ]
```

```
lambdaPlotB[lam_,startX_,skipCycles_,showCycles_] :=
    Block[{localQuad},

        localQuad[x_] := lam x (1 - x);

        Show[
            Graphics[{
                Text[
                    SequenceForm["Lambda = ", lam],
                    Scaled[{1, 1}],
                    {1, 1}
                ],
                Line[
                    Partition[
                        NestList[
                            localQuad,
                            Nest[
                                localQuad,
                                startX,
                                skipCycles
                            ],
                            showCycles
                        ],
                        2, 1
                    ]
                ]
            }],
            AspectRatio->Automatic
        ]
    ]
```

On the second day we defined a new version of the built-in **FixedPoint** function. Our version is able to detect cycles; when it does, it returns a list of the values in the cycle. **CycleFixedPoint** takes the same arguments as **FixedPoint**, including the optional third argument that specifies a maximum number of iterations.

CycleFixedPoint can also be given options to specify what functions to use to determine if two values are "the same". The **SameTest** option specifies the function to use while testing for convergence, and the **CycleTest** option specifies what function to use when measuring the cycle length. (In numerical cases, the **CycleTest** should normally ignore a few decimal places of accuracy, to avoid problems with numerical round-off error. The **SameTest** should require a higher level of accuracy than the **CycleTest**.)

CycleFixedPoint returns a list of the elements of the limit cycle. If the function converges to a single value, a list containing a single value is returned. If the function does not converge in the maximum number of steps (if any) specified, then a single number (the last one calculated) is returned, not surrounded by list brackets.

If the option **IterationCount -> True** is included, then **CycleFixedPoint** returns a list of two elements. The first element is the normal result described above. The second element is the number of cycles required to reach this result (which will be equal to the maximum specified if the function did not converge).

```
Options[CycleFixedPoint] =
     {
          SameTest -> SameQ,
          CycleTest -> SameQ,
          IterationCount -> False
     };
```

```
CycleFixedPoint[f_, start_, options___] :=
     CycleFixedPoint[f, start, Infinity]
```

```
CycleFixedPoint[f_, start_, maxIterations_?NumberQ,
                    options___] := Block[
     {nth, twonth, counter, result, value,
      sameTest, cycleTest},

     {sameTest, cycleTest, iterationCount} =
          {SameTest, CycleTest, IterationCount} /.
                    {options} /.
                    Options[CycleFixedPoint];

     nth = f[start];
     twonth = f[f[start]];
     counter = 1;
```

```
    If[maxIterations =!= Infinity,
        While[ (!sameTest[nth, twonth]) &&
                (counter < maxIterations),
            nth = f[nth];
            twonth = f[f[twonth]];
            ++counter;
        ]
    ,
        While[!sameTest[nth, twonth],
            nth = f[nth];
            twonth = f[f[twonth]];
            ++counter;
        ]
    ];

    If[counter == maxIterations,
        result = nth
    ,
        result = {twonth};
        value = f[twonth];
        While[!cycleTest[value, twonth],
            AppendTo[result, value];
            value = f[value]
        ]
    ];

    If[TrueQ[iterationCount],
        {result, counter},
        result
    ]
]
```

We defined a function for use with **CycleFixedPoint** that compares only the first 12 or so places of a floating-point number (assuming it is close to 1):

```
FuzzySameQ[x_, y_] := Abs[x - y] < 10^-12
```

Using these functions, we made a utility function to return the limit cycle of the **quad** function, together with the number of cycles required to reach this limit:

```
lambdaCycle[lambda_, maxCycles_] :=
    CycleFixedPoint[
        lambda # (1 - #)&,
        0.5,
        maxCycles,
        CycleTest -> FuzzySameQ,
        IterationCount -> True
    ]
```

We also defined a function for making orbit diagrams quickly. First, we needed some utility functions:

```
VeryFuzzySameQ[x_, y_] := Abs[x - y] < 10^-6
```

```
VeryVeryFuzzySameQ[x_, y_] := Abs[x - y] < 10^-5
```

```
sloppyLambdaCycle[lambda_, maxCycles_] :=
    CycleFixedPoint[
        lambda # (1 - #)&,
        0.5,
        maxCycles,
        SameTest -> VeryFuzzySameQ,
        CycleTest -> VeryVeryFuzzySameQ,
        IterationCount -> True
    ]
```

The **orbitDiagram** function can use either **sloppyLambdaCycle** (defined above), or **sloppyLambdaCycleC** (defined in the following section).

```
orbitDiagram[lower_, upper_, divisions_,
            useC_, options___] :=
Block[
    {lambda},
    ListPlot[
        Flatten[
            Table[
                cycle = First[
                    If[useC,
                        sloppyLambdaCycleC,
                        sloppyLambdaCycle
                    ][lambda, 500]
                ];
                Map[
                    {lambda, #}&,
                    If[Head[cycle] === List,
                        Take[
                            cycle,
                            Min[
                                Length[cycle],
                                128
                            ]
                        ],
                        NestList[
                            N[lambda # (1-#)]&,
                            cycle, 128]
                    ]
                ],
            {lambda, lower, upper,
             (upper-lower)/divisions}
            ],
            1
        ],
        options,
        PlotStyle -> PointSize[0.001]
    ]
]
```

■ For Those Who Want To Know

Starved for speed, we resorted to writing some fast C language code to compute limit cycles. The following *Mathematica* function exactly simulates the **lambdaCycle** function implemented above. The only difference is that while **lambdaCycle** is written entirely in *Mathematica* and will run on any *Mathematica* system, **lambdaCycleC** is designed to execute an external C program to do the actual iteration. This means it will work only if that external program has been compiled and is available on the same computer *Mathematica* is running on.

You can use the two interchangeably: The only difference is that one runs about 500 times faster. You get to guess which one that is.

```
lambdaCycleC[lambda_, maxCycles_] :=
    Get[StringJoin[
        "!LambdaCycle/lambdaCycle ",
        ToString[N[lambda]],
        " ",
        ToString[N[maxCycles]]
    ]]
```

This function assumes that the following C function (which should work on any normal UNIX system) is available as an executable, called `lambdaCycle`, in a subdirectory called `LambdaCycle`. The subdirectory should be located where *Mathematica* can find it (typically in the home directory of the person running *Mathematica*).

```
/*
 * Lambda Cycle calculator.
 *
 * This program takes two arguments on its command line:
 * First, a value of lambda (floating point number between
 * zero and 4, typically).  Second, a maximum number of
 * iterations before giving up (20000 or so is reasonable).
 * It prints to standard output a Mathematica list containing
 * two elements.  The first element is a list of the values
 * in the limit cycle of the iterated quadratic map
 * x1 - lambda * x0 * (1 - x0), or a single number if the
 * iteration failed to converge.  The second element is
 * a number giving how many cycles were required for the
 * iteration to converge.
 */
```

```
main(argc, argv)
int     argc;
char    **argv;
{
    double  lambda;
    double  x, y;
    double  z;
    long    counter;
    long    maxCycles;
    double  slop = 0.00000000000001;

    sscanf(argv[1], "%lf", &lambda);
    sscanf(argv[2], "%ld", &maxCycles);

    x = 0.5;
    y = 0.5;
    counter =  maxCycles;

    do
        {
        x = lambda * x * (1 - x);
        y = lambda * y * (1 - y);
        y = lambda * y * (1 - y);
        --counter;
        }
    while ((x != y) & (counter != 0));

    printf("{");

    if (counter == 0)
        printf("%lg", y);
    else
        {
        printf("{");

        x = y;
        printf("%lg", y);
        y = lambda * y * (1 - y);
        while (z = x - y, (z > slop) || (z < -slop))
            {
            printf(", %lg", y);
            y = lambda * y * (1 - y);
            }
```

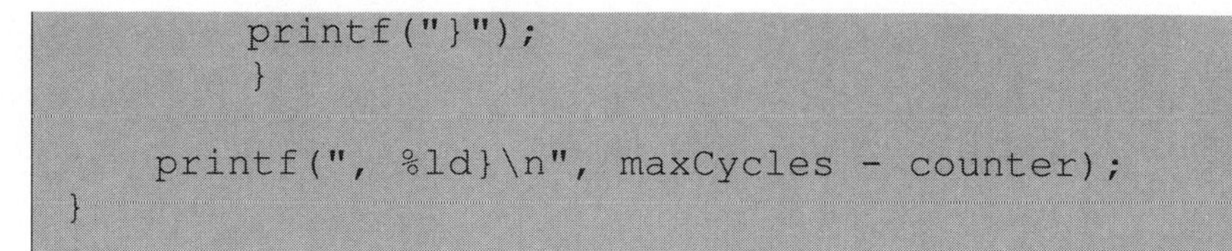

```
        printf("}");
        }

    printf(", %ld}\n", maxCycles - counter);
}
```

The **sloppyLambdaCycle** function is simulated by the following function (which works just like **lambdaCycleC**):

```
sloppyLambdaCycleC[lambda_, maxCycles_] :=
    Get[StringJoin[
        "!LambdaCycle/sloppyLambdaCycle ",
        ToString[N[lambda]],
        " ",
        ToString[N[maxCycles]]
    ]]
```

This function uses the following C program:

```
/*
 * Sloppy Lambda Cycle calculator.
 *
 * The program does the same thing as the Lambda Cycle
 * program, except that the tests for convergence are
 * very sloppy.  It is used for making plots where the
 * resolution of the plot limits what can be seen anyway.
 */

main(argc, argv)
int     argc;
char    **argv;
{
    double  lambda;
    double  x, y;
    double  z;
    long    counter;
    long    maxCycles;
    double  slopA = 0.000001;
    double  slopB = 0.00001;

    sscanf(argv[1], "%lf", &lambda);
    sscanf(argv[2], "%ld", &maxCycles);
```

```
    x = 0.5;
    y = 0.5;
    counter =  maxCycles;

    do
        {
        x = lambda * x * (1 - x);
        y = lambda * y * (1 - y);
        y = lambda * y * (1 - y);
        --counter;
        }
    while ((z = x - y, (z > slopA) || (z < -slopA)) &
            (counter != 0));

    printf("{");

    if (counter == 0)
        printf("%lg", y);
    else
        {
        printf("{");

        x = y;
        printf("%lg", y);
        y = lambda * y * (1 - y);
        while (z = x - y, (z > slopB) || (z < -slopB))
            {
            printf(", %lg", y);
            y = lambda * y * (1 - y);
            }

        printf("}");
        }

    printf(", %ld}\n", maxCycles - counter);
}
```

References

1) Mitchell Feigenbaum, *Universal Behavior of Nonlinear Systems*, Los Alamos Science, Summer, 1 (1980).
2) Martin Gardner in one of his Scientific American articles.
3) James Gleick, *Chaos, Making a New Science*, Viking, 1978.
4) H.-O. Peitgen and P. H. Richter, *The Beauty of Fractals*, Springer-Verlag, 1986.
5) Robert L. Devaney, *Chaos, Fractals, and Dynamics*, Addison-Wesley, 1990.
6) Michael Barnsley, *Fractals Everywhere*, Academic Press, 1988.

Chapter Eight
Fractals When Least Expected

Jerry and Theo discover that fractals are everywhere, and end up making some very beautiful pictures.

■ Dialog

Theo: In Chapter 3 we looked at nested **Sin** functions and we made plots, varying either the starting value or the number of **Sin** functions or both. Something we haven't done is look at what happens if we use a complex number as the starting point.

Jerry: Let's try an example of that:

```
Sin[1.2 + 0.3 I]
0.974296 + 0.110345 I
```

So if you start with a complex number, the value of the **Sin** function is also a complex number. How can we make a plot of such a function? If we want to make a plot of a complex function in the complex plane, we are going to have to plot four separate variables: the real and imaginary parts of the input value, and the real and imaginary parts of the function value. We can use the x and y axes of a three dimensional plot for the real and imaginary parts of the input, but what can we do with the real and imaginary parts of the function value? Since we are handicapped by our three-dimensional reality, we have to figure out some way to plot two variables using only one more dimension in space (the z axis).

Theo: We could use color as the fourth variable.

Jerry: So, the real and imaginary parts of the function will be plotted using the height of the surface for one part, and the color of the surface for the other part?

Theo: Right. It turns out that usually the most useful thing to do is the make the height represent the absolute value of the function, and the color represent the complex phase angle of the function.

Jerry: You sound as if this is a well known thing. Is this technique widely used?

Theo: Variations of it are. Many people who use *Mathematica* find that it's very easy to plot functions this way, so it has been gaining in popularity. In the past, making colored plots was generally quite difficult compared to how it's done in *Mathematica*. The only serious problem remaining with colored plots is how to print them. With current technology, color printing, whether single pages or large quantities, is quite expensive. Many scientific journals don't print graphics in color.

Jerry: Perhaps I should explain a bit about absolute values and phase angles. A complex number can be thought of as a point in a plane whose x-coordinate is its real part and y-coordinate its imaginary part. If you think of the complex number as a point, then the absolute value of the number is its distance from the origin. The complex phase angle of the number is the angle that a vector from the origin to the point makes with the real axis. Here's an illustration of this:

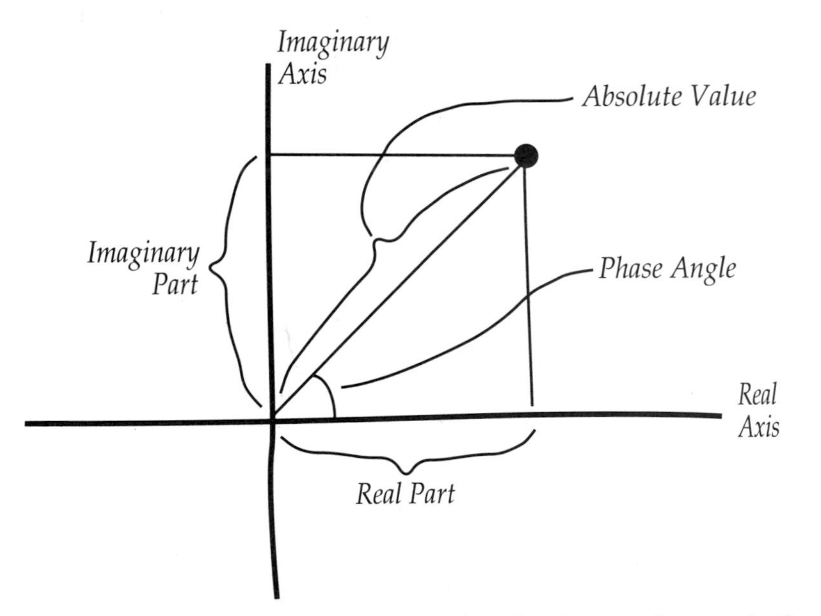

So, converting a complex number into its absolute value and phase angle is very much like converting a point in the plane into polar coordinates.

How are we going to use this to plot the function?

Theo: Well, the absolute value part is easy. In fact, we can make a plot right now, just by plotting the absolute value of the **Sin** function:

```
Plot3D[Abs[Sin[x + y I]], {x, -Pi, Pi}, {y, -Pi, Pi}];
```

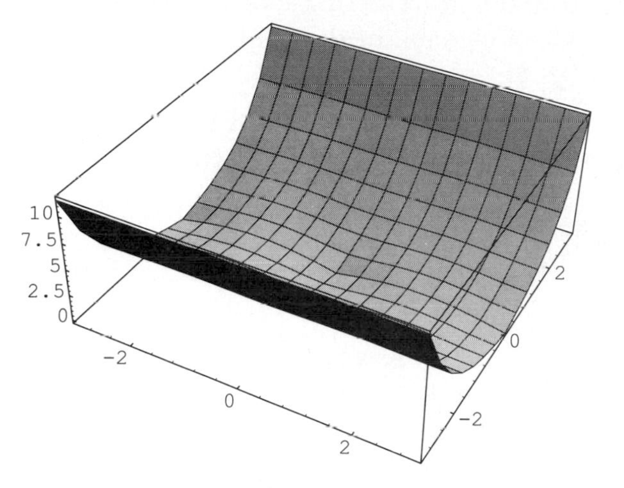

This plot doesn't include all the information from the function, however. We can see how far the value is away from the origin, but we can't see in what direction. A very nice way to show this is to map the phase angle of the function onto the color wheel.

Jerry: In other words, as the angle moves around the circle, the color of the surface will go from red to green to blue and back to red again. I like this idea. How do we tell *Mathematica* to do this?

Theo: Fortunately there is a function called **Hue** which does exactly this. Both **Hue[0]** and **Hue[1]** represent pure red -- hue goes in a circle. Starting at zero, the the hue moves from red to yellow to green to cyan to blue to magenta, and back to red.

There is a *Mathematica* function called **Arg** which calculates the phase angle (in radians) of a complex number:

```
Arg[1.2 + 0.3 I]
0.244979
```

Arg always returns an angle between **-Pi** and **Pi**. To use this as a parameter to **Hue** we need to convert it into a number between zero and 1. We can do that by adding **Pi** and then dividing by **2Pi**:

```
Hue[N[(Pi + Arg[1.2 + 0.3 I]) / (2 Pi)]]
Hue[0.53899]
```

Jerry: Could we make picture of what each point in the complex plane would look like using this coloring scheme?

Theo: Sure. The way you add mathematically defined color to a plot is by using a list as the first argument to the **Plot3D** function. The first element of the list becomes the height of the surface, and the second element of the list becomes the color of the surface. Let's try just plotting **x + y I**:

```
Plot3D[
    {
        Abs[x + y I],
        Hue[N[(Pi + Arg[x + y I]) / (2 Pi)]]
    },
    {x, -2, 2},
    {y, -2, 2},
    AxesLabel -> {"Re", "Im", "Abs[z]"},
    PlotPoints -> 15
];
```

See picture on color plates page ii.

Jerry: I see that as we move away from the origin, the surface gets higher. If we move in a straight line away from the origin, the color stays the same. If we move in a circle a constant distance from the origin, the height stays the same, but the color changes. Now I see why it's such a nice thing that the color wheel is a wheel. Just as a circle comes back to where it started from, so the color wheel comes back. It fits perfectly.

Theo: It really is a perfect match. Before we go on, let's define a function we can use to translate a single complex number into a list of absolute value and color. We can use this function whenever we want to plot a complex function:

```
maxZ = N[10^30];
nPi = N[Pi];
ntwoPi = N[2Pi];
complexToZandColor[z_] := Block[{nz, az},
    nz = N[z];
    az = Abs[nz];
    {
        Min[az, maxZ],
        Hue[(nPi+If[nz==0, 0, Arg[nz / az]])/(ntwoPi)]
    }
    ]
```

This function is rather more complicated than might seem necessary, because it is designed to take care of a number of problems at once. First, we define some numerical constants. This avoids having to recalculate the numerical value of **Pi** many times. Second, we clip the range of the absolute value to less than 10^30. This is to prevent **Plot3D** from rejecting points with huge absolute values as unplottable (you'll see why we need this later). Third, we take care of the problem of what color to use if the absolute value is zero. For a point actually on the origin the phase angle is undefined (a zero length vector could be pointing in any direction). **Arg** refuses to return a value in this case. I've just arbitrarily decided that if the absolute value is zero we'll pretend that the phase is also zero.

Jerry: We could use this function to make the same plot as before, like this:

```
Plot3D[
    complexToZandColor[x + y I],
    {x, -2, 2},
    {y, -2, 2},
    AxesLabel -> {"Re", "Im", "Abs[z]"}
];
```

See picture on color plates page ii.

Theo: Now we can try that plot of **Sin[z]** again, this time using this function to color the surface:

```
Plot3D[
    complexToZandColor[Sin[x + y I]],
    {x, -Pi, Pi},
    {y, -Pi, Pi}
];
```

See picture on color plates page ii.

Jerry: Clearly we were missing something in the earlier version of this picture!

Theo: Let's start experimenting more with nested **Sin** functions in the complex plane. (I should warn people that some of the following plots take quite a long time to produce on 1990-style desktop computers.) We've seen **Sin[z]** in the plot above. It looks sort of boring. Let's try **Sin[Sin[z]]**:

```
Plot3D[
     complexToZandColor[Sin[Sin[x + y I]]],
     {x, -Pi, Pi},
     {y, -Pi, Pi},
     PlotPoints -> 50
];
```

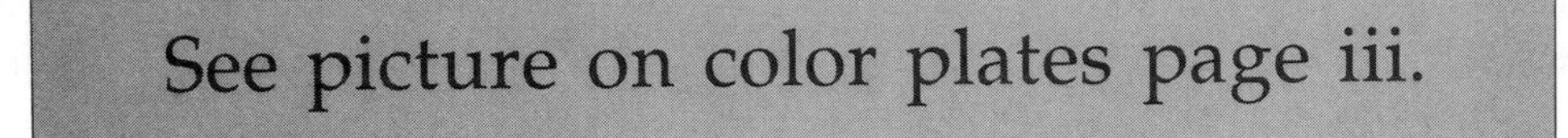

Jerry: It looks like the absolute value of **Sin[Sin[z]]** gets very big when **z** moves off in the imaginary direction. It also looks like the phase is starting to do some interesting things. It moves around the circle several times on the way to those large peaks. I wonder what **Sin[Sin[Sin[z]]]** will look like:

```
f[x_] := Sin[Sin[Sin[x]]]
Plot3D[
     complexToZandColor[f[x + I y]],
     {x, -Pi, Pi},
     {y, -Pi, Pi},
     PlotPoints -> 100,
     PlotRange -> {0, 10^6}
     ];
```

See picture on color plates page iv.

Theo: Now that's *really* interesting! Let's see what we can notice about this plot. First, **Sin[Sin[Sin[z]]]** gets *very* big as you move off in the imaginary direction (notice that the plot range is from zero to a million, and the peaks are still cut off). Let's look at the value right in the middle of one of those big peaks, at **z = Pi + Pi I**:

```
Sin[Sin[Sin[Pi + Pi I]]]//N
```

```
                22506
-1.6055795685 10      I
```

Very big indeed. Second, we notice that the phase gets all twisted around near the peaks. It seems to start oscillating faster and faster as you move up the peak. Third, and perhaps most interesting, we notice that the peaks seem to be fluted. They have a definite rib-like structure. Unfortunately, it's hard to see what they really look like because they are so hugely tall. In a situation like this, the **Log** function often comes in handy. By plotting the **Log** of a function instead of the function itself, you can compress a large range of values into a much smaller range. It turns out that this function varies so much that we have to use the **Log** of the **Log** to get it into a form where we can see the whole thing at once. To do this we define a new version of our **complexToZandColor** function:

```
nPi = N[Pi];
ntwoPi = N[2Pi];
nE = N[E];
complexToLogLogZandColor[z_] := Block[{nz, az},
     nz = N[z];
     az = Abs[nz];
     {
          Log[Log[az + nE]],
          Hue[(nPi +
               If[nz == 0,
                    0,
                    Arg[nz / az]
               ])/(ntwoPi)
          ]
     }
     ]
```

Jerry: Wait a minute, what's that "**+ nE**" doing in there?

Theo: It's there because **Log[z]** is sometimes negative (for values of **Abs[z]** less than **E**), and if **Log[Abs[z]]** is negative then **Log[Log[Abs[z]]]** would be complex, which gets us right back where we started. By adding **E** we can be sure that **Log[Abs[z] + E]** will never be negative (since **Abs[z]** is always positive).

We can use this new function to produce this plot:

```
f[x_] := Sin[Sin[Sin[x]]]
Plot3D[
    complexToLogLogZandColor[f[x + I y]],
    {x, -Pi, Pi},
    {y, -Pi, Pi},
    PlotPoints -> 100,
    PlotRange -> All
    ];
```

See picture on color plates page v.

Jerry: Now we can really see the ribbed peaks. I'm surprised by those ribs.

Theo: So am I. It's not what I expected (although I'm not sure what I *did* expect). Let's push on to **Sin[Sin[Sin[Sin[z]]]]**:

```
f[x_] := Sin[Sin[Sin[Sin[x]]]]
Plot3D[
    complexToLogLogZandColor[f[x + I y]],
    {x, -Pi, Pi},
    {y, -2, 2},
    PlotPoints -> 100
    ];
```

See picture on color plates page vi.

Jerry: Now it looks like the peaks are even more ribbed. In fact, it looks like there are big ribs which themselves have smaller ribs. You don't suppose....

Theo: I certainly do! There's a fractal in there somewhere! Just think, maybe if we nest another **Sin** function in, then there will ribs upon ribs upon ribs. If we nested an infinite number of **Sin** functions, then there would be an infinite number of levels of ribs. In other words, no matter how closely we looked at a particular rib, we would always see that it had more ribs itself. That's what a fractal is: No matter how close you look, it always has more ribs.

Unfortunately, **Sin[Sin[Sin[Sin[Sin[z]]]]]**, the next plot we might want to make, is almost impossible. The function gets so huge so fast that it's really hard to see anything. We need to rethink our whole approach.

The problem with these functions is that they get very big very quickly. Maybe we can turn that around and use it to our advantage. If you look at the plots, the ribs seem to stay pretty close to zero, while everything around them zooms off into the sky. A reasonable guess might be that as you nest more and more **Sin** functions the value along these ribs actually moves steadily towards zero. (This is true, for example, along the biggest rib of all, the real axis. We know from Chapter 3 that if you nest **Sin** functions and look only at real starting values, the output gets smaller and smaller the more **Sin** function you nest.) Another reasonable guess might be that for any points not on one of the ribs the value zooms steadily towards infinity as you nest more and more **Sin** functions.

If this is the case, then the basic choice is between points on the surface that move towards zero as you nest **Sin** functions, and points on the surface that move towards infinity. (Those of you familiar with the famous Mandelbrot set are probably salivating already.)

A bit of analysis is enough to convince me, at least, that if, as you progressively nest **Sin** functions, the imaginary component of the value ever gets very large, then the value is going to keep getting bigger forever.

On the other hand, if you have nested a whole bunch of **Sin** functions and the value is still relatively small, it's probably going to stay that way.

So, to see what the function looks like for a really large number of **Sin** functions, we don't necessarily have to calculate all the **Sin** functions for all the values. All we have to do for each value is to start nesting **Sin** functions until the imaginary component of the value either gets very big, in which case we can stop, or until we nest enough **Sin** functions to convince us that the value is probably going to stay small.

An interesting number might be how many **Sin** functions we have to nest before the value gets bigger than a certain threshold beyond which we don't think we have to calculate. Here's a function to tell us this. (Note that I've arbitrarily decided that if the absolute value of the imaginary part gets bigger than 10000, or we reach 100 **Sin** functions without the value getting that big, then we're through):

```
numberOfSinsToEscape[z_] := Block[
     {count, nz = N[z]},
     For[
          count = 0,
          (Abs[Im[nz]] < 10000) && (count < 100),

          nz = Sin[nz];
          ++count
     ];
     count
]
```

Using this function we can make the following plot:

```
Plot3D[
     numberOfSinsToEscape[x + I y],
     {x, -2Pi, 2Pi}, {y, -2Pi, 2Pi},
     PlotPoints -> 50
];
```

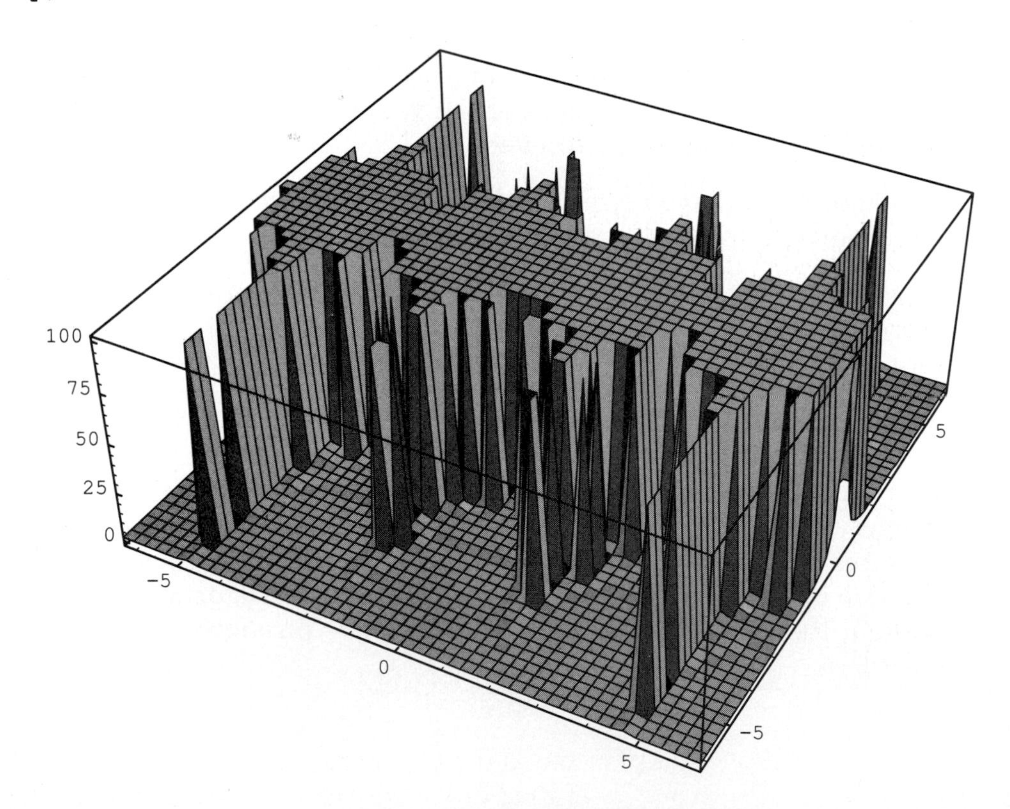

Jerry: Let's understand what this plot is. The points on this surface whose heights are near zero are those that take only a small number of **Sin** functions before they get very big. So they represent the huge peaks we saw in the other plots. The points whose heights are near 100 are those that never got big enough, so they represent the furrows in the ribs. I can sort of see how this plot fits with the other plots, but it's not easy.

Theo: You're right. We need to plot many more points to see enough detail. But really, there's no need to use three-dimensional plots anymore. There's just one parameter we are plotting, namely the number of iterations before the value runs away. We could just make a two-dimensional plot where we color each spot according to how many iterations it took. Let's use white for those points which stay small, and black for those which get big very fast. Shades of gray can represent values that get big more slowly. To make such a plot, all we have to do is to replace the word **Plot3D** with **DensityPlot** in the command above. Here's the plot:

```
DensityPlot[
     numberOfSinsToEscape[x + I y],
     {x, -2Pi, 2Pi}, {y, -2Pi, 2Pi},
     PlotPoints -> 50
];
```

6
4
2
0
−2
−4
−6
−6 −4 −2 0 2 4 6

It's still sort of hard to see. The problem is that we need vastly more points in our plot, so we can make out the details of the border between the white and the black areas. We need some heavy artillery to calculate enough points to make out the details. The technique we will use is discussed in the Functions That We Defined section. If you want to know how this technique works, or if you want to generate your own fractals, you should read the section. Otherwise, all you need to know is that the functions we define in that section make plots just like the one above, only faster, provided you are running on a UNIX computer and have the appropriate external programs available. The following command duplicates the plot above but with 500 by 500 samples points instead of 50 by 50:

```
fractalPlot[
    sinFractalC[{{-7, 7}, {-7, 7}}, 500],
    {0, 50}];
```

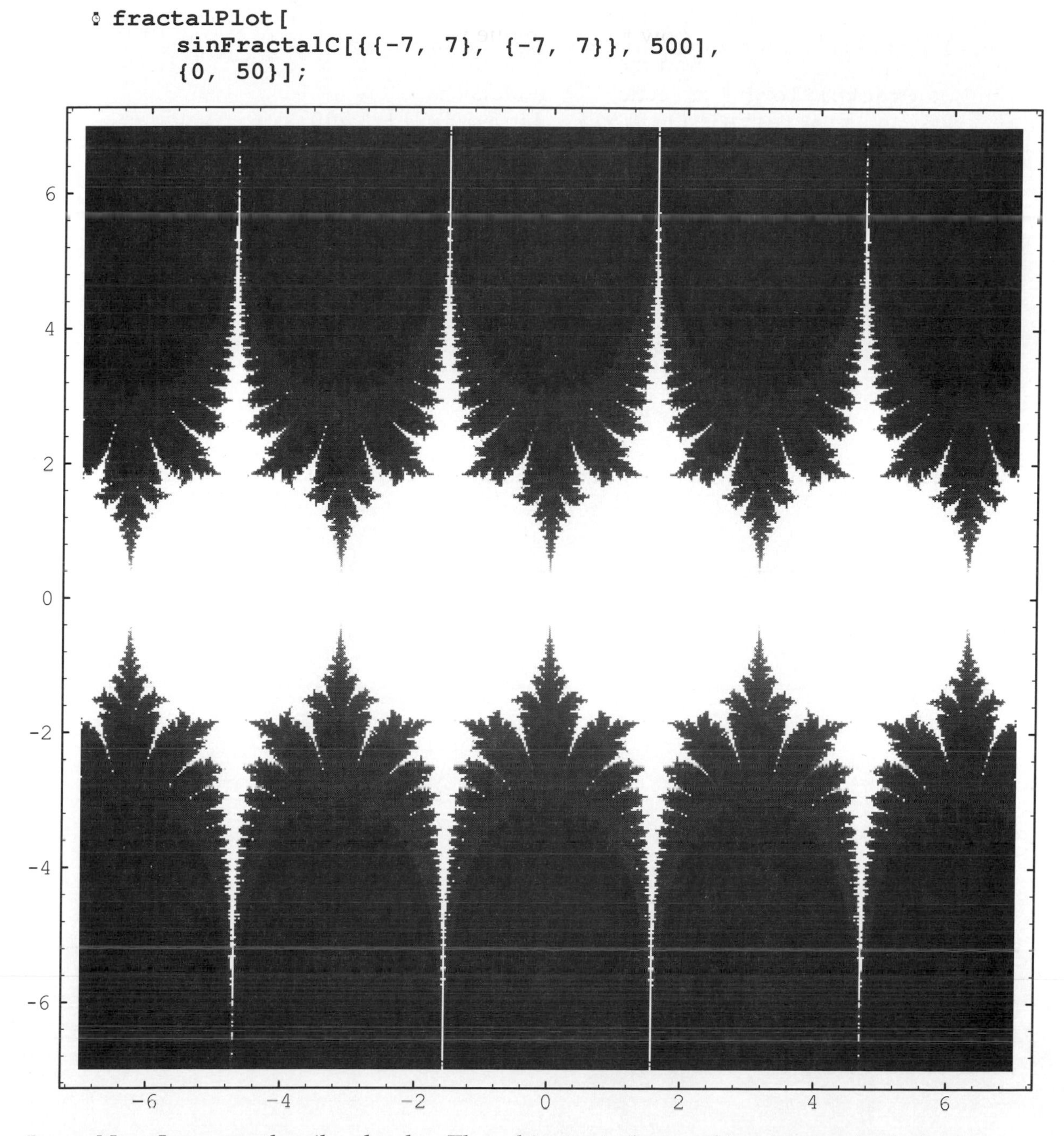

Jerry: Now I can see the ribs clearly. The white tentacles reaching into each black section are the low points of the ribs. And Whoa!, it certainly does look like a fractal!

Theo: If it's a fractal, then we should be able to blow up a small section and see even more tentacles. Let's try magnifying one of the small tentacles:

```
fractalPlot[
    sinFractalC[{{0, 1}, {1.3, 2.7}}, 500],
    {0, 20}];
```

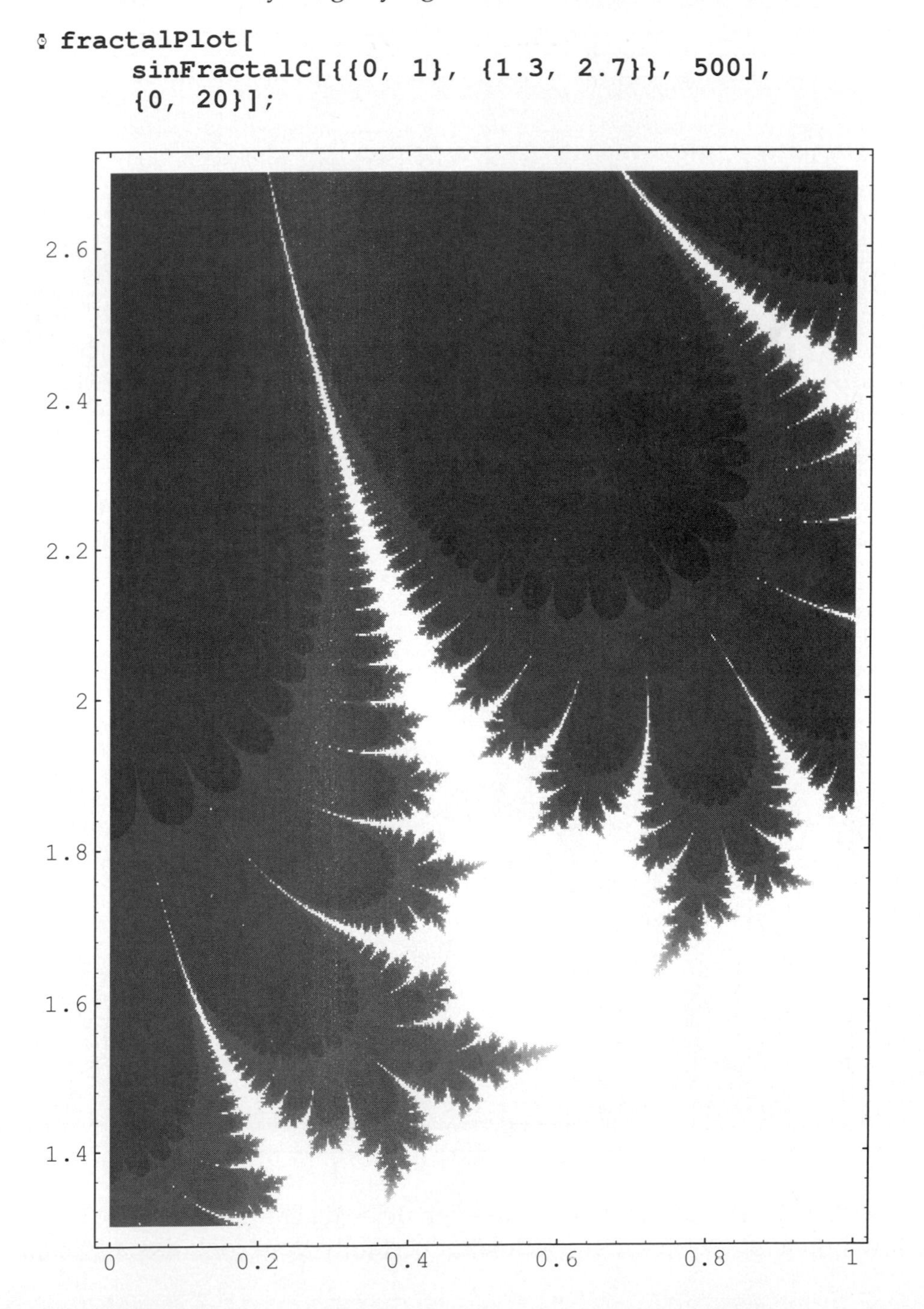

And we can zoom in again:

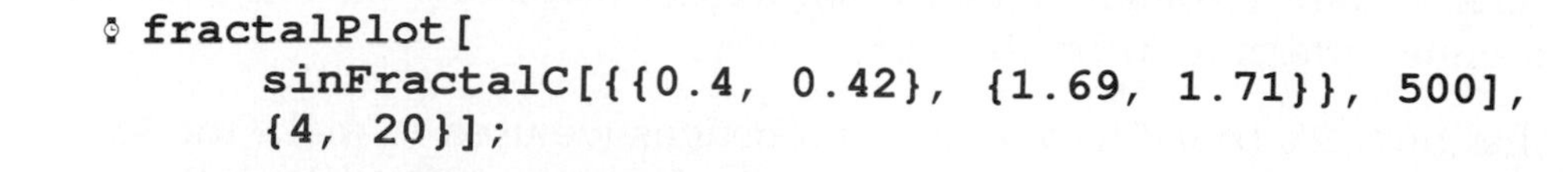

```
fractalPlot[
    sinFractalC[{{0.4, 0.42}, {1.69, 1.71}}, 500],
    {4, 20}];
```

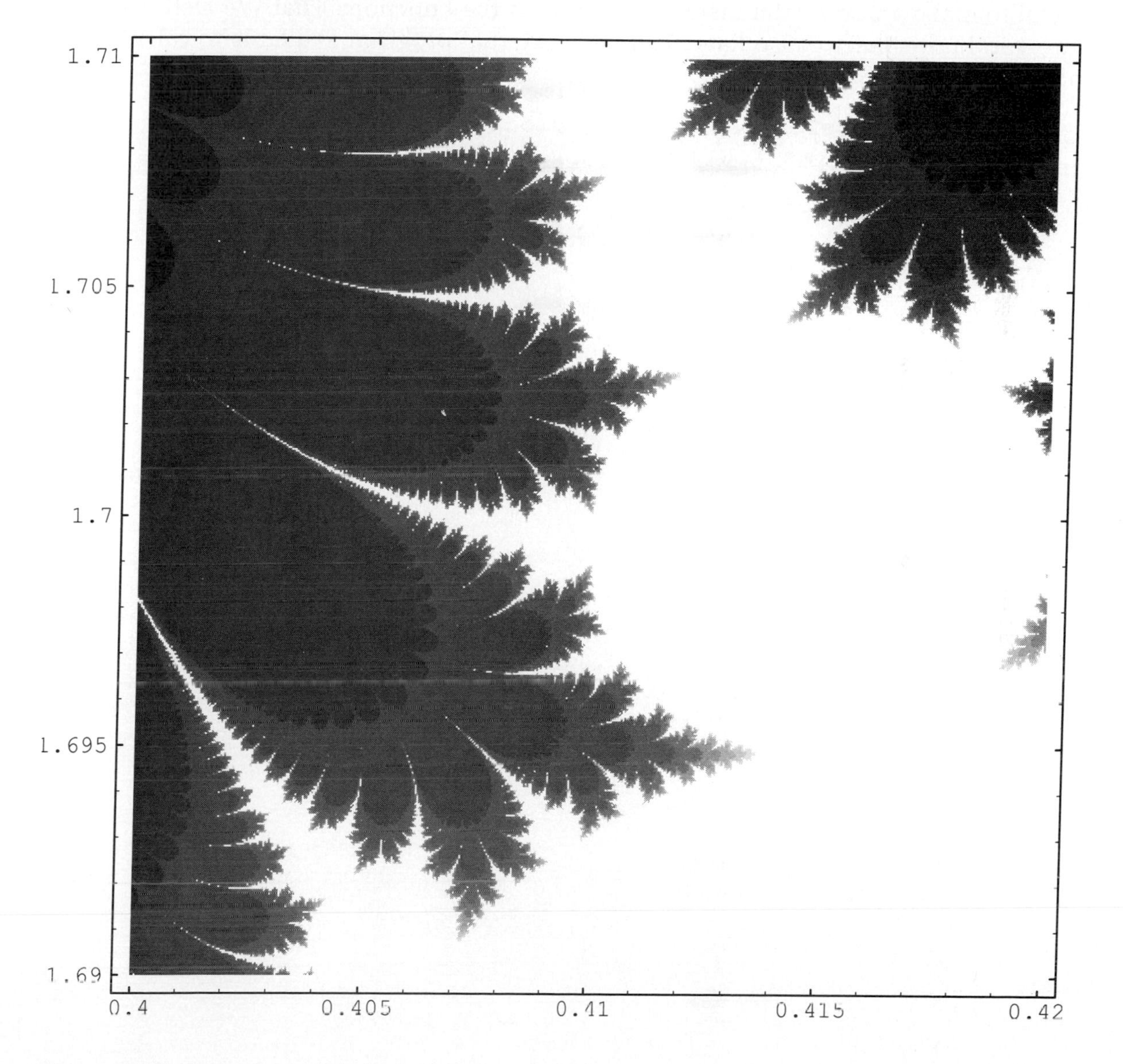

Theo: Well, I'm sold. This is *definitely* a fractal, no doubt about it. We've zoomed in by a factor of more than 100, and it still looks the same.

Jerry: So, do you suppose that if we nest **Cos** functions instead of **Sin** functions we would also get a fractal? I remember that when we did that along the real axis, the **Cos** behaved quite differently from the **Sin**.

Theo: Let's find out. It's trivial to modify the functions we used to make the **Sin** fractal to make a **Cos** fractal instead (if you read the Functions That We Defined section, you know that we also have a suitable external program available to calculate this fractal very quickly):

```
numberOfCossToEscape[z_] := Block[
     {count, nz = N[z]},
     For[
          count = 0,
           (Abs[Im[nz]] < 10000) && (count < 100),

          nz = Cos[nz];
          ++count
     ];
     count
]
```

```
fractalPlot[
     cosFractalC[{{-7, 7}, {-7, 7}}, 500],
     {0, 20}];
```

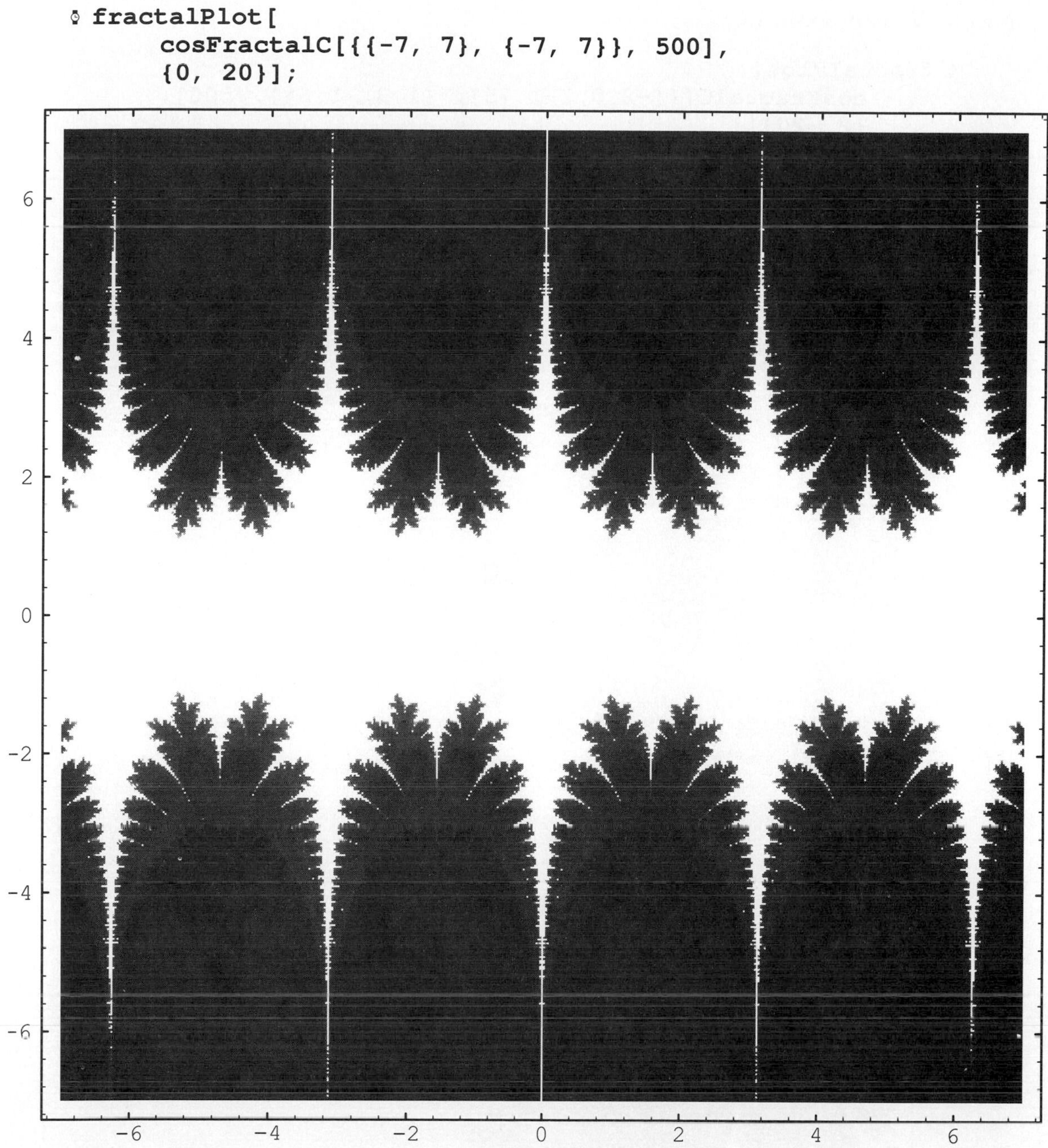

Nice! We can zoom in:

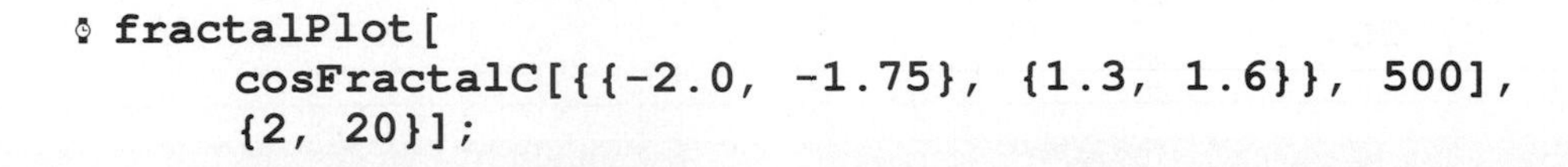

```
fractalPlot[
    cosFractalC[{{-2.0, -1.75}, {1.3, 1.6}}, 500],
    {2, 20}];
```

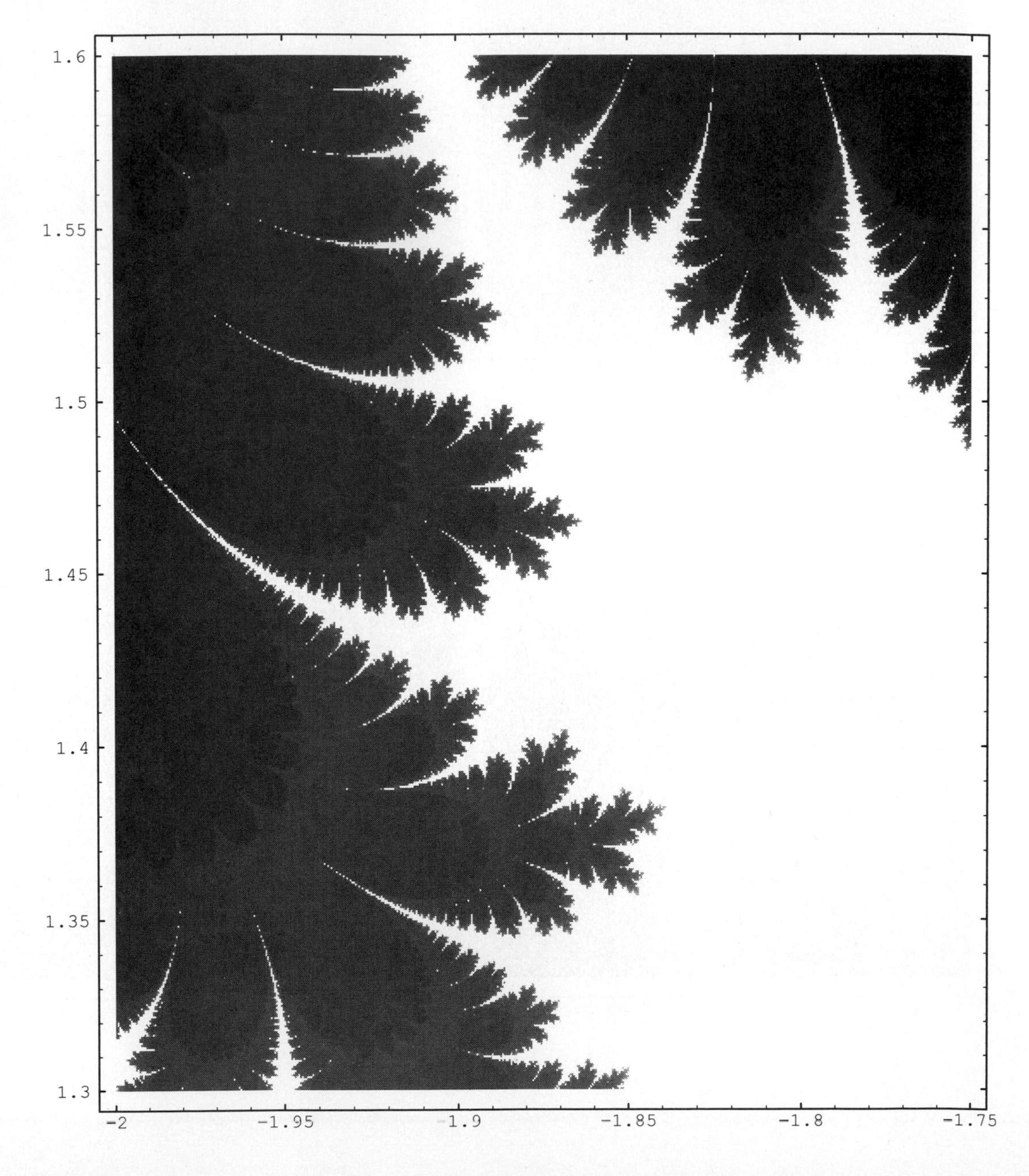

Similar, but different from the `Sin` fractal. These fractals are clearly very self-similar. They look a lot like some Julia sets I've seen.

Jerry: Shouldn't these pictures be in color? Everyone else's fractals are in color!

Theo: Sure, we could make them in color if we wanted to (by adding the option `ColorFunction -> Hue` to the `ListDensityPlot` used in the definition of `fractalPlot`), but I prefer them the way they are. They're dignified, not gaudy.

After we finished this chapter, I happened to notice an exercise in one of Devaney's fractal books **[1]** that was exactly this problem.

Jerry: Devaney also has created some excellent videotapes. The first one, "Chaos, Fractals and Dynamics: Computer Experiments in Mathematics" **[2]** would be of interest to anyone reading this book. His second videotape "Transition to Chaos: The Orbit Diagram and the Mandelbrot Set" **[3]** has spectacular animations of orbit diagrams like ours.

Functions That We Defined

Theo: In this chapter we defined two functions for translating a complex number into a height and a color. One function made the height equal to the absolute value of the number; the other function made it equal to the log of the log of the absolute value:

```
maxZ = N[10^30];
nPi = N[Pi];
ntwoPi = N[2Pi];
complexToZandColor[z_] := Block[{nz, az},
    nz = N[z];
    az = Abs[nz];
    {
        Max[Min[az, maxZ], -maxZ],
        Hue[(nPi+If[nz == 0, 0, Arg[nz / az]])/(ntwoPi)]
    }
]
```

```
nPi = N[Pi];
ntwoPi = N[2Pi];
nE = N[E];
complexToLogLogZandColor[z_] := Block[{nz, az},
     nz = N[z];
     az = Abs[nz];
     {
          Log[Log[az + nE]],
          Hue[(nPi+If[nz == 0, 0, Arg[nz / az]])/(ntwoPi)]
     }
]
```

We also defined a function to count the number of **Sin** functions you have to nest around a given starting point before the imaginary part of the result gets very large:

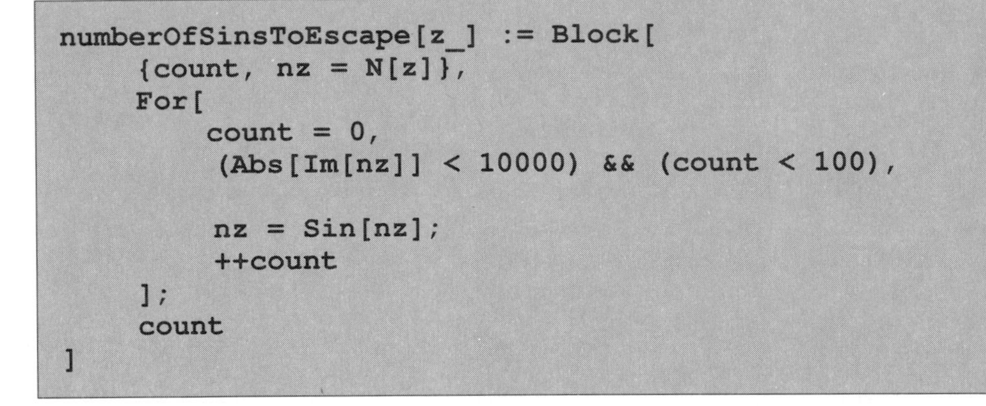

```
numberOfSinsToEscape[z_] := Block[
     {count, nz = N[z]},
     For[
          count = 0,
          (Abs[Im[nz]] < 10000) && (count < 100),

          nz = Sin[nz];
          ++count
     ];
     count
]
```

■ For Those Who Want To Know

This section describes how we can calculate the fractal plots in this chapter very quickly using some complicated techniques. You can skip it if you like.

Theo: Unfortunately *Mathematica* is not very good at iterating simple functions a huge number of times. It's sort of slow. To calculate really large numbers of points, we can write a special program in, for example, the C programming language. C is a very low-level language, and it is possible to compile very efficient programs in C. I've written such a program for calculating the points in this plot, and I've defined some *Mathematica* functions that allow us to pretend that my C function is actually a built-in *Mathematica* function.

Here's how it works. We want a *Mathematica* function called **`sinFractal`**, which can give us a matrix of values representing the points in a plot like the one above. For convenience, we might also want the function to return the range of z values that those points represent. We can define the function in two different ways. First, we can define a version, which we'll call **`sinFractalM`**, which uses only *Mathematica* functions. Second, we can define a version called **`sinFractalC`** which has exactly the same arguments and produces exactly the same result, but does it by calling a C program behind our backs. Here's the first version:

```
sinFractalM[{{ReMin_, ReMax_}, {ImMin_, ImMax_}},
            steps_] :=
    {
        {{ReMin, ReMax}, {ImMin, ImMax}},
        Table[
            numberOfSinsToEscape[x + y I],
            {y, ImMin, ImMax, (ImMax - ImMin)/steps}
            {x, ReMin, ReMax, (ReMax - ReMin)/steps},
        ]
    }
```

This function returns a list with two elements. The first element is just the range specification exactly as we gave it to the function. Later we'll use this for drawing tick marks correctly. The second element of the list is the actual matrix of values. Here's how we define the same function using an external C program:

```
sinFractalC[{{ReMin_, ReMax_}, {ImMin_, ImMax_}},
             steps_] :=
    {{{ReMin, ReMax}, {ImMin, ImMax}},
        Get[StringJoin[
            "!sinfrac ",
            ToString[ReMin], " ",
            ToString[ReMax], " ",
            ToString[ImMin], " ",
            ToString[ImMax], " ",
            ToString[steps]
        ]]
    }
```

The **"!sinfrac "** means execute the external C program named **sinfrac**. (Note that this works only when the *Mathematica* kernel is running on a UNIX system and you have the **sinfrac** C program correctly compiled and installed. The **sinfrac** program produces a matrix in *Mathematica*-compatible input form, which is read back in by the **Get** command. All the **ToString** functions tell **sinfrac** the range to calculate over. In the end, this function returns the same thing as the pure *Mathematica* version does, just a whole lot faster. Well over a hundred times faster, in fact.

To look at these matrices of values, let's define a function called **fractalPlot**, which takes a list like the one returned by **sinFractal** and plots it in a pleasing way Here it is:

```
fractalPlot[
        {{{ReMin_, ReMax_}, {ImMin_, ImMax_}}, matrix_},
        {minCount_, maxCount_}] :=
Block[{},
    ListDensityPlot[matrix,
        PlotRange -> {minCount, maxCount},
        MeshRange -> {{ReMin, ReMax},
                      {ImMin, ImMax}},
        Mesh -> False,
        AspectRatio -> Automatic];
    ]
```

In addition to the list from **sinFractal**, this function takes a second argument that specifies how the gray levels in the plot should be scaled. The **minCount** is the number of nested **Sin** functions that will be colored black, and the **maxCount** is the number of nested **Sin** functions that will be colored white.

We can use **`fractalPlot`** with either the *Mathematica* or the C version of **`sinFractal`**.

Here is the function for the cosine-fractal:

```
cosFractalC[{{ReMin_, ReMax_}, {ImMin_, ImMax_}},
                  steps_] :=
    {{{ReMin, ReMax}, {ImMin, ImMax}},
        Get[StringJoin[
            "!cosfrac ",
            ToString[ReMin], " ",
            ToString[ReMax], " ",
            ToString[ImMin], " ",
            ToString[ImMax], " ",
            ToString[steps]
        ]]
    }
```

Here is the C program used by the functions above:

```
#include <math.h>

typedef struct
    {
    double  re;
    double  im;
    } complex;
#define TRUE    1
#define FALSE   0

extern complex complexSin(complex z);

main(argc, argv)
int     argc;
char    **argv;
{
    complex     min, max;
    int         steps;
    complex     delta;
    complex     z;
    complex     z1;
    int         count;
    int         firstCol;
    int         firstRow;
```

```
if (argc != 6)
    {
    printf("usage: sinfrac Rmin Rmax Imin Imax npts\n");
    return;
    }

sscanf(argv[1], "%lf", &min.re);
sscanf(argv[2], "%lf", &max.re);
sscanf(argv[3], "%lf", &min.im);
sscanf(argv[4], "%lf", &max.im);
sscanf(argv[5], "%d", &steps);

delta.re = (max.re - min.re) / steps;
delta.im = (max.im - min.im) / steps;

printf("{");
firstRow = TRUE;
for (z.im = min.im; z.im <= max.im; z.im += delta.im)
    {
    if (firstRow)
        firstRow = FALSE;
    else
        printf(",\n");

    printf("{");
    firstCol = TRUE;

    for (z.re = min.re; z.re <= max.re; z.re += delta.re)
        {
        for(count = 1, z1 = z; count < 100; ++count)
            {
            z1 = complexSin(z1);
            if (fabs(z1.im) > 10000)
                break;
            }
        if (firstCol)
            firstCol = FALSE;
        else
            printf(",");
        printf("%d", count);
        }

    printf("}");
    }
```

```
    printf("}\n");
}

complex complexSin(complex z)
{
    complex z1;

    z1.re = (exp(z.im) + exp(-z.im)) * sin(z.re) * 0.5;
    z1.im = -(exp(-z.im) - exp(z.im)) * cos(z.re) * 0.5;

    return(z1);
}
```

For **cosFractalC**, replace the function `complexSin` in the above program with the one below:

```
complex complexCos(complex z)
{
    complex z1;

    z1.re = (exp(z.im) + exp(-z.im)) * cos(z.re) * 0.5;
    z1.im = (exp(-z.im) - exp(z.im)) * sin(z.re) * 0.5;

    return(z1);
}
```

References

1) Chaos, Fractals, and Dynamics, *Robert L. Devaney*, Addison-Wesley, 1990, p 172.
2) Chaos, Fractals and Dynamics: Computer Experiments in Mathematics (Videotape), *Robert L. Devaney*, Science Television, 212-569-8079.
3) Transition to Chaos: The Orbit Diagram and the Mandelbrot Set (Videotape), *Robert L. Devaney*, Science Television, 212-569-8079.

Chapter Nine
Plotting and Graphics

Theo and Jerry eschew obfuscation enlighteningly.

■ Dialog–First Day: 2D Plots

Jerry: A chapter on plotting and graphics evokes my Indian Ocean reaction **[1]**. So many possibilities!

Theo: I think we need a chapter that explains *Mathematica*'s basic graphics functions in a simple straightforward way, without engaging in flights of fancy or unnecessary complications.

Jerry: It sounds like a manual to me: one of those books that tells you everything that is correct, but causes you to have no interest in ever doing any of it. Yuck!

Theo: No, no, it will be OK. We'll keep it short and explain just those things that everyone is burning to know. Also, we'll do lots of referring to the actual manual **[2]**. Anywhere we don't explain a function, people can look it up in the manual.

Let's start with two-dimensional plotting. The first command to know is **Plot**, and it works like this:

```
Plot[Cos[x], {x, -2Pi, 2Pi}];
```

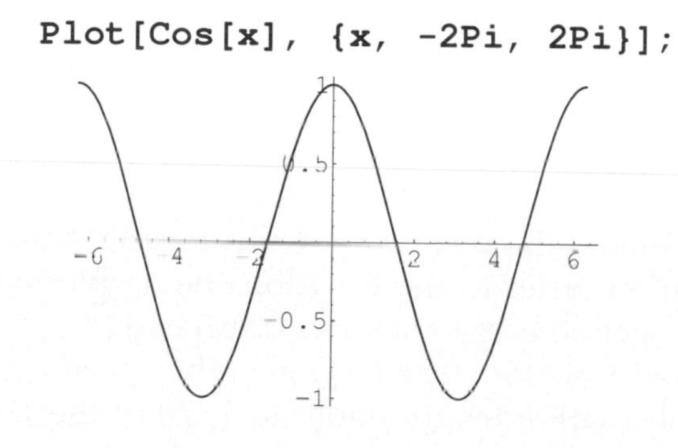

The first argument to **Plot** is the function to be plotted, and the second argument is a list containing the variable to be used and the end points of the range of values over which the variable runs.

If you want to plot more than one function at the same time, you can give **Plot** a list of functions as its first argument:

```
Plot[{Sin[x], Cos[x]}, {x, -2Pi, 2Pi}];
```

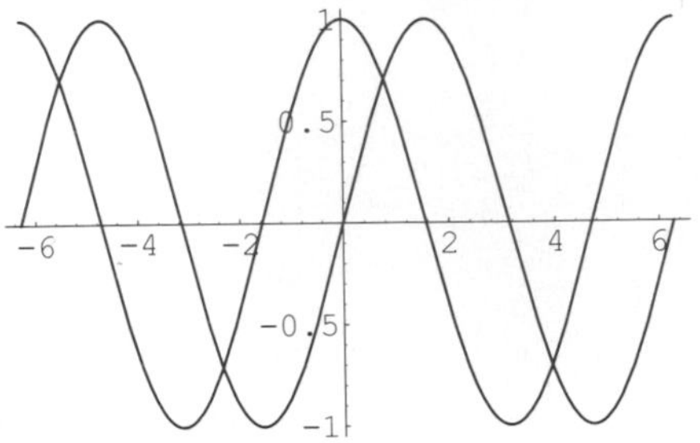

Jerry: How can we tell which line is which? Is there some way to make them look different?

Theo: Yes. You can use the **PlotStyle** option to do this. It's a bit strange, but here's an example showing how to turn the second one into a dashed line:

```
Plot[{Sin[x], Cos[x]}, {x, -2Pi, 2Pi},
    PlotStyle -> {
        Dashing[{}],
        Dashing[{1/20}]
    }];
```

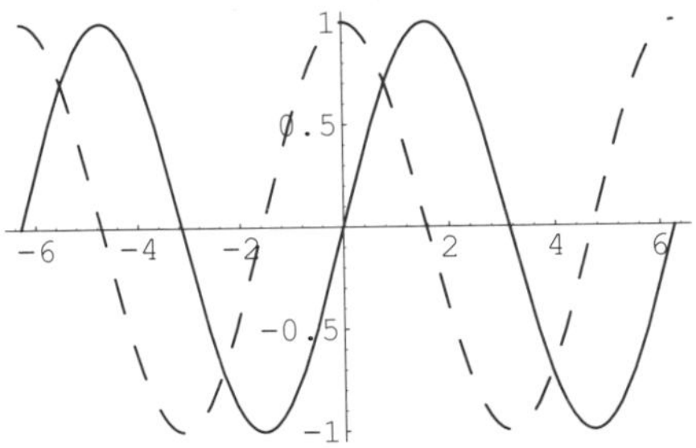

PlotStyle takes a list of graphics modifiers like **Dashing**, **GrayLevel**, **RGBColor**, or **Thickness**, and applies them in order to the functions being plotted. In this example, the first function, **Sin[x]**, gets the specification **Dashing[{}]** (which means no dashing), and the second function, **Cos[x]**, gets the specification **Dashing[{1/20}]** (which means dashes with a length equal to 1/20 of the width of the plot).

You can have more than one modifier per function by using a list of lists. For example, this command turns the first curve blue and the second one red, at the same time giving them different dashing patterns and thicknesses:

```
Plot[{Sin[x], Cos[x]}, {x, -2Pi, 2Pi},
    PlotStyle -> {
     {Thickness[1/40],RGBColor[0,0,1],Dashing[{}]},
     {Thickness[1/80],RGBColor[1,0,0],Dashing[{1/20}]}
    }];
```

See picture on color plates page vii.

Jerry: This last example is excellent! All these damn little curly brackets have an important purpose in life, but I appreciate it only when I'm trying to do something complicated. When I'm doing simple things, it's a *big pain*.

Without sub-lists, there would be no way to tell **Plot** to associate more than one modifier with each function.

Theo: Often people want the two axes in a plot to have the same scaling (i.e., they want lines with a slope of one to be drawn as 45° lines on the screen, and circles to be drawn as circles instead of ellipses). You can do this with the option **AspectRatio -> Automatic**:

```
Plot[{x, Sin[x], Cos[x]}, {x, -2Pi, 2Pi},
    AspectRatio -> Automatic];
```

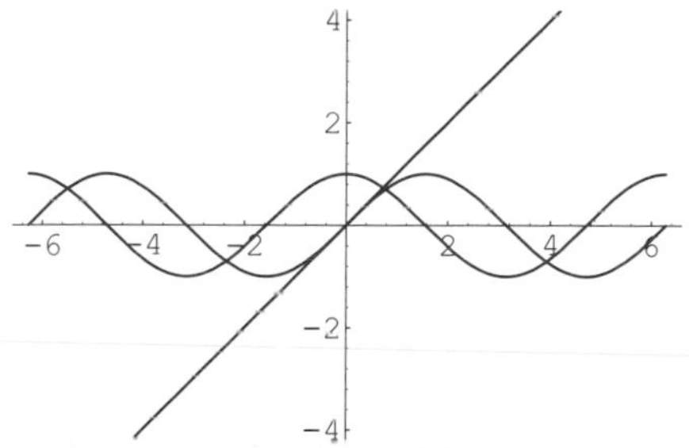

Jerry: Suppose I want to add some dots to this plot?

Theo: You can use the **Prolog** and **Epilog** options to add lists of graphics objects to a plot. Among the graphics objects you can add are **Point**, **Line**, **Rectangle**, **Polygon**, **Circle**, and **Disk**. You can also use modifiers like **PointSize**,

Thickness, **GrayLevel**, and **RGBColor** to affect these objects. Here's an example:

```
Plot[{Sin[x], Cos[x]}, {x, -2Pi, 2Pi},
    Epilog -> {
        PointSize[1/20],
        Point[{-7Pi/4, Sin[-7Pi/4]}],
        Point[{-3Pi/4, Sin[-3Pi/4]}],
        Point[{Pi/4, Sin[Pi/4]}],
        Point[{5Pi/4, Sin[5Pi/4]}]
    }];
```

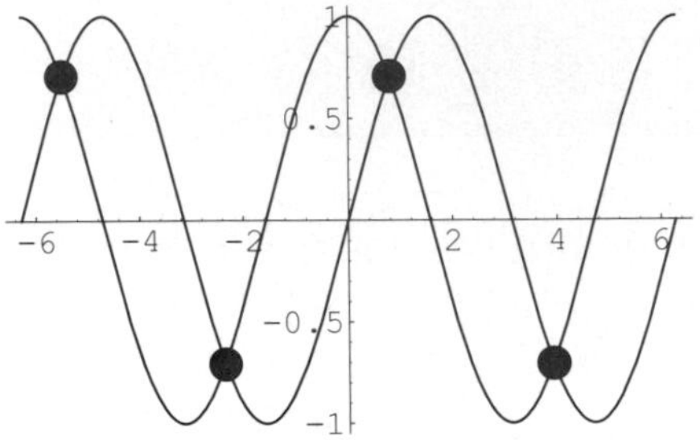

Jerry: What's the difference between the **Prolog** and **Epilog** options?

Theo: Prolog adds its list of objects *behind* the plot (drawn first), while **Epilog** adds them *in front of* the plot (drawn last). This makes a difference only if the objects overlap and are of different colors. You can have both options in the same plot, if you like.

Jerry: What if I want to plot *mostly* dots? Is there a more direct way of doing that?

Theo: ListPlot takes a list of points and plots them. For example, we can plot a few points like this:

```
ListPlot[{
    {2, 7},
    {1, Sqrt[2]/7},
    {5, 4},
    {1/10, 3}
}];
```

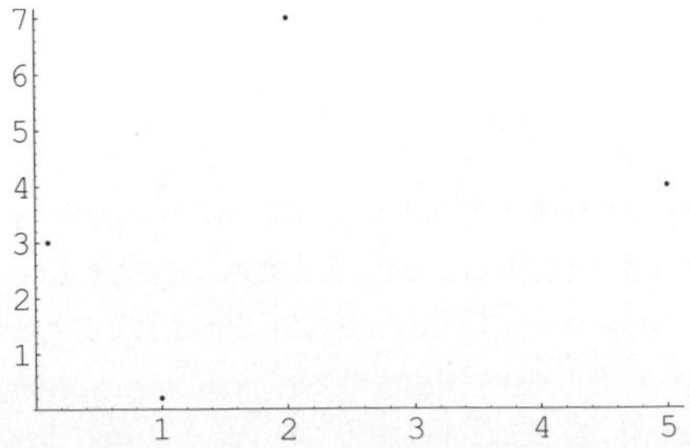

Jerry: How do we make these dots big enough to see?

Theo: By using the `PlotStyle` option, which works the same way as it does for `Plot`:

```
ListPlot[{
    {2, 7},
    {1, Sqrt[2]/7},
    {5, 4},
    {1/10, 3}
}, PlotStyle -> PointSize[1/40]];
```

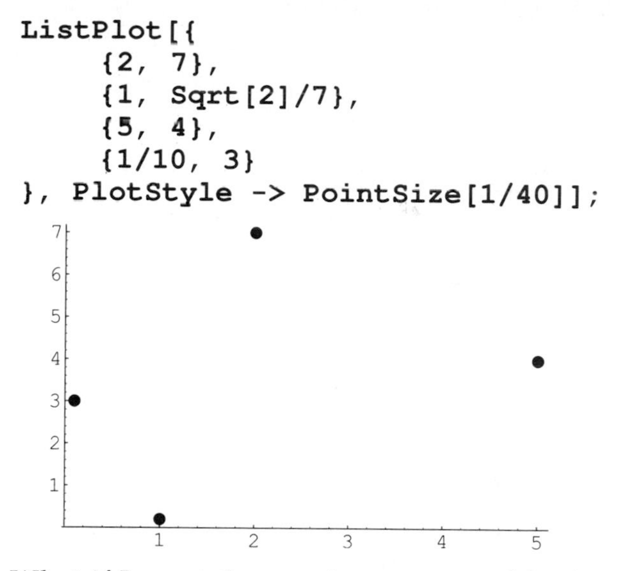

Jerry: What if I want these points connected by lines instead of drawn as dots?

Theo: The option `PointJoined -> True` will do that:

```
ListPlot[{
    {2, 7},
    {1, Sqrt[2]/7},
    {5, 4},
    {1/10, 3}
}, PlotJoined -> True];
```

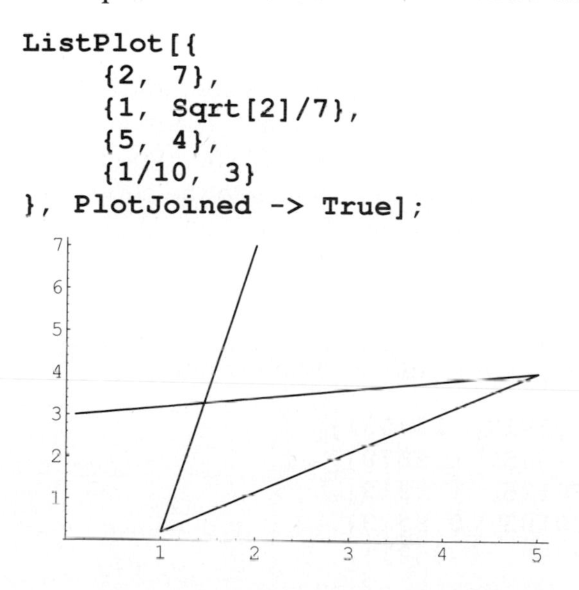

If you give **ListPlot** a list of numbers instead of a list of pairs, it automatically spaces them out horizontally:

```
ListPlot[{5, 4, 2, 7, 2, 3},
          PlotStyle -> PointSize[1/40]];
```

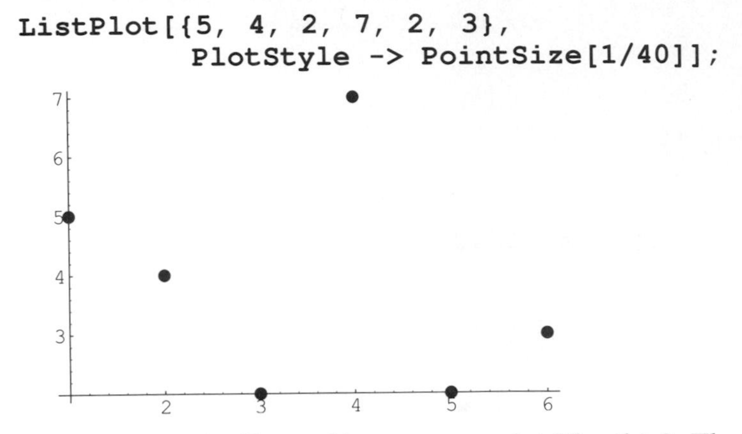

Jerry: How do I read off coordinates on a plot like this? That is, I want to point to a place on the plot and ask what its x and y coordinates are.

Theo: If you're reading the electronic edition in a full version of *Mathematica* (not *MathReader*), you select the plot by clicking on it. Then you hold down the Command key and point at the plot. The coordinates of the mouse (in the coordinate system of the plot) will be displayed at the bottom of the window.

Jerry: Very nice! What if I want to remember the coordinates? Do I have to write them down on paper?

Theo: Oh no! Too middle-of-the-20th-century. If you click while holding down the command key, you will get a small dot. You can click as many times as you like, or hold down the mouse button and drag to get a line of dots. When you're done, use the Copy command in the Edit menu to copy the coordinates to the clipboard. You can then paste them anywhere you like. For example, here is a coordinate-pair copied from the plot above:

```
{4.1729, 3.09409}
```

If you select more than one point, you get a list of coordinate-pairs:

```
{{4.98429, 1.7931}, {5.13917, 1.7931},
{5.33278, 1.6678}, {5.52638, 1.6678},
{5.68126, 1.5425}, {5.68126, 1.2919},
{5.83615, 1.0412}, {5.99103, 0.8533},
{6.10719, 0.6027}, {6.3008, 0.3521},
{6.45568, 0.2267}}
```

■ Dialog–Second Day: 3D Plots

Theo: `Plot3D` is a wonderful command.

Jerry: Says who? Everybody *loves* 3D plots, but are they really important?

Theo: That's not the point. `Plot3D` makes lovely pictures. If you want them to mean something, fine, but I don't really care one way or the other.

`Plot3D` works like `Plot`, except that you name a second variable and its range. Here we go:

```
Plot3D[Sin[x + Sin[y]], {x, 0, 4Pi}, {y, 0, 4Pi}];
```

See picture on color plates page vii.

Jerry: This is not the most beautiful plot in the world. How do we improve it?

Theo: First we need to increase the number of divisions. The `PlotPoints` option can be used to do this:

```
Plot3D[Sin[x + Sin[y]], {x, 0, 4Pi}, {y, 0, 4Pi},
    PlotPoints -> 30];
```

See picture on color plates page viii.

Now that we have a nice colored surface, we might like to remove the black grid lines. We can do that by adding the option `Mesh -> False` (I've increased the plot points even more, to make it look really smooth):

```
Plot3D[Sin[x + Sin[y]], {x, 0, 4Pi}, {y, 0, 4Pi},
    PlotPoints -> 50, Mesh -> False];
```

See picture on color plates page viii.

When printing in black-and-white, you might want to tell *Mathematica* to use only gray levels, instead of colors. You can do that by using the option **Lighting -> False**:

```
Plot3D[Sin[x + Sin[y]], {x, 0, 4Pi}, {y, 0, 4Pi},
    PlotPoints -> 30, Lighting -> False];
```

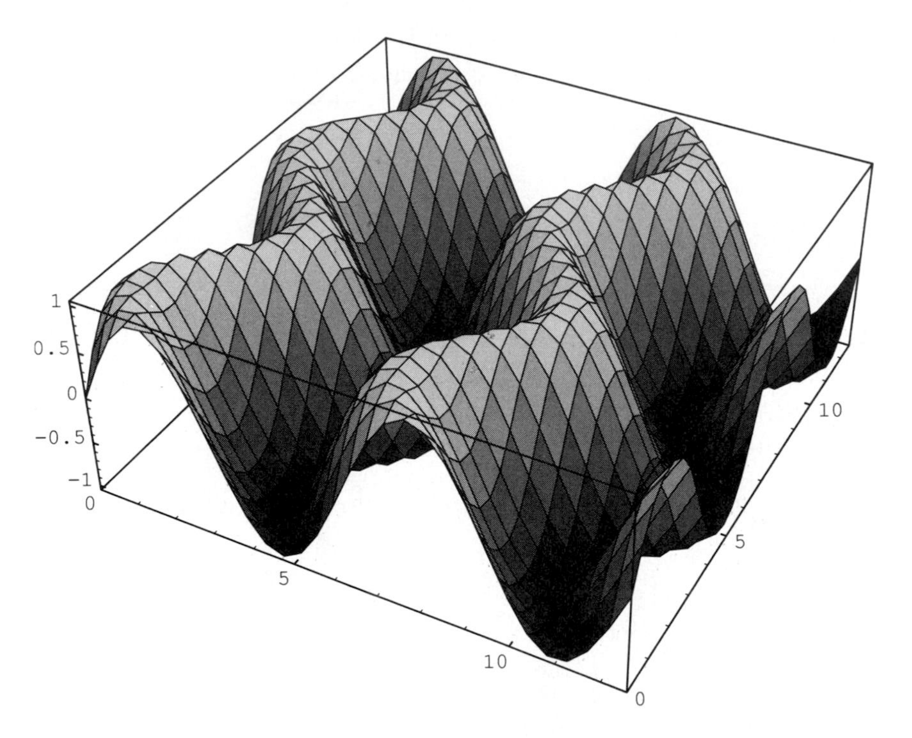

Higher areas are lighter, lower areas are darker. (For those who have used *Mathematica* in the past, note that this used to be the default way of making 3D plots. The default has been changed to color.)

Jerry: Suppose I want to see what this looks like from another viewpoint? Maybe lower and more in front.

Theo: You do this as follows. First, type or copy the **Plot3D** command you want to use. Just to the left of the last closing bracket, type a comma. It should look like this:

```
Plot3D[Sin[x + Sin[y]], {x, 0, 4Pi}, {y, 0, 4Pi},
    PlotPoints -> 30, ];
```

Then use the 3D View Point Selector command in the Prepare Input sub-menu of the Action menu. You will get a dialog box with a neat little cube you can rotate with the mouse (using the scroll bars, or by dragging the cube itself). When you have the viewpoint where you want it, click the Paste button in the dialog box, then the OK button. It should look like this:

```
Plot3D[Sin[x + Sin[y]], {x, 0, 4Pi}, {y, 0, 4Pi},
    PlotPoints -> 30,
    ViewPoint->{0.047, -3.078, -1.405}];
```

See picture on color plates page ix.

Of course, you can type in the **ViewPoint** option manually, if you like.

Jerry: When do we get to see this rotating in real-time?

Theo: That depends on what you mean by real-time. If you just want to see an animation showing this figure spinning, you can do that without too much trouble. In the command below, the option **SphericalRegion -> True** tells *Mathematica* to scale each individual plot the same way, regardless of what the view point is. (People familiar with older versions of *Mathematica* may wonder why this option is now necessary: It's a bit subtle, but generally it's necessary to include this option any time you are making a 3D animation where the view point changes in the course of the animation.)

```
Do[
    Plot3D[Sin[x + Sin[y]], {x, 0, 4Pi}, {y, 0, 4Pi},
        PlotPoints -> 30,
        SphericalRegion -> True,
        Axes -> None,
        ViewPoint->{2 Cos[t], 2 Sin[t], 1.3}
    ],
    {t, 0, 2Pi-2Pi/36, 2Pi/36}
]
```

See picture on color plates page x.

By using the horizontal scroll bar while the animation is running, you can make it spin back and forth at any speed you like.

Jerry: OK, but by "rotating in real-time" I mean using the mouse to pick up and rotate a 3D graph any way I like. What you did above allows me to see one preprogrammed rotation. Seeing the graph from an angle not in the animation requires recalculating, which takes a long time. When will I be able to rotate it in any direction I want?

Theo: Late 1991. It's not a hard thing to do, but it's hard to do in a way that works on all the different platforms that *Mathematica* runs on.

Dialog–Third Day: Parametric Plots, 2D and 3D

Jerry: I assume **ParametricPlot** exists so we can graph non-functions in two dimensions. Likewise **ParametricPlot3D** should let us graph non-functions in three dimensions.

Theo: Sounds right to me. Making a circle, for example, is very hard using **Plot**, but easy using **ParametricPlot**. You give **ParametricPlot** a list of two function as its first arguments. They give the **x** and **y** values, respectively:

```
ParametricPlot[{Cos[t], Sin[t]}, {t, 0, 2Pi}];
```

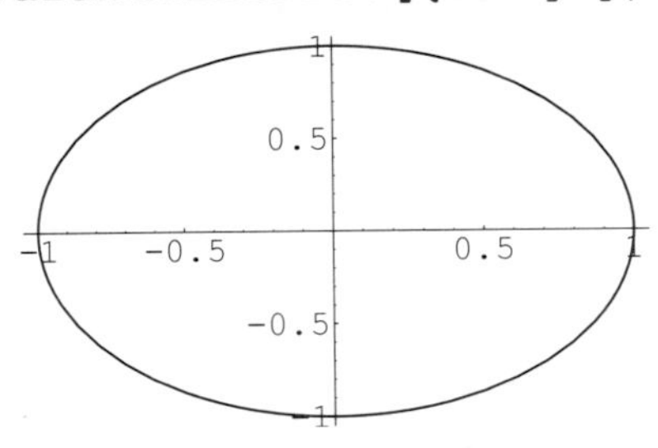

Jerry: Awful-looking circle! This sounds like a good place to use the **AspectRatio -> Automatic** option (which works with **ParametricPlot** the same way it does with **Plot**):

```
ParametricPlot[{Cos[t], Sin[t]}, {t, 0, 2Pi},
    AspectRatio -> Automatic];
```

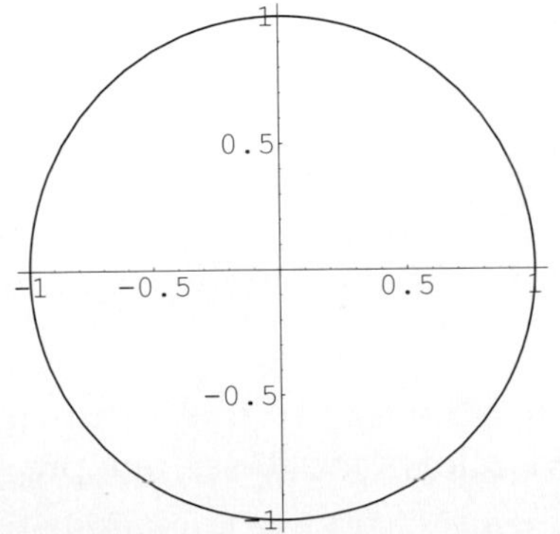

Lovely.

Theo: If you want to plot more than one parametric expression, you can give **ParametricPlot** a list of lists of functions, like this:

```
ParametricPlot[{
        {Cos[t], Sin[t]},
        {5 Cos[t]/6, Sin[t]},
        {Cos[t], 5 Sin[t]/6},
        {Cos[3t]/2, Sin[5t]/2}
    },
    {t, 0, 2Pi},
    AspectRatio -> Automatic];
```

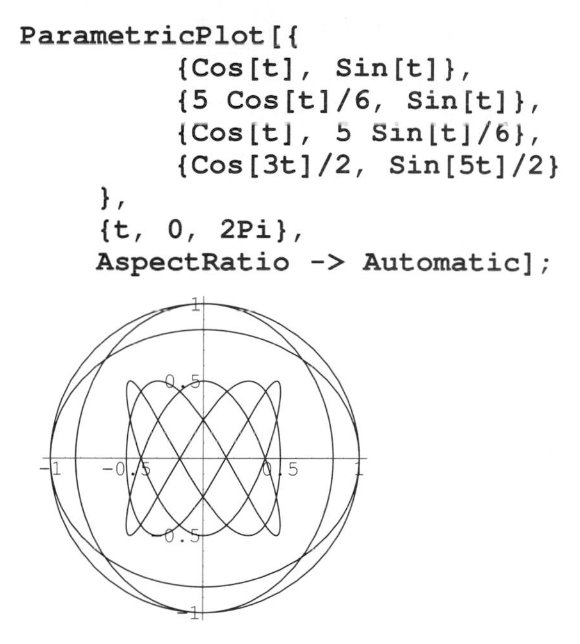

Jerry: Looks a bit complicated at first, but it has a certain excitement to it. I guess the first pair in the list makes the circle, the next two pairs make ellipses, and the last pair makes the Lissajous figure in the middle. (Although we have **t** go from **0** to **2Pi** in each case, we can, of course, use what ever end points we like.)

Theo: ParametricPlot3D is to **ParametricPlot** what **Plot3D** is to **Plot**. Its first argument is a list of three functions that give the x, y, and z coordinates of a curve. Here is an example:

```
ParametricPlot3D[{Cos[t], Sin[t], t/5}, {t, 0, 4Pi}];
```

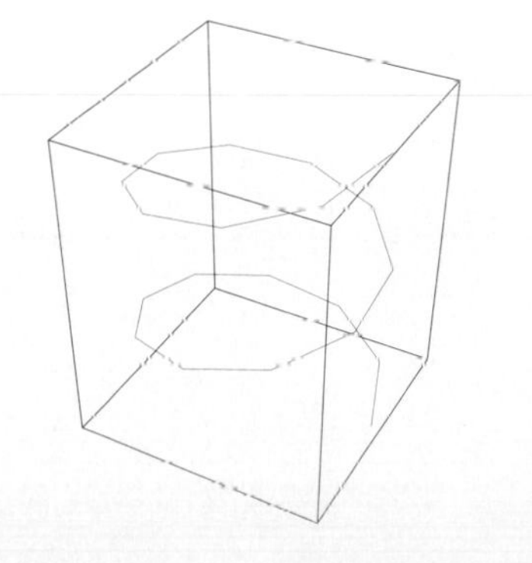

Jerry: This looks a bit ratty. Can we smooth out the curve using the **`PlotPoints`** option?

Theo: Yes:

```
ParametricPlot3D[{Cos[t], Sin[t], t/5}, {t, 0, 4Pi},
        PlotPoints -> 50];
```

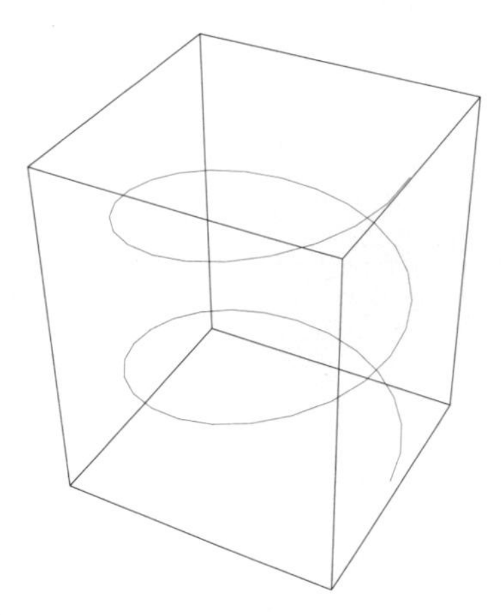

Here's another nice plot:

```
ParametricPlot3D[
    {t Cos[t], t Sin[t], -2t}, {t, 0, 20Pi},
    PlotPoints -> 200];
```

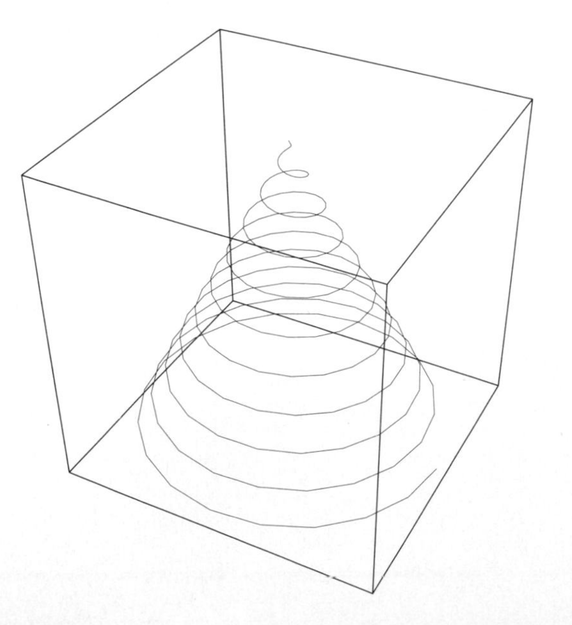

Jerry: Can we combine two spacecurves? Let me try it:

```
ParametricPlot3D[{
        {t Cos[t], t Sin[t], 2t},
        {t Cos[t] + 150, t Sin[t], 2t}
    },
    {t, 0, 20Pi},
    PlotPoints -> 200];
```

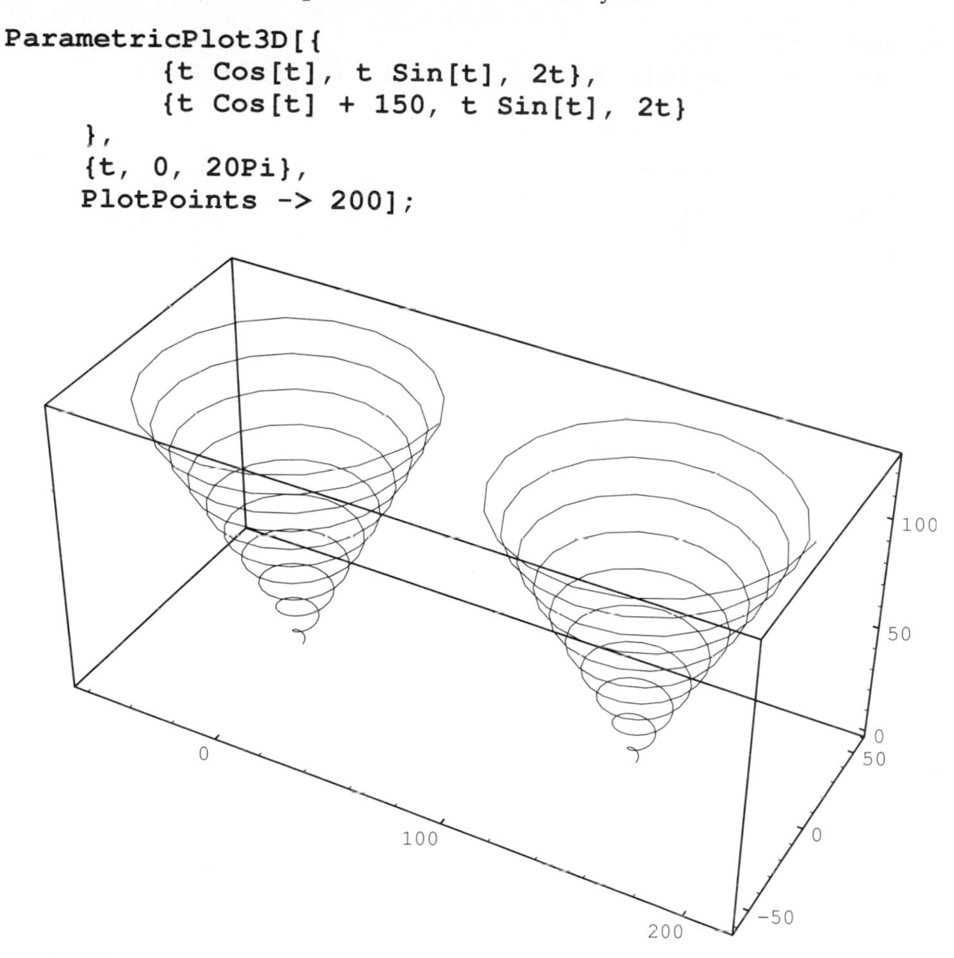

Nice, that's what I was expecting.

Theo: Remarkably, you can also use **ParametricPlot3D** to make parametric surfaces, just by adding another variable-range specification. Here's a cylinder:

```
ParametricPlot3D[
    {Cos[u], Sin[u], v},
    {u, 0, 2Pi}, {v, 0, 2}];
```

See picture on color plates page ix.

And here's a sphere:

```
ParametricPlot3D[
    {Sin[v] Cos[u], Sin[v] Sin[u], Cos[v]},
    {u, 0, 2Pi}, {v, 0, Pi}];
```

See picture on color plates page xi.

By making the two variables run over smaller ranges, we can make a cut-away view:

```
ParametricPlot3D[
    {Sin[v] Cos[u], Sin[v] Sin[u], Cos[v]},
    {u, 0, 3Pi/2}, {v, Pi/8, Pi}];
```

See picture on color plates page xi.

Of course you can combine more than one parametric surface by giving a list of parametric expressions:

```
ParametricPlot3D[{
        {Sin[v] Cos[u], Sin[v] Sin[u], Cos[v]},
        {Cos[u]/2, Sin[u]/2, v - 1}
    },
    {u, 0, 3Pi/2}, {v, 0, Pi}];
```

See picture on color plates page xii.

Jerry: *(Gaping)* Two 3D objects intersecting, *and* both cut-away so we can see the space between the surfaces, *and* we see where they cross! Amazing!

When can I move the mouse across these surfaces and get a read-out of the (x, y, z) coordinate as I can with 2D plots?

Theo: Probably never.

Jerry: I'm sure this answer is wrong. Some reader of this page will come up with it by 1992. Don't forget where you read this!

Theo: Hmm...

We should probably mention that using **ParametricPlot3D** is not the only way to combine different 3D objects with each other. In Chapter 11 we do a lot more with 3D graphics.

Our publisher has insisted that there be "spectacular graphics" in this book. We have resisted this, but not very hard. Using **ParametricPlot3D**, it is possible to make some rather nice animations. Here is an example:

```
Do[
    ParametricPlot3D[
        {
            Cos[u],
            Sin[u] + n 2 Cos[v],
            (1-n) (v-Pi) + n 2 Sin[(n+1) v]
        },
        {u, 0, 2Pi},
        {v, 0, 2Pi},
        PlotPoints -> 20,
        PlotRange -> {{-3, 3}, {-3, 3}, {-3, Pi}},
        Boxed -> False,
        ViewPoint->{2.668, -0.987, 1.833}
    ],
    {n, 0, 1, 0.025}
]
```

See picture on color plates page xiii.

Jerry: I need an explanation connecting these nice pictures with the (x, y, z) functions that generated them.

Theo: This animation is a transition between two objects. The starting object is a cylinder, similar to the one we made earlier. Its functions are:

```
{Cos[u], Sin[u], v-Pi}
```

The ending object is a variation where I have added a sine component to the z function, and another one to the y function.

```
{Cos[u], Sin[u] + 2 Cos[v], 2 Sin[2 v]}
```

It's sort of like a Lissajous figure in the y-z plane, with a self-intersecting tube thrown in.

To make the animation, I mixed these two sets of functions together, with the animation parameter, **n**, determining how much of each set to add. When **n = 0** at the start of the animation, only the first set is included. When **n = 1** at the end, only the last set.

Dialog–Fourth Day: Contour and Density Plots

Jerry: Who needs all this contour and density stuff? What can we do with these that we haven't done already?

Theo: All sorts of things! Contour plots are made with the **ContourPlot** command, which takes the same arguments as **Plot3D**. Here's an example:

```
ContourPlot[Sqrt[9 - x^2] y, {x, -3, 3}, {y, -3, 3}];
```

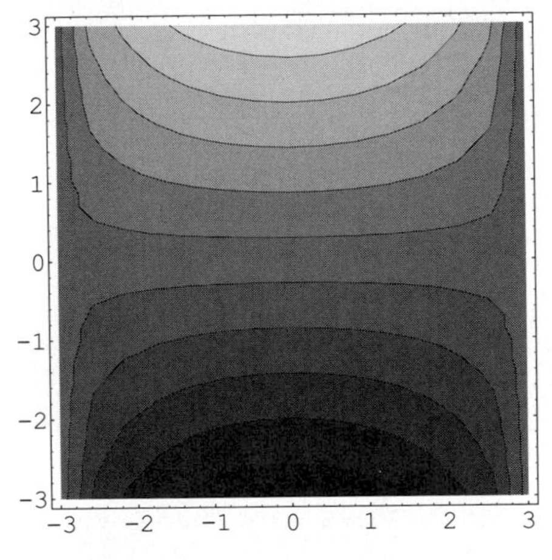

Jerry: The dark part must be the valley (where the sun never shines), and the light part must be the tops of the mountain, where the buffalo roam (oh, sorry: where the sun always shines).

Theo: Right. And the curves are drawn so that each one stays at the same height.

If you don't want the shading, you can use the option **ContourShading -> False**. If you want the contour lines to be smoothed (which often makes them look nicer), you can use the option **ContourSmoothing -> True**.

You can use the option **ColorFunction -> f** to change the way that space between contour lines is colored. The function **f** is a function that converts a number into a color. It is applied to the value of each contour in the contour plot to determine the color to make it.

One useful function to use with **`ColorFunction`** is **`Hue`**. **`Hue`** takes a number and converts it into a color whose hue depends on the value of the number. **`Hue[0]`** gives red, as does **`Hue[1]`**. Numbers in between give other colors (see Chapter 8 for more information). Numbers between 1 and 2 give the same colors as numbers between 0 and 1 (in other words, the colors go in a circle, completing one revolution for each unit increase in the number). Here is an example:

```
ContourPlot[Sqrt[9 - x^2] y, {x, -3, 3}, {y, -3, 3},
    ColorFunction -> Hue];
```

See picture on color plates page xiv.

Jerry: Quite garish. I think I prefer the gray scale version of this plot. Now, what about density plots?

Theo: Density plots are similar to contour plots, except that there are no level lines, just a rectangular grid:

```
DensityPlot[Sqrt[9 - x^2] y, {x, -3, 3}, {y, -3, 3}];
```

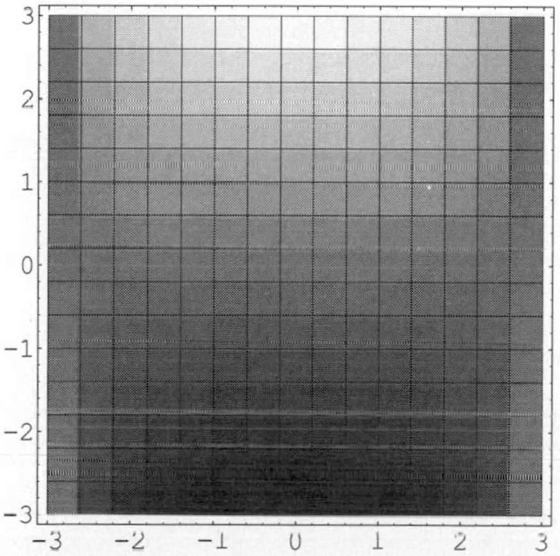

Jerry: This is obviously inferior to the contour plot version. Why would anyone want to use **`DensityPlot`**?

Theo: Ah, we'll explore that question in great detail in Chapter 12. There are many good reasons to use **`DensityPlot`**, not the least of which is that it's the fastest of all *Mathematica*'s 3D graphics commands. For example, here's a nice plot that would be hard to make any other way (I've added the option **Mesh -> False** to remove the grid lines, and **PlotPoints -> 100** to smooth out the picture):

```
DensityPlot[Sin[x + Sin[y]], {x, 0, 8Pi}, {y, 0, 8Pi},
    Mesh -> False, PlotPoints -> 100];
```

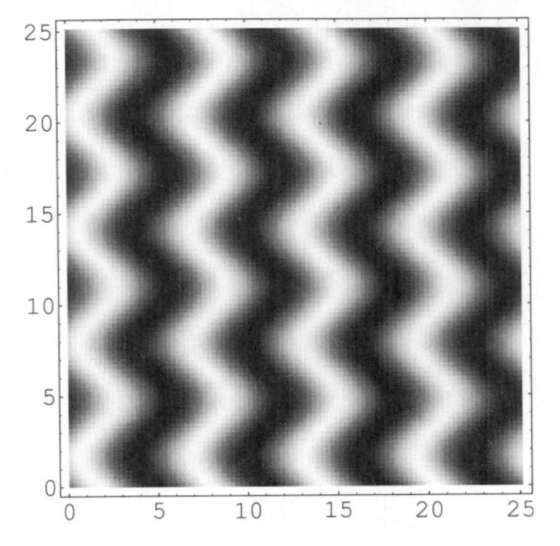

Even though this had to calculate 10,000 sample points (100 by 100), it didn't take an unreasonably long time, and it doesn't use up an unreasonably large amount of memory, as **Plot3D** would have.

Jerry: What do you mean "unreasonable"? Two years and 237MB on a 350MHz HAL9000?

Theo: No, no, unreasonable on smallish computers. It's a minute or two on a Mac II. **Plot3D** takes a few more minutes, and uses up quite a lot more space, but it's not out of the question either. On a Mac Plus, probably only the **DensityPlot** is an option.

You can use the **ColorFunction** option with **DensityPlot** also, as in the following example:

```
DensityPlot[Sin[x + Sin[y]], {x, 0, 8Pi}, {y, 0, 8Pi},
    Mesh -> False, PlotPoints -> 100,
    ColorFunction -> Hue];
```

See picture on color plates page xiv.

There are many other options (of course), and they are all described in the *Mathematica* book.

Let's move on to something more fun. Like all other *Mathematica* graphics commands, **`DensityPlot`** and **`ContourPlot`** can be used to make animations.

Jerry: That sounds exotic. I don't think I've every seen an animated contour plot before!

Theo: Not many people have. Here's a pretty one. The function is **`Sin[x]* Sin[y]`**, with a phase shift added to both **`x`** and **`y`**. The only tricky part is the **`ColorFunction`** option. I've used a pure function (described in the *Mathematica* book) to modify the value passed to the **`Hue`** function. By adding a number that goes from zero to 1 over the course of the animation, I made the colors slowly shift around a cycle. (Astute readers will realize that I am relying on the fact that **`Hue`** produces the same cycle of colors for argument values between 1 and 2 as between zero and 1.)

```
Do[
     ContourPlot[
          Sin[x+n] Sin[y+n],
          {x, -Pi, Pi},
          {y, -Pi, Pi},
          ColorFunction -> (Hue[# + n/(2Pi)]&),
          PlotPoints -> 40
     ],
     {n, 0, 2Pi-2Pi/50, 2Pi/50}
]
```

See picture on color plates page xv.

References

1) J. Glynn and T. Gray, *Exploring Mathematics with Mathematica, Addison-Wesley*, 1991, p1.
2) S. Wolfram, *Mathematica, A System for Doing Mathematics by Computer,* Second Edition, Addison-Wesley, 1991.

Chapter Ten
Two Dimensional Graphics, Differently

In which Theo and Jerry see what it's like living in Flatland.

Dialog–First Day

Jerry: "Two Dimensional Graphics" I understand, to some extent; it's the "Differently" that has me puzzled. Why did you put that in the title of this chapter?

Theo: We do all sorts of graphics in other chapters, but in this chapter we're going to do different kinds of thing, things that would not have any place in the other chapters.

Jerry: In the other chapters we always used graphics to help see something we were interested in. The graphs always served some purpose. Are you saying that in this chapter we are going to do graphics that don't serve any purpose?

Theo: More or less. Here we are going to do graphics for their own sake. Mainly this will involve using the lower-level *Mathematica* graphics functions to build up interesting things.

Jerry: By lower level you mean `Line` and `Point`, rather than `Plot`, which would be higher level?

Theo: Right. The available primitives (in two dimensions) are `Line`, `Point`, `Rectangle`, `Circle`, `Disk`, `Polygon`, `Raster`, `RasterArray`, and `Text`, and they can be modified by `Thickness`, `PointSize`, `Dashing`, `GrayLevel`, `RGBColor`, `CMYKColor`, and `Hue`.

Jerry: Let's have some examples of these. Can I use `Polygon` to draw an 11-sided regular polygon?

Theo: Boy, you ask the strangest questions. **`Polygon`** takes a list of pairs of numbers and draws a polygon whose vertices are these points. To make a regular polygon, we need to compute the coordinates of the vertices. Fortunately, this is not too hard using the **`Table`** command (we'll assign the list to a variable so we can use it later):

```
verticesForElevengon =
   Table[{Cos[t], Sin[t]}, {t, 0, 2Pi, 2Pi/11}]
```

$$\{\{1, 0\}, \{\mathrm{Cos}[\frac{2\ \mathrm{Pi}}{11}], \mathrm{Sin}[\frac{2\ \mathrm{Pi}}{11}]\}, \{\mathrm{Cos}[\frac{4\ \mathrm{Pi}}{11}], \mathrm{Sin}[\frac{4\ \mathrm{Pi}}{11}]\},$$
$$\{\mathrm{Cos}[\frac{6\ \mathrm{Pi}}{11}], \mathrm{Sin}[\frac{6\ \mathrm{Pi}}{11}]\}, \{\mathrm{Cos}[\frac{8\ \mathrm{Pi}}{11}], \mathrm{Sin}[\frac{8\ \mathrm{Pi}}{11}]\},$$
$$\{\mathrm{Cos}[\frac{10\ \mathrm{Pi}}{11}], \mathrm{Sin}[\frac{10\ \mathrm{Pi}}{11}]\}, \{\mathrm{Cos}[\frac{12\ \mathrm{Pi}}{11}], \mathrm{Sin}[\frac{12\ \mathrm{Pi}}{11}]\},$$
$$\{\mathrm{Cos}[\frac{14\ \mathrm{Pi}}{11}], \mathrm{Sin}[\frac{14\ \mathrm{Pi}}{11}]\}, \{\mathrm{Cos}[\frac{16\ \mathrm{Pi}}{11}], \mathrm{Sin}[\frac{16\ \mathrm{Pi}}{11}]\},$$
$$\{\mathrm{Cos}[\frac{18\ \mathrm{Pi}}{11}], \mathrm{Sin}[\frac{18\ \mathrm{Pi}}{11}]\}, \{\mathrm{Cos}[\frac{20\ \mathrm{Pi}}{11}], \mathrm{Sin}[\frac{20\ \mathrm{Pi}}{11}]\},$$
$$\{1, 0\}\}$$

Jerry: **`{Cos[t], Sin[t]}`** looks to me like a parametric expression that, left to itself, would probably produce a circle. Let's try making a parametric plot of it, to see if I'm right (we'll use the option **`AspectRatio -> Automatic`** to make the circle come out looking like a circle, instead of like a flattened ellipse):

```
ParametricPlot[{Cos[t], Sin[t]}, {t, 0, 2Pi},
   AspectRatio -> Automatic];
```

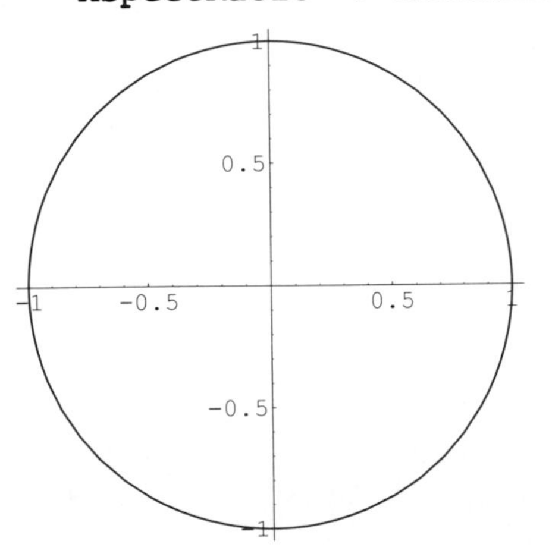

We can plot the points in the list you made, to see how they compare to this circle:

```
ListPlot[verticesForElevengon,
    AspectRatio -> Automatic];
```

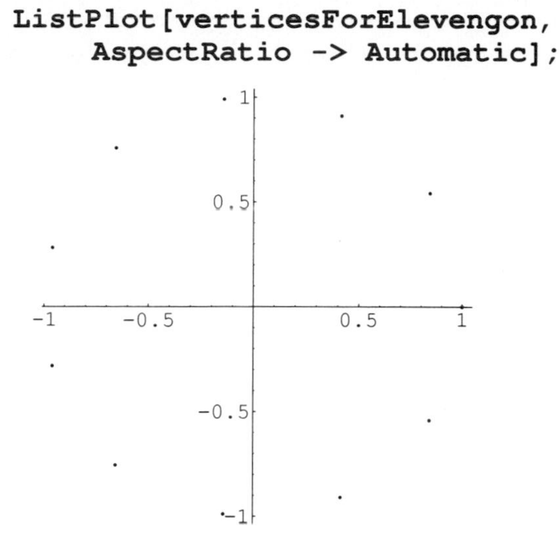

How can we use your list of numbers to draw an actual polygon?

Theo: We need to use two functions, **`Show`** and **`Graphics`**. Their roles are a little confusing. **`Show`** takes one argument, and, optionally, a list of options. The first argument is a list of "graphics objects". A "graphics object" is an expression of the form **`Graphics[`*`list`*`]`**, where ***`list`*** is a list of graphics primitives, such as the ones we talked about above. If ***`list`*** has only one element, you don't have to put it inside list brackets.

Putting this all together, here is how we draw a polygon from our list of points:

```
Show[Graphics[Polygon[verticesForElevengon]],
    AspectRatio -> 1];
```

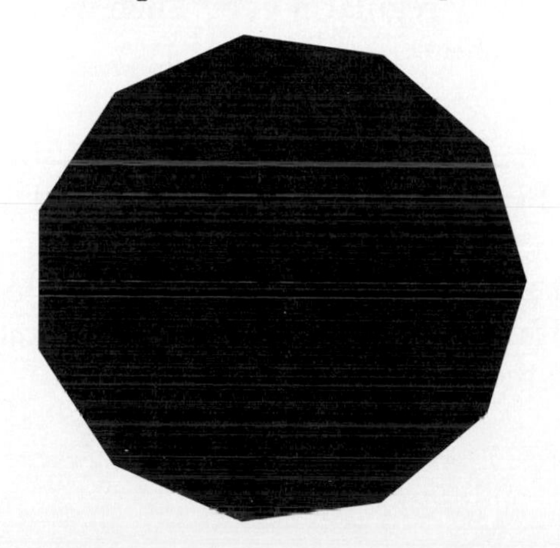

Jerry: Wonderful! How did it get filled?

Theo: That's just in the nature of **Polygon**. It draws filled polygons. If you want an unfilled polygon, you need to use the **Line** command (after all, an unfilled polygon is really just a line drawn around its outside).

```
Show[Graphics[Line[verticesForElevengon]],
    AspectRatio -> 1];
```

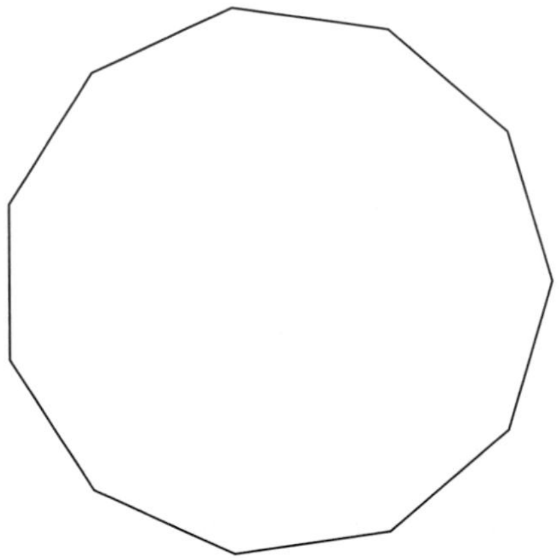

Jerry: Believe it or not, many people think of this last figure as the "polygon", and the picture before that as the polygon and its interior.

Theo: Well, yes, but it makes a certain amount of sense to have **Line** mean a line, and **Polygon** mean the whole area occupied by the polygon.

Jerry: I'd be inclined to draw a seven-sided regular polygon superimposed on this one.

Theo: I suggest we write a function to generate any regular n-gon. Then we can superimpose as many as you like. Here is the function:

```
RegularNGon[n_] :=
    Line[
        Table[{Cos[t], Sin[t]}, {t, 0, 2Pi, 2Pi / n}]
    ]
```

If we replaced the **Line** function with a **Polygon** function, it would generate filled polygons instead of outlined ones. We can draw both a 7-gon and an 11-gon at the same time by making a list of the two inside the **Graphics** function:

```
Show[
    Graphics[{
        RegularNGon[7],
        RegularNGon[11]
    }],
    AspectRatio -> Automatic
];
```

Jerry: We see mostly squares, hexagons, and octagons in the world of regular polygons. Rarely do we get to see a 7 or 11 sided polygon, much less the two of them superimposed. Let's splurge. How about superimposing all the regular polygons up to 50 sides?

Theo: We can use the **Table** command to make a table of the first 50 regular polygons, and give it to **Show**:

```
Show[
    Graphics[Table[
        RegularNGon[m],
        {m, 1, 50}
    ]],
    AspectRatio -> Automatic
];
```

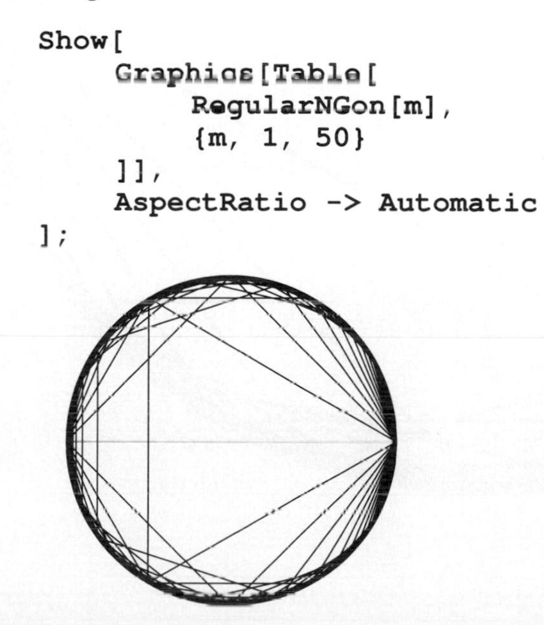

Jerry: This would be much nicer if it were bigger, so we could see all the lines.

Theo: In the Notebook Front End we can use the mouse to drag a graphic out as large as we like (having a big monitor helps, but even on a small monitor you can drag the graphic out big, and then look at parts of it by scrolling around). Here is the result of making it somewhat bigger:

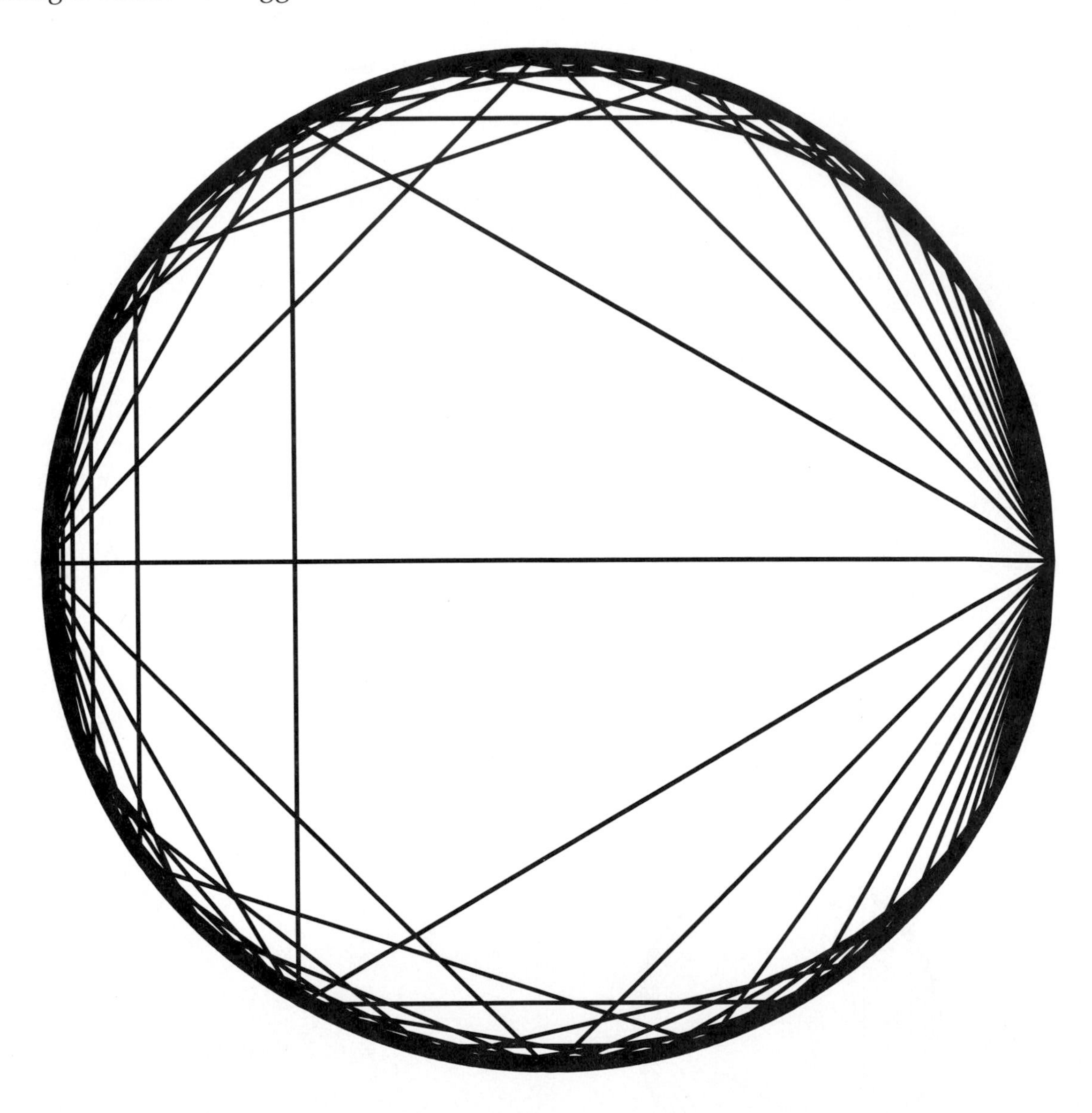

Jerry: This is bigger, but we can't really see much more detail, because the lines have gotten thicker as well. Can we make the lines thin without making the plot any smaller?

Theo: The "Make Lines Thin" option in the Graph menu of the Notebook Front End can be used to make all the lines in a graphic thinner, like this:

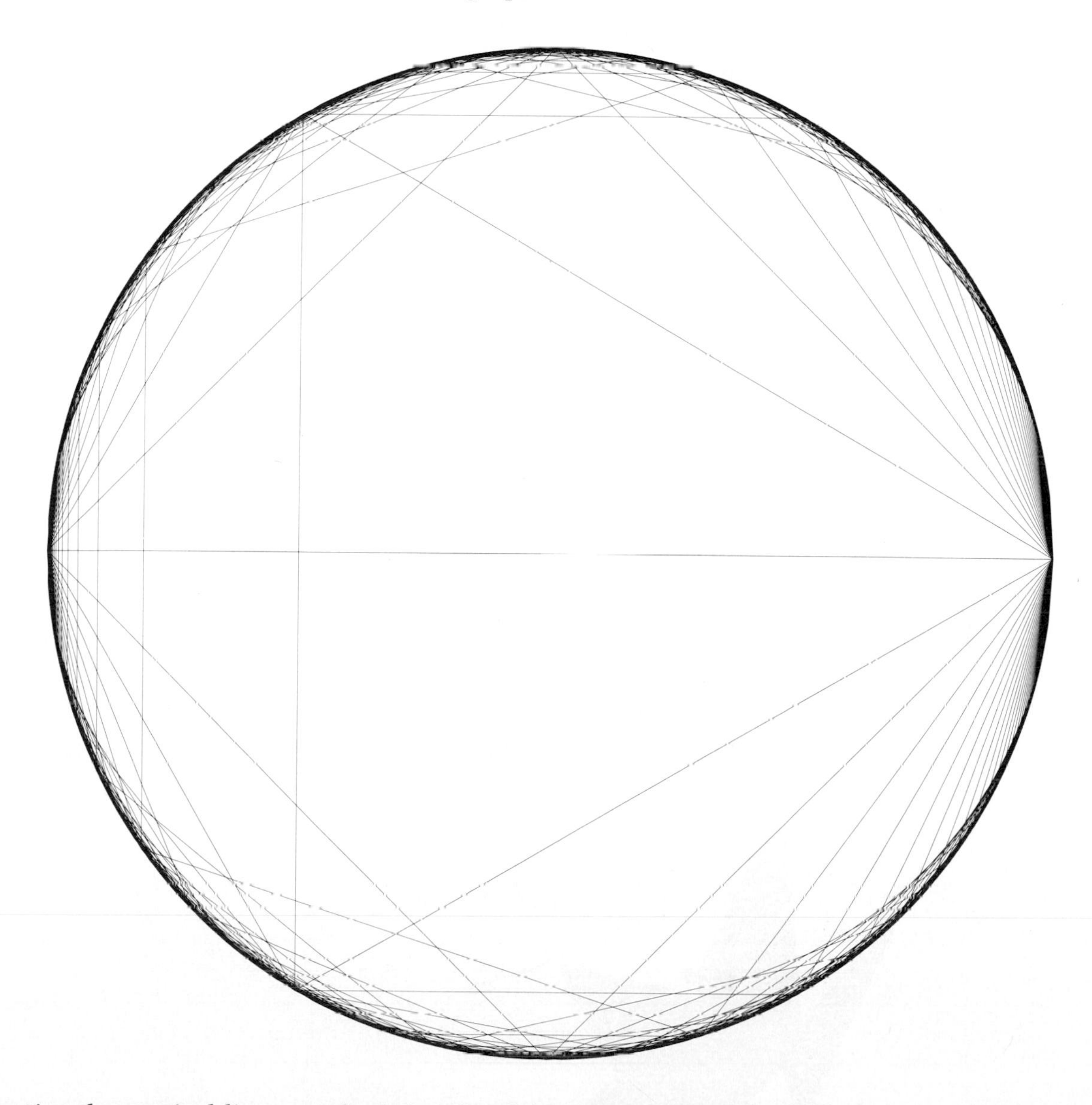

I notice the vertical lines on the left side of this figure seem to converge in a very logarithmiclike way. We could probably calculate exactly how they are spaced, but let's move on to some more graphics, since this is a graphics chapter. I would like to make some stars.

Jerry: Go for it!

Theo: Here it is!

```
star[n_] :=
     Polygon[
          Flatten[
               Table[
                    {
                              {Cos[t],         Sin[t]        },
                         2 {Cos[t + Pi/n], Sin[t] + Pi/n}
                    },
                    {t, 0, 2Pi, 2Pi / n}
               ],
               1
          ]
     ]
```

There is one funny thing about this function. I use **Table** to generate a table of pairs of points, the first on a circle of radius one and the second on a circle of radius two. Unfortunately, I can't just pass this table to **Polygon**, because **Polygon** expects a list of points (pairs of numbers), not a list of pairs of points. So, I have to use the **Flatten** function, which changes the list of pairs of points (pairs of pairs of numbers) into a list of points (pairs of numbers). Unfortunately, if I use **Flatten** in the normal way, it will reduce the list to a list of numbers. By giving **Flatten** a second argument, **1**, I tell it to flatten only the first level of the list, leaving the points as pairs.

This idea may take a while to understand. Here's what we get:

```
Show[Graphics[star[5]], AspectRatio -> 1];
```

Oops! This is not what I would consider a satisfactory star.

Jerry: But I think it's great. Don't throw it away.

Theo: Yes, it's a nice looking thing. The reader may want to explore what happens for other values of **n** in this function. I, however, am going to fix the function. I wanted the points on the outer circle to be rotated by half the distance between points. I did this by putting in a phase shift of **Pi/n** (where **n** is the number of points on the star). Unfortunately, I misplaced a square bracket, which caused the phase shift to happen only for the x coordinate. The fixed function looks like this:

```
star[n_] :=
     Polygon[
          Flatten[
               Table[
                    {
                           {Cos[t],          Sin[t]        },
                         2 {Cos[t + Pi/n], Sin[t + Pi/n]}
                    },
                    {t, 0, 2Pi, 2Pi / n}
               ],
               1
          ]
     ]

Show[Graphics[star[10]], AspectRatio -> 1];
```

Jerry: So, you've got two concentric circles with 10 points equally distributed on each, but one circle is rotated by half the distance between consecutive points. Then you connect a point from the inner circle to the next point on the outer circle, and then to the next point on the inner circle, and so on.

Theo: Right.

Now that we have a star, let's think of what we could change about this star. I can think of three variables: the number of points, the ratio between the inner and outer radii, and the amount by which we rotate the circles relative to each other. Normally we rotate them by exactly half the spacing of the points. If we were to rotate by a different amount, we would get a twisted star. Here is a function where each of these variables is adjustable (**twist** is the amount by which both the inner and outer circles are rotated, in opposite directions):

```
star[n_, radius_, twist_] :=
    Polygon[
        Flatten[
            Table[
                {
                                {Cos[t - twist],
                                 Sin[t - twist]},
                        radius {Cos[t + twist + Pi/n],
                                 Sin[t + twist + Pi/n]}
                },
                {t, 0, 2Pi, 2Pi / n}
            ],
            1
        ]
    ]
```

Let's try it with some random choices:

```
Show[Graphics[star[15, 3, 0.5]], AspectRatio -> 1];
```

Jerry: Can we change one variable in a consistent manner, and see the results as an animation?

Theo: Of course. We can use a **Do** loop to vary the twisting parameter between zero (inner and outer points lined up radially), and **Pi** (lined up again, but twisted one full circle around).

```
Do[
    Show[Graphics[star[10, 2, twist]],
        AspectRatio -> 1,
        PlotRange -> {{-2, 2}, {-2, 2}}],
    {twist, 0, Pi - Pi/100, Pi/100}
]
```

(The printed version includes fewer steps, for space reasons.)

This looks very nice on paper, and even nicer when animated on the screen. It's almost like a kaleidoscope. All the interesting patterns inside the star are caused by the algorithm used to fill self-intersecting polygons. It's not always clear what the "inside" of such a polygon is. The rule used is called the "even-odd" rule. To determine if a given spot is "inside", you draw a line from that spot all the way outwards. If that line crosses the boundary of the polygon an odd number of times, the spot is "inside". If it crosses an even number of times, it's "outside". This is a somewhat arbitrary rule, but it works, and it can be implemented efficiently on computers.

An alternate rule, called the non-zero winding rule, is used by PostScript printers. This rule is rather more sophisticated, and tends to correspond much better with what people think of as "inside", but when these figures are printed out, they don't look nearly as interesting. It is possible to tell a PostScript printer to use the "even-odd" rule instead, which is how we printed these figures in this book. (For PostScript aficionados: The way to do this is to use the PostScript command "`eofill`" instead of "`fill`".)

■ Dialog–Second Day

Jerry: I recently met Ron Avitzur, the author of the Macintosh program *Milo,* a charming and exciting semi-symbolic algebra program with an amazing front end. Before I had a chance to ask, he volunteered that his favorite 2D graph is **`Sin[Tan[x]] - Tan[Sin[x]]`**. *Milo* produced an excellent picture, as I'm sure *Mathematica* will:

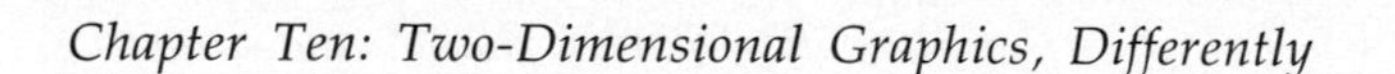


```
Plot[Sin[Tan[x]] - Tan[Sin[x]], {x, -2Pi, 2Pi}];
Plot::notreal:
   Sin[Tan[x]] - Tan[Sin[x]]
     does not evaluate to a real number at x=-4.71239.
```

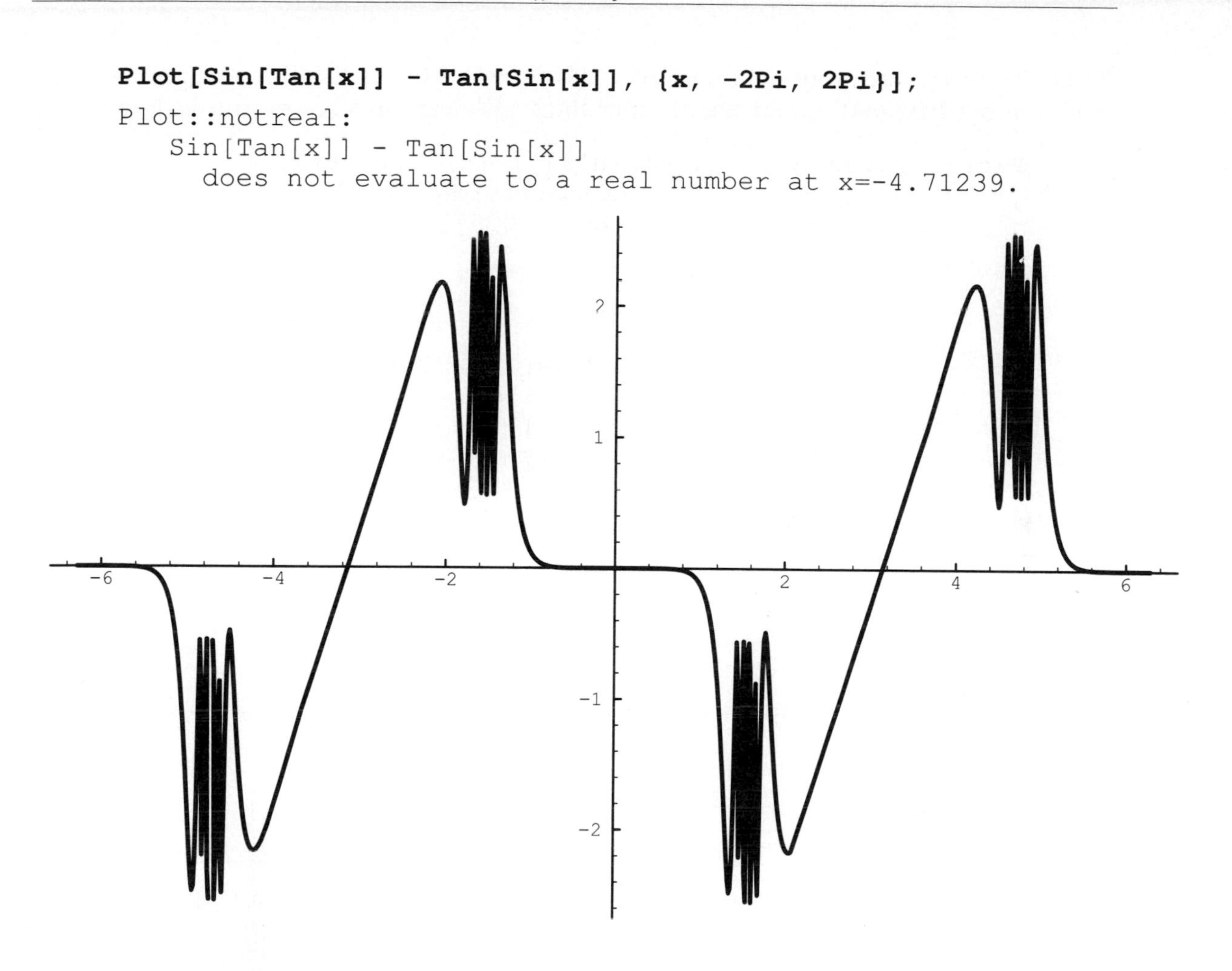

Theo: To bring out more detail, let's tell *Mathematica* to start with more points, subdivide more frequently, and use thinner lines. We can also zoom in a little bit:

```
Plot[Sin[Tan[x]] - Tan[Sin[x]], {x, -Pi, Pi},
    PlotPoints -> 100, PlotDivision -> 30,
    PlotStyle -> Thickness[0.0005]];
```

```
Plot::notreal:
   Sin[Tan[x]] - Tan[Sin[x]]
     does not evaluate to a real number at x=-1.5708.
```

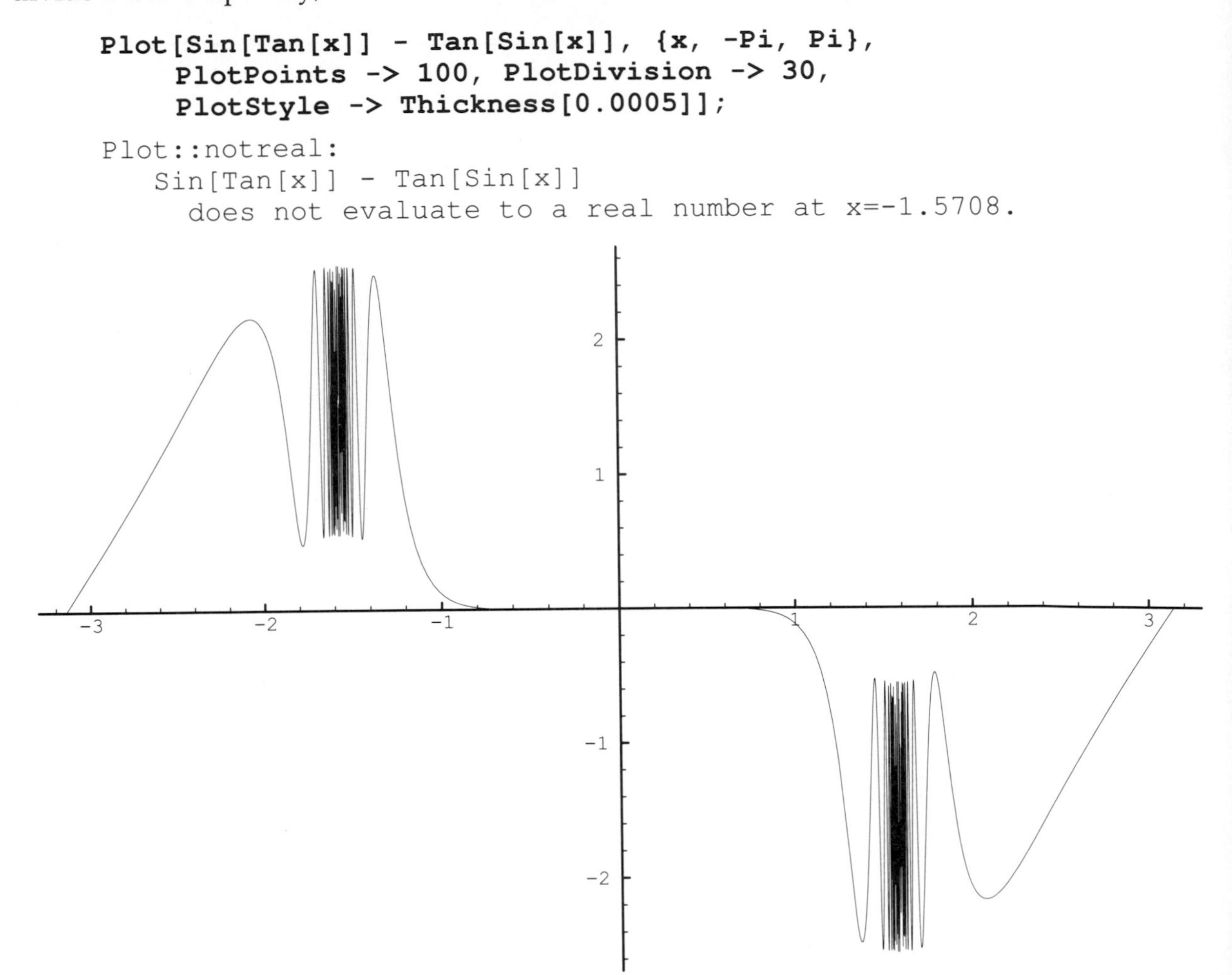

Jerry: I think this plot would be spectacular if it were about 3 feet wide. Can we do that?

Theo: Of course. To enlarge any graphic, click somewhere in the middle of it. This will give you a bounding box with eight small resizing handles on it. Click and drag the bottom right handle off to the right (the aspect ratio of the plot will be preserved automatically, so you don't have to worry about distorting it). If you drag outside the window, the graph will be scrolled automatically, allowing you to drag it out as large as you like.

Jerry: If I wanted a graph 3 feet wide, how would I know when it's big enough?

Theo: While you are dragging a graphic, its current size is displayed in the small status area at the bottom of the window. The width and height are given in "points", which are 1/72 of an inch. To make a graphic 3 feet wide, drag it until the width is 72x12x3 = 2592 points.

I made a copy of the graphic above and enlarged it to this size. I also used the **Make Lines Thin** command in the Graph menu to make all the lines in the plot be drawn one pixel wide. This is often a good idea with very large plots.

In the paper edition of this book, we include a small excerpt from the whole graphic, all of which would occupy many pages.

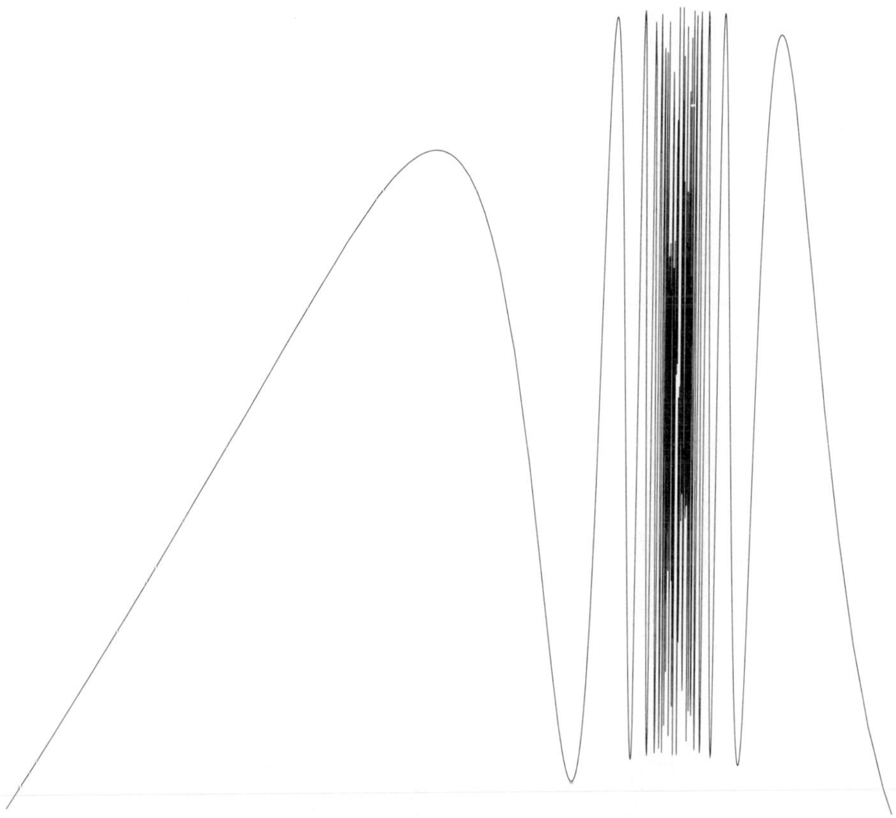

Jerry: This gives the reader a feeling for what can be done when you have *Mathematica* available, and can manipulate the graphs on your computer screen. Do we gain or lose anything when we go from the computer screen to the printed page?

Theo: For plots like this one, we gain a lot going to the printed page: This book was printed at a resolution of about 2400 dots per inch, compared with a resolution of about 72 dots per inch on the screen. That means that our 3 foot wide version (which is about 2500 dots wide on screen) has about the same total resolution as a 1 inch square printed version.

Jerry: So in this case people reading the electronic version must look in the paper book to see the really good picture. Until now, I thought the electronic edition was always better than the paper edition.

Theo: Because *Mathematica* produces all its graphics in resolution-independent PostScript form, they will generally look better when printed than on-screen. In the chapter about density graphics (see Chapter 12) we will find a counter-example to this general rule.

Functions That We Defined

Theo: We defined a simple function for generating regular polygons:

```
RegularNGon[n_] :=
    Line[
        Table[{Cos[t], Sin[t]}, {t, 0, 2Pi, 2Pi / n}]
    ]
```

A simple variation of it can be used to generate filled polygons:

```
FilledRegularNGon[n_] :=
    Polygon[
        Table[{Cos[t], Sin[t]}, {t, 0, 2Pi, 2Pi / n}]
    ]
```

We also defined two versions of a star-generating function. The first version generates a simple star with **n** points:

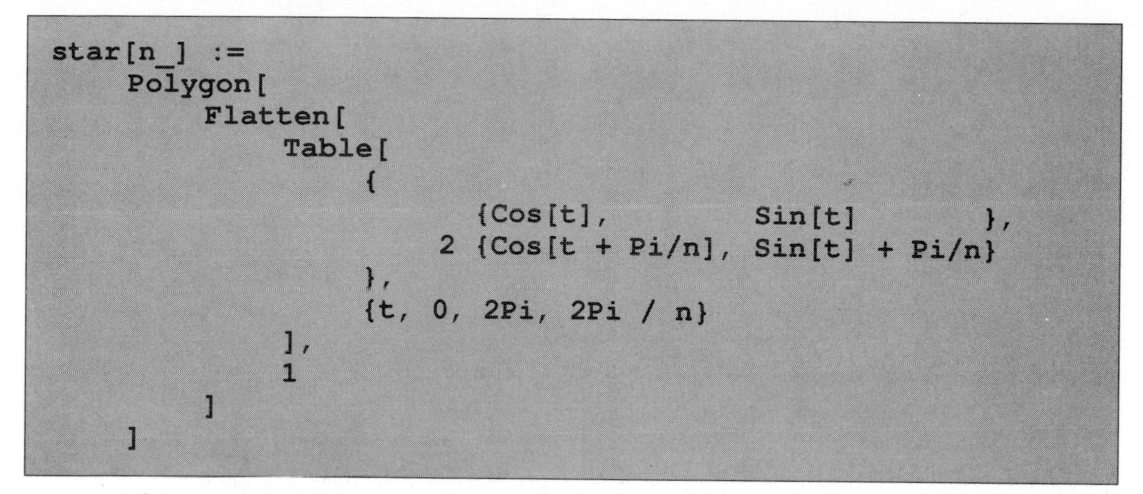

```
star[n_] :=
    Polygon[
        Flatten[
            Table[
                {
                        {Cos[t],          Sin[t]        },
                    2 {Cos[t + Pi/n], Sin[t] + Pi/n}
                },
                {t, 0, 2Pi, 2Pi / n}
            ],
            1
        ]
    ]
```

The more sophisticated version allows us to vary the ratio of the inner to outer radius of the star, and the amount by which the inner and outer points of the star are rotated relative to each other:

```
star[n_, radius_, twist_] :=
    Polygon[
        Flatten[
            Table[
                {
                            {Cos[t - twist],
                             Sin[t - twist]},
                    radius {Cos[t + twist + Pi/n],
                             Sin[t + twist + Pi/n]}
                },
                {t, 0, 2Pi, 2Pi / n}
            ],
            1
        ]
    ]
```

Chapter Eleven
Three-Dimensional Graphics, Differently

Theo and Jerry branch out into higher dimensions.

■ Dialog

Theo: Every book about *Mathematica* has a chapter on 3D graphics.

Jerry: Maybe another one isn't needed?

Theo: I don't think there's any harm in having one, just so long as it's *completely different* from the others.

Jerry: Most of the 3D graphics I've seen on *Mathematica* show surfaces; I'm interested in what happens in the interior. Some years ago in The Math Program we built regular polyhedra out of big dowel rods and rubber bands. Since the sticks were about 18 inches long, we needed to work in my backyard. It was easy to make tetrahedrons, octahedrons, but icosahedrons were quite difficult. After a time, we were successful and began stringing thread from corner to corner, to represent the diagonals. There seemed to be a shape emerging inside the icosahedron, but it was hard to isolate, because the thread was so thin. John Walker and I thought to put aluminum foil around the sections of the diagonals which seemed to be part of the shape being formed. When we stood back, and the sun reflected off the aluminum foil, it was clear there was a dodecahedron in there. Could we do anything like that in *Mathematica*?

Theo: Are you suggesting that *Mathematica* might NOT be able to simulate strung up tinfoil in your backyard? Surely you don't doubt it?

Jerry: You're right, I asked the wrong question! Let me try again: Could we do anything like that in *Mathematica* in less than two weeks?

Theo: No problem. Let's start by making a list of the vertices of an icosahedron. We could get this list from the standard package `Polyhedra.m` by Roman Maeder, like this:

```
Needs["Graphics`Polyhedra`"];
Show[Graphics3D[Icosahedron[]]];
```

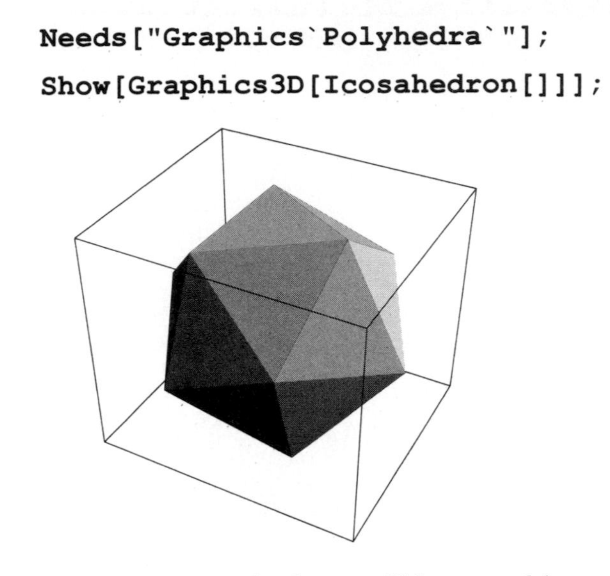

Jerry: I think this icosahedron will be a problem. In my backyard we had trouble seeing the structure of the icosahedron clearly until we hung it from one vertex.

Theo: Unfortunately, Roman's vertices seem to be rotated in a funny way, and it's not obvious how to rotate them to the orientation you want. It's easier to make up our own list of vertices from scratch (see the Functions That We Defined section below). Here are our twelve vertices (the columns are x, y, and z coordinates):

```
icosahedronVertices//TableForm
0           0           0
0.850651    0           0.525731
0.262866    0.809017    0.525731
-0.688191   0.5         0.525731
-0.688191   -0.5        0.525731
0.262866    -0.809017   0.525731
0.688191    0.5         1.37638
-0.262866   0.809017    1.37638
-0.850651   0           1.37638
-0.262866   -0.809017   1.37638
0.688191    -0.5        1.37638
0           0           1.90211
```

Jerry: Where is the origin relative to these coordinates, and can we see a picture of these vertices?

Theo: The origin is at the bottom vertex. In other words, the icosahedron is balanced perfectly on the origin.

Jerry: So, (0, 0, 1.90211) is the top vertex?

Theo: Right. We can plot dots at these vertices like this:

```
Show[Graphics3D[{PointSize[0.03],
    Map[Point, icosahedronVertices]}]];
```

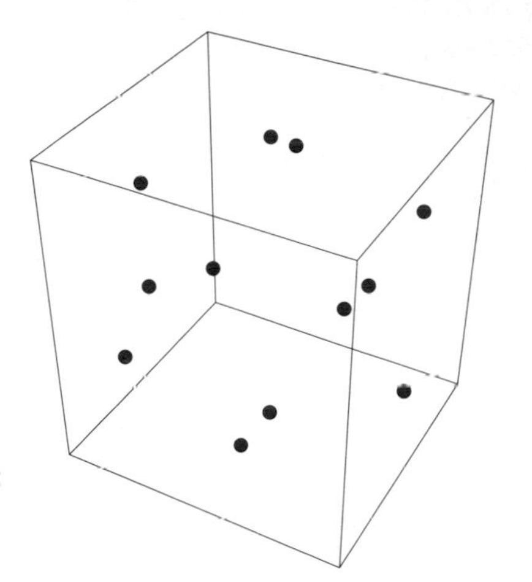

This is not very intelligible. We need to connect these vertices together with lines. To do that, I've written a function (**sortedDiagonals**) that takes a list of points (like these vertices) and makes lines from every point to every other point. It then sorts them into lists according to their length. The following command will generate all the lines, and assign them to a variable (I put a semicolon at the end of the command so we don't actually see the list of lines, because it is very long). To help us see what's going on, **sortedDiagonals** prints a list of the unique lengths of lines it found (note that this is *not* the same as what's assigned to the **icosahedronLines** variable, it's just an informative printout).

```
icosahedronLines = sortedDiagonals[icosahedronVertices];
{1., 1.61803, 1.90211}
```

Jerry: So this says that if you draw lines from every vertex to every other vertex, you will get lines of three different lengths. The shortest ones must be the edges, and the longest ones must be lines between opposite vertices. I wonder what the middle ones are?

Theo: I'll assign the three groups of lines to three separate variables, so we can deal with them easily:

```
icosahedronEdges = icosahedronLines[[1]];
icosahedronMiddles = icosahedronLines[[2]];
icosahedronOpposites = icosahedronLines[[3]];
```

We can plot each of these sets of lines separately:

```
Show[Graphics3D[icosahedronEdges]];
Show[Graphics3D[icosahedronMiddles]];
Show[Graphics3D[icosahedronOpposites]];
```

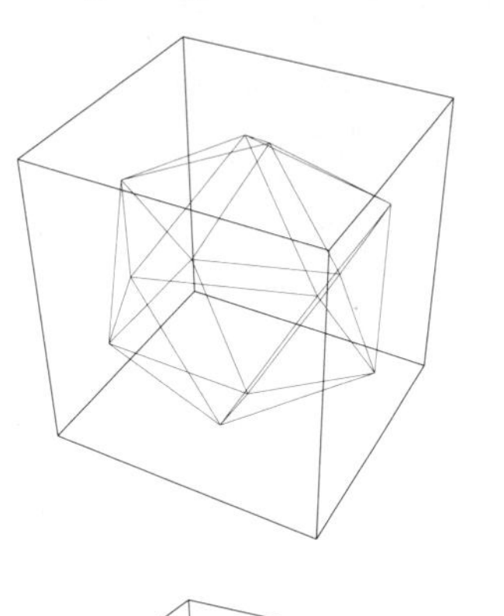

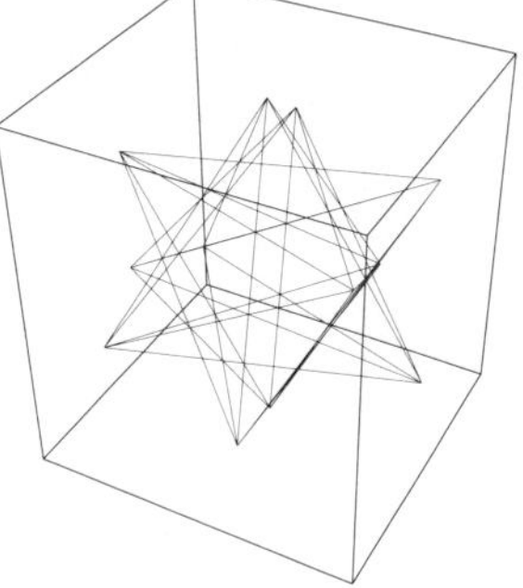

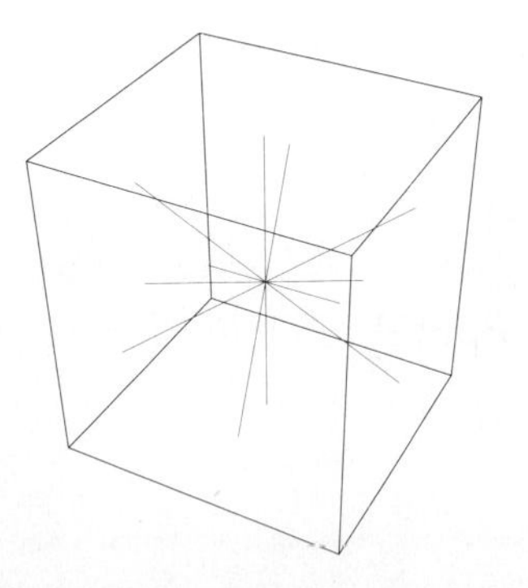

Jerry: When I picture an icosahedron hanging from a vertex, I see four parallel planes. The top and bottom planes each go through a single vertex....

Theo: Wait a minute! I'm totally confused already. We're going to need some pictures of these planes. It's a bit messy, but I've written a function (see the Functions That We Defined section below) to draw these four planes, one at a time:

```
doFourPlanesAnimation[]
```

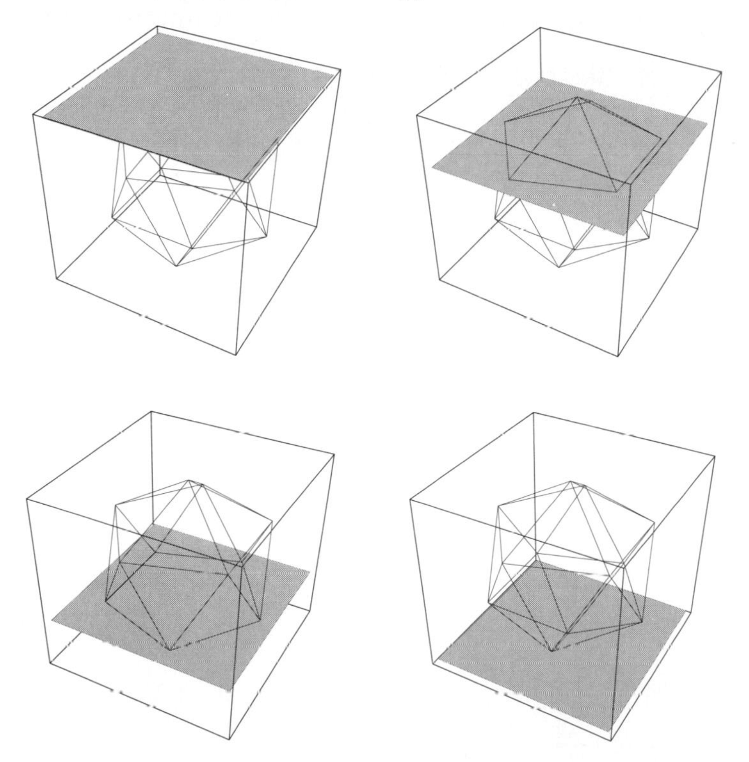

Jerry: Right. As I was saying, the top and bottom planes each go through a single vertex. The middle two planes each go through a pentagon of vertices. We can use these planes to understand why there are lines of three different lengths between the vertices.

Starting at the top vertex, we can draw five lines to points on the second plane from the top. These lines are all the same length, and are edges of length 1.

Also starting at the top, we can draw five more lines to points on the third plane from the top. These lines are also all the same length, and are the middle-length diagonals (1.618 long).

From the top we can draw one line to the bottom plane, forming one of the opposite diagonals of length 1.902.

Theo: If we are to get a dodecahedron out of these diagonals, it's clearly going to have to be using the middle-length ones, because they are the only ones that intersect each other in enough places.

Jerry: Looking at the picture of these diagonals above, I can't even begin to see a dodecahedron. It's just like in my backyard before we added the aluminum foil.

Theo: Clearly we need to have the diagonals change color each time they intersect with each other. That way we can emphasize the inside portions that (you claim) form a dodecahedron. It's not that easy to figure out when two lines intersect in three dimensions, but I've written a function that does it.

Using this function and several others that use the intersection points to pick apart and re-color the lines, we can produce the following striking picture:

```
coloredDiagonals = splitAndColor[icosahedronMiddles];
Show[Graphics3D[coloredDiagonals]];
```

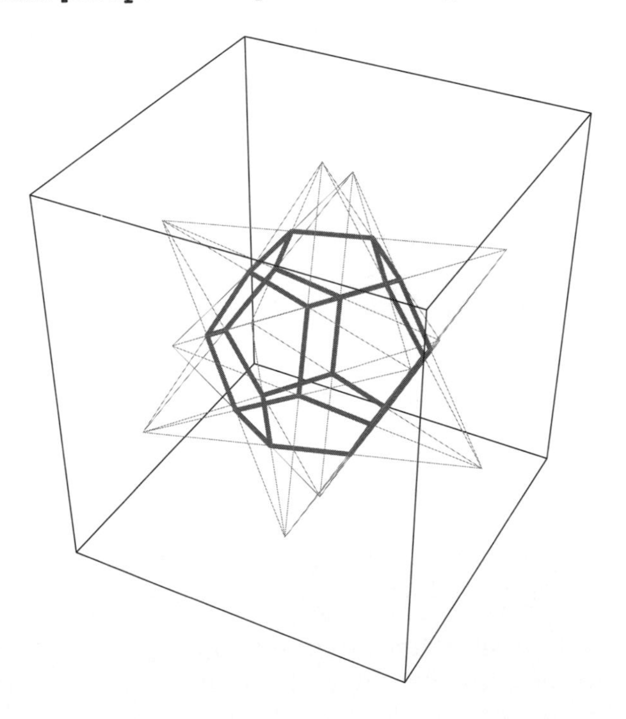

Jerry: This is a little confusing, although there certainly is no doubt about the dodecahedron! I can't really make out where all the diagonals are going.

Theo: Let's add the edges of the icosahedron in black:

```
icosahedronPicture = Show[Graphics3D[{
    coloredDiagonals,
    Thickness[0.005], GrayLevel[0],
    icosahedronEdges
}]];
```

See picture on color plates page xvi.

Jerry: OK, now the only thing left to do is to hang it from a tree and spin it. The hanging part we can do just by looking at it from lower down, but what about the spinning part? I have an MS-DOS program called AcroSpin by David Parker which manages to take 3D lines and rotates them in real time. Even on a 12MHz AT the effect is wonderful. Can we use AcroSpin?

Theo: We could translate the *Mathematica* graphics into AcroSpin format if we wanted to. If we were using *Mathematica* on a Silicon Graphics or RS/6000 computer we could use their fancy graphics programs to spin the object. We can't yet rotate graphics in real time on a Macintosh, but we can do the next best thing, which is to generate a fixed sequence of frames and look at them.

```
Do[
    Show[
        icosahedronPicture,
        ViewPoint -> {Cos[t], Sin[t], 0.3},
        SphericalRegion -> True,
        Boxed -> False
    ],
    {t, 0, 2Pi/5 - 2Pi/150, 2Pi/150}
]
```

(In the printed edition we included only one frame, because the other 29 look pretty much the same printed.)

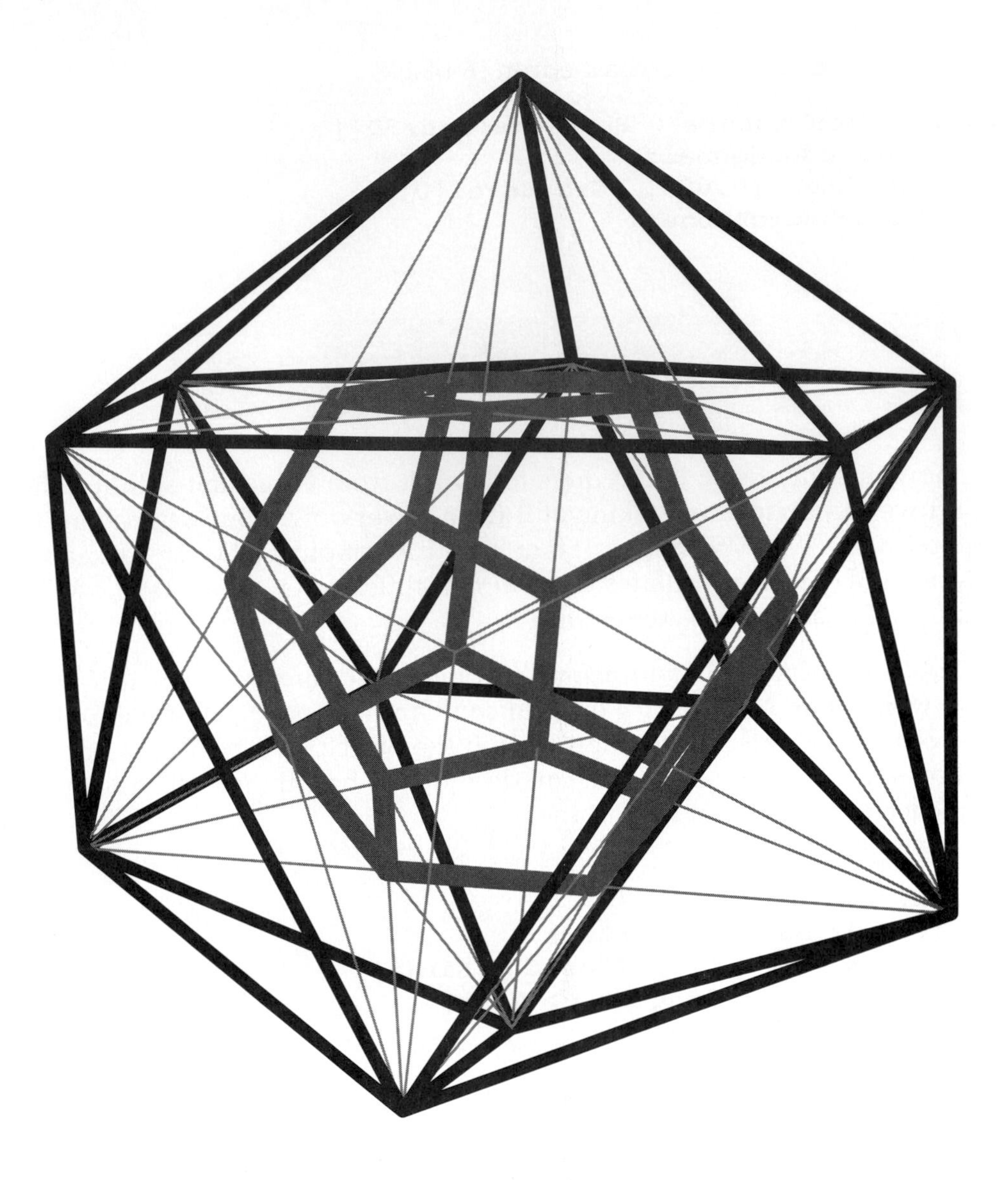

Functions That We Defined

Calculating the Icosahedron Vertices

An icosahedron consists of four parallel planes of vertices: one containing a single vertex, two containing pentagons of vertices, and a fourth containing a single vertex.

First we calculate the radius of the two pentagons:

```
radius = Sin[54 Degree]/Sin[72 Degree]//N
0.850651
```

Next the height separating the plane containing a single vertex from the plane containing the first pentagon:

```
height1 = Sqrt[1-radius^2]
0.525731
```

Now the distance from the center of one pentagon to the middle of one of its edges (which is the radius of the circle inscribed in the pentagon):

```
inscribedRadius = radius Sin[54 Degree]//N
0.688191
```

Next the height separating the two planes of pentagons:

```
height2 = Sqrt[3/4 - (radius - inscribedRadius)^2]
0.850651
```

From these values we can calculate the vertices of the icosahedron:

```
icosahedronVertices = Flatten[N[{
     {{0, 0, 0}},
     Table[
          {radius Cos[t], radius Sin[t], height1},
          {t, 0, 2Pi-2Pi/5, 2Pi/5}
     ],
     Table[
          {radius Cos[t+2Pi/10], radius Sin[t+2Pi/10],
           height1 + height2},
          {t, 0, 2Pi-2Pi/5, 2Pi/5}
     ],
     {{0, 0, height1 + height2 + height1}}
}], 1]
```

```
{{0, 0, 0}, {0.850651, 0, 0.525731},
  {0.262866, 0.809017, 0.525731},
  {-0.688191, 0.5, 0.525731},
  {-0.688191, -0.5, 0.525731},
  {0.262866, -0.809017, 0.525731},
  {0.688191, 0.5, 1.37638},
  {-0.262866, 0.809017, 1.37638},
  {-0.850651, 0, 1.37638},
  {-0.262866, -0.809017, 1.37638},
  {0.688191, -0.5, 1.37638}, {0, 0, 1.90211}}
```

■ Diagonal Lines

This function takes a list of vertices and returns a list of lists of lines. Each sublist of the main list contains lines of the same length. The main list is sorted in order of increasing line length. The first element of the list will be a list of lines between neighboring vertices, the second element will be a list of next-to-closest vertices, etc.

```
lineLength[Line[{a_, b_}]] := N[Sqrt[(b-a) . (b-a)]]
```

```
sortedDiagonals[vertices_] := Block[
     {lines, lengths, uniqueLengths, lengthsAndLines},

     lines = Flatten[
          Table[
               Line[Sort[vertices[[{i, j}]]]],
               {i, 1, Length[vertices]},
               {j, i+1, Length[vertices]}
          ]
     ];

     lengths = Map[lineLength, lines];
     lengths = Floor[0.5 + 1000000 lengths /
                    lengths[[1]]];
     uniqueLengths = Union[lengths];
     Print[N[lineLength[lines[[1]]] uniqueLengths /
                    1000000]];

     lengthsAndLines = Transpose[{lengths, lines}];
     Table[
          Map[
               Last,
               Select[
                    lengthsAndLines,
                    (#[[1]] == uniqueLengths[[i]])&
               ]
          ],
          {i, 1, Length[uniqueLengths]}
     ]
]
```

■ Intersections

```
crossingPoints[Line[{a1_, b1_}], Line[{a2_, b2_}]] :=
Block[
    {t1, t2, line1, line2, denom},

    If[(a1 === a2) || (b1 == b2),
        Return[{}]
    ];

    d1 = (b1 - a1);
    d2 = (b2 - a2);

    denom = (d2[[1]] d1[[2]] - d1[[1]] d2[[2]]);

    If[Abs[denom] < 0.000001,
        t1 = 0;
        t2 = 0
    ,
        t1 = -(((a1[[2]] - a2[[2]]) d2[[1]]) / denom) +
            ((a1[[1]] - a2[[1]]) d2[[2]]) / denom;
        t2 = -(((a1[[2]] - a2[[2]]) d1[[1]]) / denom) +
            ((a1[[1]] - a2[[1]]) d1[[2]]) / denom;
    ];

    line1 = a1 + (b1 - a1) t1;
    line2 = a2 + (b2 - a2) t2;

    If[Abs[line1[[3]] - line2[[3]]] > 0.00001,
        {},
        {line1}
    ]
]
```

```
splitOneAtCrossings[Line[{a1_, b1_}], lines_] :=
Block[
    {points},

    points = Join[
        {a1},
        Apply[Join, Map[
            crossingPoints[Line[{a1, b1}], #]&,
            lines]
        ],
        {b1}
    ];

    Map[
        Line,
        Partition[Union[Floor[
            1000000000 points + 0.5]] / 1000000000.,
            2, 1
        ]
    ]
]
```

```
splitAtCrossings[lines_] :=
    Map[splitOneAtCrossings[#, lines]&, lines]
```

```
splitAndColor[lines_] :=
    Map[
        {
            Thickness[0.001], RGBColor[0.5, 0.5, 0.5],
            #[[1]],
            Thickness[0.008], RGBColor[1, 0, 0],
            #[[2]],
            Thickness[0.001], RGBColor[0.5, 0.5, 0.5],
            #[[3]]
        }&,
        splitAtCrossings[lines]
    ];
```

```
splitAndColorThick[lines_] :=
    Map[
        {
            Thickness[0.001],
            SurfaceColor[RGBColor[1, 1, 1]],
            #[[1]],
            Thickness[0.001],
            SurfaceColor[RGBColor[1, 0, 0]],
            #[[2]],
            Thickness[0.001],
            SurfaceColor[RGBColor[1, 1, 1]],
            #[[3]]
        }&,
        splitAtCrossings[lines]
    ];
```

■ Four Planes Animation

This function produces the animation of the four planes going through an icosahedron.

```
aPlane[height_] :=
    Show[Graphics3D[{icosahedronEdges,
        Polygon[{
            {-1, -1, height}, {1, -1, height},
            {1, 1, height}, {-1, 1, height}
        }]
    }]];
```

```
doFourPlanesAnimation[] :=
    (
    aPlane[height1 + height2 + height1];
    aPlane[height1 + height2];
    aPlane[height1];
    aPlane[0];
    )
```

Chapter Twelve
Density Plots

One of the less appreciated Mathematica functions gets its day.

■ Dialog

Theo: People don't use **DensityPlot** nearly enough. It has an undeserved reputation as a dull, uninteresting command. Perhaps plots like the following give it this reputation:

```
DensityPlot[x y, {x, -3, 3}, {y, -3, 3}];
```

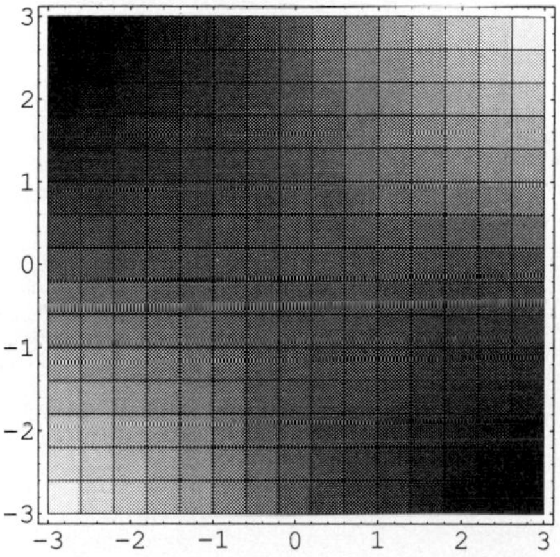

3
2
1
0
−1
−2
−3
−3 −2 −1 0 1 2 3

Jerry: I agree that this is a dull density plot. The dullness is clearly caused by a lack of wigglyness. If we put in a **Sin** function or two, things will get more interesting.

Theo: Let's try one:

```
DensityPlot[Sin[x y], {x, -3, 3}, {y, -3, 3}];
```

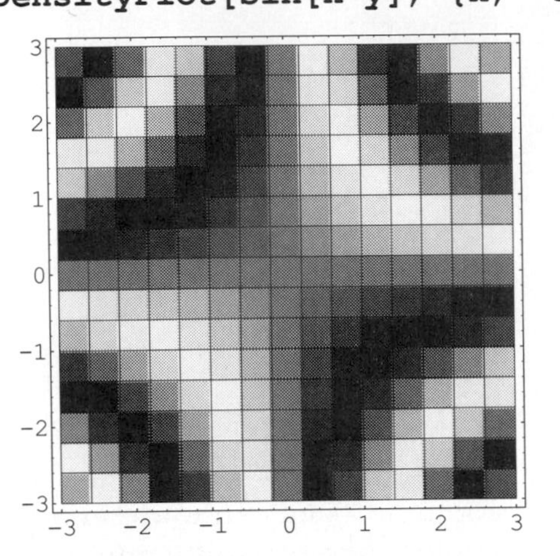

This is getting better, but clearly has a ways to go. Let's try increasing the number of sample points from the default of 15 to 50, widening the plot range, and removing the grid lines:

```
DensityPlot[Sin[x y], {x, -6, 6}, {y, -6, 6},
    PlotPoints -> 50, Mesh -> False];
```

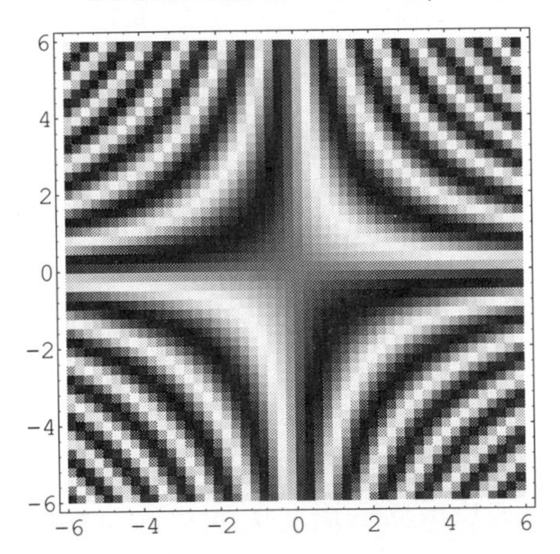

Jerry: This is starting to look good. Let's really crank up the plot points, and widen the plot range even more:

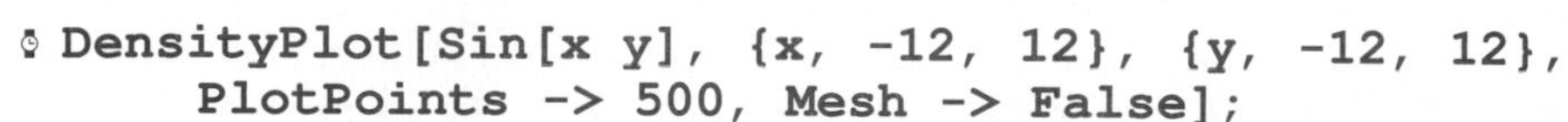

```
DensityPlot[Sin[x y], {x, -12, 12}, {y, -12, 12},
    PlotPoints -> 500, Mesh -> False];
```

Dramatic improvement!

Theo: Is that all you can say? This is a *spectacular* plot! Too bad it takes so long to calculate (about an hour on a Mac II). I wonder what happens if we try similar functions. For example, let's try `Sin[x/y]` instead of `Sin[x y]`:

```
DensityPlot[Sin[x/y], {x, -10, 10}, {y, -5, 5},
PlotPoints->500, Mesh->False];
```

Jerry: Fantastic! It looks a lot like satin sheets. It has such a three-dimensional quality, except for the busy part across the center horizontal. That part looks like many of the pictures you see in the fractal books.

Theo: The bubbles you see are not really there. They are Moiré patterns caused by the interaction of the lines with the 500x500 grid of points we plotted. Actually, there are infinitely many ever-more-closely-spaced lines radiating out from the origin, all bunched up against the horizontal axis.

Jerry: This talk of lines in the plot of `Sin[x/y]` is confusing. Where are these lines coming from?

Theo: There are radial lines in the plot because the ratio `x/y` is a constant along any line radiating out from the origin. Along any line from the origin the color (gray level) will be constant. Lines with different slopes will have different colors. So, we see a series of lines that get thinner and thinner as `y` gets smaller.

You can also think about a vertical line cut through the plot. Along a vertical line, the `x` value is constant. If we pick `x = 1`, for example, the function becomes `Sin[1/y]`, which is a wave that oscillates faster and faster as `y` gets smaller.

Jerry: All this was accomplished by the first variation we tried. Let's see some more!

Theo: Instead of just making a few more variations, let's make *all* possible variations. (See **[1]** for more information.) To avoid spending an infinite length of time at this task, we can make the following restrictions on the meaning of "variation":

> **1)** We must use `Sin`, `x`, and `y` exactly once in each function.
>
> **2)** We can use as many multiplications, divisions, and 1's as we want.
>
> **3)** We can't use any other functions, including addition and subtraction.

There are exactly 14 different functions we can make, given these restrictions. Here they are:

```
Sin[x y];
Sin[x/y];
Sin[1/(x y)];
x Sin[y];
x Sin[1/y];
Sin[y] / x;
Sin[1/y] / x;
x / Sin[y];
x / Sin[1/y];
1 / (x Sin[y]);
1 / (x Sin[1/y]);
1 / Sin[x y];
1 / Sin[x/y];
1 / Sin[1/(x y)];
```

Here are the plots, one after the other. **Plot3D**? Who needs it!

```
DensityPlot[Sin[1/(x y)], {x, -2, 2}, {y, -2, 2},
  PlotPoints->500, Mesh->False];
```

```
DensityPlot[x Sin[y], {x, -12, 12}, {y, -12, 12},
  PlotPoints->500, Mesh->False];
```

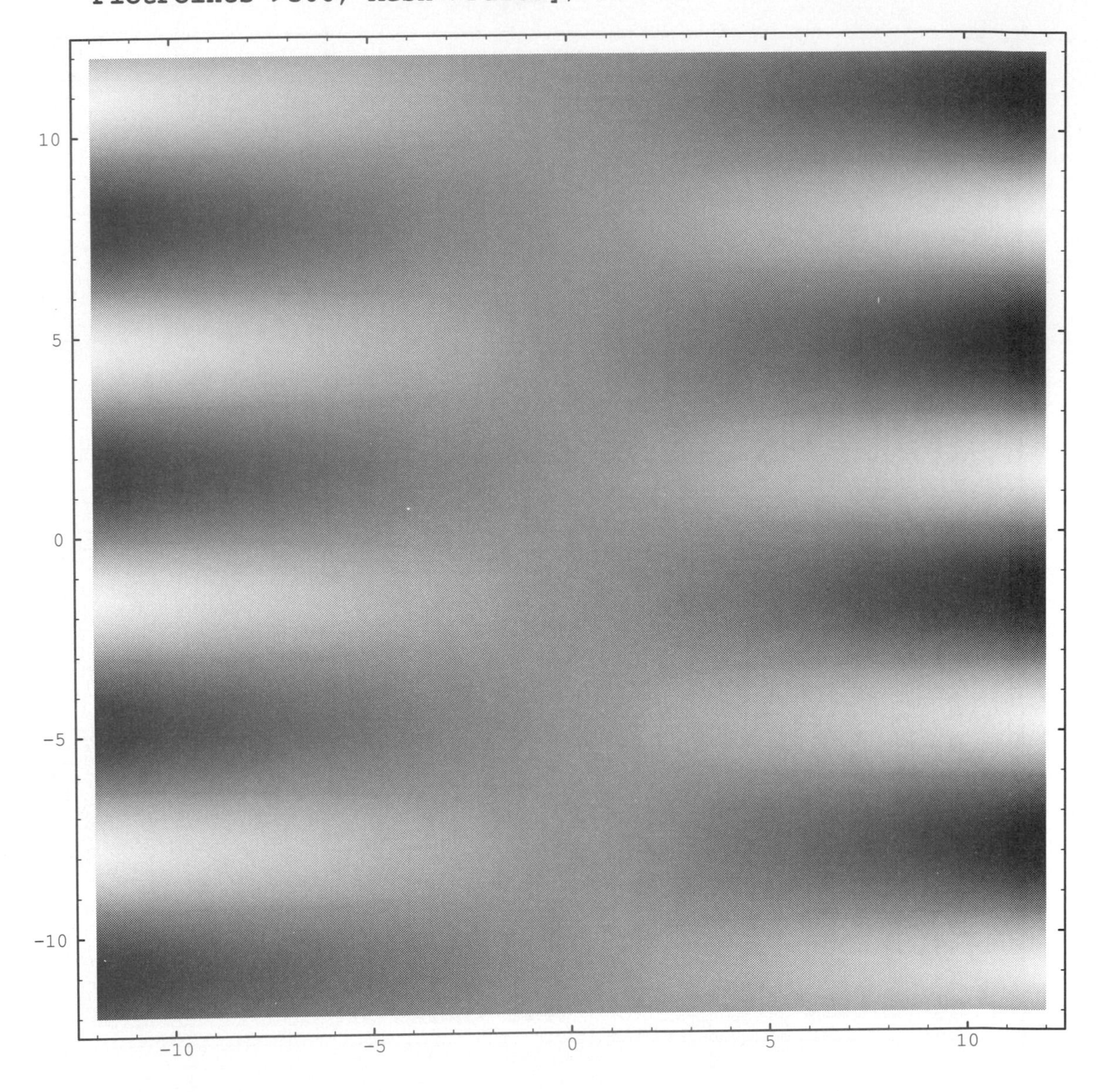

```
DensityPlot[x Sin[1/y], {x, -12, 12}, {y, -5, 5},
 PlotPoints->500, Mesh->False];
```

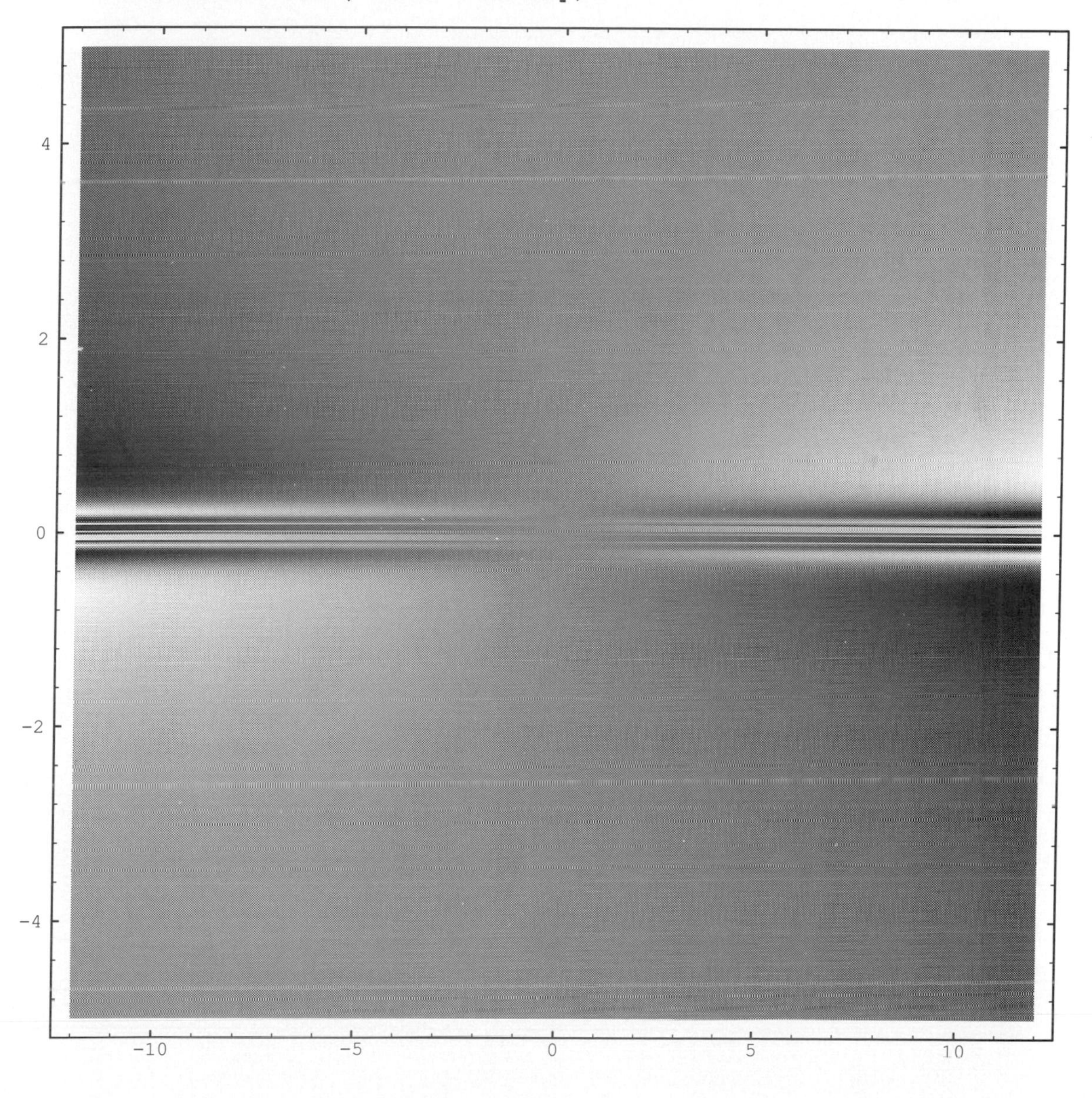

```
DensityPlot[Sin[y] / x, {x, -12, 12}, {y, -12, 12},
PlotPoints->500, Mesh->False];
```

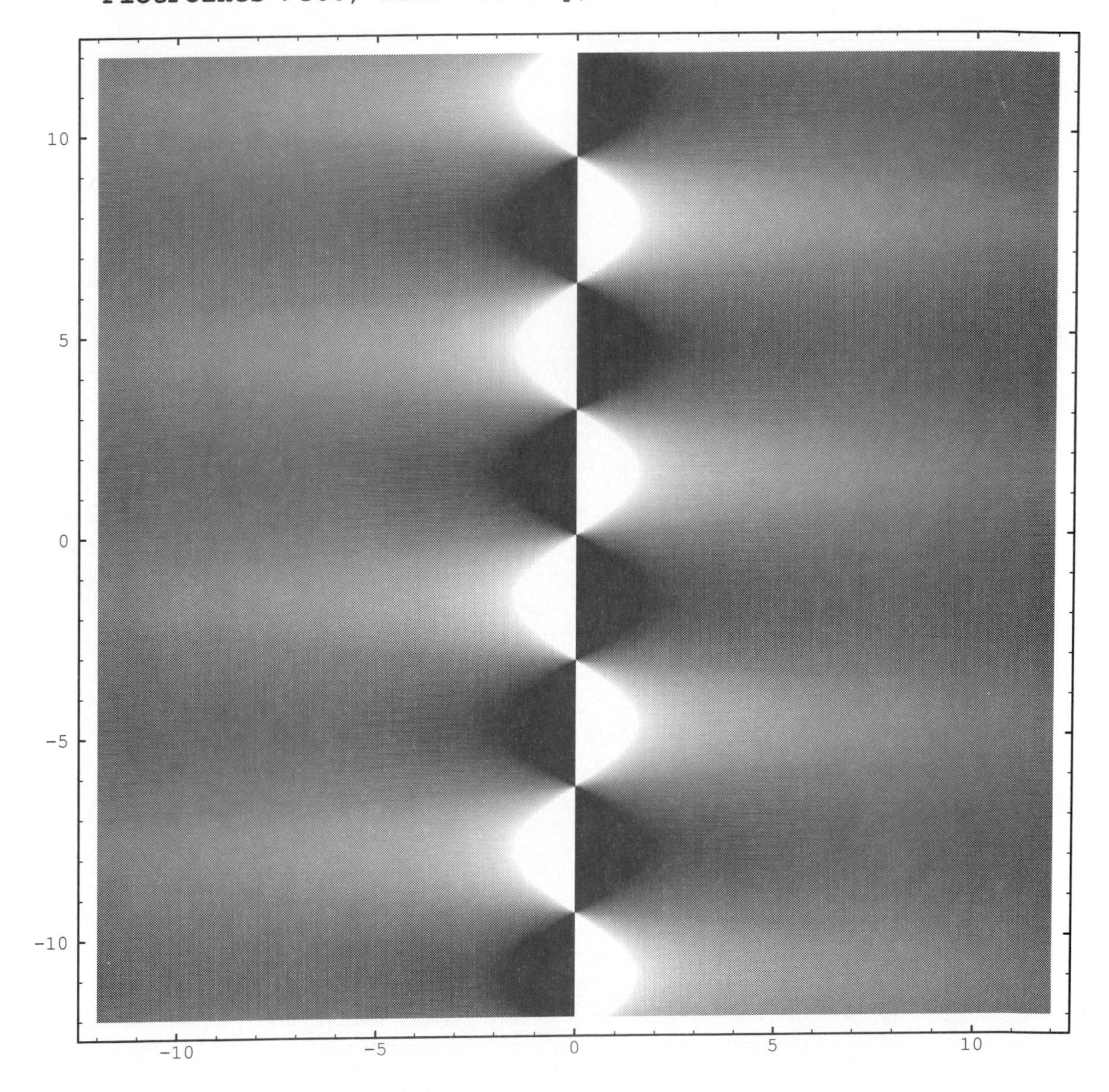

```
DensityPlot[Sin[1/y] / x, {x, -12, 12}, {y, -5, 5},
PlotPoints->500, Mesh->False];
```

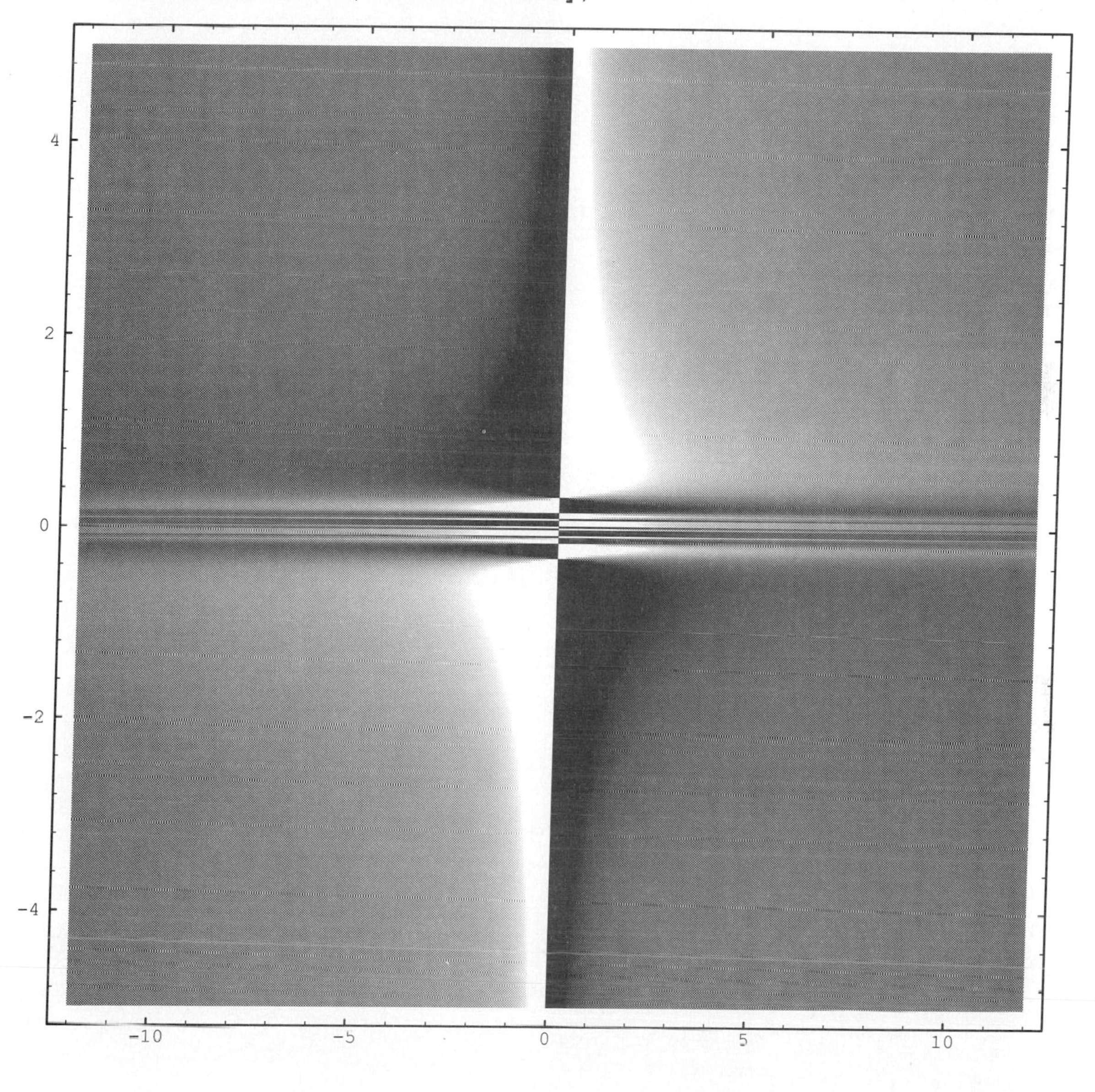

```
DensityPlot[x / Sin[y], {x, -12, 12}, {y, -12, 12},
 PlotPoints->500, Mesh->False];
```

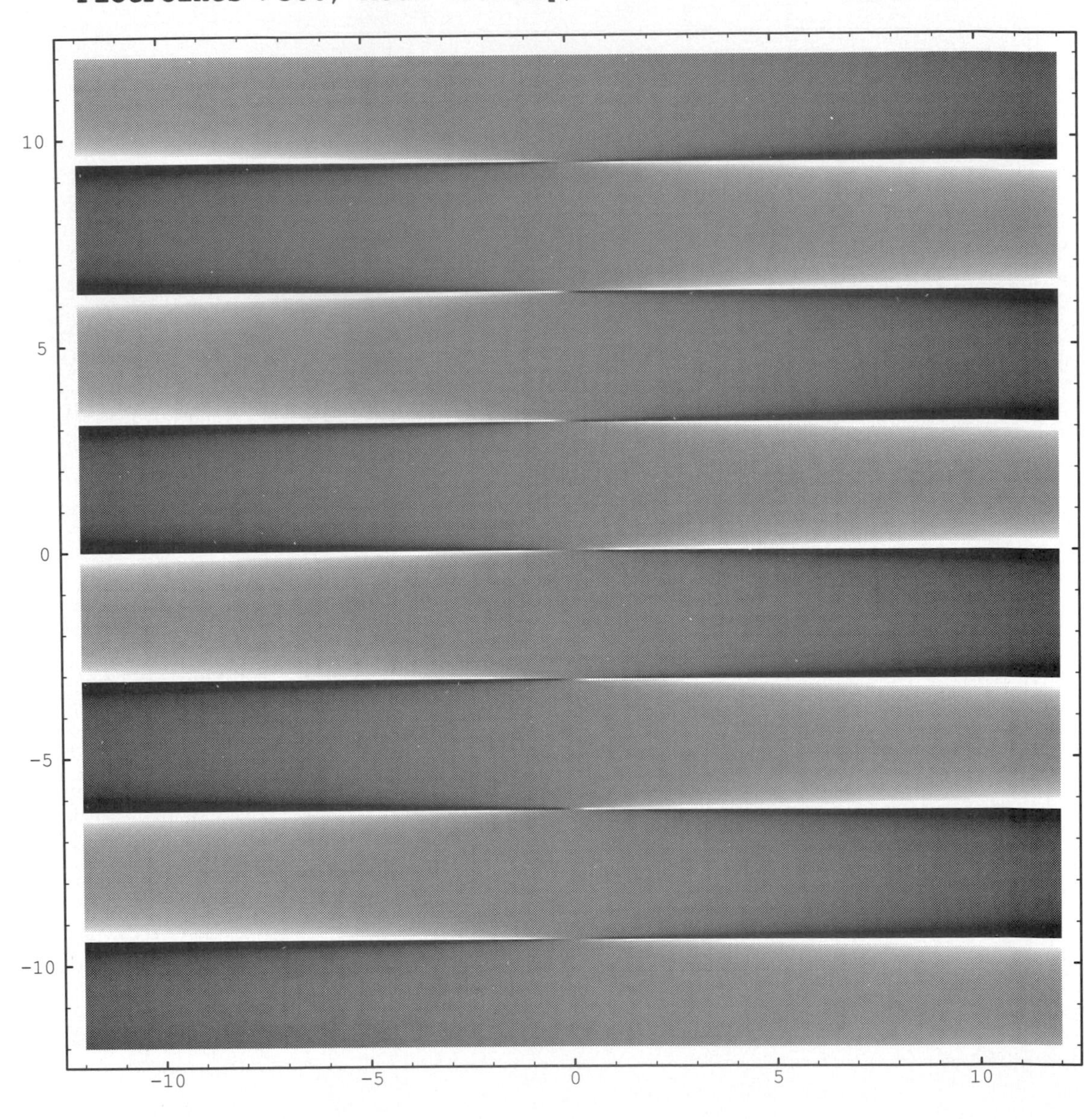

```
DensityPlot[x / Sin[1/y], {x, -12, 12}, {y, -5, 5},
  PlotPoints->500, Mesh->False];
```

```
DensityPlot[1 / (x Sin[y]), {x, -12, 12}, {y, -12, 12},
PlotPoints->500, Mesh->False];
```

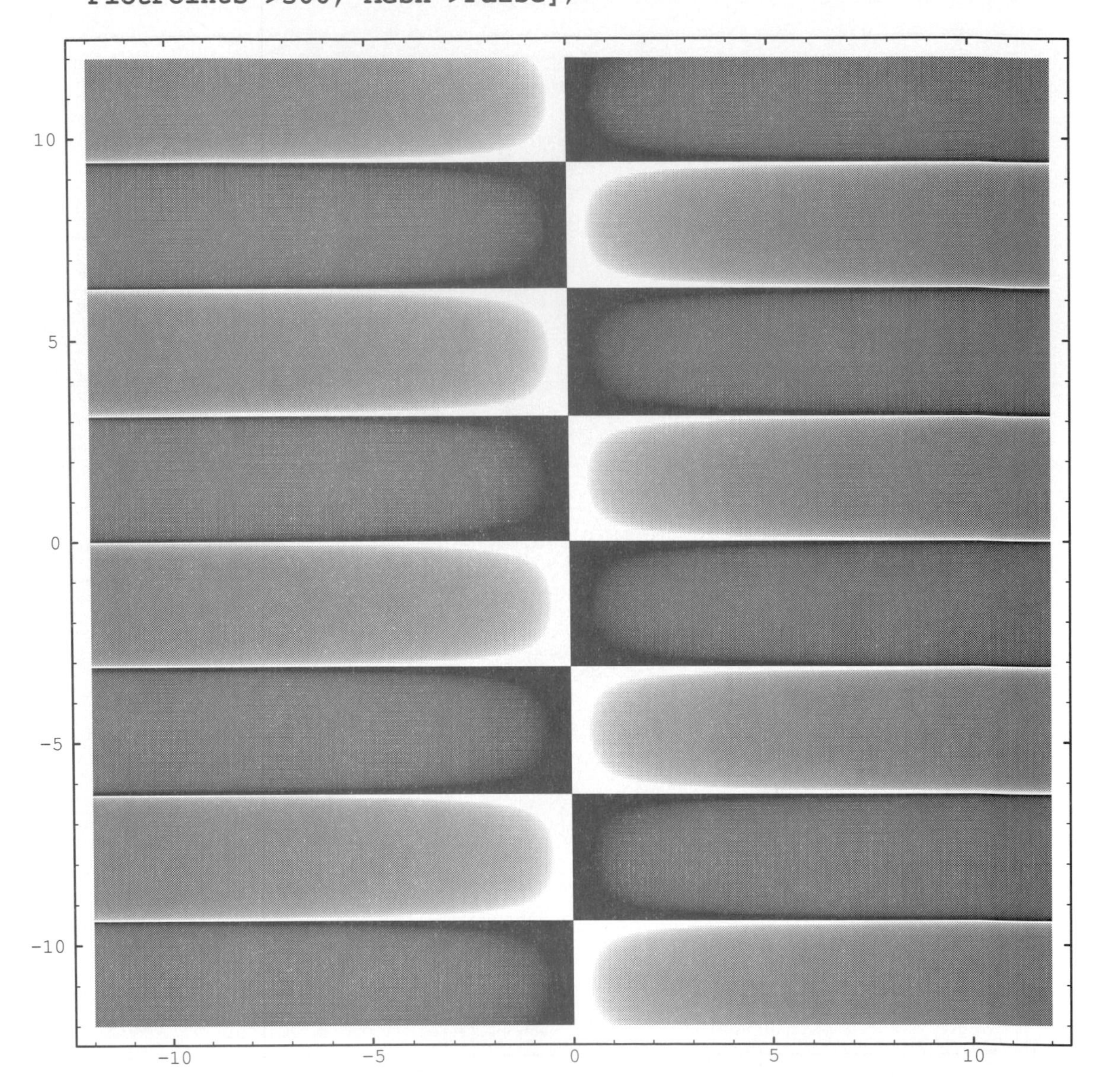

```
DensityPlot[1 / (x Sin[1/y]), {x, -12, 12}, {y, -5, 5},
 PlotPoints->500, Mesh->False];
```

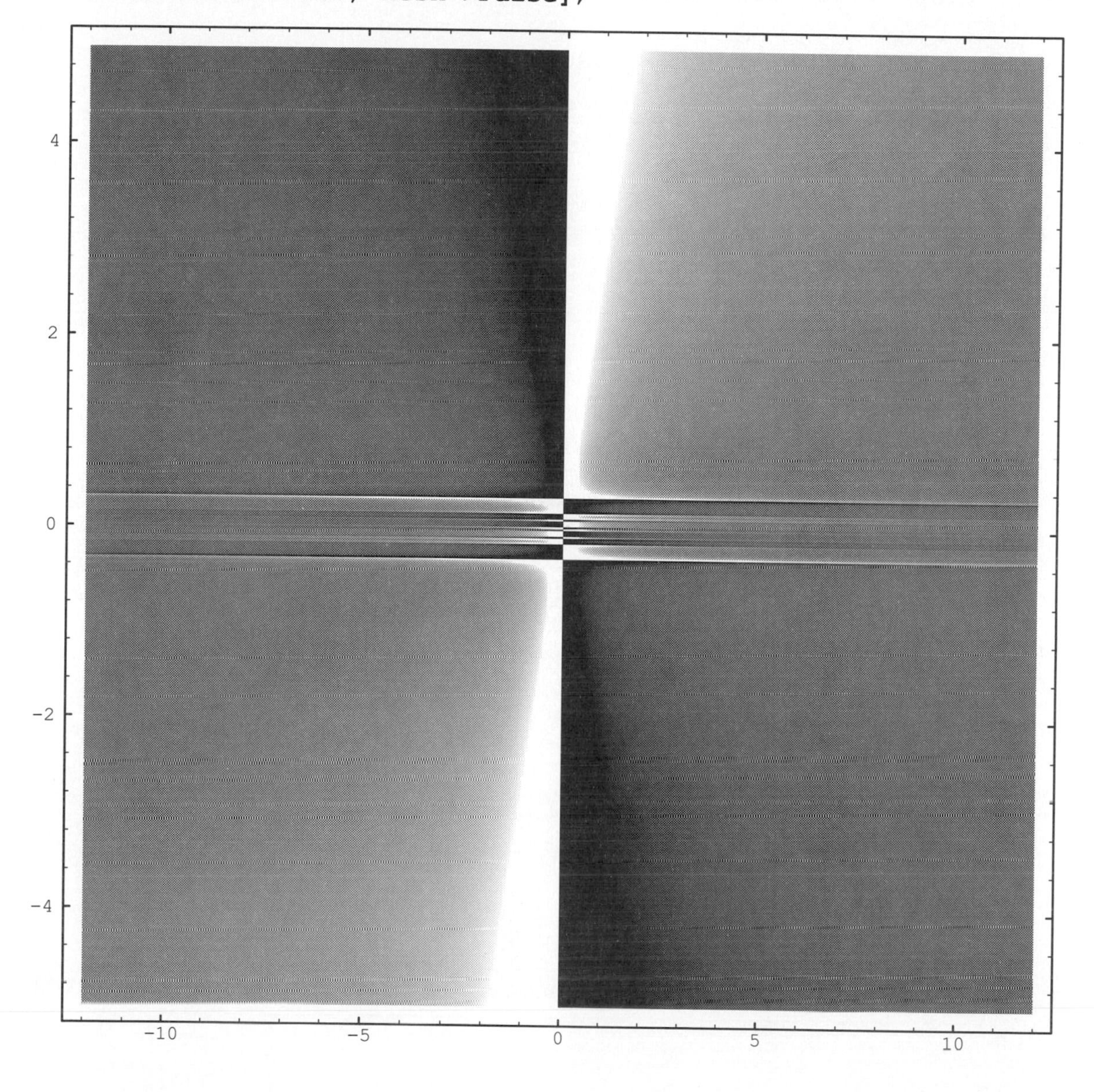

```
DensityPlot[1 / Sin[x y], {x, -12, 12}, {y, -12, 12},
PlotPoints->500, Mesh->False];
```

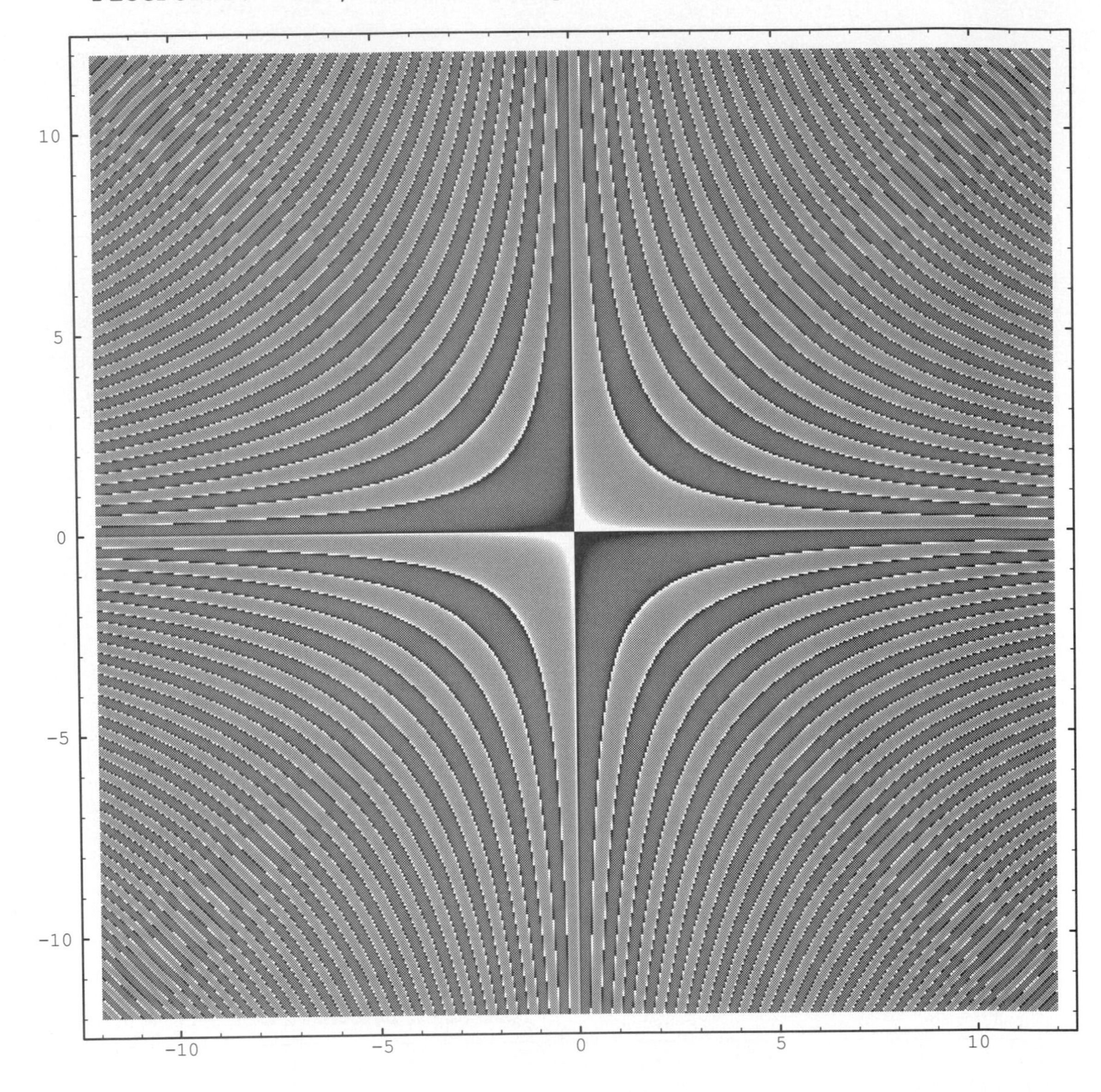

```
DensityPlot[1 / Sin[x/y], {x, -12, 12}, {y, -5, 5},
PlotPoints->500, Mesh->False];
```

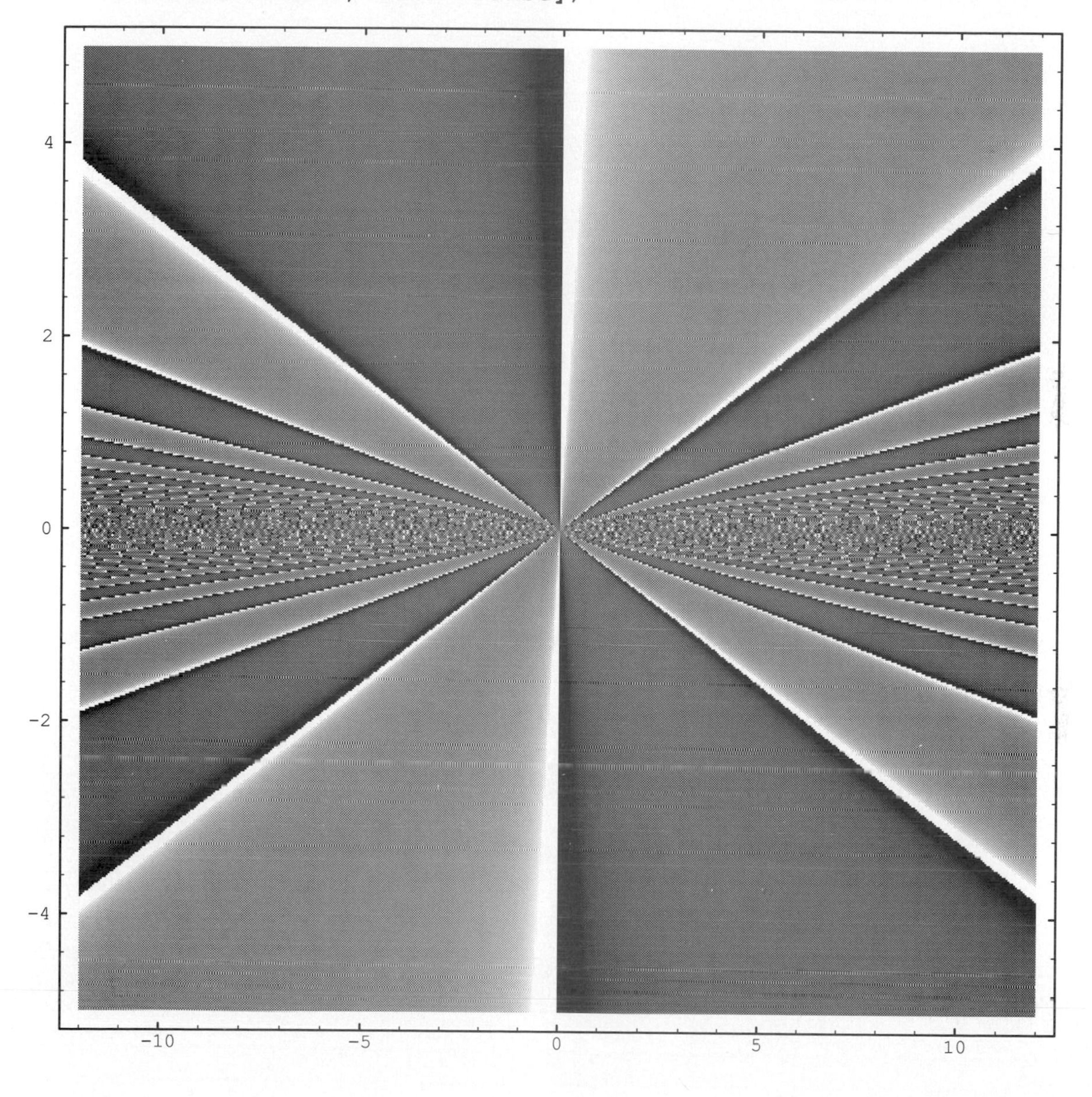

```
DensityPlot[1 / Sin[1/(x y)], {x, -2, 2}, {y, -2, 2},
PlotPoints->500, Mesh->False];
```

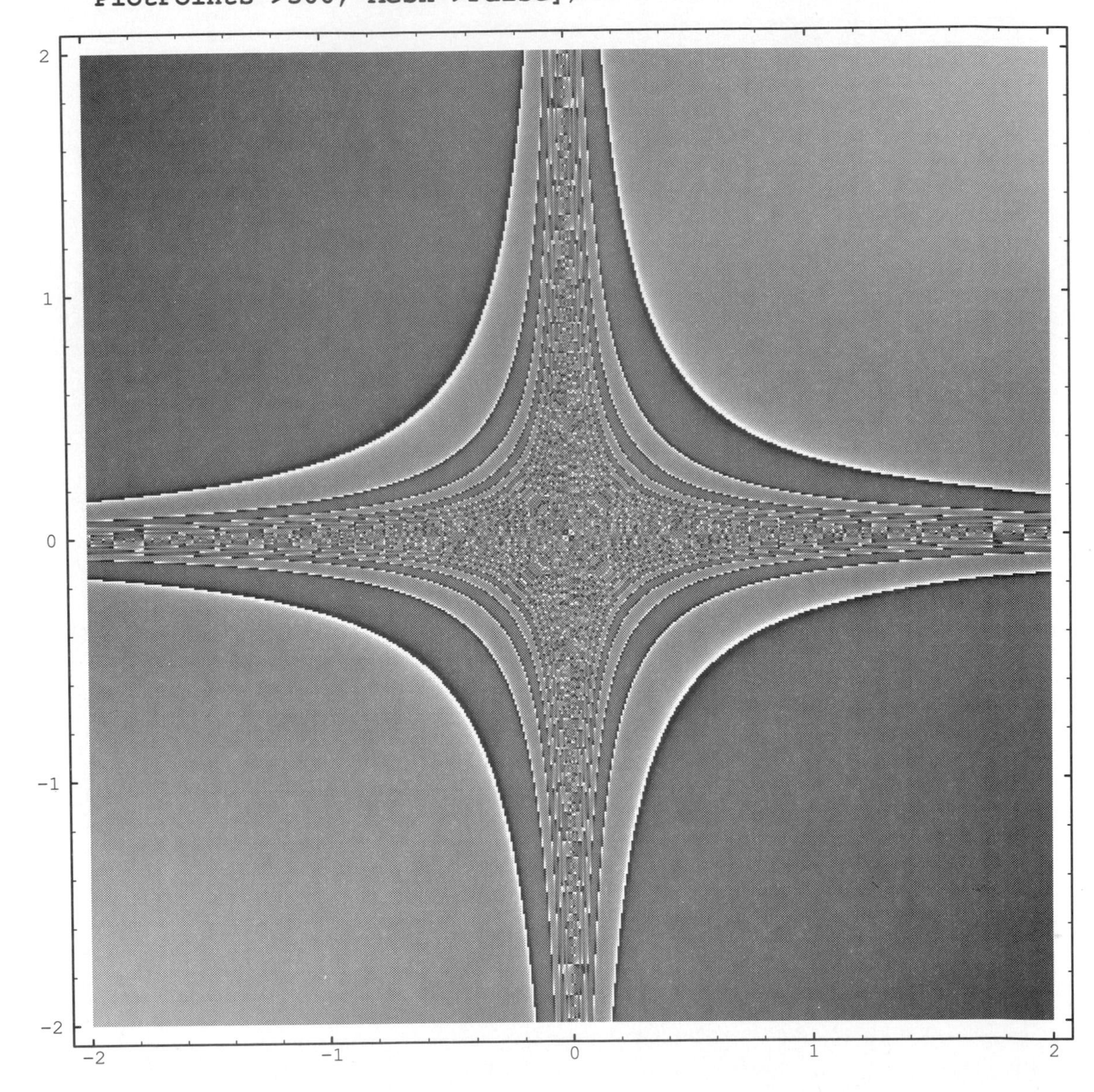

Theo: I think these are some of the most striking plots in the whole book, particularly when they are viewed on a good quality color monitor.

Jerry: I notice that they lose a lot of their effect in print. I always thought that graphics looked better in print than on screen, but this set seems to be a counterexample.

Theo: Yes it's true that most graphics look better when printed. That's because most graphics have lines and polygons in them. A high-resolution printer is able to draw lines without the "jaggies" that you see on screen. These plots, on the other hand, consist entirely of smooth, continuously varying gray levels.

Printers are not very good at rendering smooth gray levels, because they have to use half-tone screens (meshes of differently-sized black dots) to represent different levels of gray. A typical color monitor, on the other hand, can display any of 256 different levels of gray in any pixel. This allows the monitor to display very smooth, continuous variations of gray.

Jerry: Why do you keep saying "color monitor" in the same sentence with "gray levels"? Are we talking about color or what?

Theo: Well, as far as monitors are concerned, gray levels are colors too. A typical color monitor can display 256 different colors, and you get to choose which colors those are. When this Notebook is displayed on such a monitor, *Mathematica* automatically chooses to have all those colors be different levels of gray.

Better monitors (known as "24-bit" or "32-bit" monitors) are able to display 16 million different colors in any pixel, but for plots likes the ones in this chapter this doesn't make much difference, since the human eye can't detect more than about 256 different gray levels anyway.

There are also gray-level-only monitors, which can't display colors. Typically these work just like color monitors, except that your only choice is to have levels of gray as your "colors".

Jerry: What happens if you don't have such a fancy (or, as you put it, "typical") color monitor? How are these levels of gray represented on a Mac Plus monitor?

Theo: If you have fewer than 256 different levels of gray available, *Mathematica* uses "dither patterns" to approximate the grays. This is similar to what a printer does. By using patterns of black and white dots, you can approximate any gray level you like, but it doesn't look nearly as good as when you actually draw that gray level.

Some monitors can display 4 or 16 levels of gray, and on those monitors it is possible to construct mixtures of the closest available gray levels to approximate the desired one. This generally looks better than black-and-white, but not as good as 256 colors.

Jerry: Well, it certainly is nice to see these plots on a color monitor. I don't think I would be nearly as impressed if I saw them only on paper.

Theo: It's a good thing this book comes with an electronic copy, so that everyone has the opportunity to see the pictures on screen.

■ References

1) The Seven Faces of Sin, Theodore W. Gray, *The Mathematica Journal*, Summer 1990, Volume 1, Number 1, p36.

Chapter Thirteen
Making Posters

Theo and Jerry find out what fast printers are really *for.*

Dialog

Jerry: I remember some late nights before *Mathematica* was introduced to the world when large posters were being assembled from laser-printed sheets generated by *Mathematica*. It seemed like a good idea at the time, since many people hadn't seen these 3-D surfaces, and certainly not as *big* as those posters. Can we show people how to make their own posters?

Theo: Of course. For the *Mathematica* product announcement I wanted to have the largest stellated icosahedron in the world, so I put in the necessary features to allow *Mathematica* to print it out. It's very simple: *Mathematica* prints out many sheets of paper, each with a small area of your poster on it. You glue the sheets together.

Jerry: Do I need a special set up, or lots of memory, or a special printer to make posters?

Theo: No, *Mathematica* and a regular Macintosh with 4MB of memory, or a standard NeXT computer, is all you need. Any PostScript-compatible printer will do. Non-PostScript printers (dot-matrix printers, and some laser printers) may have more trouble with very large posters, so it's best to stick with PostScript printers.

Jerry: Does the construction of posters require tremendous technical skill, patience, and time?

Theo: It does require quite a bit of time for very large posters, and a certain amount of skill in cutting and gluing sheets of paper. In Appendix P I describe in complete detail the exact procedure for making posters of any size.

Jerry: I've read your description, and it radiates an air of natural authority that can come only from experience. One or two questions come to mind: Why are we doing so much cutting before we paste? Why not use the edges of the paper, which are very accurately cut, rather than an edge we cut ourselves?

Theo: Unfortunately, there are two reasons we can't just glue the pages edge-to-edge onto a backing. First, laser printers don't print in an accurate position on the page. The print can shift by as much as 1/8 of an inch from page to page. Second, laser printers can't print right up to the very edge of the paper. Typically there is at least a 3/16 inch border than can't be printed. There are various reasons for this. Printers need to grip the sheet of paper, and they typically do this with two rollers on the outside edges of the paper. Also, since they can't print in an accurate location on the page, if they tried to print right up to the edge, they would probably miss anyway. Dot matrix printers dare not print right up to the edge because if they went off the edge their print wires could be damaged by the lack of paper to run into.

So it is necessary to trim each page to the exact edge of the printed area. This generally doesn't take as long as the gluing phase anyway.

Jerry: Could you recommend a project for first-time poster makers?

Theo: I would suggest the tried-and-true stellated icosahedron. The following commands will generate this figure:

```
Needs["Graphics`Polyhedra`"];

Show[Graphics3D[Stellate[Icosahedron[]]]];
```

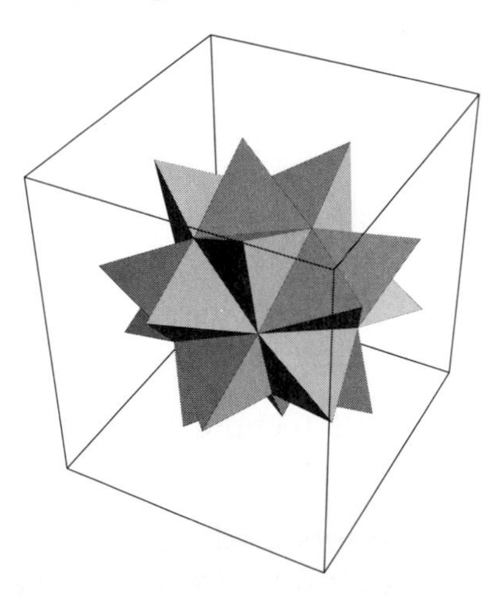

This looks fine as a 3x2 page or a 4x5 page poster. Appendix P describes how to proceed from here. Many of the graphics in this book will work well as posters, or you can make up your own. You can also use a similar construction technique to turn animations into long strips of paper. Just print one frame per page, and glue the pages side by side.

Chapter Fourteen
Fourier for Sound

Jerry and Theo explain the Fourier transform as it applies to sounds.

■ Dialog

Jerry: Plot four or five random points for me!

Theo: OK. I'll make a list of some random points.

```
theList = {{1, 0.794}, {2, 0.833}, {3, 0.019},
           {4, 0.269}, {5, 0.964}}
{{1, 0.794}, {2, 0.833}, {3, 0.019}, {4, 0.269},
  {5, 0.964}}
```

We can plot the points:

```
plot = ListPlot[theList];
```

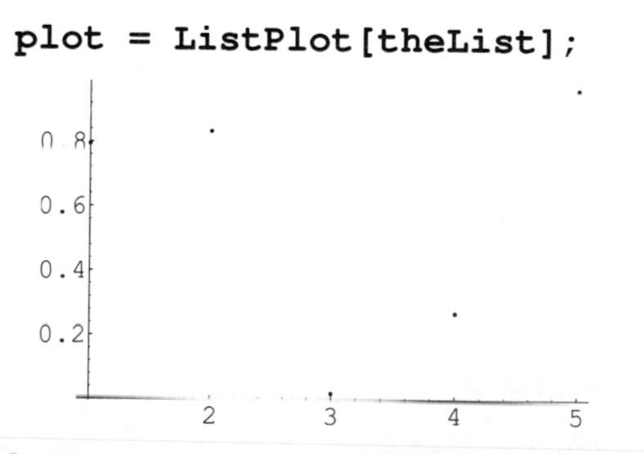

Jerry: Suppose we want a graph to go through these points. I'd like to try this in a very primitive and intuitive way. I'll start with a sine curve and then adjust it to get as close as I can by eye. Let's put in `Sin[x]` on top of that:

```
Plot[Sin[x], {x, 0, 2Pi},
        Evaluate[AddPoints[theList]]];
```

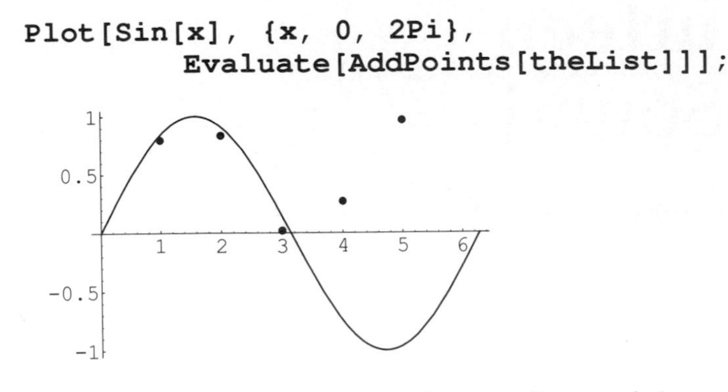

It looks to me that we come quite close on three of the points, and miss two of the points by quite a lot. The question is, how do I adjust **Sin[x]** to come closer to all the points. What do you think?

Theo: I would suggest a Fourier transform, but you probably wouldn't like that. Instead we could try doing a least-squares fit to the points, using sine functions at different frequencies as our fitting functions.

Jerry: These are nice ideas, but they will undermine my attempt to do this in a "primitive and intuitive" way.

Theo: Well, I think you will find that fitting sine functions to lists of points is not a particularly intuitive thing to do, except in hand-picked cases. Let's try fitting two sine functions (plus a constant) to these points.

The **Fit** function takes three arguments: first the list of data points we want to fit, second a list of the individual terms whose coefficients we want to calculate, and third the name of the variable used in those terms.

In this example, we use **{1, Sin[x], Sin[2x]}** as our second argument, which means we want to fit a function of the form:

```
a 1 + b Sin[x] + c Sin[2x]
```

Fit is going to tell us the best values for **a**, **b**, and **c**.

```
Fit[theList, {1, Sin[x], Sin[2x]}, x]
0.581806 + 0.0570782 Sin[x] - 0.125882 Sin[2 x]
```

```
Plot[%, {x, 0, 2Pi}, PlotRange -> {-1, 1},
        Evaluate[AddPoints[theList]]];
```

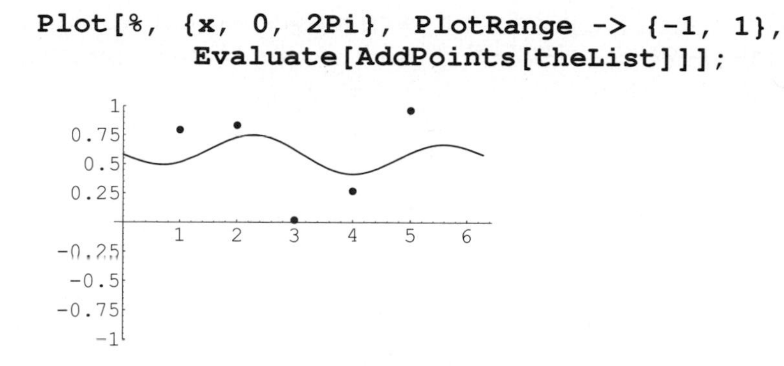

Jerry: It looks like a good first try. I can think of many ways to add more functions to improve our fit. We might try fitting to `{1, Sin[x], Sin[2x], Sin[4x], Sin[8x]}`, or the same thing for powers of 3, or some mixture of sines and cosines. And, what about tangents?

Theo: NO! Let's try to maintain some sort of dignity.

Jerry: That's the trouble with young people.

All right, let's just try the following:

```
Fit[
    theList,
    {1, Sin[x], Sin[2x], Sin[3x], Sin[4x], Sin[5x]},
    x
]
0.536598 - 0.133158 Sin[x] + 0.181244 Sin[2 x] +
  0.0625401 Sin[3 x] + 0.324963 Sin[4 x] -
  0.460676 Sin[5 x]
```

```
Plot[%, {x, 0, 2Pi}, PlotRange -> {-1.5, 1.5},
        Evaluate[AddPoints[theList]]];
```

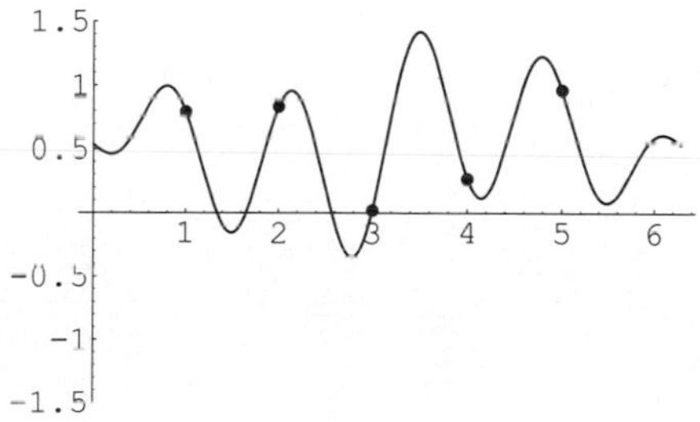

Theo: This curve does go through the points, but even a five-year-old would say it's much too wiggly a way to connect those points. I think we might do better if we add in some cosines instead of more sines:

```
Fit[
    theList,
    {1, Sin[x], Sin[2x], Cos[x], Cos[2x]},
    x
]
0.578062 + 0.281948 Cos[x] - 0.338676 Cos[2 x] +
  0.0537554 Sin[x] - 0.134799 Sin[2 x]

Plot[%, {x, 0, 2Pi}, PlotRange -> {-1, 1},
        Evaluate[AddPoints[theList]]];
```

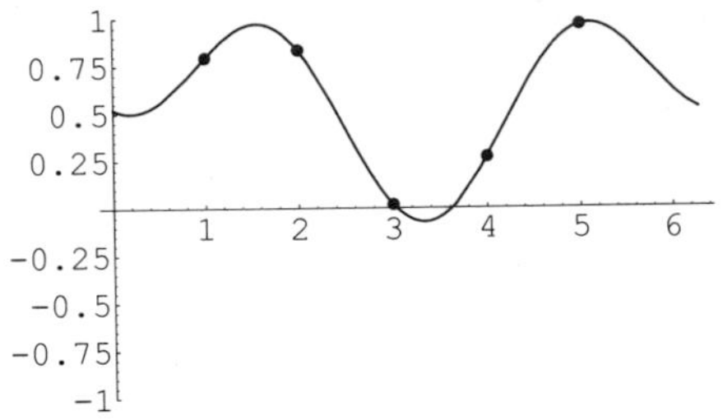

```
1
0.75
0.5
0.25
   1  2  3  4  5  6
-0.25
-0.5
-0.75
-1
```

This goes through all the points, and it's also nice and smooth. It looks just about right.

Jerry: OK, so now you've done it automatically. Can we try to understand why **`Fit`** gave us the coefficients it did?

Theo: I don't think so. I claim that selecting the coefficients in front of the sine and cosine functions is not an intuitive business. Let's plot the individual functions used to build up this fit:

```
Plot[{
        0.57806,
        0.28194 Cos[x],
        -0.33867 Cos[2 x],
        0.05375 Sin[x],
        -0.13479 Sin[2 x]
    },
    {x, 0, 2Pi},
    PlotRange -> {-1, 1},
    Evaluate[AddPoints[theList]]
];
```

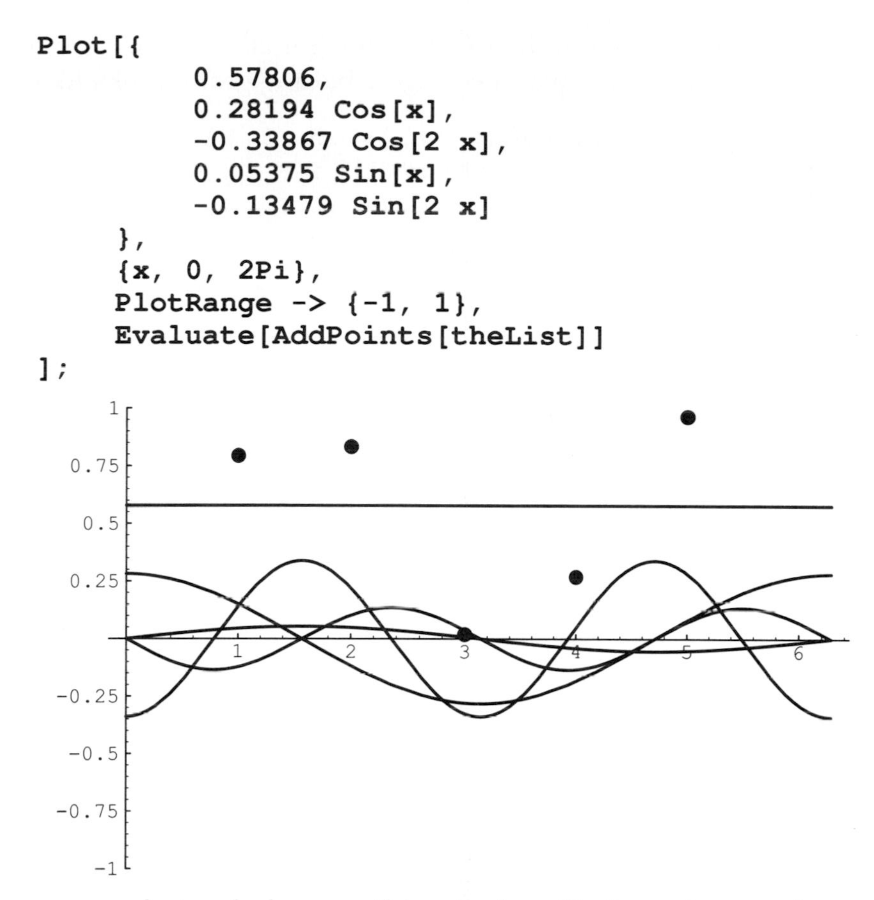

These curves obviously have nothing to do with the points, and yet when added together the sum passes through the points.

Jerry: I think you're being too pessimistic. Couldn't we get the best fit for each of the terms individually (which is very intuitive), and then add together these individual best fits?

Theo: Hmm... Well we could try, although I'm really quite sure it won't work.

This is the function we get by applying **Fit** to each term individually:

```
Fit[theList, {1.0}, x] +
    Fit[theList, {Sin[x]}, x] +
    Fit[theList, {Sin[2x]}, x] +
    Fit[theList, {Cos[x]}, x] +
    Fit[theList, {Cos[2x]}, x]
0.5758 + 0.0825233 Cos[x] - 0.75845 Cos[2 x] +
  0.0985434 Sin[x] - 0.0625055 Sin[2 x]
```

The coefficients are quite different from the function we got above, when we did a single fit to all the terms at the same time. Let's see what the graph looks like:

```
Plot[%, {x, 0, 2Pi}, PlotRange -> {-1.5, 1.5},
        Evaluate[AddPoints[theList]]];
```

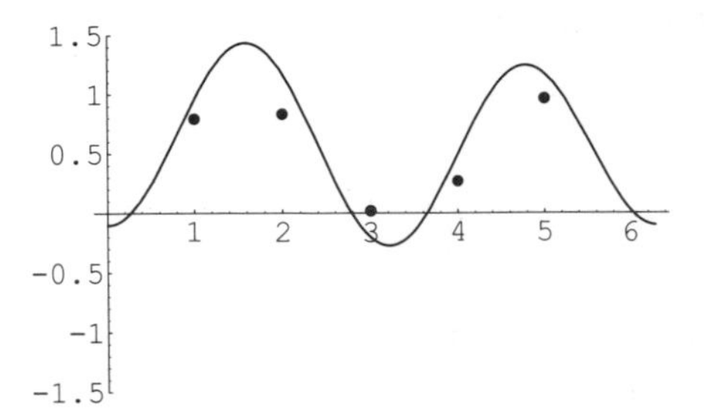

Jerry: Not very good. So, if I try to fit the terms separately to the points, I get poor answers, and the sum of these poor answers is a poor answer. In the earlier case, we got a sum which was an excellent answer, but when we plotted the pieces separately, they were terrible as fits. This could be a definition for "counterintuitive". Very persuasive.

Theo: Now that we've given up on this intuitive nonsense, we could get on to Fourier transforms.

Jerry: I remember looking at the description of the **Fourier** function in the *Mathematica* book. It didn't help me understand it.

Theo: Nor me. Fortunately, we have access to Jerry Keiper (one of the authors of *Mathematica*), who was able to provide us with a new, improved version of **Fourier** which gives its answers in a more intelligible form (see Appendix A).

The function **NicerFourier** takes a list of numbers and a variable name and returns a sum of **Sin** and **Cos** functions. Unfortunately, we have a list of points (x, y)–not a list of numbers. Furthermore, **NicerFourier** assumes that the list of numbers you give it is evenly spaced over a range of x values from 0 to 2Pi.

Fortunately, I've defined a function (see Appendix A) that takes any list of points and redistributes their x values over the desired range. Let's make a new list from the old one:

```
newList = Resample[theList]

{{0, 0.794}, {1.25664, 0.833}, {2.51327, 0.019},
  {3.76991, 0.269}, {5.02655, 0.964}}
```

I've defined another function, **YValuesOnly**, that takes a list of points and returns a list of their y values. Here's what **YValuesOnly** does:

```
YValuesOnly[newList]
{0.794, 0.833, 0.019, 0.269, 0.964}
```

Now we can use the **NicerFourier** function on this list, with **x** as our variable:

```
NicerFourier[YValuesOnly[newList], x]
0.5758 + 0.446523 Cos[x] - 0.228323 Cos[2 x] -
  0.108614 Sin[x] + 0.0643057 Sin[2 x]
Plot[%, {x, 0, 2Pi}, PlotRange -> {-1, 1},
        Evaluate[AddPoints[newList]]];
```

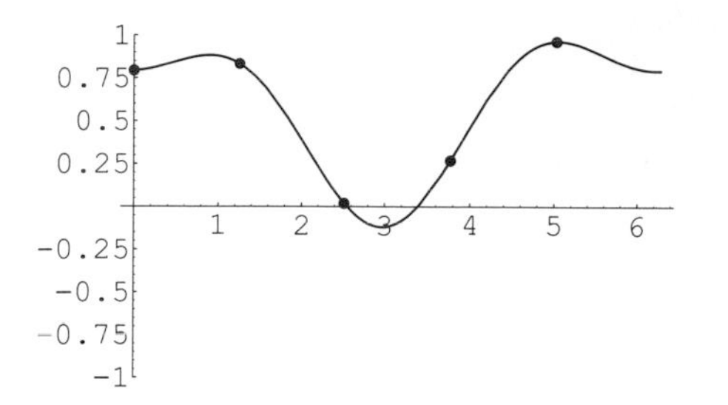

Jerry: Is this the same answer as we would get from **Fit**? Let's find out:

```
Fit[newList, {1, Sin[x], Sin[2x], Cos[x], Cos[2x]}, x]
0.5758 + 0.446523 Cos[x] - 0.228323 Cos[2 x] -
  0.108614 Sin[x] + 0.0643057 Sin[2 x]
```

So we get exactly the same answer from **Fit** as we did from **NicerFourier**. Who needs Fourier transforms if **Fit** does just as well?

Theo: In principle you don't, but you will find that, when you have a lot of data points, **Fit** will take an incredibly long time but **NicerFourier** will get the job done very efficiently.

Let's try **NicerFourier** on some numbers traced in using the graphical input feature. In the Notebook Front End, I selected a graph (like the one above) by clicking on it, held down the Command key (on a Macintosh or NeXT) or the Alt key (On an MS-DOS computer), clicked inside the graph, and dragged the mouse around while holding down the mouse button. This selected a series of points in the graph. Then I used the Copy command to copy the coordinates of the points into the clipboard, typed **theList2 =** into a new cell, and used the Paste command to paste the coordinates into the cell.

Here is the cell I got:

```
theList2 = {{0.0693451, 0.422046}, {0.170573, 0.521354},
{0.37303, 0.595836}, {0.575487, 0.670317},
{0.777943, 0.719972}, {0.9804, 0.744799},
{1.18286, 0.719972}, {1.38531, 0.670317},
{1.48654, 0.571009}, {1.53716, 0.4717},
{1.63838, 0.372391}, {1.63838, 0.273083},
{1.73961, 0.148947}, {1.79023, -1.61802*10^-05},
{1.99268, -1.61802*10^-05}, {1.99268, -0.0993249},
{2.09391, -0.198634}, {2.14453, -0.297942},
{2.34698, -0.273115}, {2.3976, -0.173806},
{2.60005, -0.0993249}, {2.80251, -0.0993249},
{3.00497, -0.148979}, {3.05558, -0.248288},
{3.15681, -0.347597}, {3.30865, -0.446905},
{3.46049, -0.546214}, {3.66295, -0.571041},
{3.86541, -0.546214}, {4.01725, -0.446905},
{4.21971, -0.347597}, {4.42216, -0.273115},
{4.574, -0.173806}, {4.67523, -0.0744977},
{4.87769, -1.61802*10^-05}, {5.02953, 0.0992925}};
```

Jerry: These points don't seem to be evenly spaced in the x direction. We won't be able to do the Fourier transform until they are adjusted so as to be evenly spaced between 0 and 2Pi. This could take weeks.

Theo: Ah, but you forget that I wrote a function, **Resample**, which automatically does this to any list of points. Here we go:

```
newList2 = Resample[theList2]

{{0, 0.422046}, {0.174533, 0.521354},
  {0.349066, 0.595836}, {0.523599, 0.670317},
  {0.698132, 0.719972}, {0.872665, 0.744799},
  {1.0472, 0.719972}, {1.22173, 0.670317},
  {1.39626, 0.571009}, {1.5708, 0.4717},
  {1.74533, 0.372391}, {1.91986, 0.273083},
  {2.0944, 0.148947}, {2.26893, -0.0000161802},
  {2.44346, -0.0000161802}, {2.61799, -0.0993249},
  {2.79253, -0.198634}, {2.96706, -0.297942},
  {3.14159, -0.273115}, {3.31613, -0.173806},
  {3.49066, -0.0993249}, {3.66519, -0.0993249},
  {3.83972, -0.148979}, {4.01426, -0.248288},
  {4.18879, -0.347597}, {4.36332, -0.446905},
  {4.53786, -0.546214}, {4.71239, -0.571041},
  {4.88692, -0.546214}, {5.06145, -0.446905},
  {5.23599, -0.347597}, {5.41052, -0.273115},
  {5.58505, -0.173806}, {5.75959, -0.0744977},
  {5.93412, -0.0000161802}, {6.10865, 0.0992925}}
```

The Fourier transform of this list is:

```
fitFunction[x_] = NicerFourier[YValuesOnly[newList2], x]
```

```
0.044121 + 0.263892 Cos[x] + 0.0477955 Cos[2 x] +
  0.00570191 Cos[3 x] - 0.0421527 Cos[4 x] +
  0.0268599 Cos[5 x] - 0.0020689 Cos[6 x] +
  0.0205209 Cos[7 x] - 0.00483375 Cos[8 x] +
  0.00689647 Cos[9 x] + 0.0115607 Cos[10 x] +
  0.00504275 Cos[11 x] + 0.00758603 Cos[12 x] +
  0.00918261 Cos[13 x] + 0.000642832 Cos[14 x] +
  0.00809095 Cos[15 x] + 0.00767693 Cos[16 x] +
  0.00139302 Cos[17 x] + 0.4646 Sin[x] +
  0.199329 Sin[2 x] - 0.0480275 Sin[3 x] +
  0.036602 Sin[4 x] + 0.0122735 Sin[5 x] +
  0.0155285 Sin[6 x] + 0.000159089 Sin[7 x] +
  0.0172027 Sin[8 x] + 0.00137931 Sin[9 x] +
  0.0107668 Sin[10 x] + 0.00565466 Sin[11 x] +
  0.0011944 Sin[12 x] + 0.00267302 Sin[13 x] +
  0.00136303 Sin[14 x] - 0.000247546 Sin[15 x] +
  0.00568507 Sin[16 x] - 0.00201664 Sin[17 x]
```

Jerry: Oh, look at that! What tells it how far to go? Why don't we have a **`Sin[96x]`** term?

Theo: Ah, well that's another one of the advantages of **`NicerFourier`** over **`Fit`**. **`NicerFourier`** knows that it needs as many degrees of freedom (terms) as there are sample points. Let's see how many points there are in the list:

```
Length[newList2]
```

```
36
```

In our Fourier transform, we have: one constant term, 17 sine terms, and 17 cosine terms, which adds up to 35—just about right.

Jerry: I think we need to plot the points and the expression, and see if they match up.

```
Plot[fitFunction[x], {x, 0, 2Pi},
          Evaluate[AddPoints[newList2]]];
```

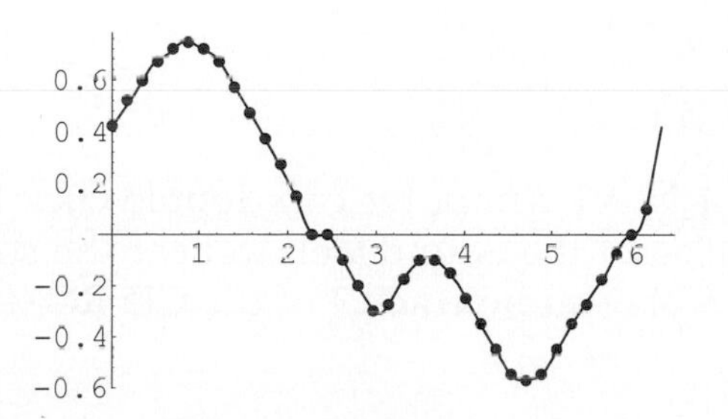

Jerry: Wonderful!

I notice that all the sine and cosine terms in this function have a period of 2Pi, or Pi, or Pi/2, etc. Does this mean that if we plot the function over a wider range of x values we will see the same shapes repeated?

```
Plot[fitFunction[x], {x, 0, 6Pi},
        Evaluate[AddPoints[newList2]]];
```

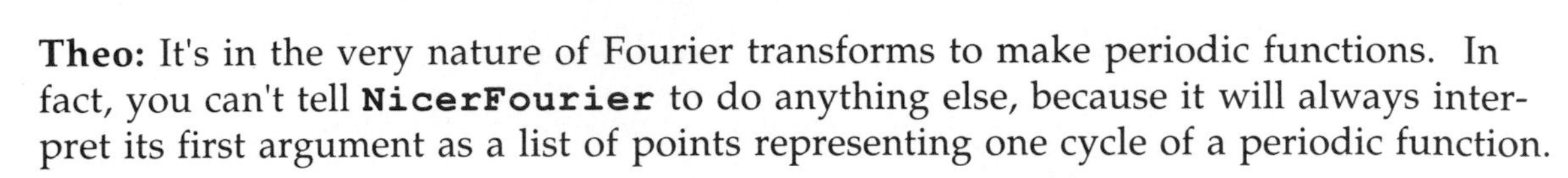

Theo: It's in the very nature of Fourier transforms to make periodic functions. In fact, you can't tell **NicerFourier** to do anything else, because it will always interpret its first argument as a list of points representing one cycle of a periodic function.

Jerry: Looking at the plot of three cycles of our function, I am struck by how much it looks like a sound wave. Can we hear it?

Theo: What do you think this is, some kind of cheap pocket calculator? Of course we can hear it! The following command generates a few seconds of sound from this waveform:

```
Play[fitFunction[2000 t], {t, 0, 1}];
```

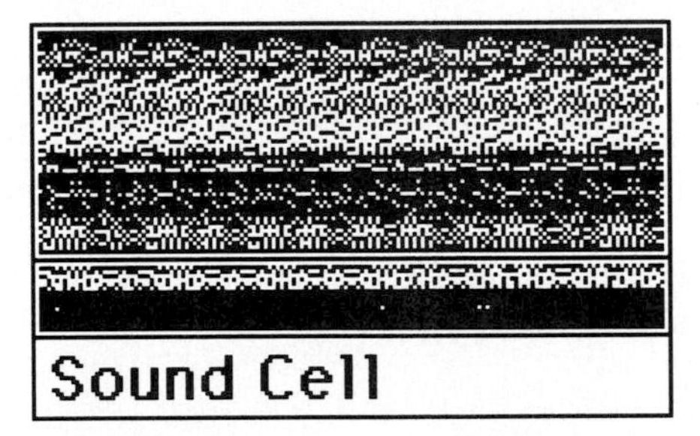

(Play track 3 of the CD to hear this sound.)

People reading this book on a Macintosh or NeXT computer can double-click the small sound icon at the top of the Cell bracket of the Sound Cell to hear the sound. Others can use a regular audio CD player to play audio track 3 of the CD-ROM in the back of their copies of the book.

Jerry: This is amusing, but I'd like to see (and hear) something more pleasant. I'd like to see the waveform for middle C on your piano.

Theo: Well, we can try, though my piano hasn't been tuned lately...

I don't know if the cord on my microphone will reach far enough.... OK, it seems to have worked. I recorded the sound on the Macintosh using Farallon's MacRecorder device.

Here's the recorded note, copied from SoundEdit and pasted into *Mathematica*. (Readers of the electronic version can double-click the sound icon at the top of the cell bracket to hear the note. Others can play track 4 of the CD-ROM version on their audio CD player.)

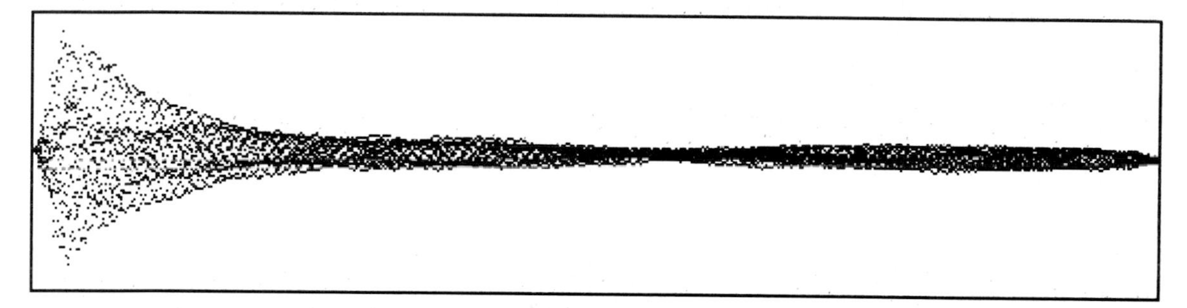

Using the SoundEdit program, I expanded the recorded waveform until I could see the individual cycles. I selected one cycle, copied it from SoundEdit, and pasted it into the following cell:

Then I used the Convert To InputForm command in the Notebook Front End to convert the sound into *Mathematica* commands. After a bit of editing to extract the list of numbers representing the waveform, I got the following cell:

```
theList3 = {0.25781, 0.24219, 0.17188, 0.09375, 0.03906, -0.02344,
-0.09375, -0.14844, -0.15625, -0.12500, -0.08594, -0.04688, 0.00000,
0.05469, 0.12500, 0.17188, 0.17969, 0.12500, 0.02344, -0.10156,
-0.19531, -0.25781, -0.26562, -0.22656, -0.14844, -0.06250,
0.03125, 0.14844, 0.24219};
```

Jerry: I'm incredibly impressed by what you just did. You took two different programs, SoundEdit and *Mathematica*, from two different companies, and made them look like they were custom built to work together. How is this possible?

Theo: This kind of cooperation is typical in the Macintosh and NeXT worlds. Because a central dictatorship (Apple and NeXT respectively) established and enforced standards for graphics, sound, and other interchange formats, almost all applications on those computers are able to communicate with each other in sensible (not to mention powerful) ways. MS-DOS users, eat your hearts out.

Jerry: I knew you'd get that in somewhere.

Anyway, these numbers are very interesting, but they don't look like a sound to me. Can we use **`ListPlot`** to plot them?

```
ListPlot[theList3, PlotJoined -> True];
```

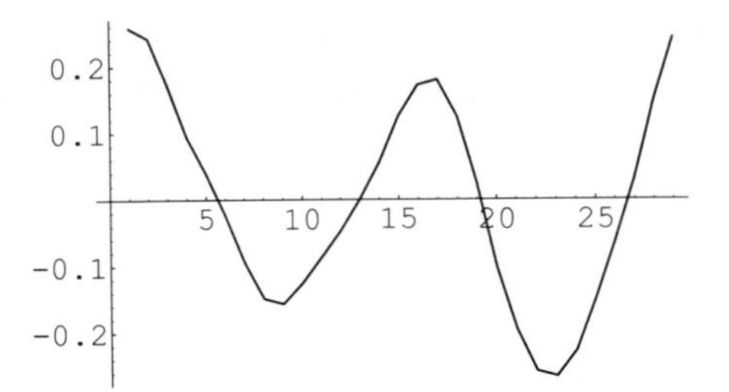

Good. How about some sines and cosines? I don't see any here.

Theo: OK, OK, give me a chance!

Now we can do the same thing to this list as we did to the others, starting with a Fourier transform:

```
fitFunction3[x_] = NicerFourier[theList3, x]
```

```
-0.0010769 + 0.0414294 Cos[x] + 0.203635 Cos[2 x] +
  0.005813 Cos[3 x] + 0.00126413 Cos[4 x] +
  0.00979687 Cos[5 x] + 0.00705703 Cos[6 x] +
  0.00117221 Cos[7 x] - 0.00246574 Cos[8 x] -
  0.00285869 Cos[9 x] - 0.00142169 Cos[10 x] -
  0.00218023 Cos[11 x] - 0.00118717 Cos[12 x] -
  0.000229629 Cos[13 x] - 0.000938074 Cos[14 x] +
  0.0345634 Sin[x] + 0.0368276 Sin[2 x] -
  0.0344975 Sin[3 x] + 0.010888 Sin[4 x] -
  0.0112036 Sin[5 x] + 0.000466984 Sin[6 x] +
  0.000474962 Sin[7 x] - 0.00107652 Sin[8 x] -
  0.000306537 Sin[9 x] + 0.0000582713 Sin[10 x] +
  0.0000982184 Sin[11 x] - 0.0000709458 Sin[12 x] +
  0.000180763 Sin[13 x] - 0.00120728 Sin[14 x]
```

Jerry: Can we graph a few cycles of this, to see what it looks like?

```
Plot[fitFunction3[x], {x, 0, 6Pi},
                    AspectRatio -> 0.33];
```

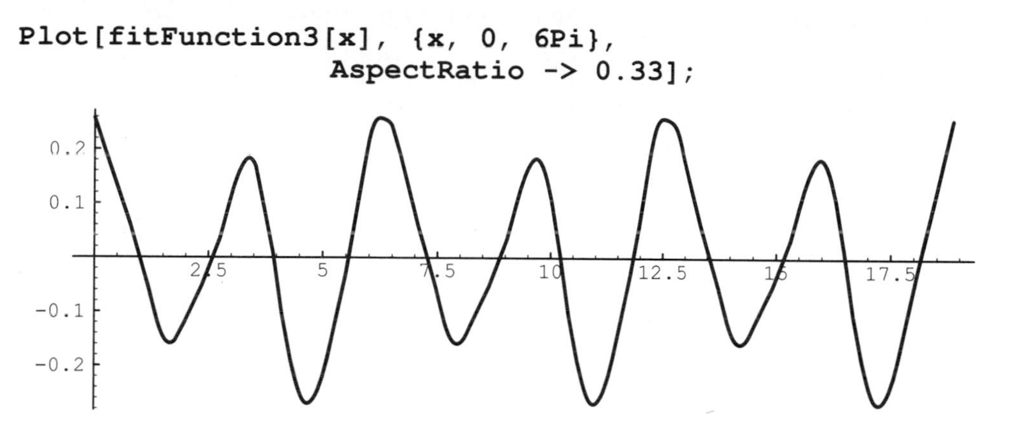

Is this graph exactly three cycles of the wave that was copied from SoundEdit of middle C on the piano?

Theo: Right. Of course, the waveform from the piano varied over the duration of the note, and this plot is just repeating one cycle picked from the middle somewhere, so it won't sound exactly like the real piano. In fact, a real piano has three strings for each of most of its notes (67 out of 88 on a grand piano, according to Joyce Mast), and each string is tuned to a slightly different frequency. We can't duplicate the subtle interplay of frequencies by repeating only a single cycle. We're making a sort of poor man's piano.

Jerry: Yes, and using about 10 million dollars' worth of computer to do it.

Theo: That's the price of using a general-purpose program.

Anyway, we can now produce a snippet of simulated piano, using this waveform:

```
Play[fitFunction3[2000 t], {t, 0, 1}];
```

Sound Cell

(Play track 5 of the CD to hear this sound.)

Jerry: Not only does this *not* sound like a piano, it doesn't even have the decency to fade out like a real piano. Can we at least get it to do that?

Theo: Looking at the outline of the note we pasted in above, we probably want a sharp rise in the loudness, followed by some sort of exponential decay. Let's try defining an envelope function to mimic this:

```
twoPi = N[2Pi];

envelope[x_] := Which[
        x < 5 twoPi, x / (5 twoPi),
        True, E^((5 twoPi - x)/(25 twoPi))
    ]
```

Jerry: What's **Which**?

Theo: No, **Who**'s on first. Oh, excuse me.

Which is a function which (OK--*that*) is very useful for defining pieced-together functions. **Which** takes a series of pairs of arguments, first a test condition, then a value. It returns the value corresponding to the first test that returned **True**. In our case, we want a function that rises linearly, then starts falling off exponentially. So, the first test is **x < 5 twoPi** (in other words, for the first 5 cycles of the waveform), and the first value is **x / (5 twoPi)**, which starts at zero and goes up to 1 just when the first limit is reached. The second test is **True**, which means always, and the second value is an exponential decay that starts at 1 and decays slowly down to near zero at the end of 100 cycles. Here's a plot of the envelope function:

```
Plot[envelope[2000 t], {t, 0, 1}];
```

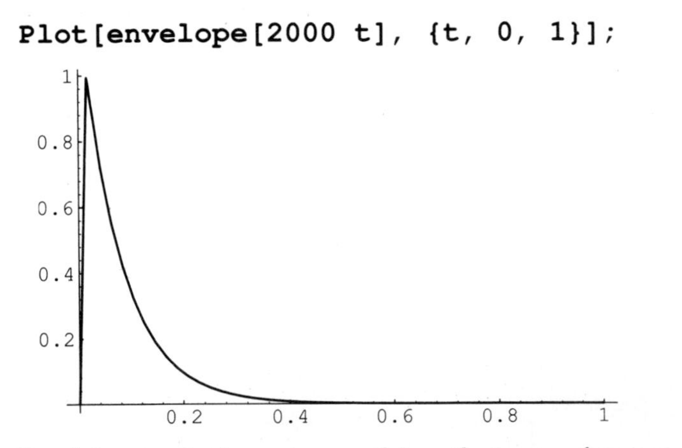

Jerry: So, this graph shows something that goes from zero to 1 quickly, in a direct manner, and then gradually fades away to zero, more quickly at first and more slowly later on. You've created this function to modify the amplitude of the single cycle, which is being repeated, to make a sound more like the original piano sound.

Theo: Right. You can use **Which** to define more complicated piece-wise-defined functions just by adding more (test, value) pairs.

All we have to do is to multiply the waveform function by the envelope function to get a new sound:

```
Play[fitFunction3[2000 t] envelope[2000 t],
     {t, 0, 1}, PlayRange -> All];
```

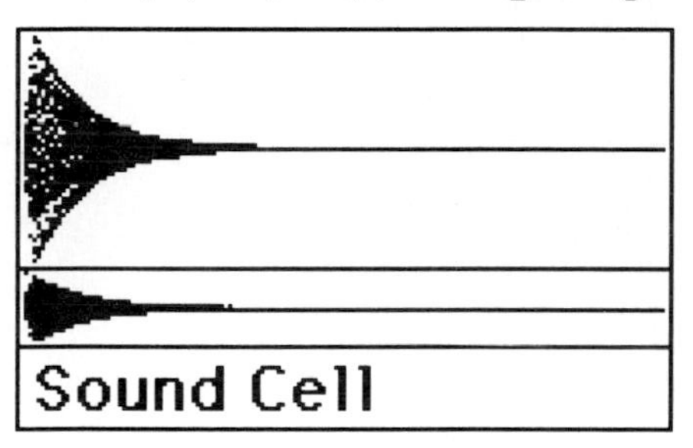

(Play track 6 of the CD to hear this sound.)

Jerry: Well, if you think that sounds like a piano, I'm glad you don't work in the music business.

Theo: The problem is that we can't do a good job simulating a piano by repeating one cycle many times. That's one reason why electronic "pianos" sounded so bad at first -- they tried to do exactly that. We have to do something a little more sophisticated.

Using the same recording of middle C, I've copied and pasted about a 1.5 second section from near the middle of the note:

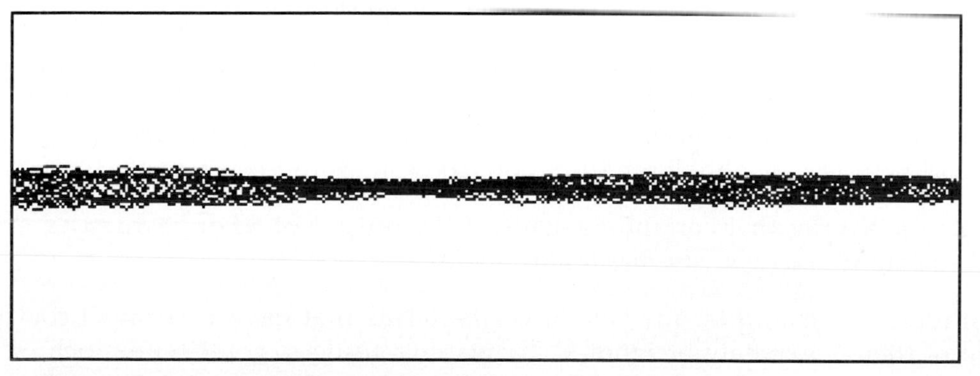

(Play track 7 of the CD to hear this sound.)

Jerry: So, instead of picking out just one cycle, you picked out what must be hundreds of cycles. Clearly this contains much more information about what the piano really sounds like.

Theo: That's the idea. We can convert the whole 1.5 seconds of sound into a *Mathematica* list, as we did the shorter sample before:

```
theList4 = {
-0.03125, -0.06250, -0.08594, -0.10156, -0.10938, -0.09375, -0.02344, 0.03125,
0.06250, 0.08594, 0.08594, 0.07031, 0.06250, 0.03125, -0.01562, -0.05469,
-0.07812, -0.07812, -0.09375, -0.10938, -0.10156, -0.03125, 0.04688, 0.06250,
0.04688, 0.05469, 0.06250, 0.03906, -0.00781, -0.04688, -0.06250, -0.08594,
-0.10938, -0.10156, -0.04688, 0.00781, 0.04688, 0.08594, 0.10156, 0.09375,
0.08594, 0.05469, 0.00781, -0.04688, -0.07031, -0.07812, -0.09375, -0.10938,
-0.10938, -0.04688, 0.03125, 0.06250, 0.04688, 0.04688, 0.06250, 0.04688,
0.01562, -0.03125, -0.05469, -0.07812, -0.10156, -0.10938, -0.06250, -0.01562,
0.02344, 0.06250, 0.08594, 0.08594, 0.07812, 0.06250, 0.02344, -0.02344,
-0.06250, -0.07031, -0.08594, -0.10938, -0.12500, -0.07812, 0.00000, 0.04688,
0.04688, 0.03906, 0.05469, 0.04688, 0.01562, -0.03125, -0.04688, -0.06250,
-0.09375, -0.10156, -0.06250, -0.02344, 0.02344, 0.06250, 0.07812, 0.08594,
0.07812, 0.07031, 0.03906, 0.00000, -0.03906, -0.05469, -0.06250, -0.08594,
-0.11719, -0.10156, -0.02344, 0.03906, 0.04688, 0.03906, 0.05469, 0.05469,
0.02344, -0.01562, -0.03906, -0.05469, -0.08594, -0.10938, -0.08594, -0.04688,
0.00000, 0.04688, 0.07031, 0.08594, 0.09375, 0.07812, 0.05469, 0.00000,
-0.04688, -0.07031, -0.07812, -0.09375, -0.13281, -0.12500, -0.04688, 0.02344,
0.03906, 0.04688, 0.05469, 0.05469, 0.03906, 0.00000, -0.03125, -0.03906,
-0.07031, -0.10938, -0.09375, -0.05469, -0.00781, 0.03125, 0.07031, 0.07812,
0.08594, 0.08594, 0.07031, 0.03125, -0.01562, -0.04688, -0.05469, -0.07031,
-0.11719, -0.12500, -0.07031, 0.00000, 0.03125, 0.03906, 0.04688, 0.05469,
0.04688, 0.00000, -0.03125, -0.03906, -0.07031, -0.10156, -0.09375, -0.07031,
-0.02344, 0.03125, 0.06250, 0.07812, 0.09375, 0.10156, 0.08594, 0.04688,
-0.00781, -0.05469, -0.05469, -0.06250, -0.10156, -0.13281, -0.09375, -0.02344,
0.02344, 0.03125, 0.03906, 0.05469, 0.05469, 0.01562, -0.02344, -0.03125,
-0.05469, -0.09375, -0.10938, -0.09375, -0.05469, -0.00781, 0.03906, 0.06250,
0.07031, 0.08594, 0.09375, 0.07031, 0.01562, -0.03125, -0.04688, -0.05469,
-0.09375, -0.13281, -0.10938, -0.03906, 0.01562, 0.03906, 0.04688, 0.06250,
0.07031, 0.03125, -0.00781, -0.02344, -0.04688, -0.07812, -0.09375, -0.08594,
-0.06250, -0.00781, 0.03125, 0.06250, 0.07031, 0.08594, 0.09375, 0.07812,
0.03125, -0.02344, -0.04688, -0.04688, -0.07812, -0.13281, -0.13281, -0.07031,
-0.00781, 0.02344, 0.03125, 0.05469, 0.07031, 0.03906, 0.00781, -0.00781,
```

This is just the smallest fraction of all the numbers: There are about 10000 in the total sample. The sound was sampled at a rate of about 7400 samples per second.

Now we could calculate a Fourier transform of this whole list, but rather than use the **NicerFourier** function we used above, I'd like to define a new function that will let us deal more effectively with this very large sample.

Instead of trying to recreate this sound exactly, let's try to extract the important elements of it and then recreate it (approximately) from a much smaller description. The output from **NicerFourier** would allow us to recreate it exactly, but the output would be unreasonably large (think of 32000 sines and cosines).

Jerry: You're saying that certain elements of the output of **NicerFourier** are not really important, and we can just ignore them?

Theo: Right. For example, any sine or cosine terms that have very small coefficients in front of them can safely be ignored: they aren't going to contribute much of anything to the final waveform. Also, if we play our waveform for many, many cycles, we don't really care if a term at a given frequency (period) is a sine or a cosine term, since this is just a phase shift, which is unimportant in the long run.

Jerry: Perhaps we want a function that combines the coefficients of each sine/cosine pair into a single number.

Theo: Right. Let's make a function that returns a list of pairs of numbers. The first number in each pair will be the frequency (this is the coefficient in front of the **`x`** in the **`Sin[n x]`** and **`Cos[n x]`** terms). The second number will be the square root of the sum of the squares of the coefficients of the corresponding sine and cosine term.

Jerry: Why not just the sum of the coefficients, or something simpler?

Theo: Think of it as the magnitude of a vector. The cosine and sine terms could be thought of as representing the x and y coordinates of a rotating vector. The coefficients define the shape of the ellipse that the vector traces out. We are interested in the length of that vector, which is the square root of x squared plus y squared.

Jerry: OK, that makes sense -- the Pythagorean theorem. So what is this new function called?

Theo: First, let's take care of one more complication. So far, we've been talking about frequency and period somewhat loosely. We have terms of the form **`Sin[x]`**, **`Sin[2x]`**, **`Sin[3x]`**, etc. Clearly these are sine waves at different frequencies, but what does that really mean? When we are talking about sounds, there is some base frequency, say 440 Hz (cycles per second) for the A above middle C. If we had a list of numbers representing one full cycle of this note, and we did a Fourier transform of the list, the **`Sin[x]`** terms would represent a sine wave at a frequency of 440 cycles per second. The **`Sin[2x]`** terms would represent a sine wave at 880 cycles per second (one octave higher), and so forth.

On the other hand, if we had a list of numbers representing two full cycles of the same note, and we took a Fourier transform of that list, the **`Sin[x]`** terms would represent a frequency of only 220Hz, while the **`Sin[2x]`** terms would be the 440Hz terms.

In general, anytime we do a Fourier transform of some sound samples, we have to re-scale the frequency depending on the length of time our sample represents. The most convenient way of specifying this is usually to say how many samples per second there are.

I've made a function, **`FrequencyAmplitudes`**, that takes two arguments, first a list of sample values, and second a number giving the sampling rate (number of samples per second). It returns a list of pairs of numbers, as described above.

Jerry: OK, let's use it on our sound!

Theo: Right, we're ready now. The sampling rate might look a bit funny, but it's what our sampler used.

```
amplitudes = FrequencyAmplitudes[theList4, 7418.1818];
```

Jerry: How come we don't see any output? I was looking forward to all those numbers!

Theo: There are such a lot of them that I thought it would be better to go directly to a plot (the semi-colon at the end of the statement suppressed the output, which is why we didn't see it). If you like, we can look at the first 10 terms:

```
Take[amplitudes, 10]
{{0.653009, 0.0000933983}, {1.30602, 0.00010179},
  {1.95903, 0.000052522}, {2.61204, 0.000075075},
  {3.26504, 0.000131353}, {3.91805, 0.0000355669},
  {4.57106, 0.0000752017}, {5.22407, 0.0000426629},
  {5.87708, 0.0000810404}, {6.53009, 0.0000668318}}
```

Jerry: This says that the magnitude of the coefficients of **`Sin[x]`** and **`Cos[x]`**, representing a frequency of **`0.653009`** Hz, is **`0.0000933983`**, etc.

Theo: Right.

We haven't mentioned the main reason that people do Fourier transforms in the first place: spectrograms. A spectrogram is a plot of basically what we've got in our list of amplitudes. The horizontal axis is frequency, and the vertical axis is amplitude at each frequency.

I've written a function to make these spectrogram plots automatically (we could use **`ListPlot`** directly on this list and get something very similar). My function makes somewhat better-looking plots. Here's what we get:

```
SpectrogramPlot[amplitudes, PlotRange -> All];
```

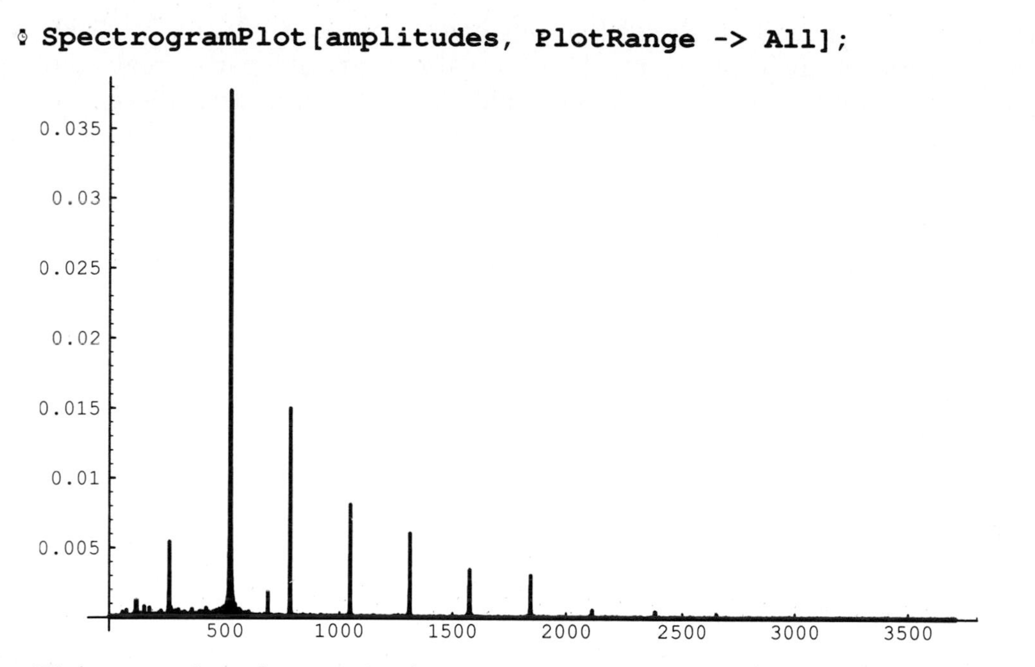

Jerry: There are a lot of points in this plot. Clearly most of the action is in the left half. Perhaps we should plot just that part.

Theo: I agree, let's plot only the first 4000 terms:

```
SpectrogramPlot[Take[amplitudes, 4000],
                PlotRange -> All];
```

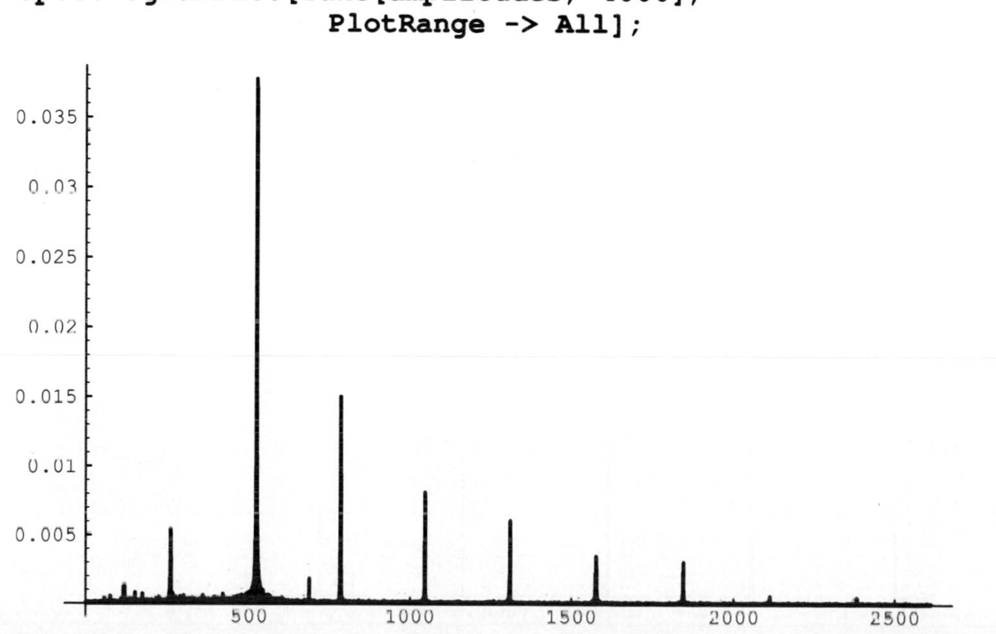

Jerry: Very pretty indeed. I can see that the lowest peak is a little above 250Hz, which makes sense for middle C at 263Hz. There are also regularly spaced peaks that must represent the various harmonics of the sound. It's interesting that the first overtone is much stronger than the base frequency.

Theo: Yes, it is. There are also thousands of points in between those peaks that are obviously of little use to us. Let's pick out just the highest of the peaks, say all those taller than 0.004, to pick a number. We can use the **`Select`** function to do this automatically:

```
pianoHarmonics =
    Select[amplitudes, (#[[2]] > 0.004)&]
```

```
{{260.551, 0.00545618}, {261.204, 0.00521308},
  {519.142, 0.00464671}, {519.795, 0.00467448},
  {520.448, 0.00563921}, {521.101, 0.00988677},
  {521.754, 0.0377572}, {522.407, 0.0126219},
  {523.06, 0.0370961}, {523.713, 0.0154129},
  {524.366, 0.010757}, {525.019, 0.00664962},
  {525.672, 0.00574432}, {526.325, 0.00495422},
  {526.978, 0.00410967}, {782.958, 0.00725495},
  {783.611, 0.0150055}, {784.264, 0.0119608},
  {784.917, 0.00937931}, {1045.47, 0.00452438},
  {1046.77, 0.00811399}, {1047.43, 0.00750902},
  {1310.59, 0.00607564}, {1311.24, 0.00495867}}
```

Jerry: Are you claiming that, out of the thousands of terms in **`amplitudes`**, only these few are really interesting?

Theo: Well, I'm claiming that we should, with luck, be able to get a reasonable approximation to the piano sound using only these terms. Let's look at the spectrogram of these terms:

```
SpectrogramPlot[pianoHarmonics,
                    AxesOrigin -> {200, 0}];
```

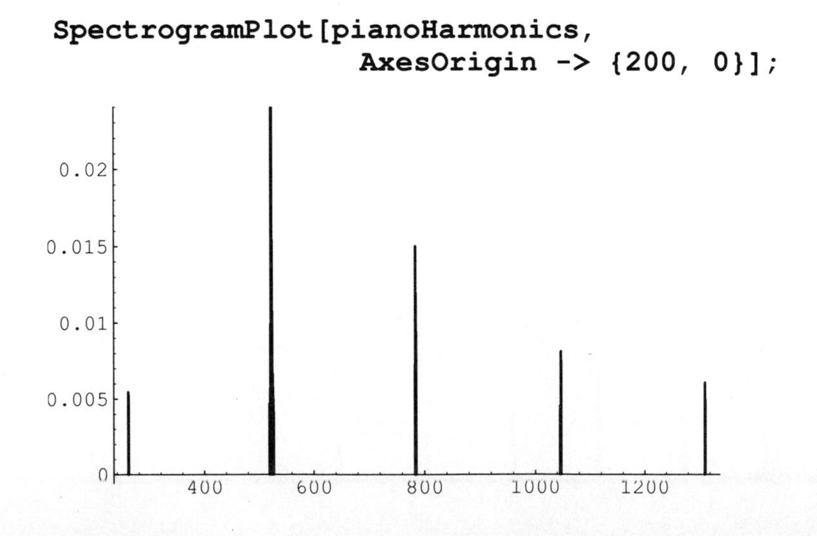

Much simpler, and much easier to handle.

Jerry: Now, how are we going to recreate the waveform from this list of frequencies and amplitudes? We have to translate this back somehow into an actual expression involving sines and cosines.

Theo: Fortunately, I've written a function to do exactly that. It takes a list of frequencies and amplitudes, and a variable name, and returns an expression. Note that the expression includes only sine functions, because we intentionally gave up all phase information. But **Synthesize** adds a random phase to each term, to help avoid having all the frequencies interfere with each other synchronously.

```
pianoWave = Synthesize[pianoHarmonics, x]
```

```
0.00545618 Sin[4.12734 + 1637.09 x] +
  0.00521308 Sin[3.94986 + 1641.19 x] +
  0.00464671 Sin[4.59927 + 3261.87 x] +
  0.00467448 Sin[3.92768 + 3265.97 x] +
  0.00563921 Sin[1.31637 + 3270.07 x] +
  0.00988677 Sin[2.96319 + 3274.18 x] +
  0.0377572 Sin[5.41702 + 3278.28 x] +
  0.0126219 Sin[0.657157 + 3282.38 x] +
  0.0370961 Sin[5.28409 + 3286.48 x] +
  0.0154129 Sin[1.11439 + 3290.59 x] +
  0.010757 Sin[3.16538 + 3294.69 x] +
  0.00664962 Sin[2.43677 + 3298.79 x] +
  0.00574432 Sin[3.86666 + 3302.9 x] +
  0.00495422 Sin[1.59677 + 3307. x] +
  0.00410967 Sin[1.47267 + 3311.1 x] +
  0.00725495 Sin[1.98001 + 4919.47 x] +
  0.0150055 Sin[3.93897 + 4923.57 x] +
  0.0119608 Sin[0.0862953 + 4927.67 x] +
  0.00937931 Sin[6.04519 + 4931.78 x] +
  0.00452438 Sin[3.14186 + 6560.87 x] +
  0.00811399 Sin[2.37139 + 6577.07 x] +
  0.00750902 Sin[0.336992 + 6581.17 x] +
  0.00607564 Sin[1.11101 + 8234.67 x] +
  0.00495867 Sin[0.923645 + 8238.78 x]
```

Jerry: Now we should be able to combine this with the envelope function we defined earlier, and make a sound!

Theo: Right.

Before we do, though, I'd like to propose a new and better envelope function. I was thinking about functions that look sort of like our envelope function, and suddenly I realized that I know a very good one. The electron probability distribution for the Hydrogen 2s orbital looks very much like what we want.

Jerry: You must be joking! That sounds like typical after-the-fact rationalization. I bet you thought up the function and then realized it was the same as the orbital function.

Theo: No, no, in fact I actually had to look up what the orbital function was, because, to my shame, I couldn't remember it. I had to modify it somewhat, and add some terms that don't exist in the real orbital, but after all is said and done, we have the following new envelope:

```
pianoEnvelope := x^0.5 (E^(-8x) + 0.08 E^(-x))
Plot[pianoEnvelope, {x, 0, 2}];
```

I'll put in a new copy of the original note, so we can compare:

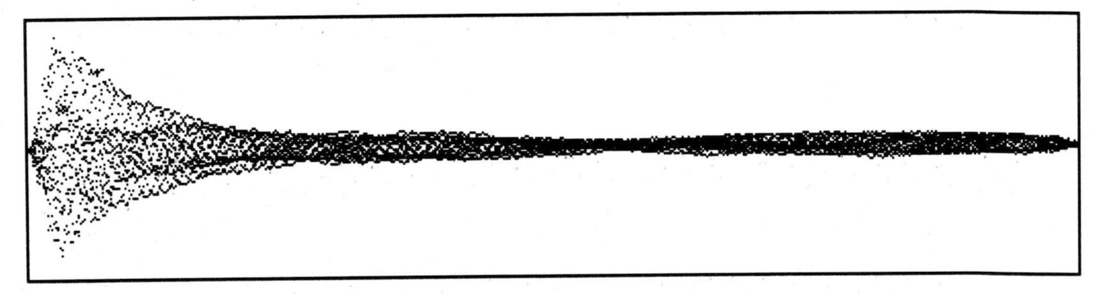

(Play track 8 of the CD to hear this sound.)

Jerry: That does look quite like the outline of the note we pasted in from SoundEdit. It's also a much simpler expression than the other envelope, which had that silly **`Which`** function in it.

Theo: Let's make a sound:

```
Play[pianoWave pianoEnvelope, {x, 0, 2},
    PlayRange -> All];
```

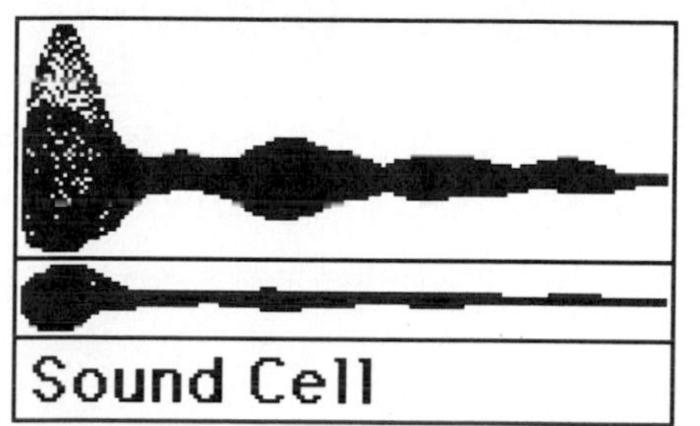

(Play track 9 of the CD to hear this sound.)

Jerry: Well! Finally something I wouldn't be ashamed to listen to! It's not perfect, but it's certainly much better than the other one.

Theo: Yes, I think it's pretty good. The main problem seems to be the too-strong beating we hear towards the end of the note. It actually sounds like it's getting a bit louder, instead of constantly softer. That's caused by the very close frequencies around each major harmonic peak. Those very similar frequencies are important, since, as we discussed, a real piano has three very slightly out of tune strings for each note. Unfortunately, we just have a little too much. We could probably improve things even more by including more frequencies, which would tend to smooth everything out more.

Jerry: It is, however, very impressive how well we have simulated the huge amount of data we started with, thousands of samples, with a very simple equation involving only twenty terms or so.

Functions That We Defined

```
AddPoints[list_] :=
    Epilog -> {PointSize[0.02], Map[Point, list]}
```

Chapter Fifteen
The Amazing Constantly Rising Tone

In which Jerry and Theo wait forever for the end of the scale.

Dialog

Theo: A few years ago I saw an amazing exhibit in the Exploratorium museum in San Francisco.

Jerry: There are many amazing exhibits in the Exploratorium. Which one did you see?

Theo: The amazing continuously rising tone exhibit. It's a small cabinet with twelve buttons arranged in a circle. Pushing a button makes a tone. If you start with the button on top, and push them in sequence clockwise, each button makes a tone higher than the one before. The amazing thing is that the button on top is also higher than the one before.

Jerry: So, you're saying that you start with one button, push a bunch of buttons and hear a higher tone each time, then push the original button again, and the tone is *even higher*? Sitting here in Urbana, this is very hard to believe.

Theo: Well, they do a lot of things in California that are hard to believe in the Midwest.

Jerry: What if you continue around two, three, or four times?

Theo: If you push buttons in the clockwise direction, every tone is higher than the one before it, no matter how many times you go around. You can start anywhere you like, and go for as long as you like. The tone rises each time.

Jerry: This sounds like the audio version of M. C. Escher's continuously rising staircase. I suppose if Escher can create such a visual illusion, then someone else can create an audio illusion. I mean, it must be an illusion, right?

Theo: Fortunately the Exploratorium considers teaching its goal, so they felt compelled to put some sort of explanation with the exhibit. I decided to ask Jim Maeder, who built the exhibit, how it works. He sent me to a paper by a Mr. Shepard.

It turns out the tones are called Shepard tones, and have been known for a long time **[1]**. Mr. Shepard has a whole theory about these things, and can make all kinds of paradoxical sounds.

Jerry: All right, let's hear these sounds.

Theo: Here they are, with no explanation (yet). Double-click the picture below (in the electronic edition). This will start a sound animation. The twelve tones will be played in order from top to bottom, and then repeated from the top again. As each tone is played, its number will be displayed at the top. Click anywhere to stop the animation.

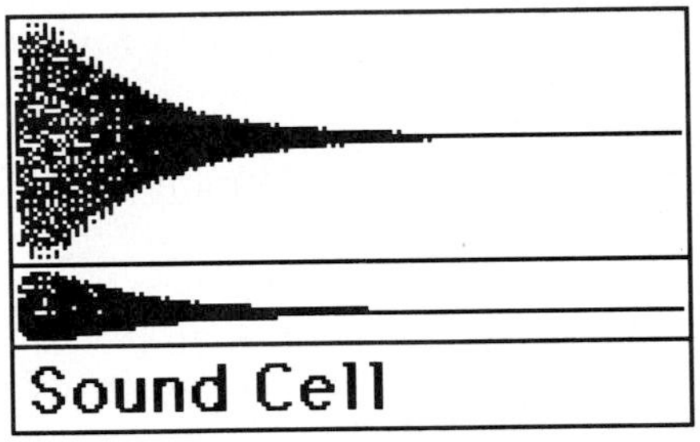

(Play track 10 of the CD to hear these sounds.)

Jerry: Fabulous. I know it must be an illusion, but I can't deny what I hear. The tones continue to rise. I wonder if we can create the same effect with stocks and bonds, or with the sales of this book.

Theo: Perhaps we should understand how this illusion works before we start using it on stocks.

Jerry: The tones are played slowly. Is this necessary for the illusion to work?

Theo: Yes. If you play the tones too quickly, you can start to hear how it works. Let's try it:

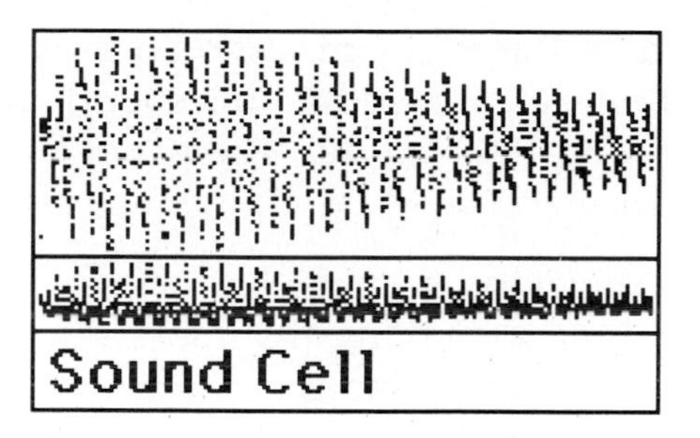

(Play track 11 of the CD to hear these sounds.)

Jerry: The illusion is destroyed! When the tones were about one second long, it was absolutely convincing. When the tones are only about 1/5 of a second, I clearly hear that the notes get lower sometimes, instead of higher.

Theo: You start to notice that although parts of a tone get higher, new lower parts get introduced. At some point your mind shifts to hearing the lower parts of the tone. Perhaps we should reveal how it's done:

As we saw in Chapter 14, a musical note or tone is usually composed of several different frequencies mixed together. How "high" the note sounds depends not only on the frequencies of these harmonics, but also how loud they are relative to each other.

Our Shepard tones consist of six different harmonics mixed together. The loudest ones are in the middle, with the highest and lowest softer. The next tone in line has all the harmonics shifted up in frequency. The tone sounds higher. At the same time the higher harmonics are made softer, and the lower ones louder. Despite this, the tone *still* sounds higher. This is the trick that makes it all work.

As you go from one tone to the next the harmonics keep getting higher, the higher harmonics keep getting softer, and the lower harmonics keep getting louder. Eventually (after a cycle of twelve tones), the highest harmonic has disappeared entirely, and a brand new one has appeared at the bottom, very softly. We are back at the starting point.

Jerry: Let's talk about how we implemented this in *Mathematica*.

Theo: Fortunately we were able to find a paper **[2]** that described some suitable parameters to use. The first thing to decide is which six harmonics to use. Let's start with middle C, 263 Hz. The next harmonic would be twice that, or 526 Hz, and so on. Here are all the harmonics we will use:

```
Table[263 2^n, {n, 0, 5}]
{263, 526, 1052, 2104, 4208, 8416}
```

The next thing to decide is how the amplitude of the harmonics should vary as their frequencies vary. The low harmonics should be soft, the middle ones loud, and the high ones soft. Clearly we want some sort of hump-like curve. The paper suggests an "inverted cosine envelope in log frequency space".

Jerry: I have no idea what an inverted cosine envelope is, and I know even less about a log frequency space. I wonder if they mean "use something that looks like a cosine curve, and have the x-axis in units of the log of the frequency".

Theo: Yes, that's exactly what they mean, but you don't get NSF grants with language like that. Let's start out with a cosine curve, and see what it looks like over our range of frequencies:

```
Plot[Cos[f], {f, 263, 8416}, PlotPoints -> 100];
```

Jerry: Clearly not good. This is oscillating wildly and confusing the plotting routines. We were supposed to use **Log[f]** anyway:

```
Plot[Cos[Log[f]], {f, 263, 8416}];
```

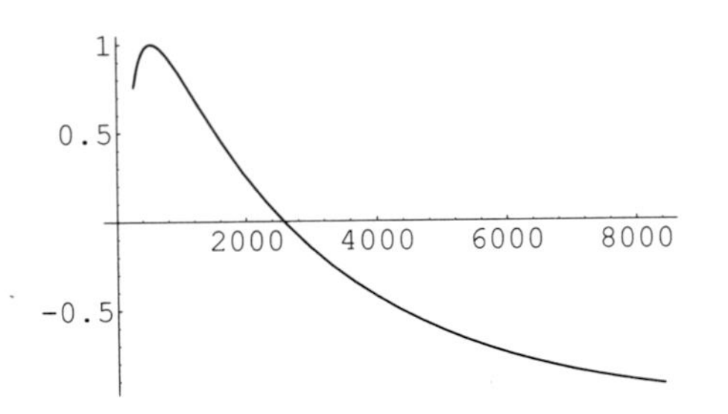

Theo: Let's remember what we are trying to get. We want a curve that is zero when f is 263 Hz. We can get that by subtracting **Log[263]** inside the cosine (which makes the cosine 1 when **f** is 263), and then subtracting the cosine from 1.

```
Plot[1 - Cos[Log[f] - Log[263]],
    {f, 263, 8416}, PlotRange -> All];
```

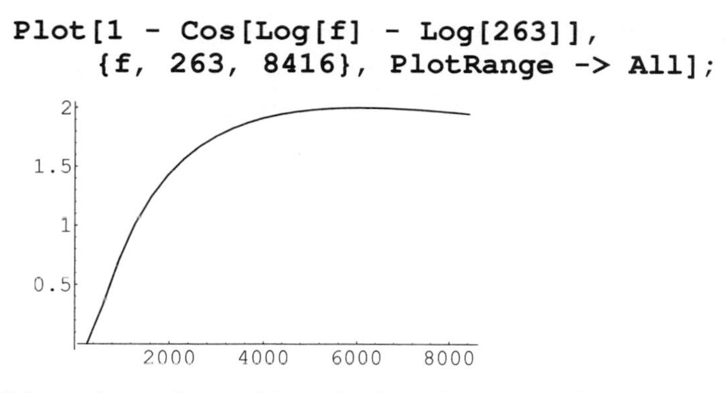

Jerry: Now the only problem is that the curve doesn't get back down to zero fast enough. We need to multiply by something that makes the argument of the cosine equal **2Pi** when **f** equals 8416. We can do that by taking the current argument (**Log[f] - Log[263]**), dividing by its value when f is 8416 (**Log[8416] - Log[263]**), and multiplying by **2Pi**:

```
Plot[1 - Cos[2Pi (Log[f] - Log[263]) /
                              (Log[8416] - Log[263])],
    {f, 263, 8416}];
```

Theo: This doesn't look like a cosine, because the x-axis is **f**, not **Log[f]**. We can use **ParametricPlot** to change the x-axis:

```
ParametricPlot[{Log[f],
    1-Cos[2Pi(Log[f]-Log[263])/(Log[8416]-Log[263])]},
    {f, 263, 8416},
    AxesOrigin -> {5.5, 0}];
```

Jerry: That's a handy little trick! By switching to parametric form, you're able to make a log-scaled x-axis.

Theo: Just one of those little things.

We can simplify this expression using the fact that **Log[a] - Log[b]** equals **Log[a / b]**.

Jerry: As long as **a** and **b** have the same sign.

Theo: Don't bother me with mathematical facts, I'm shaping curves here!

```
ParametricPlot[
    {Log[f], 1-Cos[2Pi Log[f / 263] / Log[2^5]]},
    {f, 263, 8416},
    AxesOrigin -> {5.5, 0}];
```

2
1.5
1
0.5
6 6.5 7 7.5 8 8.5 9

Let's define a function that implements this envelope:

```
harmonicAmplitude[f_] :=
    N[1 - Cos[2Pi Log[f / 263] / Log[2^5]]]
```

Jerry: Could we make a plot to show the harmonics?

Theo: Sure. We want to draw six lines, one for each harmonic. The x-coordinate will be the log of the frequency, and the y-coordinate will be the amplitude as given by our envelope. We can add these lines as an epilog to the plot command:

```
harmonicLine[f_] :=
    Line[{
        {Log[f], 0},
        {Log[f], harmonicAmplitude[f]}
    }]
```

```
ParametricPlot[
    {Log[f], 1-Cos[2Pi Log[f / 263] / Log[2^5]]},
    {f, 263, 8416},
    AxesOrigin -> {5.5, 0},
    Epilog -> Table[harmonicLine[263 2^n], {n, 0, 5}]
];
```

2
1.5
1
0.5
6 6.5 7 7.5 8 8.5 9

Jerry: I see only four lines!

Theo: The other two are there, but they have zero height, so you can't see them.

Jerry: Now we need an animation showing how these four, uh, six harmonics change from one Shepard tone to the next.

Theo: The trick is to shift all the harmonics up, while leaving the envelope in the same place relative to the frequency axis. We want to leave out the last harmonic, since once we start shifting the frequencies up, it will be outside the range of our envelope function.

```
Do[
    ParametricPlot[
        {Log[f], harmonicAmplitude[f]},
        {f, 263, 8416},
        AxesOrigin -> {5.5, 0},
        Epilog -> Table[
            harmonicLine[2^m 263 2^n], {n, 0, 4}]
    ],
    {m, 0, 11/12, 1/12}
]
```

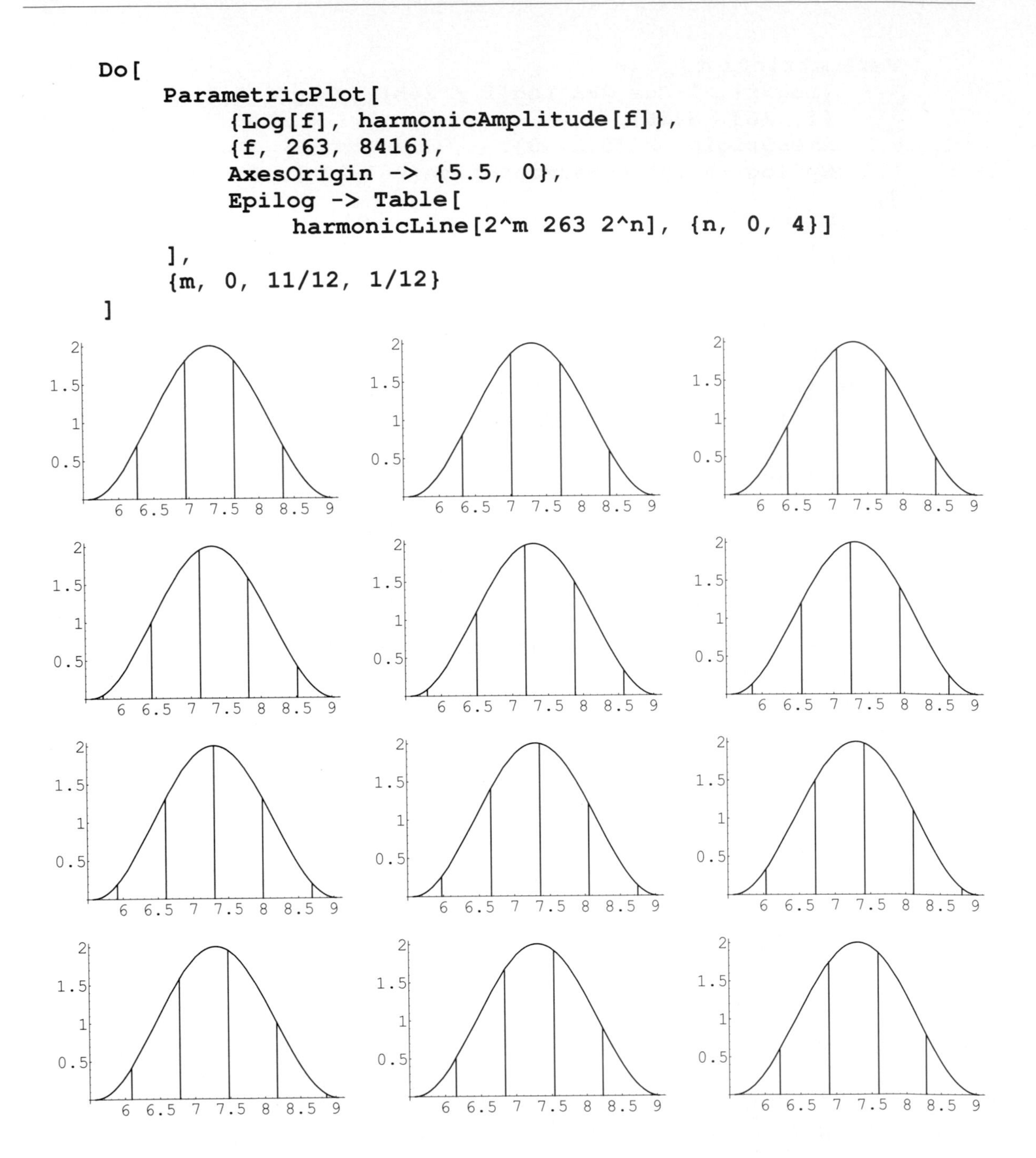

Jerry: I think we're ready to make the sounds that you so mysteriously presented at the beginning of this chapter.

Theo: The first step is to make up an amplitude envelope. This is the function that will give the tones their overall shape. Here is one I like:

```
envelope = (Log[t + 0.001] - Log[0.001]) E^(-8 t);
Plot[envelope, {t, 0, 1}];
```

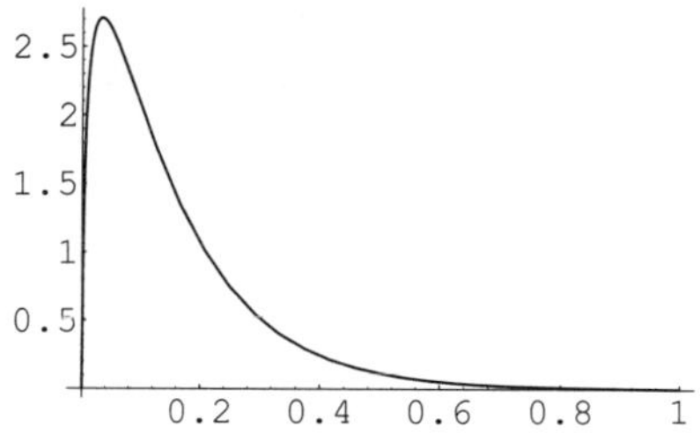

Jerry: This rises quickly and falls away slowly.

Theo: Now we need to make a function that multiplies this envelope function by the right set of harmonics to make the desired tone. The following function takes a number between 0 and 1, and produces a tone from it. If you give it increments of 1/12, it will generate notes in the standard scale.

```
tone[m_] := N[envelope Apply[
    Plus,
        Table[Sin[2.^m 263 2.^n t] *
         harmonicAmplitude[2.^m 263 2.^n], {n, 0, 4}]
    ]
]
```

We can play one of the tones like this: (The `PlayRange` was determined experimentally. You can use `PlayRange -> All` instead, but it's slower.)

```
Play[Evaluate[tone[0/12]], {t, 0, 1},
     PlayRange -> {-9, 9}];
```

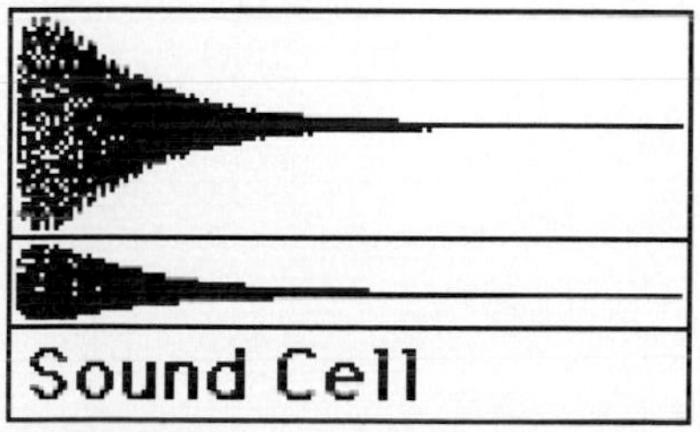

(Play track 12 of the CD to hear this sound.)

Jerry: Now we should combine this sound with the pictures!

Theo: OK. Here's a function to do that:

```
shepardPlot[m_] :=
    ParametricPlot[
        {Log[f], harmonicAmplitude[f]},
        {f, 263, 8416},
        AxesOrigin -> {5.5, 0},
        Epilog -> {
         Table[harmonicLine[2^m 263 2^n], {n, 0, 4}],
         Play[
            Evaluate[tone[m]], {t, 0, 1},
            PlayRange -> {-9, 9},
            DisplayFunction -> Identity][[1]]
        }
    ]
```

We can now use it to generate an animation with graphics *and* sound:

```
Do[shepardPlot[m/12], {m, 0, 11}]
```

(We didn't include the pictures in the printed edition, because it looks the same as the earlier one without sound. Play track 10 of the CD to hear the sounds.)

Jerry: There are lots of questions to explore here. What if the frequencies we mix are not octave harmonics? We have six harmonics: what if we use 5 or 27? Does the shape of the frequency envelope make any difference? What if it's just linear?

Theo: That's what *Mathematica* is for: experimentation. Let the readers try these ideas for themselves.

Functions That We Defined

Theo: This function takes a frequency and returns the relative amplitude that a harmonic at that frequency should have:

```
harmonicAmplitude[f_] :=
     N[1 - Cos[2Pi Log[f / 263] / Log[2^5]]]
```

This function returns a **Line** graphics primitive corresponding to a harmonic at the given frequency:

```
harmonicLine[f_] :=
     Line[{
          {Log[f], 0},
          {Log[f], harmonicAmplitude[f]}
     }]
```

This is the amplitude envelope for the tones:

```
envelope = (Log[t + 0.001] - Log[0.001]) E^(-8 t);
```

This function takes a number between 0 and 1, and returns a function that, when played, generates a tone. Going in steps of 1/12 between 0 and 1 generates tones in the normal scale:

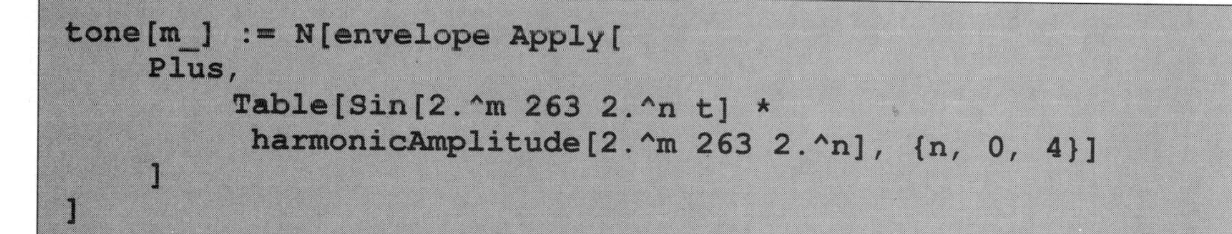

```
tone[m_] := N[envelope Apply[
     Plus,
          Table[Sin[2.^m 263 2.^n t] *
           harmonicAmplitude[2.^m 263 2.^n], {n, 0, 4}]
     ]
]
```

This function takes a number between 0 and 1 (like **tone**), and makes a plot of the harmonics with the appropriate sound associated with the plot:

```
shepardPlot[m_] :=
    ParametricPlot[
        {Log[f], harmonicAmplitude[f]},
        {f, 263, 8416},
        AxesOrigin -> {5.5, 0},
        Epilog -> {
         Table[harmonicLine[2^m 263 2^n], {n, 0, 4}],
         Play[
             Evaluate[tone[m]], {t, 0, 1},
             PlayRange -> {-9, 9},
             DisplayFunction -> Identity][[1]]
        }
    ]
```

References

1) Shepard, R. N., Circularity in Judgments of Relative Pitch, *Journal of the Acoustical Society of America*, 1964, **36**, p2345-2353.
2) Diana Deutsch, F. Richard Moore, Mark Dolson, Pitch Classes Differ with Respect to Height.

Chapter Sixteen
What Sound Is That Function in the Window?

Jerry and Theo discover what periodic functions are really *for.*

■ Dialog–First Day

Jerry: It's typical in the late twentieth century for people who have access to new technology to wander around wondering what to do with it. We aren't going to do that with the new sound technology in *Mathematica*, are we?

Theo: Of course not.

I wonder if we could use the **Play** command to listen to some functions. Any periodic function, like **Sin**, **BesselJ**, etc, is potentially a sound.

The **Play** command works like the **Plot** command. The first argument is the function to play. The second argument specifies the variable to use and the range. The range is specified in seconds, so the following command generates one second of sound. We have to multiply **x** by a large number inside the sine to get an audible frequency. Here's what a sine function sounds like:

```
Play[Sin[2000 x], {x, 0, 1}];
```

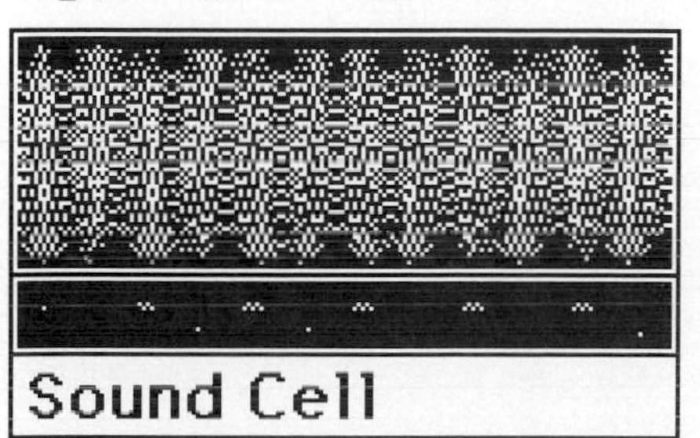

(Play track 13 of the CD to hear this sound.)

To hear this sound in the electronic edition, double-click the small speaker-like icon at the top of the cell bracket.

Jerry: What would **Sin[1000 x]** sound like?

```
Play[Sin[1000 x], {x, 0, 1}];
```

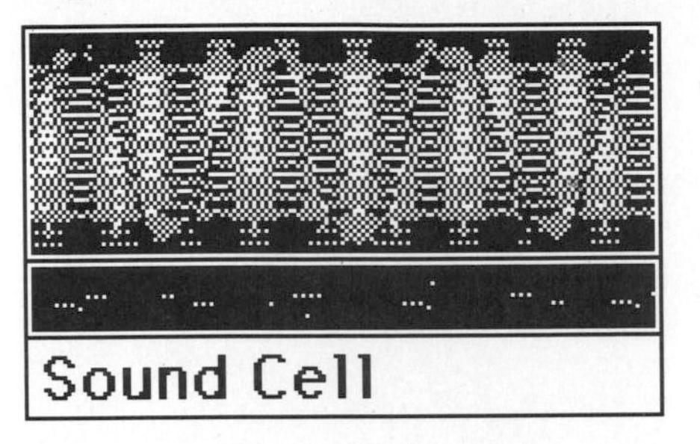

(Play track 14 of the CD to hear this sound.)

Sounds quite a bit lower to me. I'm confused by the pictures we're getting with these sounds. What are they?

Theo: By default, *Mathematica* makes up pictures like the ones above to go along with sounds. If you want to specify a different graphic to go along with a particular sound, you can use the **Epilog** option to do so. We'll probably be trying that out sometime later.

The strange dots in the default picture are an attempt to represent the waveform of the sound, but you shouldn't try to read too much into them. The idea is to have a picture that gives an overall feeling for the waveform. The picture is more or less what you would get if you replaced the word **Play** with **Plot** in the command, except that it's drawn with dots instead of connected lines.

If we play only a very short duration of this waveform, we will be able to see it better:

```
Play[Sin[1000 x], {x, 0, 0.02}];
```

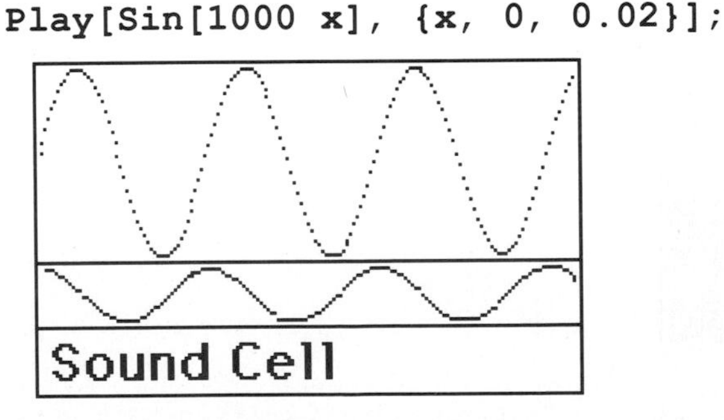

(Play track 15 of the CD to hear this sound.)

Jerry: When I listen to this sound, all I hear is a click. So, if we have a long enough duration to hear a tone, then we have too many cycles to see clearly.

Theo: Right. These pictures are particularly boring because the waveform is constant over the whole duration of the sound. More interesting sounds tend to produce more interesting pictures, although it's by no means always so.

Jerry: I'm starting to get more interested in this sound business, and I am overwhelmed by questions. How small a coefficient of `x` can we still hear? How large? Suppose we add (or subtract, or multiply, or divide) the two functions above? What combination of functions will produce that famous melody "How Much Is That Doggy In The Window?", or Beethoven's Fifth? What if we multiply the function by a constant?

Theo: If we take 20Hz to 20KHz as the maximum range of adult human hearing, `Sin[2 Pi 20 x]` is the smallest coefficient, and `Sin[2 Pi 20000 x]` the largest we could hear (with good enough speakers, and if we're not more than 25 years old, and haven't been to too many rock concerts).

Let's experiment with the combinations you suggested:

```
Play[Sin[1000 x] + Sin[2000 x], {x, 0, 1}];
```

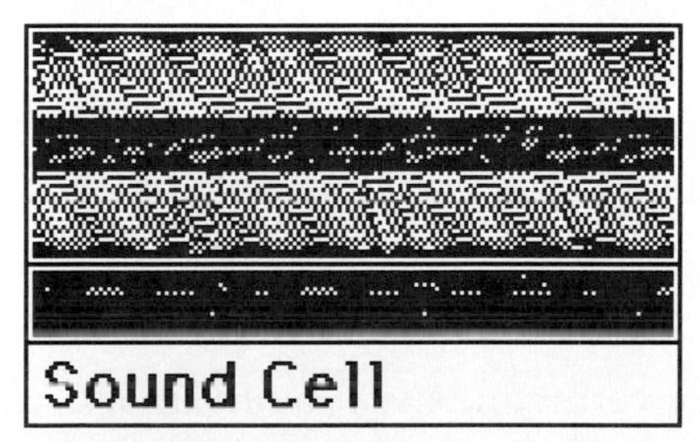


```
Play[Sin[1000 x] - Sin[2000 x], {x, 0, 1}];
```

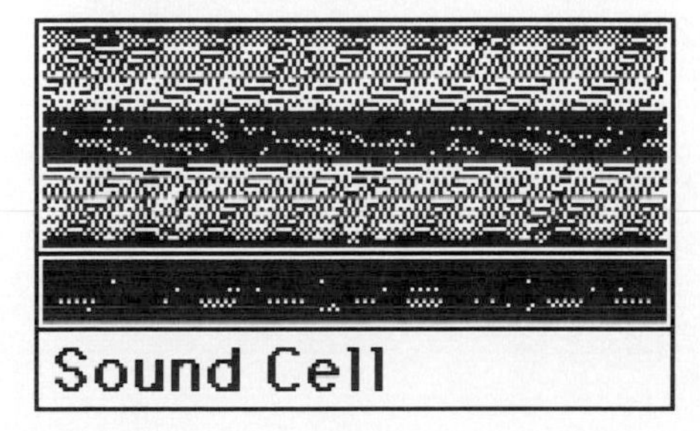

```
Play[Sin[1000 x] * Sin[2000 x], {x, 0, 1}];
```

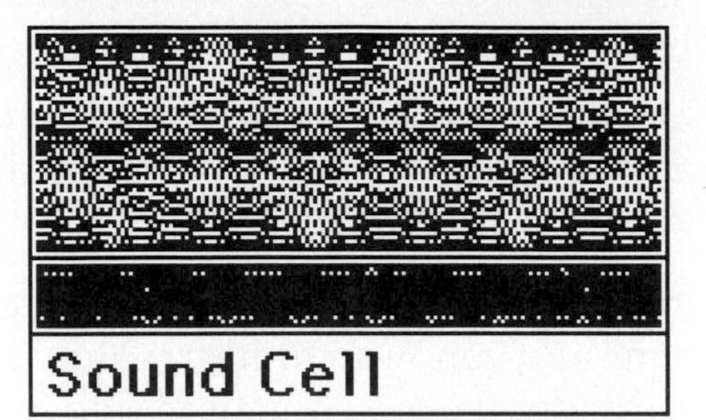


```
Play[Sin[1000 x] / Sin[2000 x], {x, 0, 1}];
```

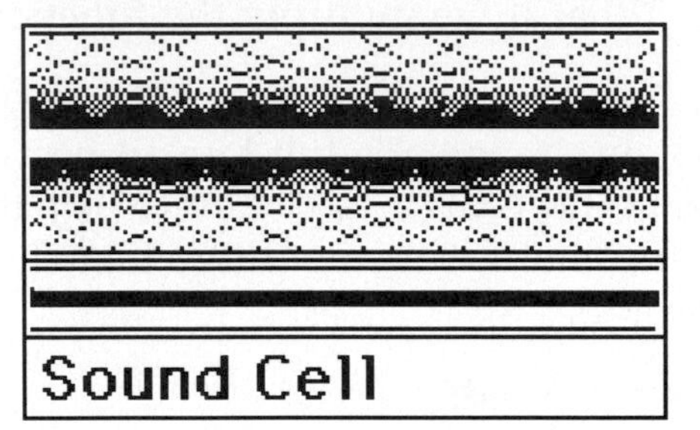

(Play track 16 of the CD to hear these four sounds.)

Jerry: So, division sounds buzzy. How do we make melodies?

Theo: Just a bunch of sine waves, nothing special. Someone could define a bunch of *Mathematica* functions that would let you type in a list of notes and chords and then hear them. I've been working on something just like that, but it's not done yet.

Jerry: Could we at least play a simple major chord?

Theo: Oh, all right. The following four **Play** commands will build up a C major chord, with the lowest note being middle-C. As they say, music is all mathematics:

```
Play[Sin[2^(0/12) 263 2 Pi x],
      {x, 0, 1}, PlayRange -> {-4, 4}];
Play[Sin[2^(0/12) 263 2 Pi x] +
      Sin[2^(4/12) 263 2 Pi x],
      {x, 0, 1}, PlayRange -> {-4, 4}];
Play[Sin[2^(0/12) 263 2 Pi x] +
      Sin[2^(4/12) 263 2 Pi x] +
      Sin[2^(7/12) 263 2 Pi x],
      {x, 0, 1}, PlayRange -> {-4, 4}];
Play[Sin[2^(0/12) 263 2 Pi x] +
      Sin[2^(4/12) 263 2 Pi x] +
      Sin[2^(7/12) 263 2 Pi x] +
      Sin[2^(12/12) 263 2 Pi x],
      {x, 0, 1}, PlayRange -> {-4, 4}];
```

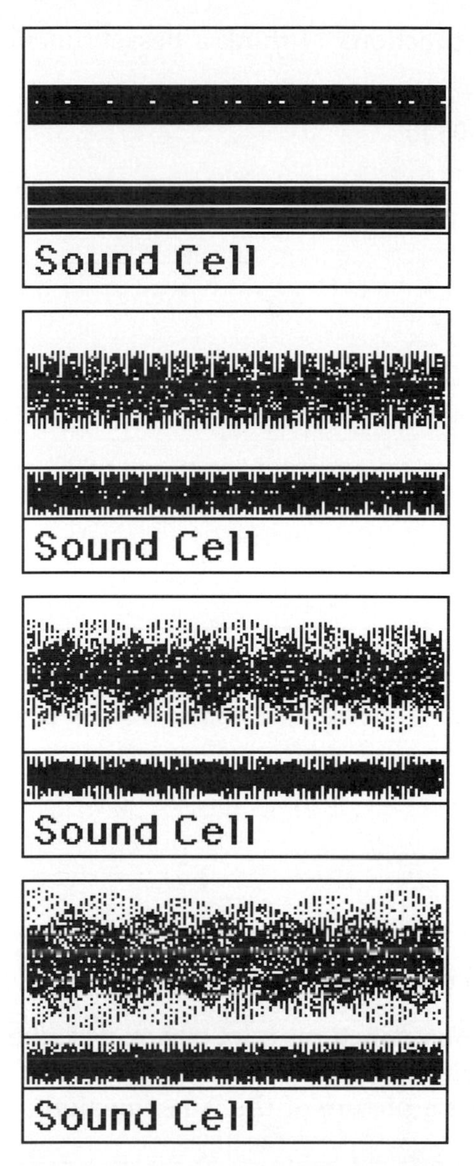

(Play track 17 of the CD to hear these sounds.)

To hear these sounds in an endless loop, double-click the first picture (right on the picture, not on the cell bracket). I've slowed down the animation (using the "slower" button in the animation control palette visible while the animation is running), so there is a slight delay between each tone.

Jerry: I think this little exercise is going to make every composer in the country run out and buy *Mathematica*. Well, maybe not *run*.

Theo: Let's listen to some more interesting functions. I think a Bessel function such as the following one might sound nice:

```
Plot[BesselJ[0, x], {x, 0, 30}];
```

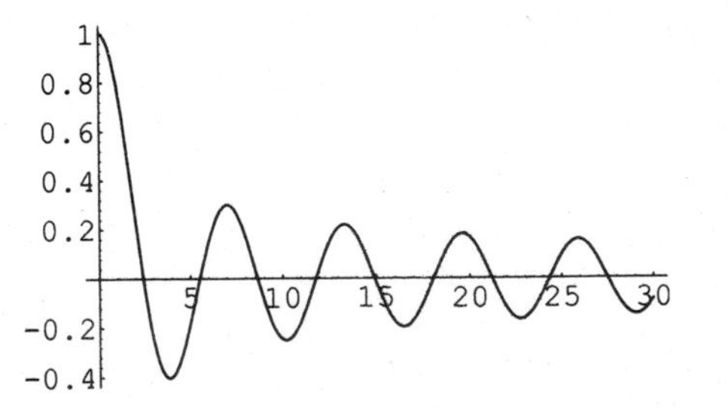

Here is the Bessel function (zeroth order) up to an argument value of 6000, as a sound:

```
Play[BesselJ[0, 2000 x], {x, 0, 3},
    PlayRange -> {-.4, 0.4}];
```

Sound Cell

(Play track 18 of the CD to hear this sound.)

Jerry: Door chimes! I was hoping for more excitement.

Theo: Back when I was a chemist we used Airy functions a lot. They describe the quantum mechanical wavefunction of a particle bouncing off a linear potential barrier. I bet they sound pretty neat too. Here's a picture of the Airy function:

```
Plot[AiryAi[x], {x, -30, 10}];
```

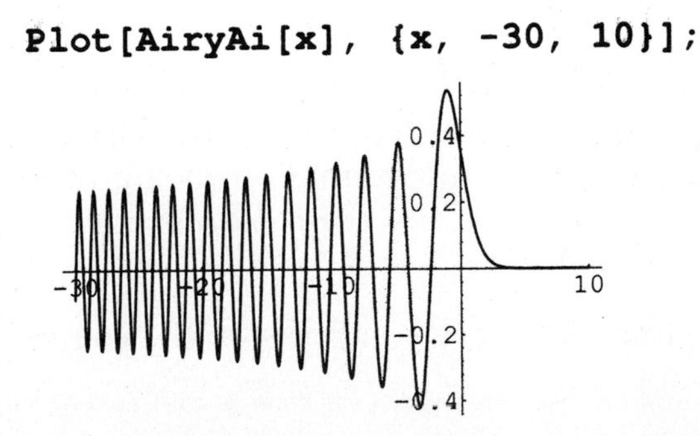

Jerry: It seems to oscillate faster and faster as you go off to the left, but stops altogether for positive values of **x**.

Theo: Right, it decays exponentially for positive **x** values. Let's listen to it from -300 up to 0:

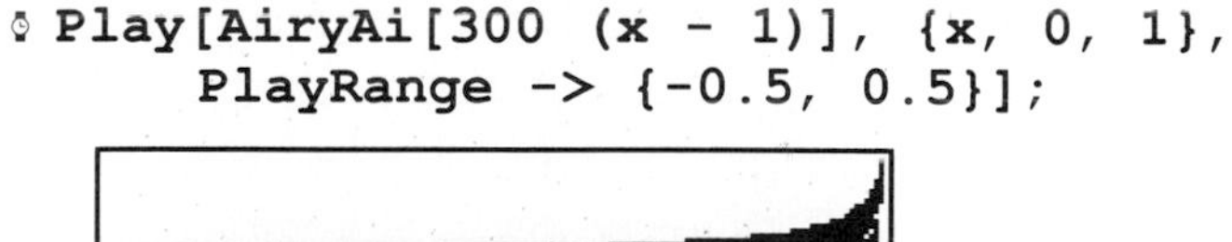

```
Play[AiryAi[300 (x - 1)], {x, 0, 1},
    PlayRange -> {-0.5, 0.5}];
```

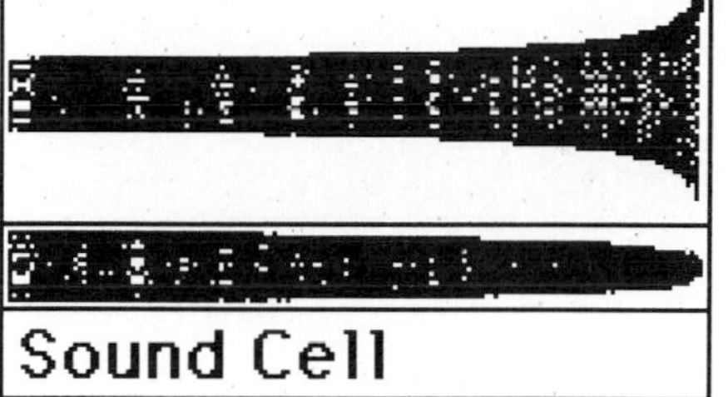

(Play track 19 of the CD to hear this sound.)

Jerry: Not bad. Sounds sort of like a horse neighing. Can we mix some of these together? How about adding one at a slightly different frequency?

```
Play[AiryAi[300 (x - 1)] + AiryAi[298 (x - 1)], {x, 0, 1},
    PlayRange -> {-0.5, 0.5}];
```

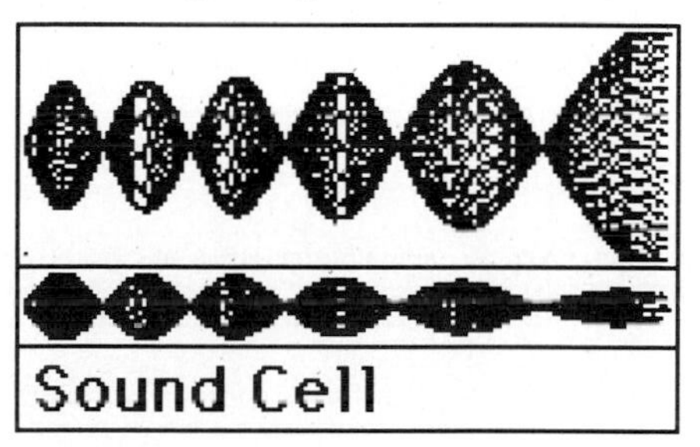

(Play track 20 of the CD to hear this sound.)

Theo: Sounds sort of like Roger Rabbit. The large-scale blobs in the picture are caused by the interaction (beating) of the two slightly different frequencies.

Jerry: I hear it as vibrato. Bravo! Encore!

Theo: OK, here's some more:

```
Play[AiryAi[300 (x - 1)] + AiryAi[285 (x - 1)], {x, 0, 1},
     PlayRange -> {-0.5, 0.5}];
```

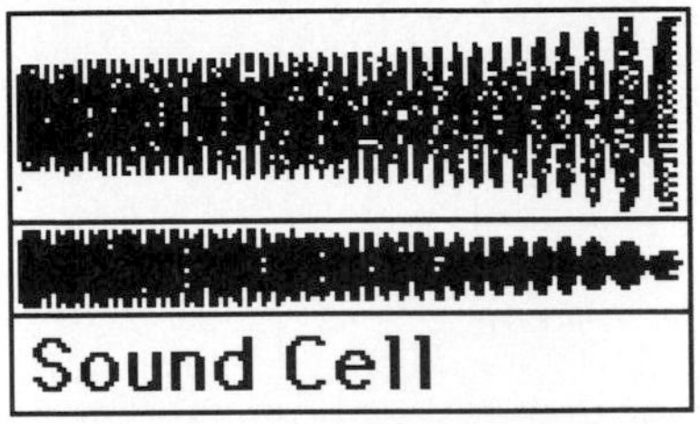

(Play track 21 of the CD to hear this sound.)

Jerry: So, first we got vibrato; now we get one of those "Oops, I'm on the wrong planet" sounds.

```
Play[Sin[100 / x], {x, 0.000001, 1},
     PlayRange -> {-1, 1}];
```

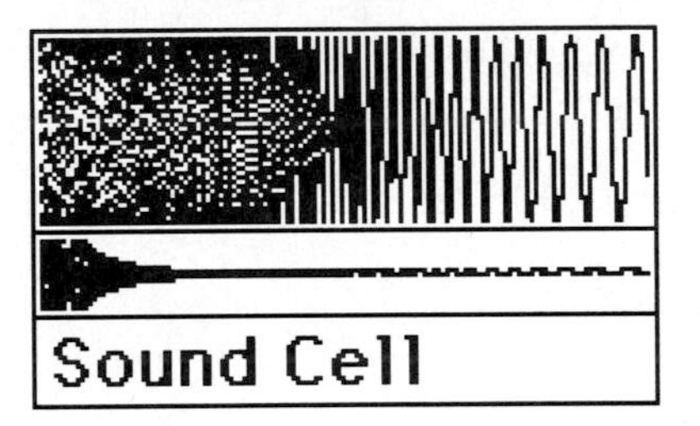

(Play track 22 of the CD to hear this sound.)

Theo: Ooo.... Ahh.... Let's hear that over a wider range, starting at a negative value. We should also increase the sampling rate (which is the same as increasing the **PlotPoints** option in a **Plot** command).

```
Play[Sin[2000 / x], {x, -3, 3},
     PlayRange -> {-1, 1},
     SampleRate -> 22254.5454];
```

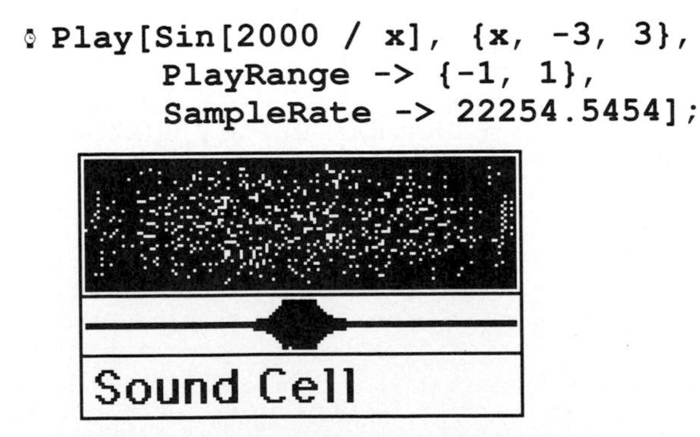

(Play track 23 of the CD to hear this sound.)

Random tree with manually specified colors for each generation of branches. From Chapter 6.

Another random tree from the same function.

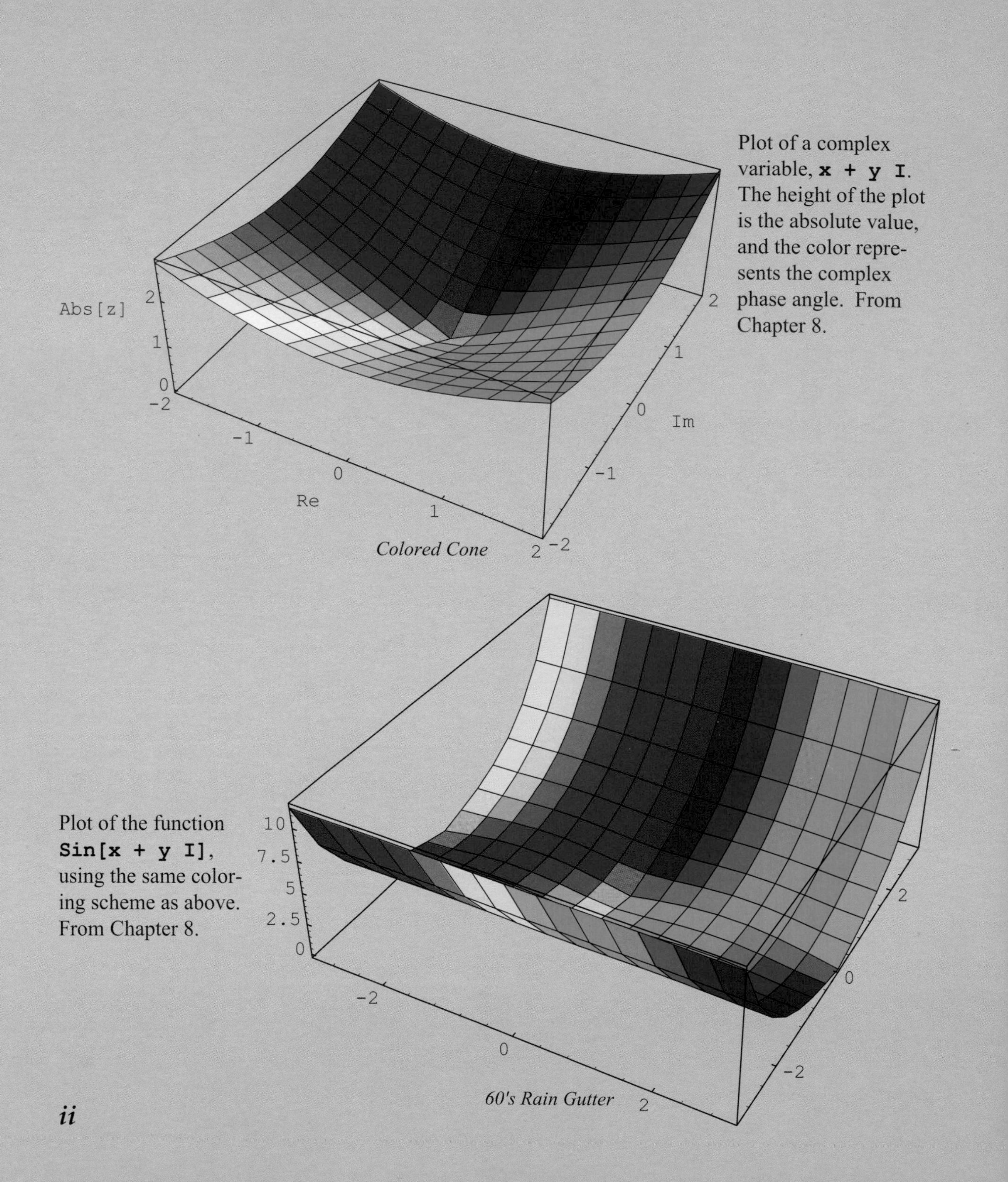

Plot of a complex variable, **x + y I**. The height of the plot is the absolute value, and the color represents the complex phase angle. From Chapter 8.

Colored Cone

Plot of the function **Sin[x + y I]**, using the same coloring scheme as above. From Chapter 8.

60's Rain Gutter

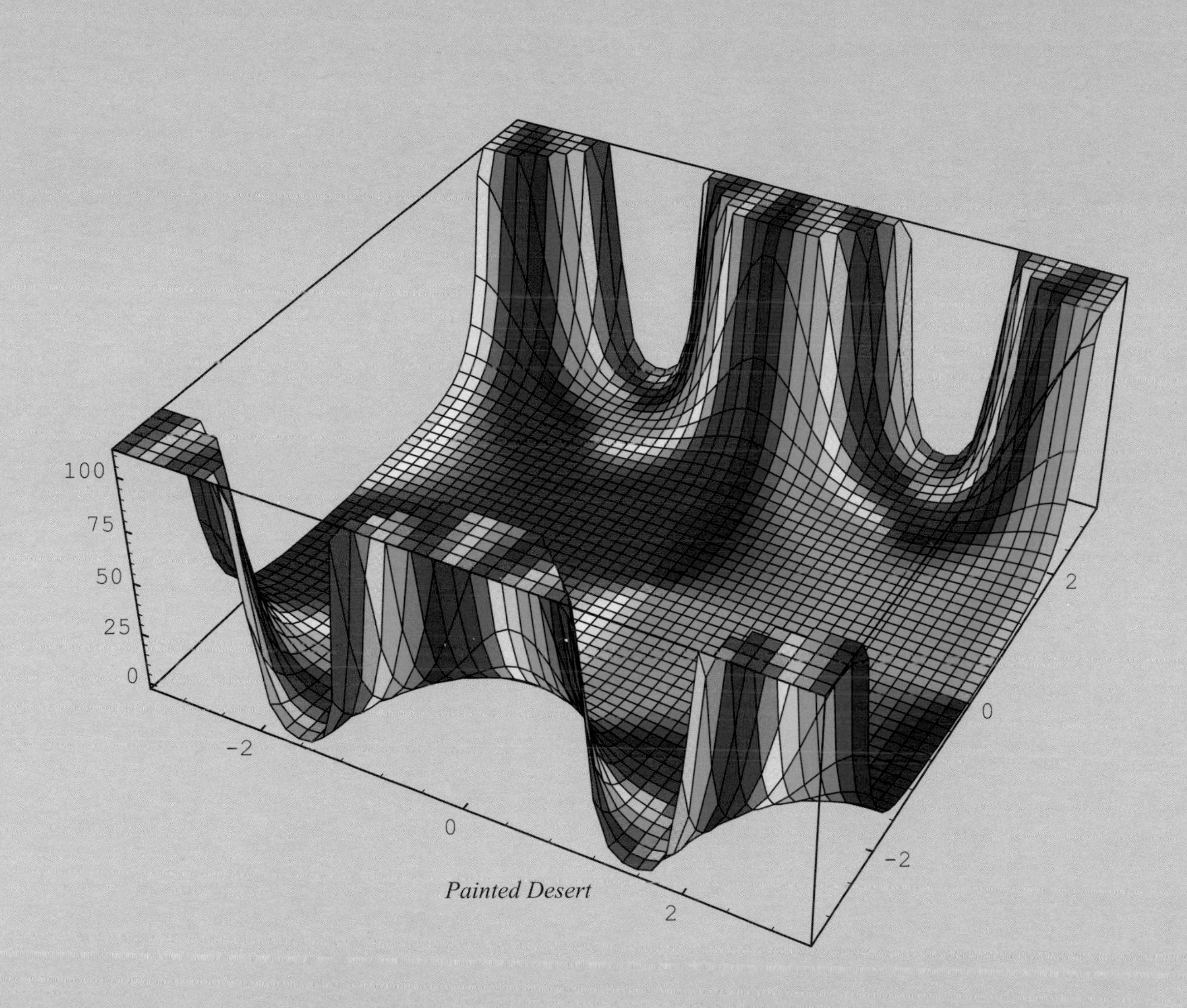

A plot of **Sin[Sin[x + y I]]**. The height of the plot is the absolute value, and the color represents the complex phase angle. The valley down the middle is along the real axis, Thc channels between the weathered buttes are where the real part is $\pi/2$ and $-\pi/2$. From Chapter 8.

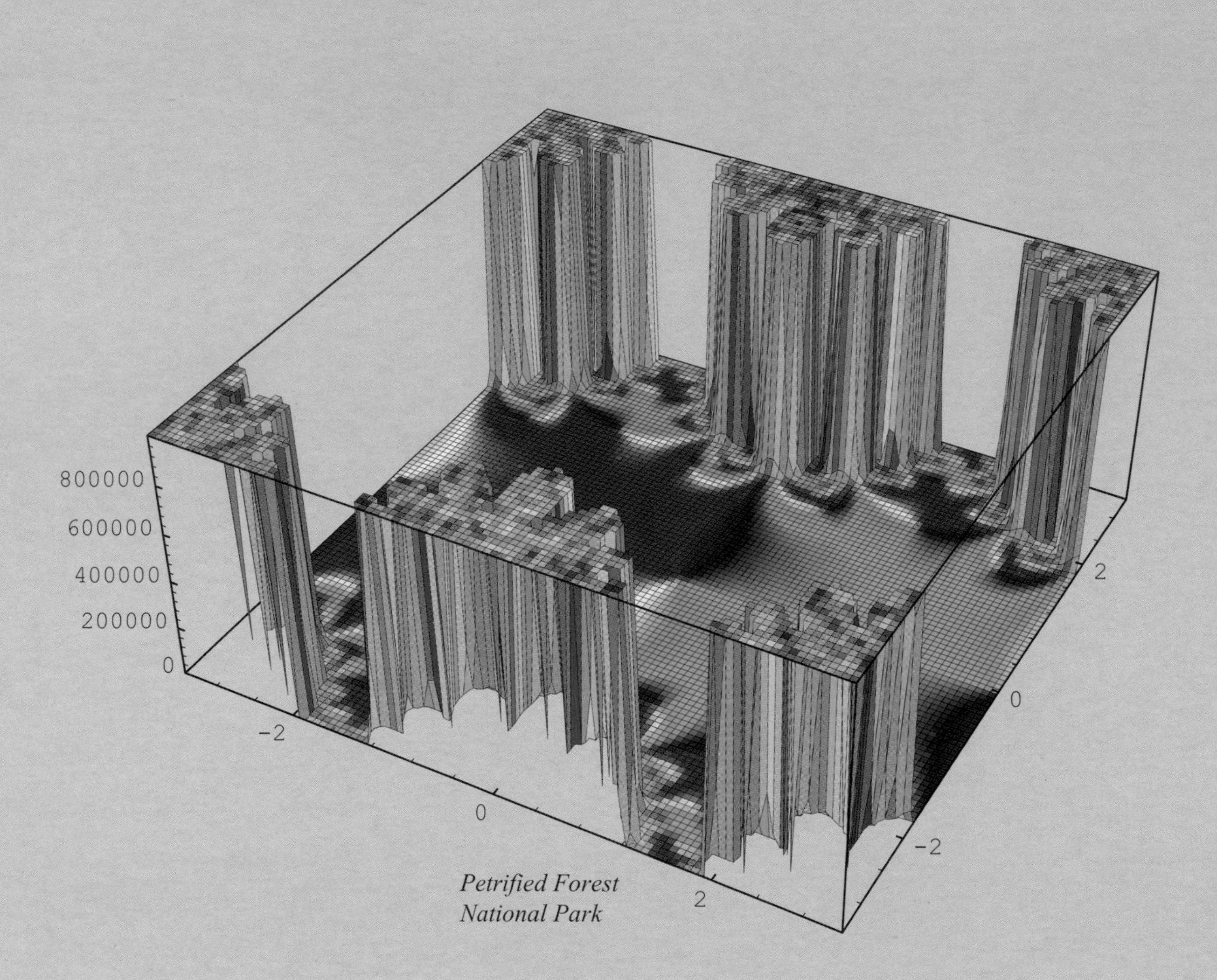

A plot of **Sin[Sin[Sin[x + y I]]]**. The sides of the peaks are so steep because the function grows very, very rapidly. From Chapter 8.

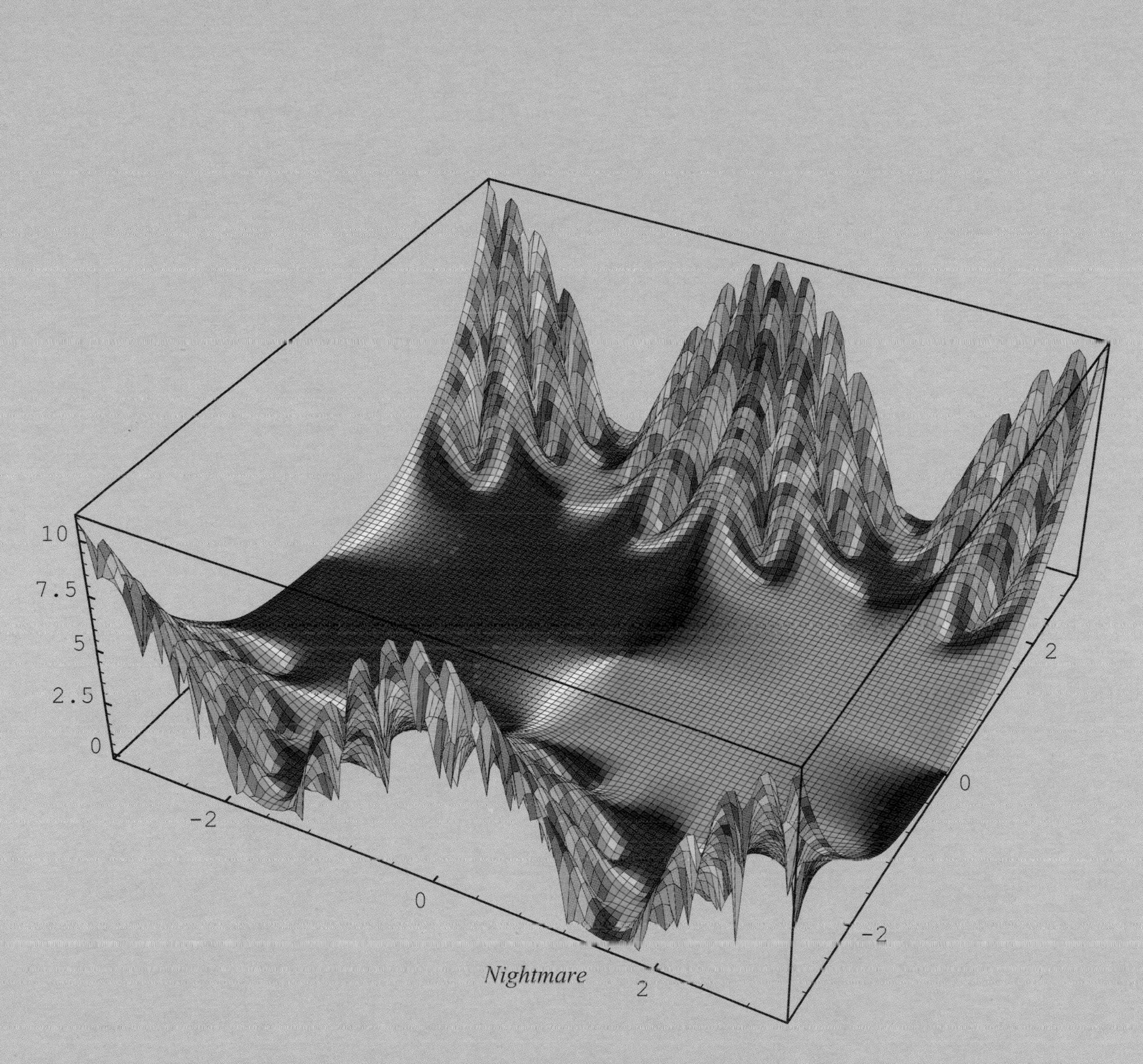

A plot of **`Log[Log[Sin[Sin[Sin[x + y I]]]]]`**. The fluted structure is highly suggestive of fractals. From Chapter 8.

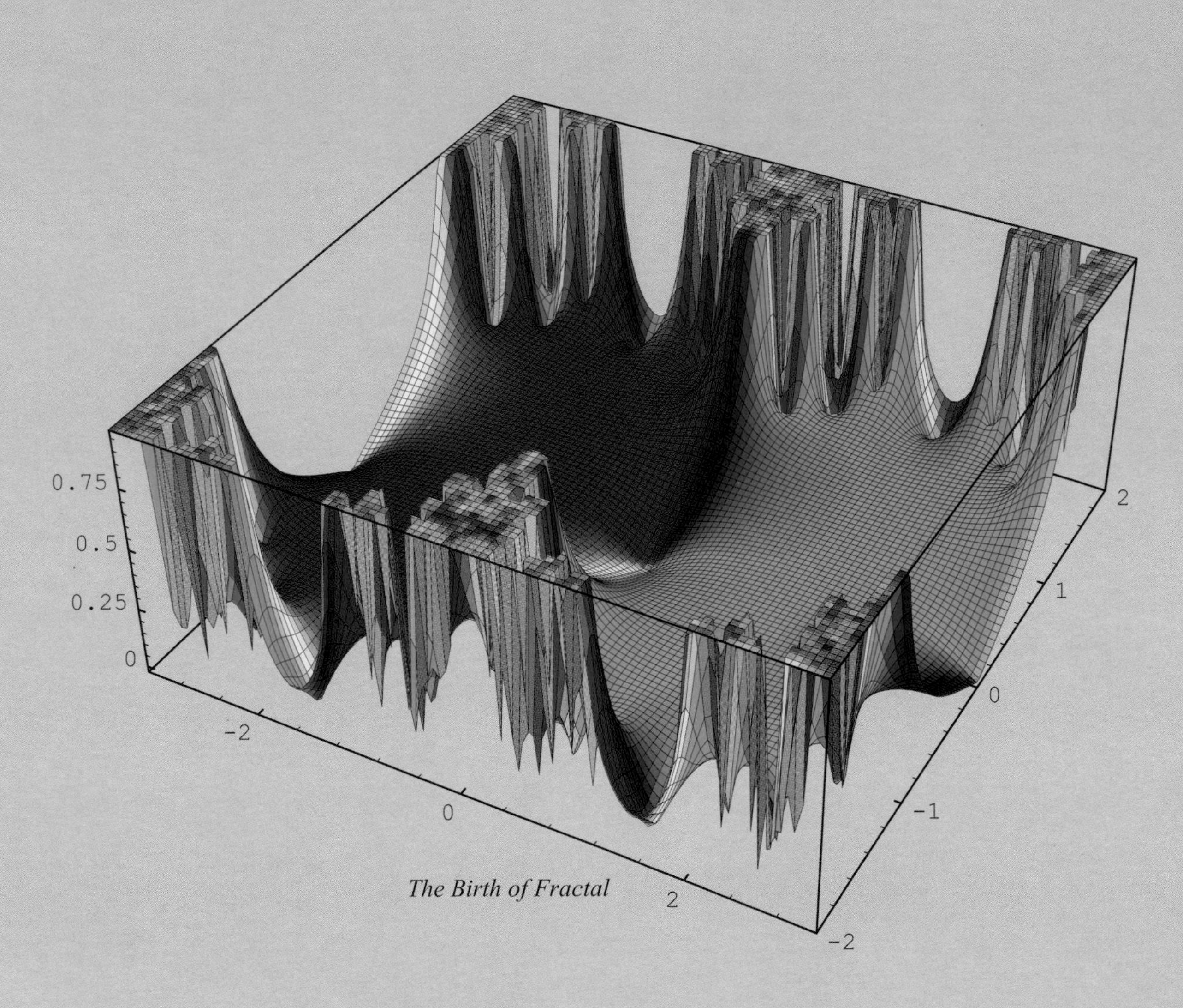

The Birth of Fractal

A plot of **`Log[Log[Sin[Sin[Sin[Sin[x + y I]]]]]]`**. The flutes seen in the last picture have now themselves become fluted, indicating that fractals are definitely in the offering. From Chapter 8.

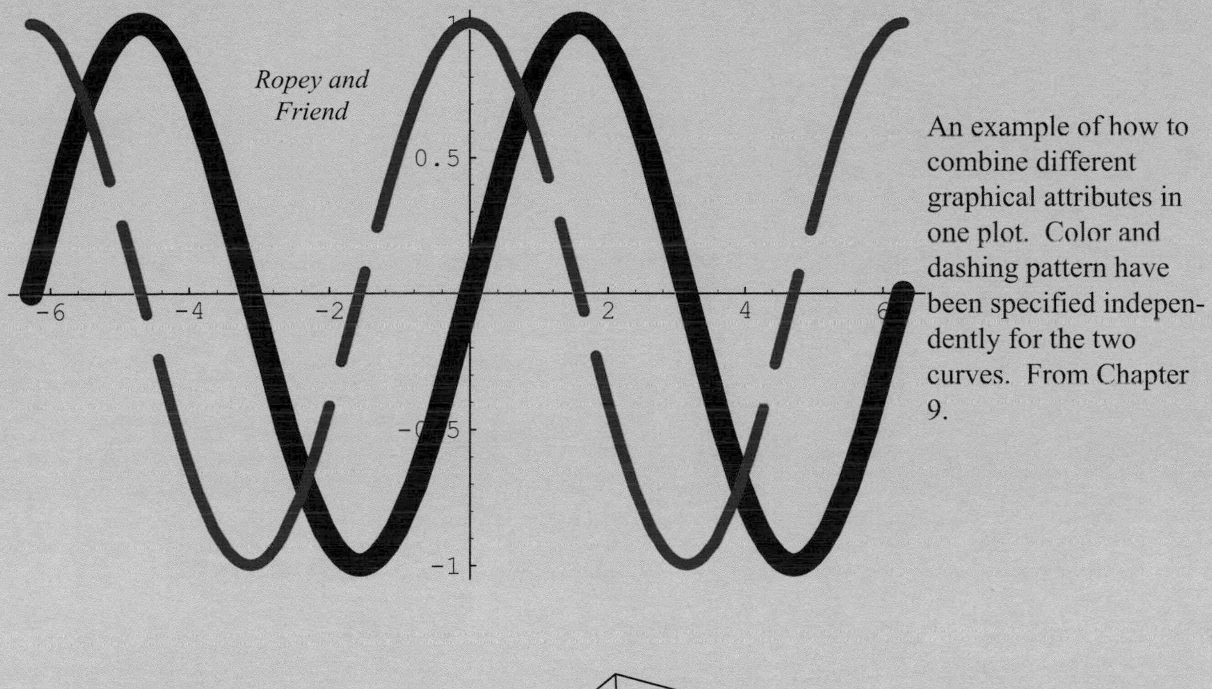

An example of how to combine different graphical attributes in one plot. Color and dashing pattern have been specified independently for the two curves. From Chapter 9.

A plot of the function **`Sin[x + Sin[y]]`**. This plot is hard to see because there are not enough subdivisions. The next plot improves the situation. From Chapter 9.

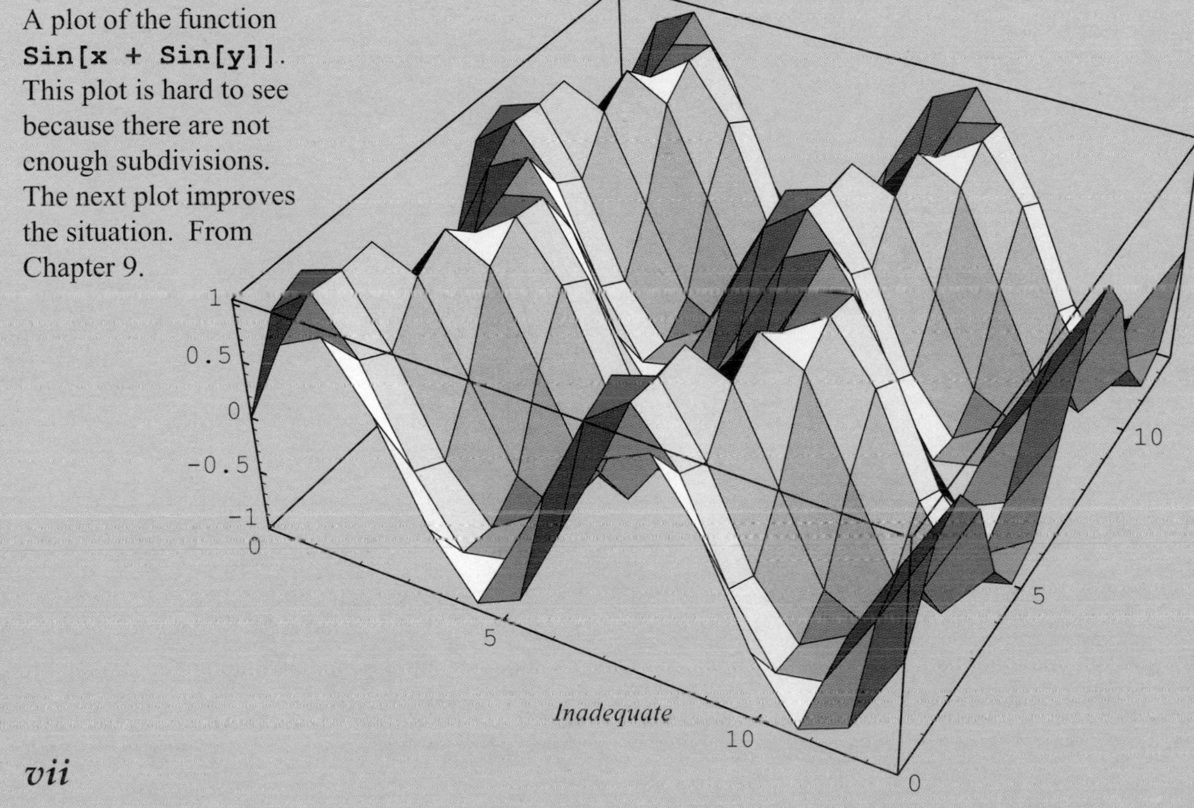

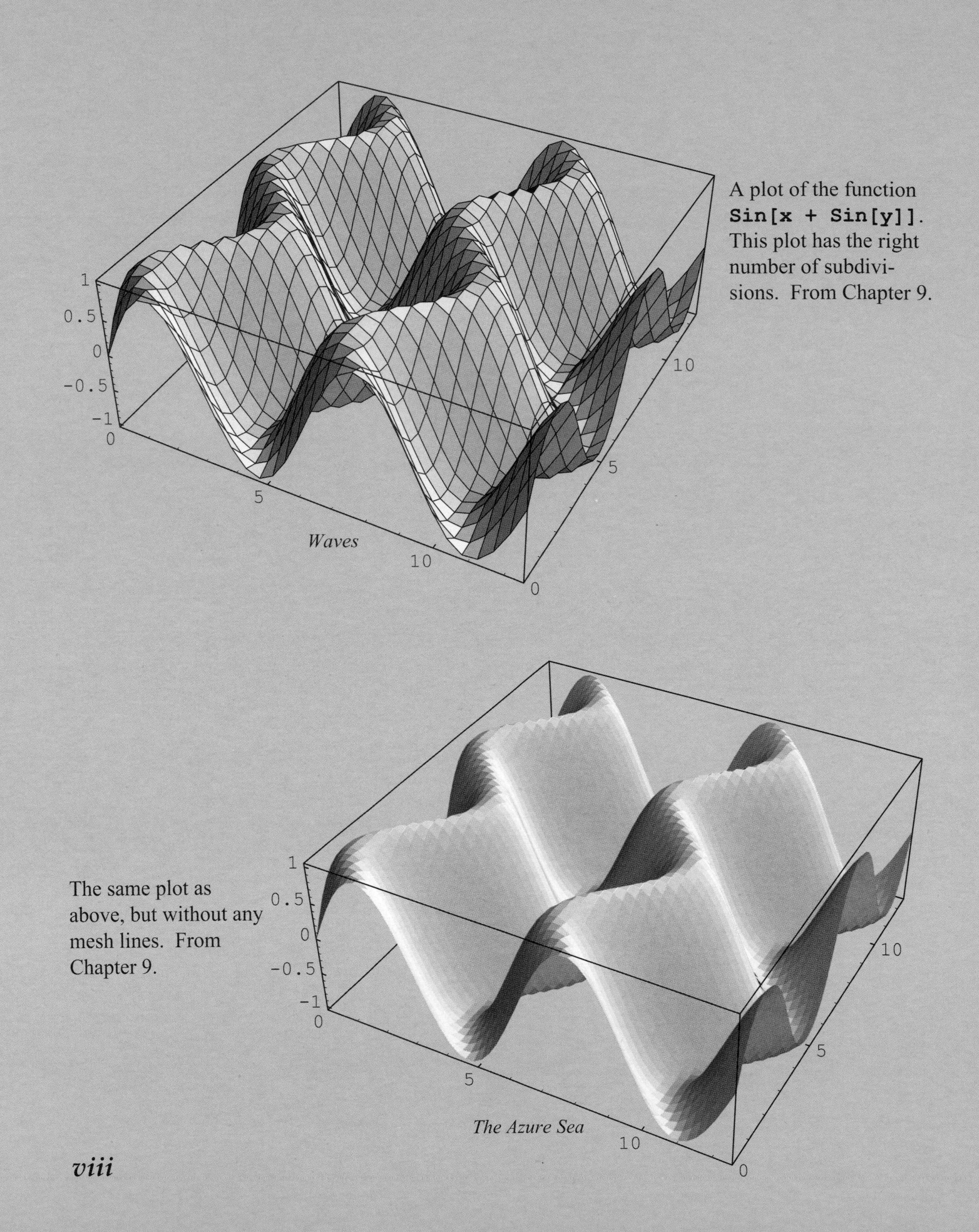

A plot of the function **Sin[x + Sin[y]]**. This plot has the right number of subdivisions. From Chapter 9.

The same plot as above, but without any mesh lines. From Chapter 9.

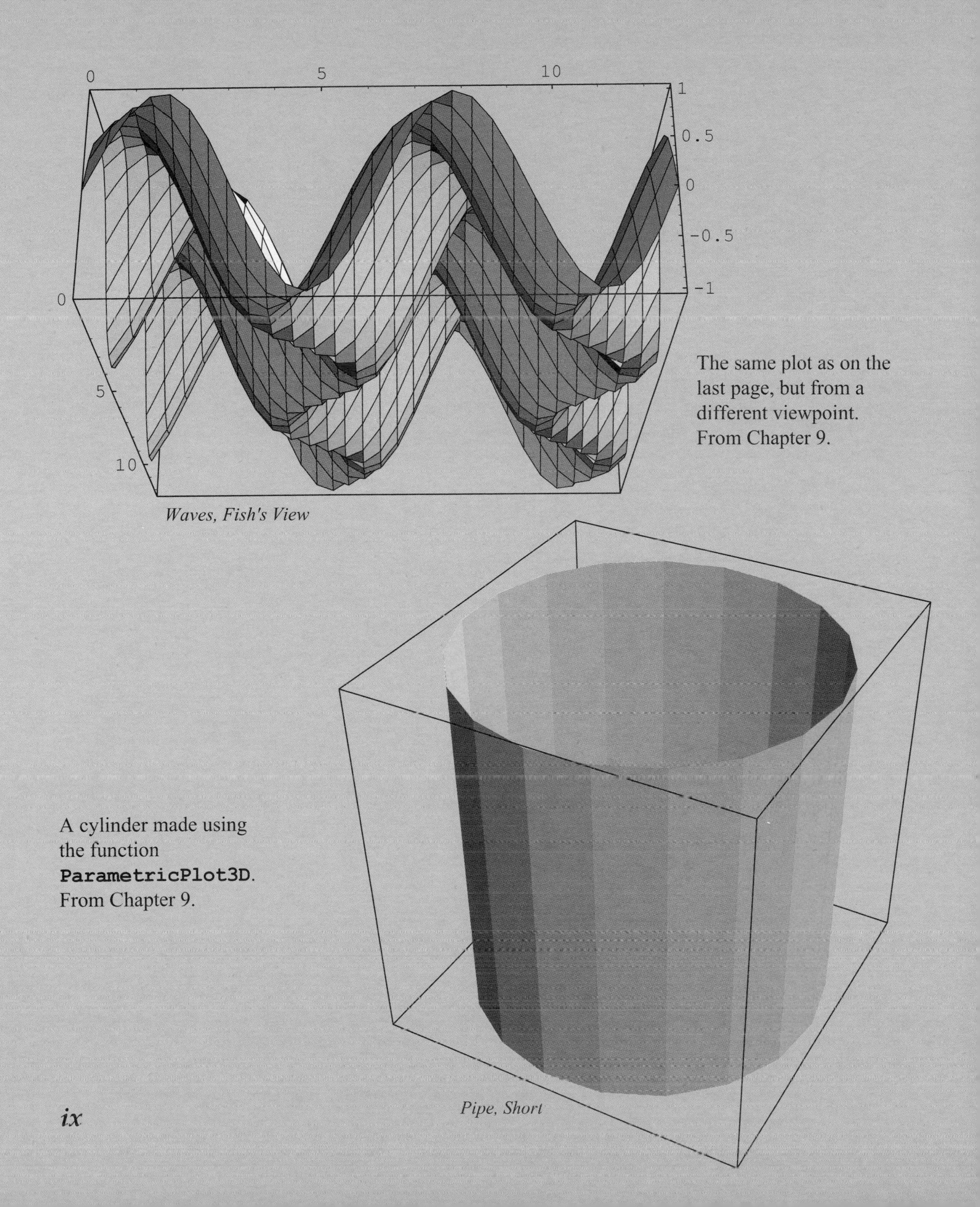

Waves, Fish's View

The same plot as on the last page, but from a different viewpoint. From Chapter 9.

Pipe, Short

A cylinder made using the function **`ParametricPlot3D`**. From Chapter 9.

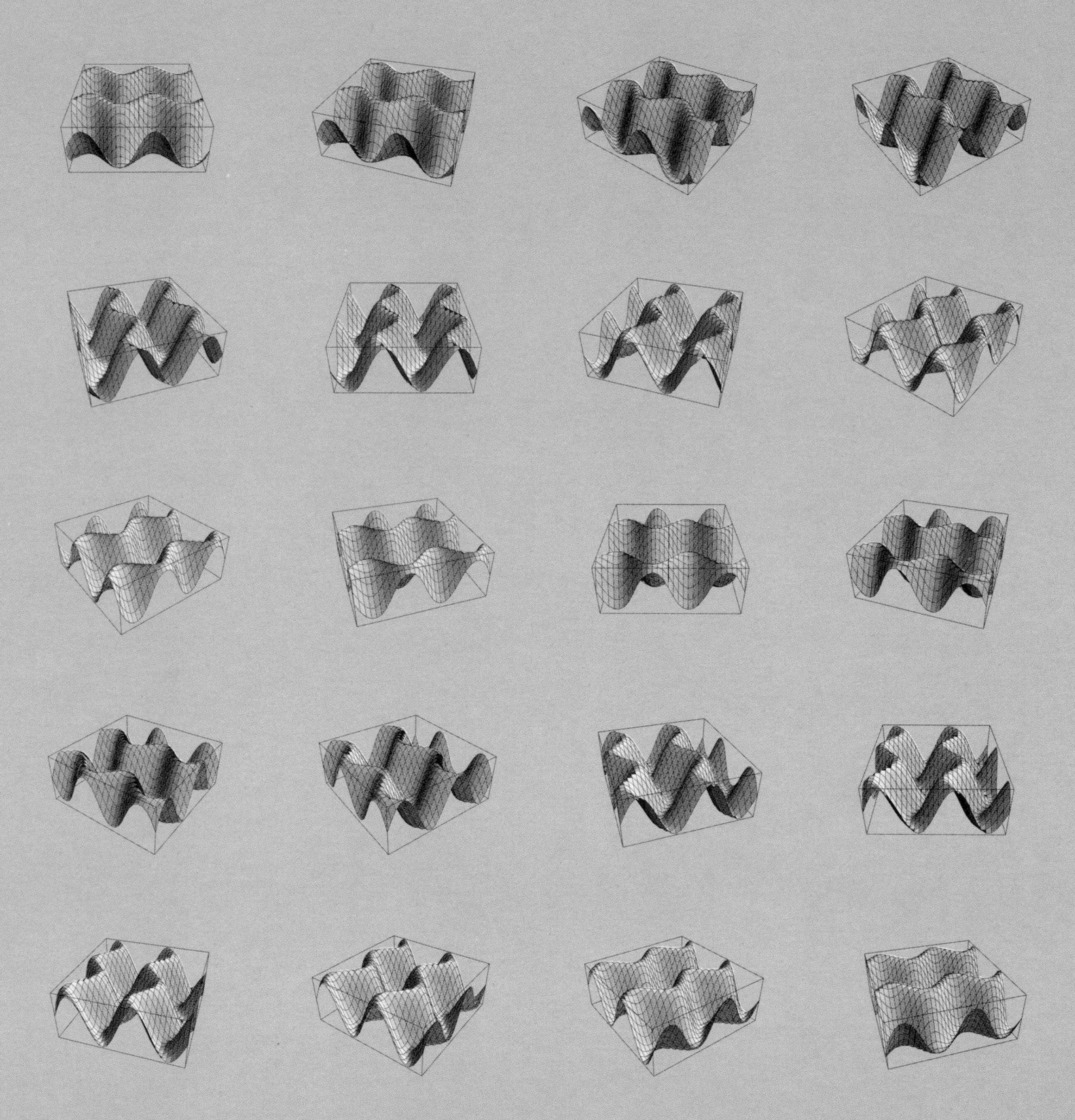

An animation of a rotating surface. From Chapter 9.

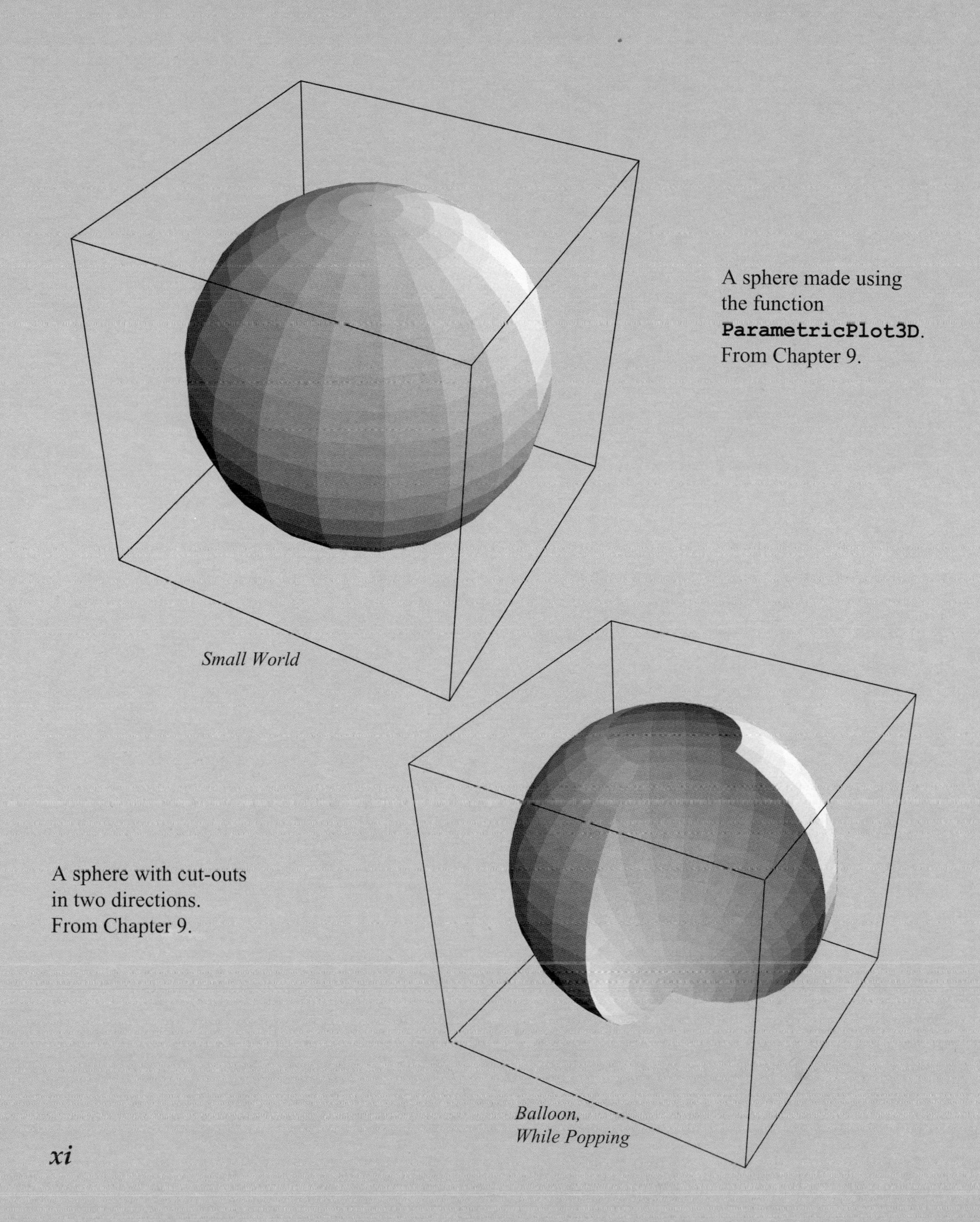

A sphere made using the function **`ParametricPlot3D`**. From Chapter 9.

A sphere with cut-outs in two directions. From Chapter 9.

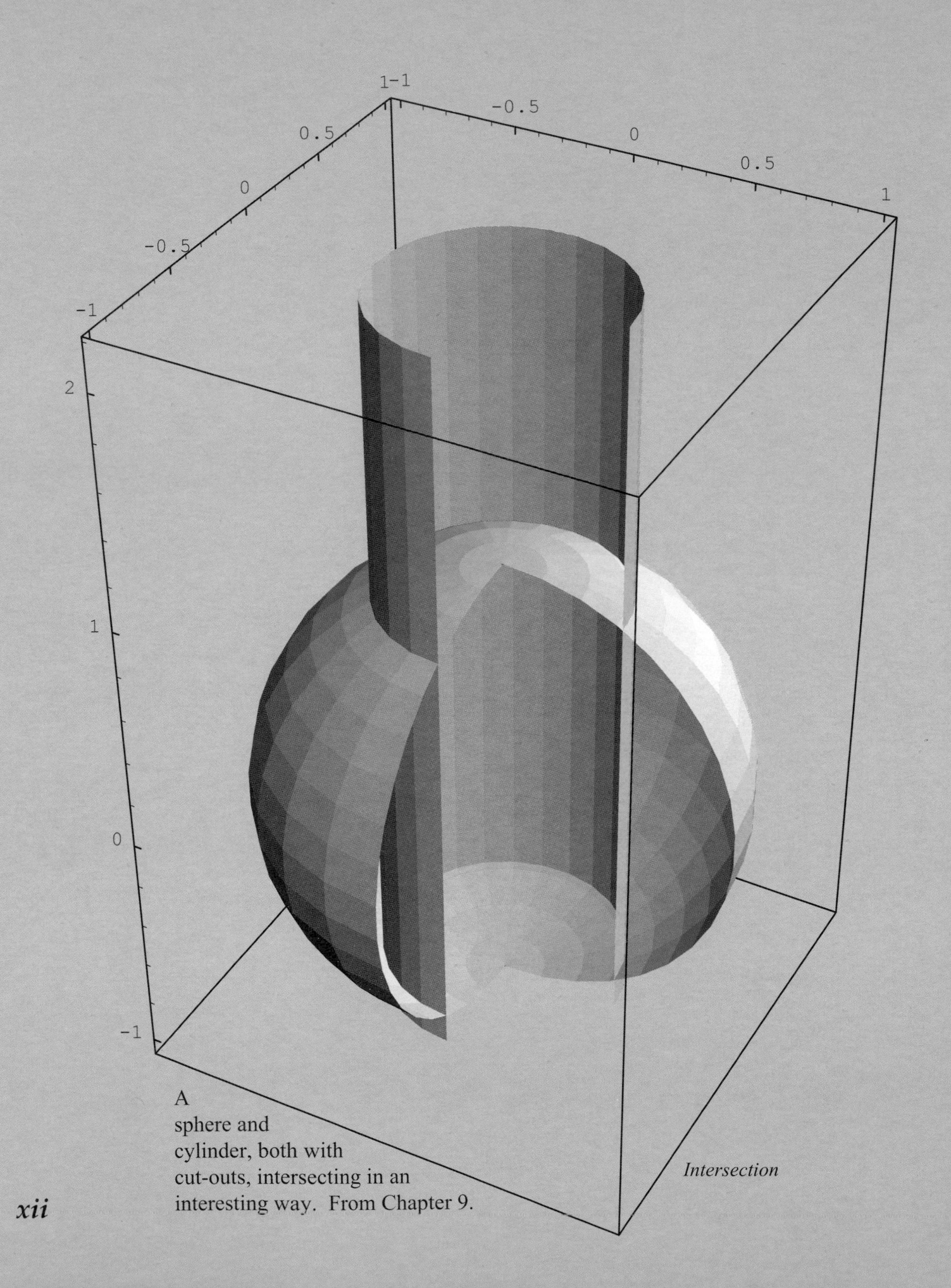

A
sphere and
cylinder, both with
cut-outs, intersecting in an
interesting way. From Chapter 9.

Intersection

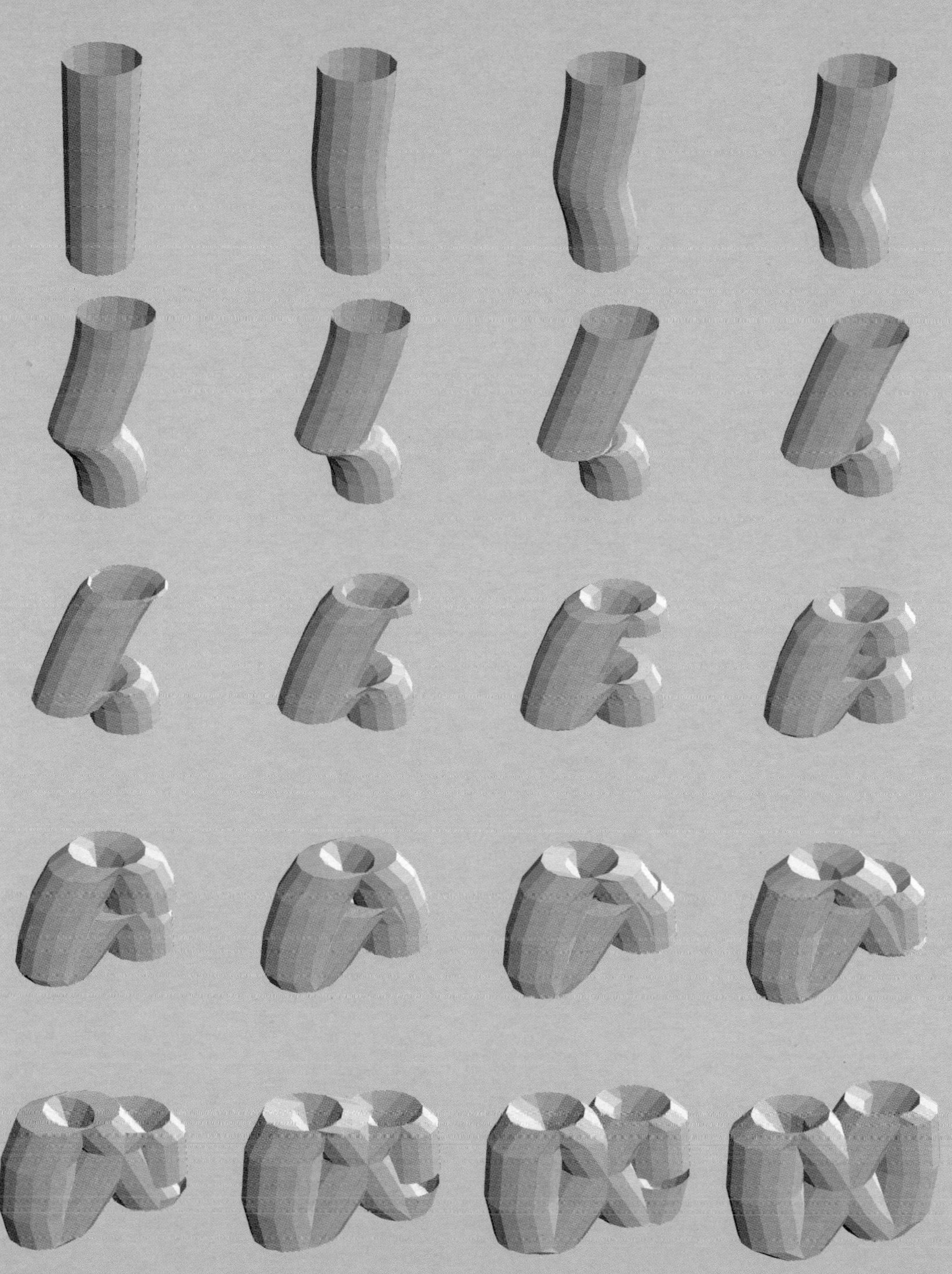

An animation of a cylinder being deformed into a self-intersecting three-dimensional Lissajous figure. From Chapter 9.

A contour plot of **Sqrt[9 - x^2] y** with the contour levels colored with the **Hue** function. From Chapter 9.

A densityplot of **Sin[x + Sin[y]]** colored with the **Hue** function. From Chapter 9.

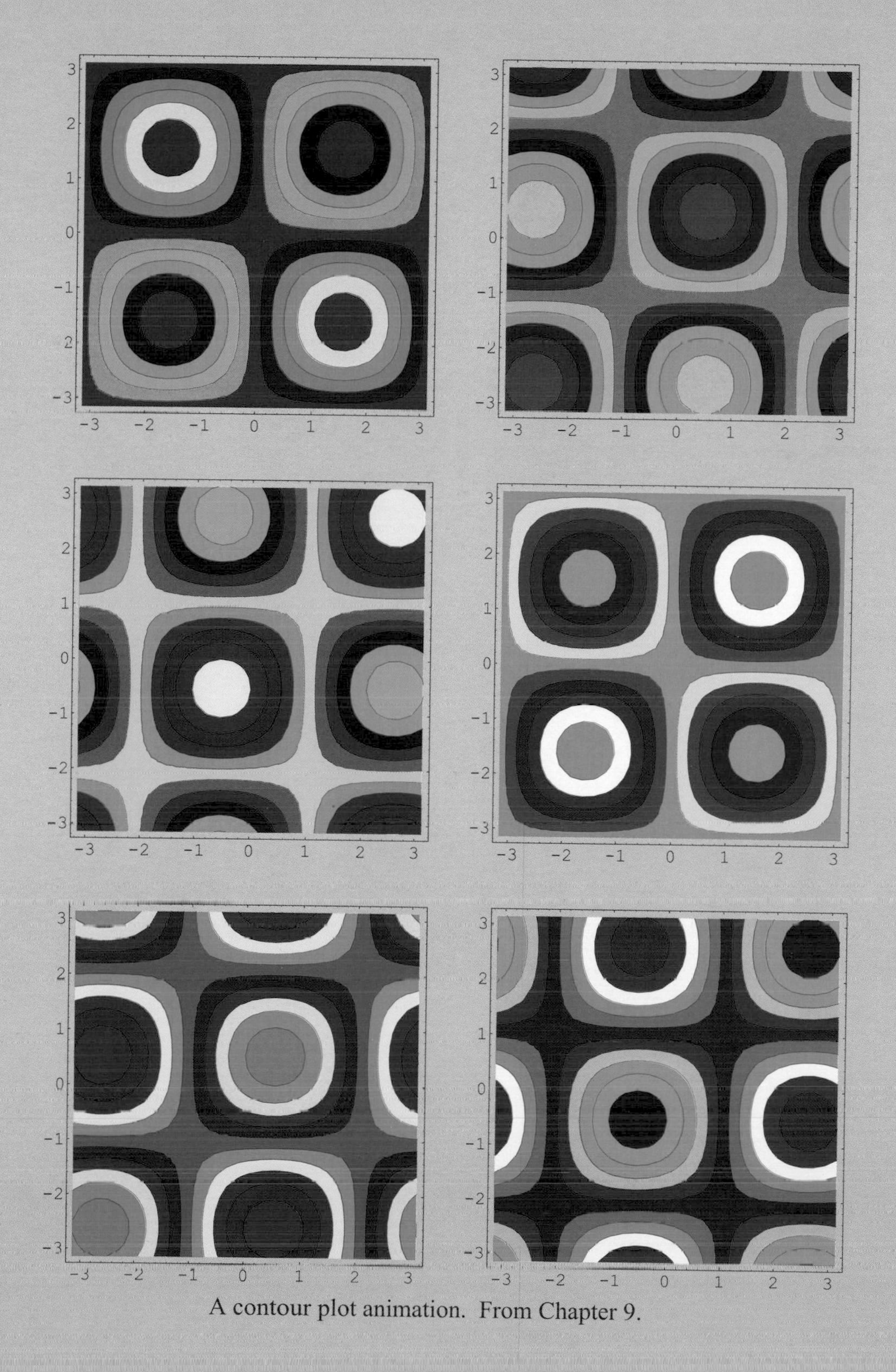

A contour plot animation. From Chapter 9.

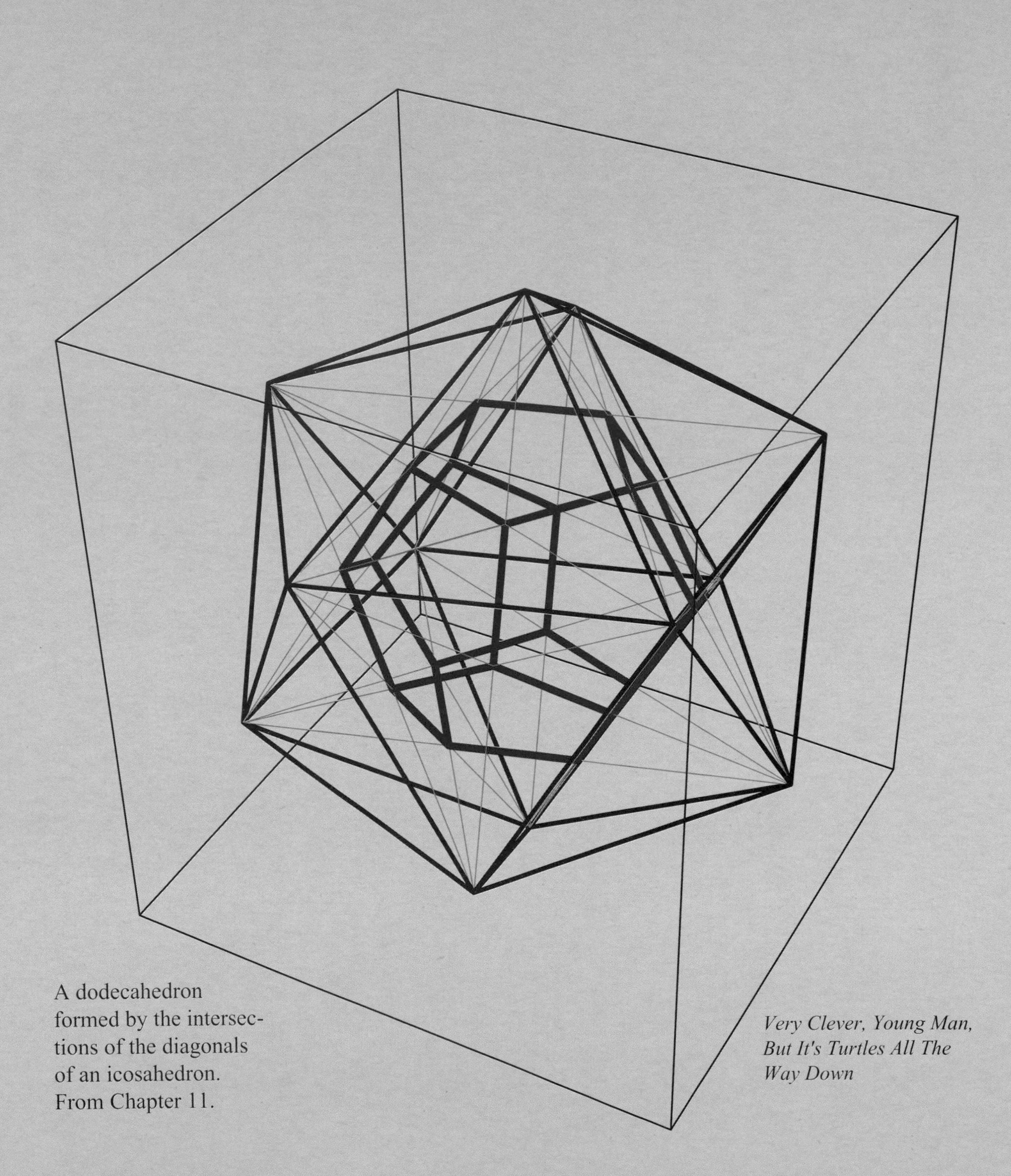

A dodecahedron formed by the intersections of the diagonals of an icosahedron. From Chapter 11.

Very Clever, Young Man, But It's Turtles All The Way Down

Jerry: Powerful! I hear static in the middle, which I assume is there for the same reason that plots of **`Sin[1/x]`** always get broken up near the origin: The function oscillates infinitely often, which is hard to plot (or to hear, for that matter).

Theo: I think at this point we should recommend that people who have not done so yet go out and get speakers to connect to their computers. Thousand-dollar computers tend to come with 50-cent speakers, which is a problem when trying to listen to dramatic, exciting sounds like the one above. A pair of sort-of-good powered speakers (Sony makes a nice line) is invaluable, and can be plugged directly into the back of a Macintosh or NeXT computer. Fifty-dollar speakers are a dramatic improvement. Headphones can also be plugged directly into the back of the Macintosh or NeXT.

Jerry: We should warn people that they probably *don't* want to connect *Mathematica* to their good stereo speakers, unless the speakers are properly protected by fuses. A few wrong functions, a heavy hand on the volume control, and you could do some serious damage.

Theo: Yes. Mathematical functions can be dangerous. Kids, don't try this at home.

Let's try using some more sophisticated, difficult-to-understand functions to make silly noises. Here's a combination of **`Sin`** and **`BesselJ`** that I quite like:

```
Play[Sin[1000 x BesselJ[0, 30 x]], {x, 0, 1},
     PlayRange -> {-1, 1}];
```

Sound Cell

(Play track 24 of the CD to hear this sound.)

Jerry: This is definitely something that the Ghostbusters® will want to use in their next movie.

Theo: By inserting an **x^2**, we can get an even more gurgly sound:

```
Play[Sin[1000 x BesselJ[0, 30 x^2]], {x, 0, 1},
    PlayRange -> {-1, 1}]
```

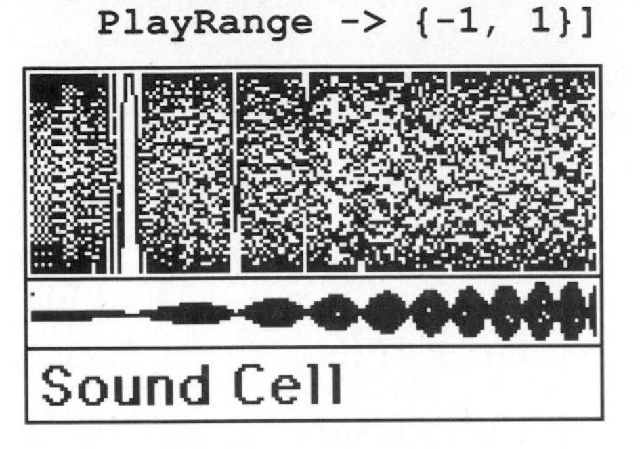

(Play track 25 of the CD to hear this sound.)

Jerry: I'm sure all the rap groups will want to use this sound on their next record.

If my students could make sounds like this one, I'm sure they would all stay at The Math Program past midnight trying to out-do each other.

Theo: I'd like to try something completely different.

There is a old problem that goes like this: Start with a number. If it's odd, multiply it by three and add one. If it's even, divide it by two. Repeat.

Jerry: Although it's never been proven, experience shows that you will always come back to 1 eventually. Can we do this in *Mathematica*?

Theo: Of course, and here is a function to do it (named after Mr. Collatz, and written by Dan Grayson, our visiting professor):

```
next[i_?OddQ] := next[i] = 3i+1
next[i_?EvenQ] := next[i] = i/2
collatz[i_] :=
    FixedPointList[next, i, SameTest -> ((#2==1)&)]
```

Let's try starting with 12:

```
collatz[12]
{12, 6, 3, 10, 5, 16, 8, 4, 2, 1}
```

How about 27:

```
collatz[27]
```

```
{27, 82, 41, 124, 62, 31, 94, 47, 142, 71, 214, 107, 322,
  161, 484, 242, 121, 364, 182, 91, 274, 137, 412, 206,
  103, 310, 155, 466, 233, 700, 350, 175, 526, 263, 790,
  395, 1186, 593, 1780, 890, 445, 1336, 668, 334, 167,
  502, 251, 754, 377, 1132, 566, 283, 850, 425, 1276,
  638, 319, 958, 479, 1438, 719, 2158, 1079, 3238, 1619,
  4858, 2429, 7288, 3644, 1822, 911, 2734, 1367, 4102,
  2051, 6154, 3077, 9232, 4616, 2308, 1154, 577, 1732,
  866, 433, 1300, 650, 325, 976, 488, 244, 122, 61, 184,
  92, 46, 23, 70, 35, 106, 53, 160, 80, 40, 20, 10, 5,
  16, 8, 4, 2, 1}
```

Jerry: Very interesting. I assume we are going to make this into a sound?

Theo: Yes. All we have to do is repeat, many times, the cycle we get for a given starting point, and use it as a waveform. Here's a function that does this. It generates the cycle for the given starting point, and then repeats this enough times to get 1000 values, which it plays as a sound:

```
playCollatz[n_] := Module[
    {digits},

    digits = collatz[n];
    digits = Flatten[
        Table[digits, {1 + 1000 / Length[digits]}]];
    digits = N[Take[digits, 1000]];

    ListPlot[Take[digits, 100], PlotJoined -> True,
        AspectRatio -> 0.2, PlotRange -> All,
        PlotLabel -> InputForm[n],
        Ticks -> None,
        Axes -> None,
        Epilog -> SampledSoundList[
                    digits / Max[digits], 7418]
    ]
]
```

Here are the sounds generated by starting at 12 and 27. Notice that the plot includes only the first 100 numbers in the repeated cycle, and that it is labeled with the starting number:

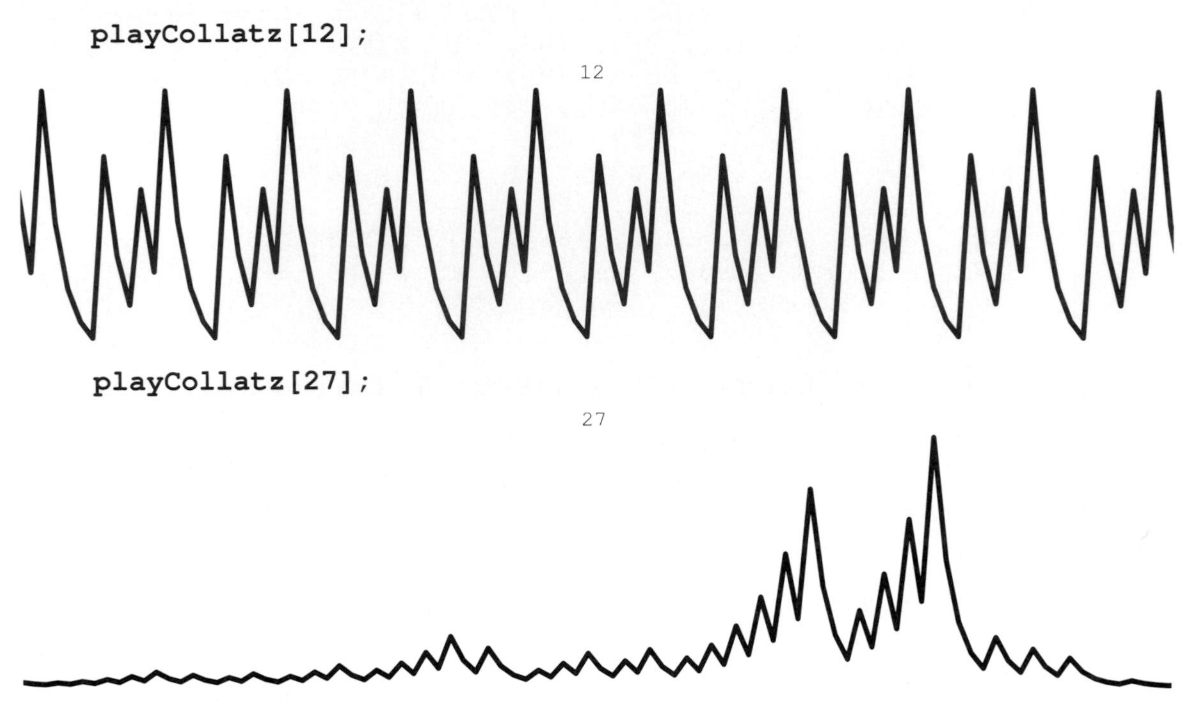

(Play track 26 of the CD to hear these two sounds.)

Jerry: There must be all sorts of different sounds we can get starting at different points. Let's do all possible starting points up to 100:

```
Do[
    playCollatz[m],
    {m, 1, 100}
]
```

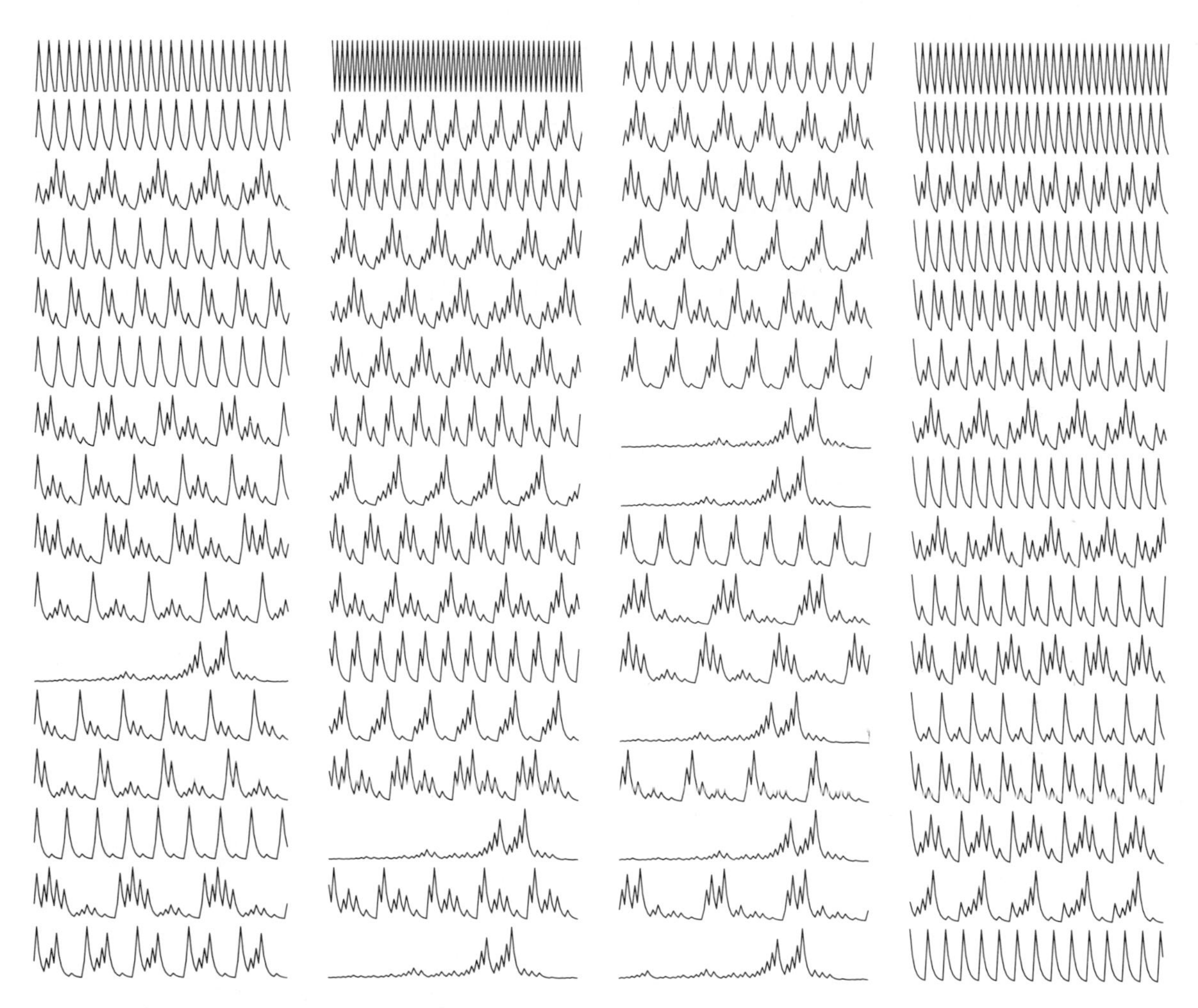

(Play track 27 of the CD to hear these sounds.)

Theo: By double-clicking the picture above (in the electronic edition), you can hear the latest in avant-garde computer music and animation. It's really quite catchy, don't you think?

Jerry: Do we have to pay royalties to Mr. Collatz, through ASCAP, for public performances of this piece? It's very attractive, in an integer sort of a way.

Theo: Not to mention mathematically informative. I notice particularly that there are many pairs of successive notes that sound the same. This is indicates that there are many pairs of successive integers that have the same cycle length, which is something I, at least, didn't know about the Collatz problem.

Jerry: Perhaps we, or our visiting professor, can figure out why these pairs are there. In the mean time, is there anything else we can do with these wonderful sounds?

Theo: Yes, there is a little-known feature of animations that is quite amusing. While an animation is running, you can click the horizontal scroll bar and drag it back and forth. As you move the scroll bar, the animation is moved forward and backward proportionally.

Jerry: You mean, if I move the scroll bar all the way to the left, I see the first frame of the animation, and as I move it to the right, it flips through the frames forward, at a speed proportional to how fast I'm moving the scroll bar?

Theo: Right. The great thing is that each time it displays a new frame, it also plays the corresponding sound. So, by moving the scroll bar back and forth, you can play little tunes using these sounds. If you move the cursor up and down without moving it horizontally, it will repeat the same note.

Jerry: What if I want to skip some notes?

Theo: If you click the pause button in the animation control palette (it's the one with two vertical bars in it), you can then click and release anywhere in the gray area of the horizontal scroll bar to move directly to that frame of the animation.

Jerry: Oh, this is very exciting. Could we make a slide trombone?

Theo: Oh, yes, let's do that!

Dialog–Second Day

Theo: To make a slide trombone in *Mathematica*, we need to draw a sequence of pictures, each of which is a trombone with the slide in a different position. Associated with each picture there will need to be a different sound.

Jerry: How on earth are we going to draw a trombone in *Mathematica*?

Theo: Well, let's look in my handy Educorp CD-ROM. Yes, here is a picture of a trombone, and it's even listed under "trombone" in the clip-art section, so I didn't have to look for it. Here's the picture that I found:

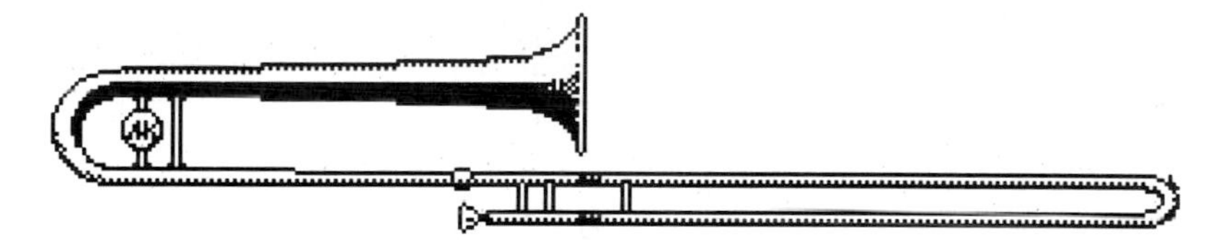

Jerry: But this is just a picture that you pasted in. We are going to have to modify it using *Mathematica* commands. How will we do that?

Theo: The first step is to trace this bit map image, and convert it into a set of lines and arcs. I used Canvas® on the Macintosh, but almost any drawing program would have worked. Here's what I got:

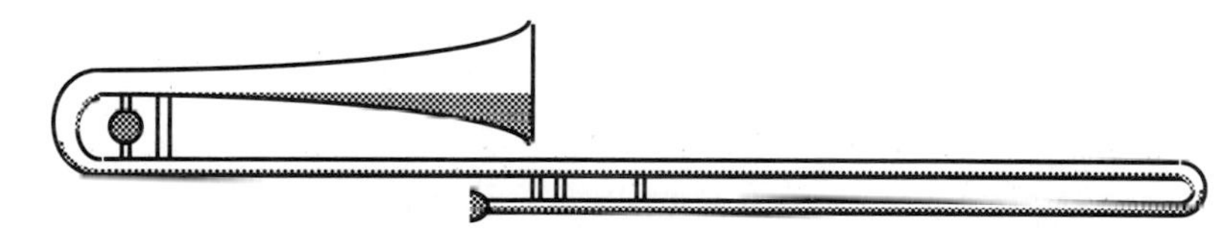

Jerry: Not bad, but this is still just a picture!

Theo: Yes, but we can use the Convert To InputForm command to convert this picture into a list of *Mathematica* graphics primitives. Here's what we get:

```
Show[Graphics[{
Thickness[0.000000], GrayLevel[0.50000],Line[{{2.380623, -1.173010}, {2.380623, -1.173010}}],
Polygon[{{0.041523, 0.069204},{0.415225, 0.069204},{0.415225, 0.110727},{0.041523, 0.110727},{0.041523, 0.069204}}],
Thickness[0.003460],GrayLevel[0.00000],Circle[{0.979239, 0.032872}, {0.022491, 0.024222},{266.0 Degree, 450.0 Degree}],
GrayLevel[1.00000],Disk[{0.977509, 0.031142}, {0.012111, 0.010381},{267.0 Degree, 450.0 Degree}],
GrayLevel[0.00000],Circle[{0.977509, 0.031142}, {0.012111, 0.010381},{267.0 Degree, 450.0 Degree}],
Line[{{0.375433, 0.008651}, {0.974048, 0.008651}}],Line[{{0.375433, 0.019031}, {0.974048, 0.019031}}],
Line[{{0.974048, 0.039792}, {0.043253, 0.039792}}],Line[{{0.974048, 0.053633}, {0.046713, 0.053633}}],
Thickness[0.000000],GrayLevel[0.50000],Line[{{0.044983, 0.055363}, {0.044983, 0.055363}}],
Thickness[0.003460],Line[{{0.375433, 0.012111}, {0.974048, 0.012111}}],
Thickness[0.000000],Line[{{0.972318, 0.013841}, {0.972318, 0.013841}}],
Thickness[0.003460],Line[{{0.970588, 0.043253}, {0.043253, 0.043253}}],
Thickness[0.000000],Line[{{0.041523, 0.044983}, {0.041523, 0.044983}}],
Thickness[0.003460],Circle[{0.974048, 0.029412}, {0.017301, 0.017301},{0.0 Degree, 90.0 Degree}],
Thickness[0.000000],Line[{{0.041523, 0.044983}, {0.041523, 0.044983}}],
Thickness[0.003460],Circle[{0.979239, 0.031142}, {0.019031, 0.019031},{266.0 Degree, 366.0 Degree}],
Thickness[0.000000],Line[{{0.041523, 0.044983}, {0.041523, 0.044983}}],
Disk[{0.359862, 0.013841}, {0.017301, 0.013841},{270.0 Degree, 450.0 Degree}],
Thickness[0.003460],GrayLevel[0.00000],Circle[{0.361592, 0.015571}, {0.017301, 0.013841},{270.0 Degree, 450.0 Degree}],
Line[{{0.413495, 0.039792}, {0.413495, 0.022491}}],Line[{{0.420415, 0.039792}, {0.420415, 0.019031}}],
Line[{{0.434256, 0.039792}, {0.434256, 0.019031}}],Line[{{0.441177, 0.039792}, {0.441177, 0.019031}}],
Line[{{0.503460, 0.039792}, {0.503460, 0.019031}}],Line[{{0.510381, 0.039792}, {0.510381, 0.019031}}],
Circle[{0.041523, 0.086505}, {0.038062, 0.044983},{86.0 Degree, 274.0 Degree}],
GrayLevel[1.00000],Disk[{0.043253, 0.081315}, {0.019031, 0.025952},{275.0 Degree, 447.0 Degree}],
GrayLevel[0.00000],Circle[{0.043253, 0.081315}, {0.019031, 0.025952},{87.0 Degree, 275.0 Degree}],
Thickness[0.000000],GrayLevel[0.50000],Line[{{0.508651, 0.020761}, {0.508651, 0.020761}}],
Thickness[0.003460],Circle[{0.043253, 0.070934}, {0.034602, 0.027682},{180.0 Degree, 270.0 Degree}],
Thickness[0.000000],Line[{{0.508651, 0.020761}, {0.508651, 0.020761}}],
Thickness[0.003460],Circle[{0.046713, 0.083045}, {0.024222, 0.029412},{86.0 Degree, 186.0 Degree}],
GrayLevel[1.00000],Disk[{0.043253, 0.067474}, {0.368512, 0.043253},{0.0 Degree, 90.0 Degree}],
GrayLevel[0.00000],Circle[{0.043253, 0.067474}, {0.368512, 0.043253},{0.0 Degree, 90.0 Degree}],
GrayLevel[1.00000],Disk[{0.043253, 0.174741}, {0.368512, 0.043253},{0.0 Degree, 270.0 Degree}],
GrayLevel[0.00000],Circle[{0.043253, 0.174741}, {0.368512, 0.043253},{270.0 Degree, 360.0 Degree}],
Line[{{0.413495, 0.167820}, {0.413495, 0.070934}}],Line[{{0.060554, 0.105536}, {0.060554, 0.053633}}],
Line[{{0.067474, 0.105536}, {0.067474, 0.053633}}],Line[{{0.091696, 0.105536}, {0.091696, 0.053633}}],
Line[{{0.102076, 0.105536}, {0.102076, 0.053633}}],Thickness[0.000000],GrayLevel[0.50000],
Line[{{0.100346, 0.055363}, {0.100346, 0.055363}}],Disk[{0.064014, 0.081315}, {0.015571, 0.015571}],
Thickness[0.003460],GrayLevel[0.00000],Circle[{0.064014, 0.081315}, {0.015571, 0.015571}]}],
AspectRatio->0.17301,PlotRange->{{0.00000, 1.00000}, {0.00000, 0.17301}}];
```

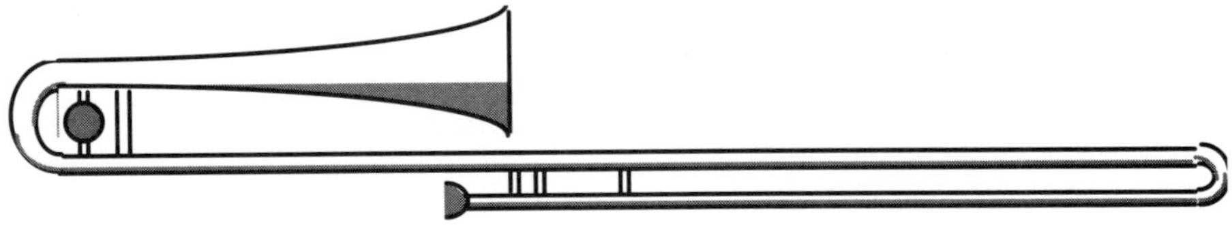

Jerry: That's amazing. The result of evaluating this load of gibberish is a trombone!

Theo: Wonders of modern science.

We are going to want to move the slide of this trombone back and forth, so we need to locate which coordinates in the "load of gibberish" correspond to the x-coordinates of the end of the slide. Fortunately, this is not too hard, because we can look for any x-coordinates larger than about 1/2; the slide is the only thing that far to the right.

I've done this, and made up a function based on the text above. It's in the Functions That We Defined section, if you want to look at it. The function is called **`trombone`**, and it takes one argument, which is a number between 0 and 1 that specifies how far the slide should be extended. Here are some examples (the **AspectRatio** and **PlotRange** options are copied from the text above):

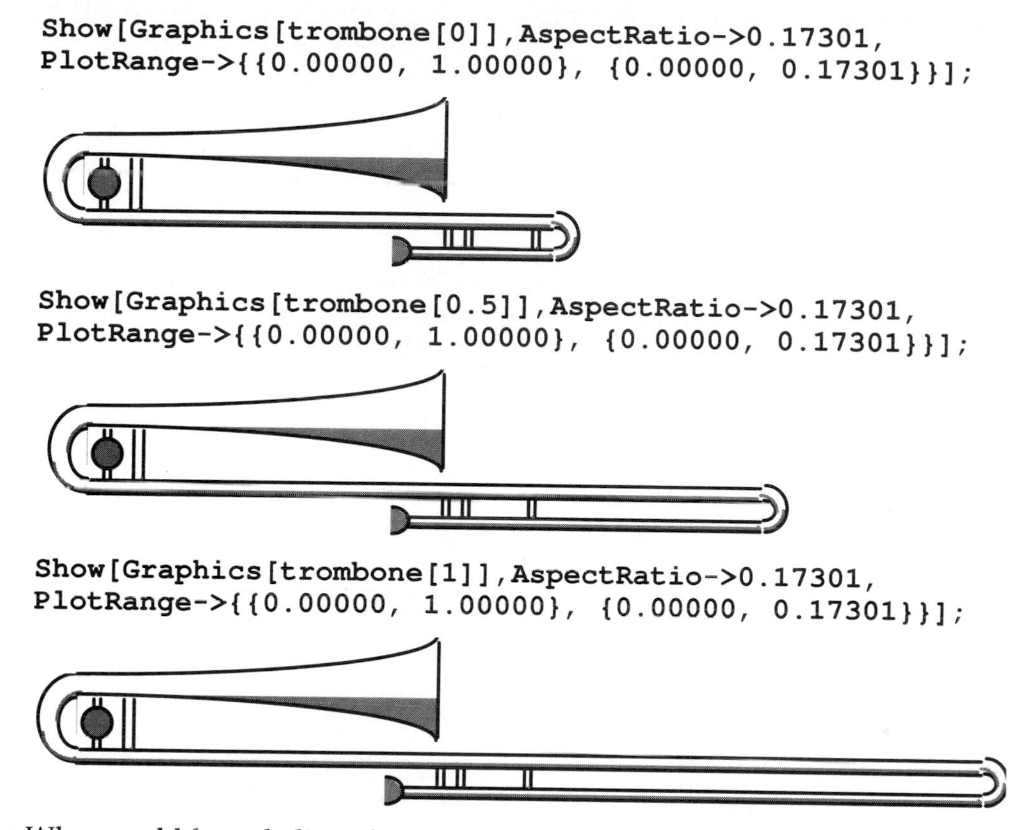

```
Show[Graphics[trombone[0]],AspectRatio->0.17301,
PlotRange->{{0.00000, 1.00000}, {0.00000, 0.17301}}];
```

```
Show[Graphics[trombone[0.5]],AspectRatio->0.17301,
PlotRange->{{0.00000, 1.00000}, {0.00000, 0.17301}}];
```

```
Show[Graphics[trombone[1]],AspectRatio->0.17301,
PlotRange->{{0.00000, 1.00000}, {0.00000, 0.17301}}];
```

Jerry: Who would have believed it! All we have to do is add a sound to each picture.

Theo: Right. Let's start at middle-C, and put in all the notes of the scale up three octaves. That will give us 36 images. Middle-C is about 263 Hz, so to get a note at middle-C we use:

```
Play[Sin[263 2 Pi x],
     {x, 0, Round[263 0.1] / 263},
     SampleRate -> 7418.1818];
```

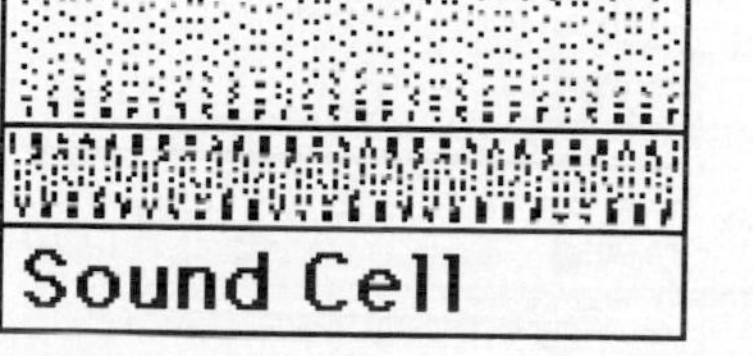

(Play track 28 of the CD to hear this sound.)

To combine this sound with a picture, we make a list of the graphics object (the result of the **trombone** function) and the sound object (the result of the **Play** function), and then use **Show** to display it. The **PlayFunction** option is used to tell **Play** not to play the sound right away, but just produce the sound object. Here it is:

```
Do[
    Show[
        Graphics[{
            freq = 263 2.0^n;
            trombone[2^n / 2^3],
            Play[Sin[freq 2 Pi x],
                {x, 0, Round[freq 0.1] / freq},
                SampleRate -> 7418.1818,
                DisplayFunction -> Identity][[1]]
        }],
        PlotRange ->
            {{0.00000, 1.00000}, {0.00000, 0.17301}},
        AspectRatio -> 0.17301
    ],
    {n, 0, 3, 1/12}
]
```

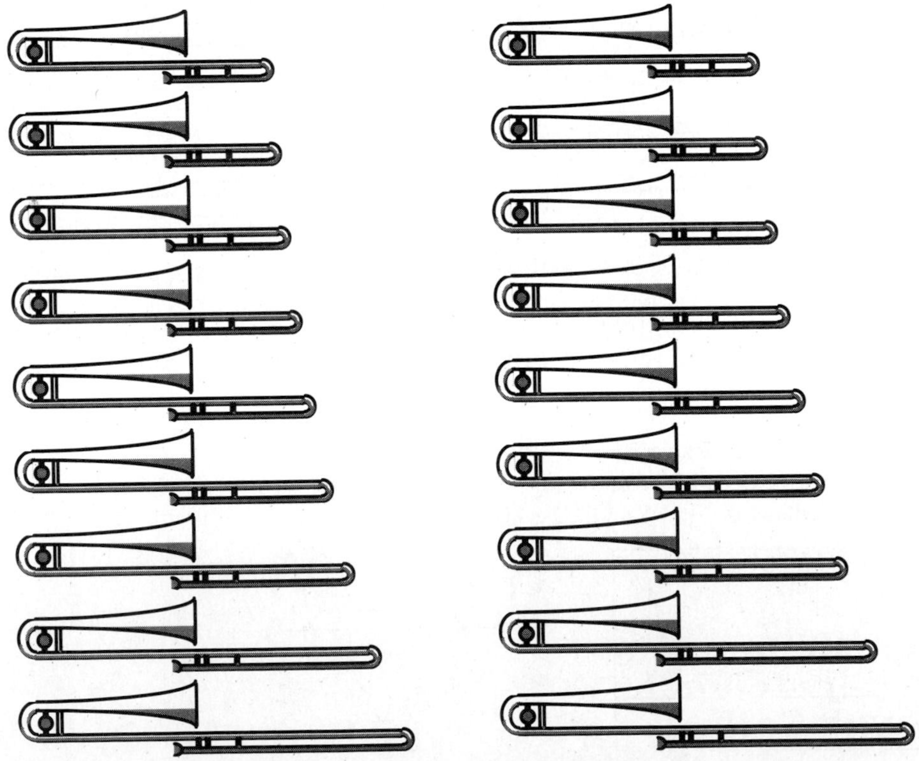

(Play track 29 of the CD to hear these sounds.)

Jerry: Incredible! Unbelievable! A *Mathematica* slide trombone!

Theo: Let's review how to play this trombone. First, double-click anywhere on the picture. This starts the animation running. You will hear rising tones as the animation runs forward. Find the horizontal scroll bar at the bottom of the window. Click and drag the "thumb" (the small box in the gray area of the scroll bar). As you move the "thumb", the trombone slide will move in and out, and appropriate sounds will be heard.

Jerry: I notice that as the slide gets longer, the sound gets higher. Isn't this contrary to the real-world trombone way of doing things? I also notice it sounds more like a marimba than a trombone. Is this a good thing?

Theo: Look, it's an electronic, *Mathematica*-based trombone. You can't expect it to conform to all sorts of pre-conceived notions about what a trombone should be.

Jerry: I get it! When cars were first invented, people made them look like horse-drawn wagons. This was clearly a mistake that we will not repeat. How would you feel if readers of this book who have *Mathematica* create a more traditional-acting and -sounding trombone?

Theo: I think it would be a waste of time. They can buy a real one for less money. They can buy a Yamaha or Korg synthesizer that sounds *much* better, also for less money. If you're going to use *Mathematica*, you ought to do something with it that it is uniquely best qualified to do. For example, a trombone that sounds like an Airy function would be an appropriate use of *Mathematica*.

■ Dialog–Third Day

Theo: Wolfram Research has produced a long, thin poster about the Riemann Zeta function. Here is part of the function on the poster:

```
Plot[RiemannSiegelZ[x], {x, 0, 85},
      AspectRatio -> 0.2, PlotPoints -> 100];
```

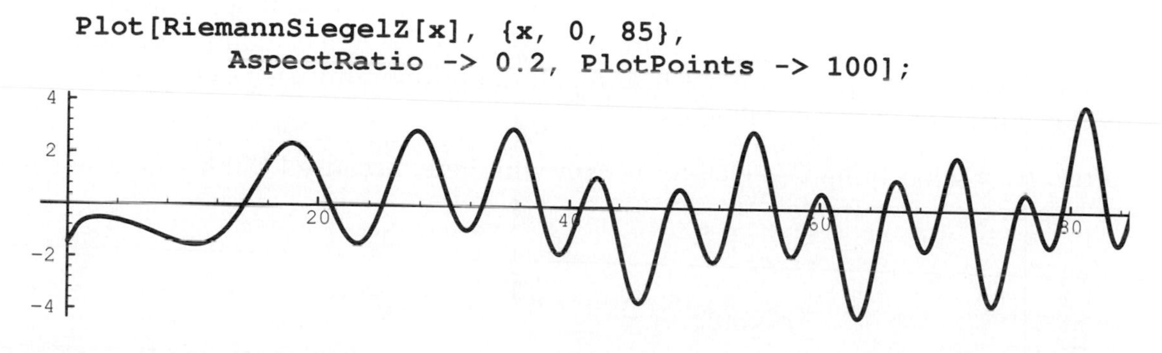

Jerry: Wait a minute, I thought you said this poster was about the Zeta function. How does **RiemannSiegelZ** connect with the Zeta function?

Theo: The poster explains it all, and you can get one by calling Wolfram Research and offering them $10. The two functions (**Zeta** and **RiemannSiegelZ**) are closely related, and in particular they have the same zeros. The one plotted above is better suited to two-dimensional plots, since **Zeta** is complex valued.

Jerry: This plot certainly looks like it might make an interesting sound! Let's try playing it:

```
Play[RiemannSiegelZ[2000 x], {x, 0, 15},
    PlayRange -> {-20, 20}, SampleRate -> 22254.5454];
```

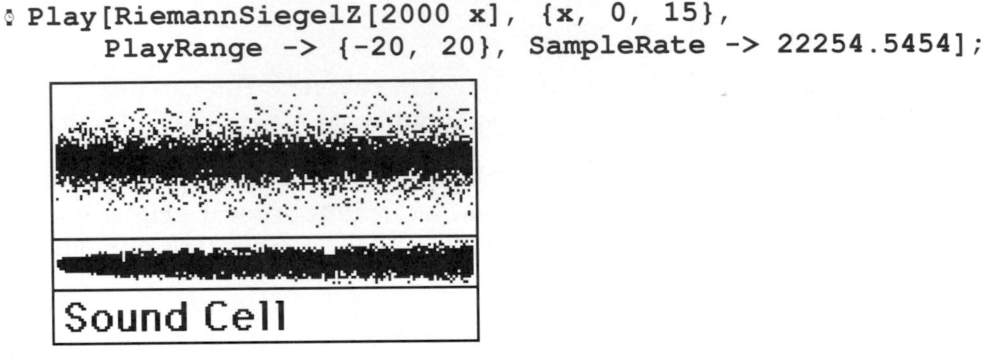

(Play track 30 of the CD to hear this sound.)

Theo: That took a pretty serious amount of time! But it was worth it. This is by far the most interesting function we've listened to yet. It sounds very rich, and grows steadily more powerful (if you have your computer connected to good speakers). It sounds a bit like some huge machine starting up.

Jerry: I read in the poster that the zeros of this function become more closely spaced in a logarithmic way as x gets bigger. That must be what causes the tone to rise quickly at first, and then more and more slowly.

Theo: I am fascinated by what might be causing the deep, rumbling, roaring sound in this function. Since the Zeta function is deeply intertwingled with prime numbers, I like to think that perhaps we are hearing all the prime numbers, crying out from the background of indifferent integers.

Jerry: It's a good thing this chapter is over with here, because I think you're starting to lose it.

Visiting Professor: Dan Grayson

We can use symbolic computation to explain the fact that some neighboring tones in the Collatz sequence have the same pitch.

Here are functions **even** and **odd** that try to tell whether a symbolic expression is even or odd. They make use of familiar rules, such as "a sum is even if its terms are both odd or both even".

```
odd[i_Integer] := OddQ[i]
even[i_Integer] := EvenQ[i]
odd [i_ j_] := odd [i] && odd [j]
even[i_ j_] := even[i] || even[j]
odd [i_+j_] := odd [i] && even[j] || even[i] && odd[j]
even[i_+j_] := even[i] && even[j] || odd [i] && odd[j]
```

Now we introduce a function for computing the Collatz sequence of an expression which makes use of the functions above.

```
Clear[next,collatz]
next[1] := End
next[End] := End
next[i_?odd] := next[i] = Expand[3i+1]
next[i_?even] := next[i] = Expand[i/2]
next[i_] := Unknown
collatz[i_] := Drop[FixedPointList[next,i],-2]
```

Now we can run it:

```
collatz[16n+5]
{5 + 16 n, 16 + 48 n, 8 + 24 n, 4 + 12 n, 2 + 6 n, 1 + 3 n}
```

This output tells us that if we start with a number of the form **16n+5** (such as 37) then its Collatz sequence begins with the numbers listed above. Eventually we get down to **3n+1**. What happens after that? Try this:

```
next[3n+1]
Unknown
```

This tells us that we can't tell what comes next in the sequence. The reason is that we don't know whether **3n+1** is even or odd. If the number **n** is odd, then **3n+1** is even, and if the number **n** is even, then **3n+1** is odd; without further information about the parity of **n**, we can't tell what to do next.

Let's construct a table that shows the Collatz sequence for numbers of the form **16n+i**, and let **i** range between **0** and **15**, so that all possibilities are covered.

```
ColumnForm[ Table[ collatz[16n+i], {i,0,15}]]
```

```
{16 n, 8 n, 4 n, 2 n, n}
{1 + 16 n, 4 + 48 n, 2 + 24 n, 1 + 12 n, 4 + 36 n, 2 + 18 n, 1 + 9 n}
{2 + 16 n, 1 + 8 n, 4 + 24 n, 2 + 12 n, 1 + 6 n, 4 + 18 n, 2 + 9 n}
{3 + 16 n, 10 + 48 n, 5 + 24 n, 16 + 72 n, 8 + 36 n, 4 + 18 n, 2 + 9 n}
{4 + 16 n, 2 + 8 n, 1 + 4 n, 4 + 12 n, 2 + 6 n, 1 + 3 n}
{5 + 16 n, 16 + 48 n, 8 + 24 n, 4 + 12 n, 2 + 6 n, 1 + 3 n}
{6 + 16 n, 3 + 8 n, 10 + 24 n, 5 + 12 n, 16 + 36 n, 8 + 18 n, 4 + 9 n}
{7 + 16 n, 22 + 48 n, 11 + 24 n, 34 + 72 n, 17 + 36 n, 52 + 108 n, 26 + 54 n, 13 + 27 n}
{8 + 16 n, 4 + 8 n, 2 + 4 n, 1 + 2 n, 4 + 6 n, 2 + 3 n}
{9 + 16 n, 28 + 48 n, 14 + 24 n, 7 + 12 n, 22 + 36 n, 11 + 18 n, 34 + 54 n, 17 + 27 n}
{10 + 16 n, 5 + 8 n, 16 + 24 n, 8 + 12 n, 4 + 6 n, 2 + 3 n}
{11 + 16 n, 34 + 48 n, 17 + 24 n, 52 + 72 n, 26 + 36 n, 13 + 18 n, 40 + 54 n, 20 + 27 n}
{12 + 16 n, 6 + 8 n, 3 + 4 n, 10 + 12 n, 5 + 6 n, 16 + 18 n, 8 + 9 n}
{13 + 16 n, 40 + 48 n, 20 + 24 n, 10 + 12 n, 5 + 6 n, 16 + 18 n, 8 + 9 n}
{14 + 16 n, 7 + 8 n, 22 + 24 n, 11 + 12 n, 34 + 36 n, 17 + 18 n, 52 + 54 n, 26 + 27 n}
{15 + 16 n, 46 + 48 n, 23 + 24 n, 70 + 72 n, 35 + 36 n, 106 + 108 n, 53 + 54 n, 160 + 162 n, 80 + 81 n}
```

Looking at the terminal numbers in each of the above sequences reveals some coincidences. For example, if we start with **2+16n** or with **3+16n**, we get to **2+9n** after 6 steps. If we start with **4+16n** or with **5+16n**, we get to **1+3n** after 5 steps. And if we start with **12+16n** or with **13+16n**, we get to **8+9n** after 6 steps. Each one of those three coincidences represents an adjacent pair of notes in the Collatz song with the same pitch, and the pattern of coincidences repeats itself after 16 notes.

Notice that I have chosen the coefficient of n above to be 16. There is no sense in taking a number like 48 (with an odd divisor of 3) as coefficient of n above, because that doesn't give us any extra information about whether numbers we encounter are even or odd. So we go to the next reasonable possibility and create a table for the numbers modulo 32.

```
ColumnForm[ Table[ collatz[32n+i], {i,0,31}]]
```

```
{32 n, 16 n, 8 n, 4 n, 2 n, n}
{1 + 32 n, 4 + 96 n, 2 + 48 n, 1 + 24 n, 4 + 72 n, 2 + 36 n, 1 + 18 n, 4 + 54 n, 2 + 27 n}
{2 + 32 n, 1 + 16 n, 4 + 48 n, 2 + 24 n, 1 + 12 n, 4 + 36 n, 2 + 18 n, 1 + 9 n}
{3 + 32 n, 10 + 96 n, 5 + 48 n, 16 + 144 n, 8 + 72 n, 4 + 36 n, 2 + 18 n, 1 + 9 n}
{4 + 32 n, 2 + 16 n, 1 + 8 n, 4 + 24 n, 2 + 12 n, 1 + 6 n, 4 + 18 n, 2 + 9 n}
{5 + 32 n, 16 + 96 n, 8 + 48 n, 4 + 24 n, 2 + 12 n, 1 + 6 n, 4 + 18 n, 2 + 9 n}
{6 + 32 n, 3 + 16 n, 10 + 48 n, 5 + 24 n, 16 + 72 n, 8 + 36 n, 4 + 18 n, 2 + 9 n}
{7 + 32 n, 22 + 96 n, 11 + 48 n, 34 + 144 n, 17 + 72 n, 52 + 216 n, 26 + 108 n, 13 + 54 n, 40 + 162 n, 20 + 81 n}
{8 + 32 n, 4 + 16 n, 2 + 8 n, 1 + 4 n, 4 + 12 n, 2 + 6 n, 1 + 3 n}
{9 + 32 n, 28 + 96 n, 14 + 48 n, 7 + 24 n, 22 + 72 n, 11 + 36 n, 34 + 108 n, 17 + 54 n, 52 + 162 n, 26 + 81 n}
{10 + 32 n, 5 + 16 n, 16 + 48 n, 8 + 24 n, 4 + 12 n, 2 + 6 n, 1 + 3 n}
{11 + 32 n, 34 + 96 n, 17 + 48 n, 52 + 144 n, 26 + 72 n, 13 + 36 n, 40 + 108 n, 20 + 54 n, 10 + 27 n}
{12 + 32 n, 6 + 16 n, 3 + 8 n, 10 + 24 n, 5 + 12 n, 16 + 36 n, 8 + 18 n, 4 + 9 n}
{13 + 32 n, 40 + 96 n, 20 + 48 n, 10 + 24 n, 5 + 12 n, 16 + 36 n, 8 + 18 n, 4 + 9 n}
{14 + 32 n, 7 + 16 n, 22 + 48 n, 11 + 24 n, 34 + 72 n, 17 + 36 n, 52 + 108 n, 26 + 54 n, 13 + 27 n}
{15 + 32 n, 46 + 96 n, 23 + 48 n, 70 + 144 n, 35 + 72 n, 106 + 216 n, 53 + 108 n, 160 + 324 n, 80 + 162 n, 40 + 81 n}
{16 + 32 n, 8 + 16 n, 4 + 8 n, 2 + 4 n, 1 + 2 n, 4 + 6 n, 2 + 3 n}
{17 + 32 n, 52 + 96 n, 26 + 48 n, 13 + 24 n, 40 + 72 n, 20 + 36 n, 10 + 18 n, 5 + 9 n}
{18 + 32 n, 9 + 16 n, 28 + 48 n, 14 + 24 n, 7 + 12 n, 22 + 36 n, 11 + 18 n, 34 + 54 n, 17 + 27 n}
{19 + 32 n, 58 + 96 n, 29 + 48 n, 88 + 144 n, 44 + 72 n, 22 + 36 n, 11 + 18 n, 34 + 54 n, 17 + 27 n}
{20 + 32 n, 10 + 16 n, 5 + 8 n, 16 + 24 n, 8 + 12 n, 4 + 6 n, 2 + 3 n}
{21 + 32 n, 64 + 96 n, 32 + 48 n, 16 + 24 n, 8 + 12 n, 4 + 6 n, 2 + 3 n}
{22 + 32 n, 11 + 16 n, 34 + 48 n, 17 + 24 n, 52 + 72 n, 26 + 36 n, 13 + 18 n, 40 + 54 n, 20 + 27 n}
{23 + 32 n, 70 + 96 n, 35 + 48 n, 106 + 144 n, 53 + 72 n, 160 + 216 n, 80 + 108 n, 40 + 54 n, 20 + 27 n}
{24 + 32 n, 12 + 16 n, 6 + 8 n, 3 + 4 n, 10 + 12 n, 5 + 6 n, 16 + 18 n, 8 + 9 n}
{25 + 32 n, 76 + 96 n, 38 + 48 n, 19 + 24 n, 58 + 72 n, 29 + 36 n, 88 + 108 n, 44 + 54 n, 22 + 27 n}
{26 + 32 n, 13 + 16 n, 40 + 48 n, 20 + 24 n, 10 + 12 n, 5 + 6 n, 16 + 18 n, 8 + 9 n}
{27 + 32 n, 82 + 96 n, 41 + 48 n, 124 + 144 n, 62 + 72 n, 31 + 36 n, 94 + 108 n, 47 + 54 n, 142 + 162 n, 71 + 81 n}
{28 + 32 n, 14 + 16 n, 7 + 8 n, 22 + 24 n, 11 + 12 n, 34 + 36 n, 17 + 18 n, 52 + 54 n, 26 + 27 n}
{29 + 32 n, 88 + 96 n, 44 + 48 n, 22 + 24 n, 11 + 12 n, 34 + 36 n, 17 + 18 n, 52 + 54 n, 26 + 27 n}
{30 + 32 n, 15 + 16 n, 46 + 48 n, 23 + 24 n, 70 + 72 n, 35 + 36 n, 106 + 108 n, 53 + 54 n, 160 + 162 n, 80 + 81 n}
{31 + 32 n, 94 + 96 n, 47 + 48 n, 142 + 144 n, 71 + 72 n, 214 + 216 n, 107 + 108 n, 322 + 324 n, 161 + 162 n,
 484 + 486 n, 242 + 243 n}
```

There are even more coincidences here! Notice that **4+32n**, **5+32n**, and **6+32n** all reach **2+9n** after seven steps. We can illustrate this with a particular value of **n**:

```
Table[Length[collatz[32*8+i]],{i,0,31}]
```

```
{9, 123, 123, 123, 30, 30, 30, 79, 30, 123, 30, 22, 30,
  30, 43, 43, 17, 30, 92, 92, 17, 17, 43, 43, 17, 43, 17,
  61, 105, 105, 105, 43}
```

The three 30's which appear near the beginning of this list are as predicted.

Functions That We Defined

Theo: The function **collatz** calculates the famous Collatz sequence for a given starting point:

```
next[i_?OddQ]  := next[i] = 3i+1
next[i_?EvenQ]  := next[i] = i/2
collatz[i_] :=
    FixedPointList[next, i, SameTest -> ((#2==1)&)]
```

The function **playCollatz** plays the sequence as an amplitude waveform:

```
playCollatz[n_] := Module[
    {digits},

    digits = collatz[n];
    digits = Flatten[
        Table[digits, {1 + 1000 / Length[digits]}]];
    digits = N[Take[digits, 1000]];

    ListPlot[Take[digits, 100], PlotJoined -> True,
        AspectRatio -> 0.2, PlotRange -> All,
        PlotLabel -> InputForm[n],
        Ticks -> None,
        Axes -> None,
        Epilog -> SampledSoundList[
                    digits / Max[digits], 7418]
    ]
]
```

The function **`trombone`** takes a number between zero and 1 and returns a **`Graphics`** object representing a trombone with its slide at a position determined by the number.

```
trombone[n_] := {Thickness[0.003460],GrayLevel[0.50000],
Line[{{2.380623, -1.173010}, {2.380623, -1.173010}}],
Polygon[{{0.042523, 0.069204},{0.415225, 0.069204},{0.415225, 0.110727},{0.042523, 0.110727},
{0.042523, 0.069204}}],GrayLevel[0.00000],
Circle[{0.979239 + n 0.45-0.45, 0.032872}, {0.022491, 0.024222}, {270 Degree, 450 Degree}],
GrayLevel[1.00000],Disk[{0.977509 + n 0.45-0.45, 0.031142}, {0.012111, 0.010381},{270 Degree,
450 Degree}],
GrayLevel[0.00000],Circle[{0.977509 + n 0.45-0.45, 0.031142}, {0.012111, 0.010381},
{270 Degree, 450 Degree}],Line[{{0.375433, 0.008651}, {0.974048 + n 0.45-0.45, 0.008651}}],
Line[{{0.375433, 0.019031}, {0.974048 + n 0.45-0.45, 0.019031}}],
Line[{{0.974048 + n 0.45-0.45, 0.039792}, {0.043253, 0.039792}}],
Line[{{0.974048 + n 0.45-0.45, 0.053633}, {0.046713, 0.053633}}],
GrayLevel[0.50000],Line[{{0.044983, 0.055363}, {0.044983, 0.055363}}],
Line[{{0.375433, 0.012111}, {0.974048 + n 0.45-0.45, 0.012111}}],
Line[{{0.972318 + n 0.45-0.45, 0.013841}, {0.972318 + n 0.45-0.45, 0.013841}}],
Line[{{0.970588 + n 0.45-0.45, 0.043253}, {0.043253, 0.043253}}],
Line[{{0.041523, 0.044983}, {0.041523, 0.044983}}],
Circle[{0.974048 + n 0.45-0.45, 0.029412}, {0.017301, 0.017301},{0.0 Degree, 90.0 Degree}],
Line[{{0.041523, 0.044983}, {0.041523, 0.044983}}],
Circle[{0.979239 + n 0.45-0.45, 0.031142}, {0.019031, 0.019031},{270 Degree, 360.0 Degree}],
Line[{{0.041523, 0.044983}, {0.041523, 0.044983}}],
Disk[{0.359862, 0.013841}, {0.017301, 0.013841},{270 Degree, 450 Degree}],
GrayLevel[0.00000],Circle[{0.361592, 0.015571}, {0.017301, 0.013841},{270 Degree, 450 Degree}],
Line[{{0.413495, 0.039792}, {0.413495, 0.022491}}],Line[{{0.420415, 0.039792}, {0.420415, 0.019031}}],
Line[{{0.434256, 0.039792}, {0.434256, 0.019031}}],Line[{{0.441177, 0.039792}, {0.441177, 0.019031}}],
Line[{{0.503460, 0.039792}, {0.503460, 0.019031}}],Line[{{0.510381, 0.039792}, {0.510381, 0.019031}}],
Circle[{0.041523, 0.086505}, {0.038062, 0.044983},{90 Degree, 270 Degree}],
GrayLevel[1.00000],Disk[{0.043253, 0.081315}, {0.019031, 0.025952},{270 Degree, 450 Degree}],
GrayLevel[0.00000],Circle[{0.043253, 0.081315}, {0.019031, 0.025952},{90 Degree, 270 Degree}],
GrayLevel[0.50000],Line[{{0.508651, 0.020761}, {0.508651, 0.020761}}],
Circle[{0.043253, 0.070934}, {0.034602, 0.027682},{180.0 Degree, 270.0 Degree}],
Line[{{0.508651, 0.020761}, {0.508651, 0.020761}}],
Circle[{0.046713, 0.083045}, {0.024222, 0.029412},{90 Degree, 180 Degree}],
GrayLevel[1.00000],Disk[{0.043253, 0.067474}, {0.368512, 0.043253},{0 Degree, 90 Degree}],
GrayLevel[0.00000],Circle[{0.043253, 0.067474}, {0.368512, 0.043253},{0 Degree, 90 Degree}],
GrayLevel[1.00000],Disk[{0.043253, 0.174741}, {0.368512, 0.043253},{0 Degree, 270 Degree}],
GrayLevel[0.00000],Circle[{0.043253, 0.174741}, {0.368512, 0.043253},{270 Degree, 360 Degree}],
Line[{{0.413495, 0.167820}, {0.413495, 0.070934}}],Line[{{0.060554, 0.105536}, {0.060554, 0.053633}}],
Line[{{0.067474, 0.105536}, {0.067474, 0.053633}}],Line[{{0.091696, 0.105536}, {0.091696, 0.053633}}],
Line[{{0.102076, 0.105536}, {0.102076, 0.053633}}],
GrayLevel[0.50000],Line[{{0.100346, 0.055363}, {0.100346, 0.055363}}],
Disk[{0.064014, 0.081315}, {0.015571, 0.015571}],
GrayLevel[0.00000],Circle[{0.064014, 0.081315}, {0.015571, 0.015571}]};
```

Chapter Seventeen
Factoring `x^n - 1`: So You Think 100 Cases Is Enough?

In which Jerry and Theo encounter a polynomial with surprising properties, and end up concluding that you just can't tell by looking **[1]**.

Dialog

Jerry: The **`Factor`** command in *Mathematica* allows us to explore a variety of cases of **`x^n - 1`** in factored form. I'm interested in this because I recently learned some amazing things about **`x^n - 1`** from my friend Wally Dodge, an excellent math teacher at New Trier High School in suburban Chicago.

Theo: Let's start factoring some examples of **`x^n - 1`**, and see what patterns we can discover.

```
Factor[x^2 - 1]
```

$(-1 + x)\ (1 + x)$

```
Factor[x^3 - 1]
```

$(-1 + x)\ (1 + x + x^2)$

```
Factor[x^5-1]
```

$(-1 + x)\ (1 + x + x^2 + x^3 + x^4)$

Jerry: We might notice many features of the factored expressions. For example, in each case the number of terms in the second factor is equal to n.

Theo: Maybe we should try some more examples, like a table of the first ten:

```
Table[{n, Factor[x^n - 1]}, {n, 1, 10}]//TableForm
```

```
1     -1 + x
2     (-1 + x) (1 + x)
                         2
3     (-1 + x) (1 + x + x )
                              2
4     (-1 + x) (1 + x) (1 + x )
                         2    3    4
5     (-1 + x) (1 + x + x  + x  + x )
                                  2              2
6     (-1 + x) (1 + x) (1 - x + x ) (1 + x + x )
                         2    3    4    5    6
7     (-1 + x) (1 + x + x  + x  + x  + x  + x )
                              2         4
8     (-1 + x) (1 + x) (1 + x ) (1 + x )
                         2         3    6
9     (-1 + x) (1 + x + x ) (1 + x  + x )
                                  2    3    4
10    (-1 + x) (1 + x) (1 - x + x  - x  + x )
                   2    3    4
        (1 + x + x  + x  + x )
```

Well, clearly that idea didn't hold up. There doesn't seem to be any pattern in the number of terms in the last factor. Perhaps the number of factors is equal to the number of prime divisors of n.

Jerry: No, that doesn't work for the case of n = 4, for example. I think it's the number of divisors of n, including non-prime divisors: 10 has factors 1, 2, 5, and 10, and that's four divisors, and there are four factors in the **x^10-1**. For 9, there are three divisors (1, 3, and 9) and **x^9 - 1** has three factors. So let me predict 15. 15 has 1, 3, 5, and 15. That's four divisors, so **x^15 - 1** will have four factors.

```
Factor[x^15 - 1]
```

```
                      2                2    3    4
(-1 + x) (1 + x + x ) (1 + x + x  + x  + x )
              3    4    5    7    8
  (1 - x + x  - x  + x  - x  + x )
```

Theo: That looks good to me. We need to check your theory a bit more thoroughly. Let's write a function that tests it for any n. First, we need to find out how many divisors a number has. Fortunately there is a function, **Divisors**, which will give us a list of all the divisors of a number.

```
Divisors[8]
```

```
{1, 2, 4, 8}
```

The function **Length** gives the number of elements in a list, and we can use it to get the number of divisors from the list of divisors.

```
Length[Divisors[8]]
4
```

Now we need the number of terms in the factored form of x^8 - 1. There is a function **FactorList**, which lists the factors of a polynomial (it's sort of like **Factor**, except that it gives a list of factors, instead of a single polynomial with those factors multiplied together):

```
FactorList[x^8 - 1]
                                2            4
{{-1 + x, 1}, {1 + x, 1}, {1 + x , 1}, {1 + x , 1}}
```

This is a list of pairs, where the first part of each pair is the factor, and the second part says how many times it appears (each of the factors of **x^8 - 1** appears just once).

```
Length[FactorList[x^8 - 1]]
4
```

This is the number of factors in **x^8 - 1**.

We can write a **Do** loop to test the first hundred cases, and print a message if Jerry is ever found to be wrong.

```
Do[If[Length[Divisors[n]] !=
              Length[FactorList[Factor[x^n - 1]]],
          Print["Jerry's Wrong!"]],
    {n, 1, 100}]
```

It certainly looks like Jerry had a pretty good idea here, since nothing got printed out. But you never know with these things. Later on, we'll have to find out for sure.

Jerry: Let's pretend that my conjecture is established, and look for other statements about **x^n - 1** that we believe are true. Looking back at our list of **x^n - 1** factored up to n = 10, I'm struck by the fact that each factored expression has an **x - 1** in it. Maybe this is always true. Can we write another test to see if this is true?

Theo: Well, we could, but I think we can probably just prove it's always true, without testing any examples.

```
Expand[(x - 1) Sum[x^m, {m, 0, 9}]]
      10
-1 + x
```

Jerry: For any n, 10 in this case, we can make up a polynomial which, when multiplied by **x - 1** gives **x^n - 1**. It's just the sum **1 + x + x^2 + ... + x^(n-1)**. So, **x^n - 1** always has **x - 1** as a factor.

Theo: Another thing we might notice about our list is that all the coefficients are either 1 or -1. I wonder if this is always true.

Jerry: Our list was for n from 1 to 10. How about trying from 69 to 71, to pick two funny numbers that come to mind.

```
Table[{n, Factor[x^n - 1]}, {n, 69, 71}]//TableForm
```

```
                         2            2     3     4     5     6     7     8     9
       (-1 + x) (1 + x + x ) (1 + x + x  + x  + x  + x  + x  + x  + x  + x  +

           10    11    12    13    14    15    16    17    18    19    20
          x   + x   + x   + x   + x   + x   + x   + x   + x   + x   + x   +

           21    22               3    4    6    7    9    10    12    13    15
          x   + x  ) (1 - x + x  - x  + x  - x  + x  - x   + x   - x   + x   -

           16    18    19    21    22    23    25    26    28    29    31
          x   + x   - x   + x   - x   + x   - x   + x   - x   + x   - x   +

           32    34    35    37    38    40    41    43    44
69        x   - x   + x   - x   + x   - x   + x   - x   + x  )

                                      2    3    4             2    3    4
70     (-1 + x) (1 + x) (1 - x + x  - x  + x ) (1 + x + x  + x  + x )

                   2    3    4    5    6             2    3    4    5    6
71       (1 - x + x  - x  + x  - x  + x ) (1 + x + x  + x  + x  + x  + x )

                   5    6    7    8    10    11    12    13    14    16    17
         (1 - x + x  - x  + x  - x  + x   - x   + x   - x   + x   - x   + x   -

           18    19    23    24                 5    6    7    8    10    11    12
          x   + x   - x   + x  ) (1 + x - x  - x  - x  - x  + x   + x   + x   +

           13    14    16    17    18    19    23    24
          x   + x   - x   - x   - x   - x   + x   + x  )

                           2    3    4    5    6    7    8    9    10    11
       (-1 + x) (1 + x + x  + x  + x  + x  + x  + x  + x  + x  + x   + x   +

           12    13    14    15    16    17    18    19    20    21    22
          x   + x   + x   + x   + x   + x   + x   + x   + x   + x   + x   +

           23    24    25    26    27    28    29    30    31    32    33
          x   + x   + x   + x   + x   + x   + x   + x   + x   + x   + x   +

           34    35    36    37    38    39    40    41    42    43    44
          x   + x   + x   + x   + x   + x   + x   + x   + x   + x   + x   +

           45    46    47    48    49    50    51    52    53    54    55
          x   + x   + x   + x   + x   + x   + x   + x   + x   + x   + x   +

           56    57    58    59    60    61    62    63    64    65    66
          x   + x   + x   + x   + x   + x   + x   + x   + x   + x   + x   +

           67    68    69    70
          x   + x   + x   + x  )
```

It looks good for your theory. Any reasonable person would have to agree. Maybe just a bit more testing. Is there any way to test without actually staring at these results? Could something be written which would test the coefficients of the factored polynomials and print something only if a non 1 or -1 coefficient appears?

Theo: Let's think about it. The first order of business is to get the list of coefficients out of the factored form. Let's start with **FactorList**:

```
FactorList[x^8 - 1]
```

```
                                   2            4
{{-1 + x, 1}, {1 + x, 1}, {1 + x , 1}, {1 + x , 1}}
```

We really don't care about the number of times that any factor showed up, so let's just extract the factors themselves. We want to make a list of just the first element of each pair in this list. Transposing this list will give us a list of all the first elements and a list of all the second elements:

```
Transpose[FactorList[x^8 - 1]]
```

$\{\{-1 + x, 1 + x, 1 + x^2, 1 + x^4\}, \{1, 1, 1, 1\}\}$

Jerry: I'm impressed with what **Transpose** just did! It must have very wide uses.

Theo: Yes, people who think **Transpose** is just for matrices are missing a lot. **Transpose** is actually one of the most useful general-purpose functions to keep in mind. Many times it will get you out of a bind.

We can now take the first element of this list:

```
First[Transpose[FactorList[x^8 - 1]]]
```

$\{-1 + x, 1 + x, 1 + x^2, 1 + x^4\}$

We would like to make a little function that does all these things for us. We'll call it **justFactorList**, because it's just like **FactorList** except that it returns just the factors themselves:

```
justFactorList[expr_] :=
    First[Transpose[FactorList[expr]]]
```

Jerry: So, now we can use the **justFactorList** function, instead of using **FactorList** together with these other functions.

Theo: Right. There is also a function, **CoefficientList**, that gives a list of the coefficients of a polynomial. For example:

```
CoefficientList[5 + 7 x + 9 x^3, x]
```

```
{5, 7, 0, 9}
```

Jerry: What's the second argument, **x**, for?

Theo: If you had a polynomial with more than one variable you might want the coefficients just of a particular variable, so the second argument tells *Mathematica* which variable you want.

Jerry: Why do I have to put in the **x**? My polynomial has only one variable in it.

Theo: Really? How do you know that it's not a polynomial in the variable **q**, that happens to have only the zero order term in it:

```
CoefficientList[5 + 7 x + 9 x^3, q]
```

```
           3
{5 + 7 x + 9 x }
```

Jerry: This says that, as a polynomial in **q**, the first coefficient is **5 + 7 x + 9 x^3**. That makes sense.

Theo: Now we can attack the list we got above (note that **CoefficientList** is "Listable", which means that if you apply it to a list of polynomials it moves inside the list and applies itself to each of the individual polynomials):

```
CoefficientList[justFactorList[x^8 - 1], x]
```

```
{{-1, 1}, {1, 1}, {1, 0, 1}, {1, 0, 0, 0, 1}}
```

Jerry: A big improvement! This list is much easier to scan. I have only to look for anything that is not 1 or -1 or 0. If I want to look at **x^4 - 1** factored, I see:

```
Factor[x^4 - 1]
```

```
                         2
(-1 + x) (1 + x) (1 + x )
```

and now in this new form, I see:

```
CoefficientList[justFactorList[x^4 - 1], x]
```

```
{{-1, 1}, {1, 1}, {1, 0, 1}}
```

Theo: Yes, but this list is really not much shorter than the polynomial itself. I think we need to make it much smaller. Let's make a list of just the unique, different factors. We start by flattening this nested list down into a simple list:

```
Flatten[CoefficientList[justFactorList[x^4 - 1], x]]
```

```
{-1, 1, 1, 1, 1, 0, 1}
```

This list has all the elements of the one above, but as members of one list, instead of three separate sublists. We can use the **Union** function, which takes a list and returns a list of the unique elements in it:

```
Union[Flatten[
    CoefficientList[justFactorList[x^4 - 1], x]]]
```

```
{-1, 0, 1}
```

There.

Jerry: OK, I agree. Let's do that on a big one, maybe x^100 - 1.

```
Union[Flatten[
    CoefficientList[justFactorList[x^100 - 1], x]]]
{-1, 0, 1}
```

Theo: Clearly we have to run this for all **n** from 1 to 100, to prove the rule. We want it to print out only if the list is not `{-1, 0, 1}` or `{-1, 1}` (which is what you get if there are no zero coefficients).

```
Do[
    test = Union[Flatten[CoefficientList[
            justFactorList[x^n - 1], x]]];
    If[
        test != {-1, 0, 1} && test != {-1, 1},
        Print[n, ") ", test]
    ],
    {n, 1, 100}
]
```

Jerry: So, no output again. Your theory looks good up to `x^100 - 1`. Is that it? My friend Wally Dodge recently suggested I check `x^105 - 1`.

Theo: Are you suggesting that something that works for the first 100 cases might not be a rule? Shocking. We better test n from 1 to 1000, just to be extra sure!

```
Do[
    test = Union[Flatten[CoefficientList[
            justFactorList[x^n - 1], x]]];
    If[
        test != {-1, 0, 1} && test != {-1, 1},
        Print[n, ") ", test]
    ],
    {n, 1, 1000}
]
105) {-2, -1, 0, 1}
165) {-1, 0, 1, 2}
195) {-2, -1, 0, 1}
210) {-2, -1, 0, 1, 2}
255) {-1, 0, 1, 2}
273) {-1, 0, 1, 2}
285) {-2, -1, 0, 1}
315) {-2, -1, 0, 1}
330) {-2, -1, 0, 1, 2}
345) {-1, 0, 1, 2}
357) {-2, -1, 0, 1}
385) {-3, -2, -1, 0, 1, 2}
```

```
390) {-2, -1, 0, 1}
420) {-2, -1, 0, 1, 2}
429) {-2, -1, 0, 1}
455) {-2, -1, 0, 1, 2}
495) {-1, 0, 1, 2}
510) {-2, -1, 0, 1, 2}
525) {-2, -1, 0, 1}
546) {-2, -1, 0, 1, 2}
555) {-2, -1, 0, 1}
561) {-1, 0, 1, 2}
570) {-2, -1, 0, 1, 2}
585) {-2, -1, 0, 1}
595) {-2, -1, 0, 1, 2, 3}
609) {-2, -1, 0, 1}
615) {-1, 0, 1, 2}
627) {-2, -1, 0, 1}
630) {-2, -1, 0, 1, 2}
645) {-2, -1, 0, 1}
660) {-2, -1, 0, 1, 2}
665) {-2, -1, 0, 1, 2, 3}
690) {-2, -1, 0, 1, 2}
705) {-1, 0, 1, 2}
714) {-2, -1, 0, 1, 2}
715) {-2, -1, 0, 1, 2}
735) {-2, -1, 0, 1}
759) {-1, 0, 1, 2}
765) {-1, 0, 1, 2}
770) {-3, -2, -1, 0, 1, 2, 3}
777) {-1, 0, 1, 2}
780) {-2, -1, 0, 1}
795) {-1, 0, 1, 2}
805) {-2, -1, 0, 1, 2}
819) {-1, 0, 1, 2}
825) {-1, 0, 1, 2}
840) {-2, -1, 0, 1, 2}
855) {-2, -1, 0, 1}
858) {-2, -1, 0, 1, 2}
897) {-2, -1, 0, 1}
910) {-2, -1, 0, 1, 2}
935) {-3, -2, -1, 0, 1, 2}
945) {-2, -1, 0, 1}
957) {-1, 0, 1, 2}
969) {-2, -1, 0, 1}
975) {-2, -1, 0, 1}
987) {-2, -1, 0, 1}
990) {-2, -1, 0, 1, 2}
```

Oh my! You know, it's enough to shake your confidence in the whole idea of experimental mathematics.

Jerry: So there are many **x^n - 1**'s whose coefficients are not just -1 or 0 or 1. (For example, when **x^935 - 1** is factored, it has coefficients -3, -2, -1, 0, 1, and 2.)

Theo: But none before **x^105- 1**!

Jerry: Very hard to believe.

I wonder if my theory that the number of terms in the factored form of **x^n - 1** is the same as the number of divisors of n is valid also only for the first 100 or so cases.

Theo: I wonder. Our friend Dan Grayson, who knows everything we are trying to find out, but won't tell, suggested that we look at the roots of the factors of **x^n - 1**. (Dan's comments appear in the Visiting Professor section at the end of this chapter.)

Jerry: Lets' start with **x^8 - 1**. Factor it:

```
Factor[x^8 - 1]

                              2           4
(-1 + x) (1 + x) (1 + x ) (1 + x )
```

So, what we're talking about are the roots of **x-1 == 0**, and **1+x == 0**, and **1+x^2 == 0**, and **1+x^4 == 0**. (*Mathematica* uses "==" to mean equality, and "=" to mean assignment to a variable.)

Theo: Right.

Jerry: Can we use **Solve** to get the roots, like this?

```
Solve[x^8 - 1 == 0, x]

                    I/4 Pi                        (3 I)/4 Pi
{{x -> 1}, {x -> E      }, {x -> I}, {x -> E            },
                    (5 I)/4 Pi
  {x -> -1}, {x -> E          }, {x ->  I},
          (7 I)/4 Pi
  {x -> E          }}
```

Theo: These are all the roots, but we can't tell from which of the factors each root came. That is, with which factor in the factored form of **x^n - 1** each root is associated. We know that all the factors together will have the same set of roots as the original polynomial, but Dan says that we should be interested in which root goes with which factor. We really need to use **Solve** on each of the factors separately. We can use the **Map** function to do this. First we need a function that will take any polynomial (like one of the factors), set it equal to zero, and solve it for **x**. Here is such a function:

```
solveIt[f_] := Solve[f == 0, x]
```

We can use **Map** to apply this function to each element in the list of factors:

```
Map[solveIt, justFactorList[x^8 - 1]]
{{{x -> 1}}, {{x -> -1}}, {{x -> I}, {x -> -I}},
  {{x -> (-1)^(1/4)}, {x -> I (-1)^(1/4)}, {x -> -(-1)^(1/4)},
   {x -> -I (-1)^(1/4)}}}
```

Jerry: Show me a simpler example of how the **Map** function works. I'm not sure I understand what just happened.

Theo: OK, we can use the **Map** function to apply the **Sin** function to each element in a list:

```
Map[Sin, {a, b, c}]
{Sin[a], Sin[b], Sin[c]}
```

Jerry: Ah.

Theo: I think I would get a better idea of what these roots are by looking at them in decimal form:

```
N[Map[solveIt, justFactorList[x^8 - 1]]]
{{{x -> 1.}}, {{x -> -1.}}, {{x -> 1. I}, {x -> -1. I}},
  {{x -> 0.707107 + 0.707107 I},
   {x -> -0.707107 + 0.707107 I},
   {x -> -0.707107 - 0.707107 I},
   {x -> 0.707107 - 0.707107 I}}}
```

Jerry: Maybe not. Wouldn't it be nice if we could see these roots plotted in the complex plane? That way we could see where they are, relative to each other.

Theo: After considerable experimentation, we have come up with the following little bit of *Mathematica*:

```
makePoints[{eqn_, size_}] :=
    Block[{t},
        t = x /. N[Solve[eqn == 0, x]];
        {
            PointSize[size],
            Map[Point, Transpose[{Re[t], Im[t]}]]
        }
    ]
```

```
sizedPlot[n_] :=
    Show[
        Graphics[{
            Circle[{0, 0}, 1],
            Line[{{-1, 0}, {1, 0}}],
            Line[{{0, -1}, {0, 1}}],
            Map[makePoints,
                Transpose[{justFactorList[x^n - 1],
                           Divisors[n] 0.1/n}]
            ]
        }],
        AspectRatio -> Automatic,
        PlotRange -> {{-1.15, 1.15}, {-1.15, 1.15}},
        PlotLabel -> n
    ]
```

Jerry: Am I supposed to understand all of this?

Theo: No, but we do explain some of it in Chapters 6, 9, and 10. It allows us to make plots like the following:

```
sizedPlot[8];
```

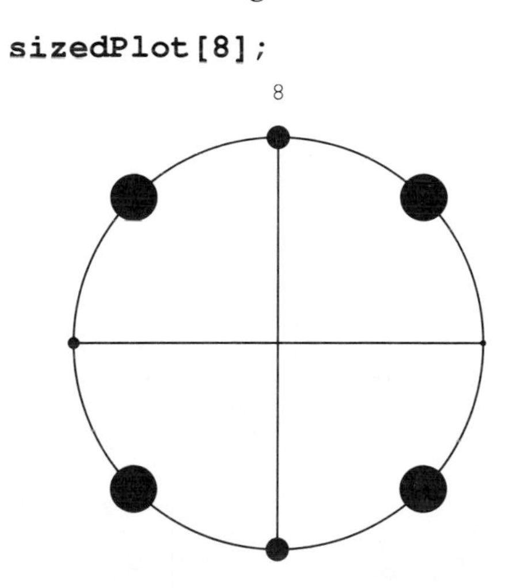

This is a plot of the roots of **x^n - 1**, for **n = 8**: The vertical line is the imaginary axis, and the horizontal line is the real axis. The circle is the unit circle, a circle of radius one.

As many of you may know, the roots of **x^n - 1** must lie about the unit circle in the complex plane, and there must be **n** of them evenly spaced. You can see the 8 roots clearly in this plot (except for the smallest dot at (1, 0), which is hard to see).

You can also see that we have plotted different roots with differently sized dots. The idea is that the size of the dot shows which term in the factored form of **x^n - 1** the root comes from. So, the smallest dot is at (1, 0), because it belongs to the first term, **x - 1**. The four biggest dots belong to the last term, **1 + x^4** which, being a quartic equation, must have four roots. There are four different sizes of dots because there are four factors of **x^8 - 1**.

We can use this function to make a table of a few values of n, to see if there is any pattern. Here is the table:

```
Do[sizedPlot[n], {n, 1, 25}]
```

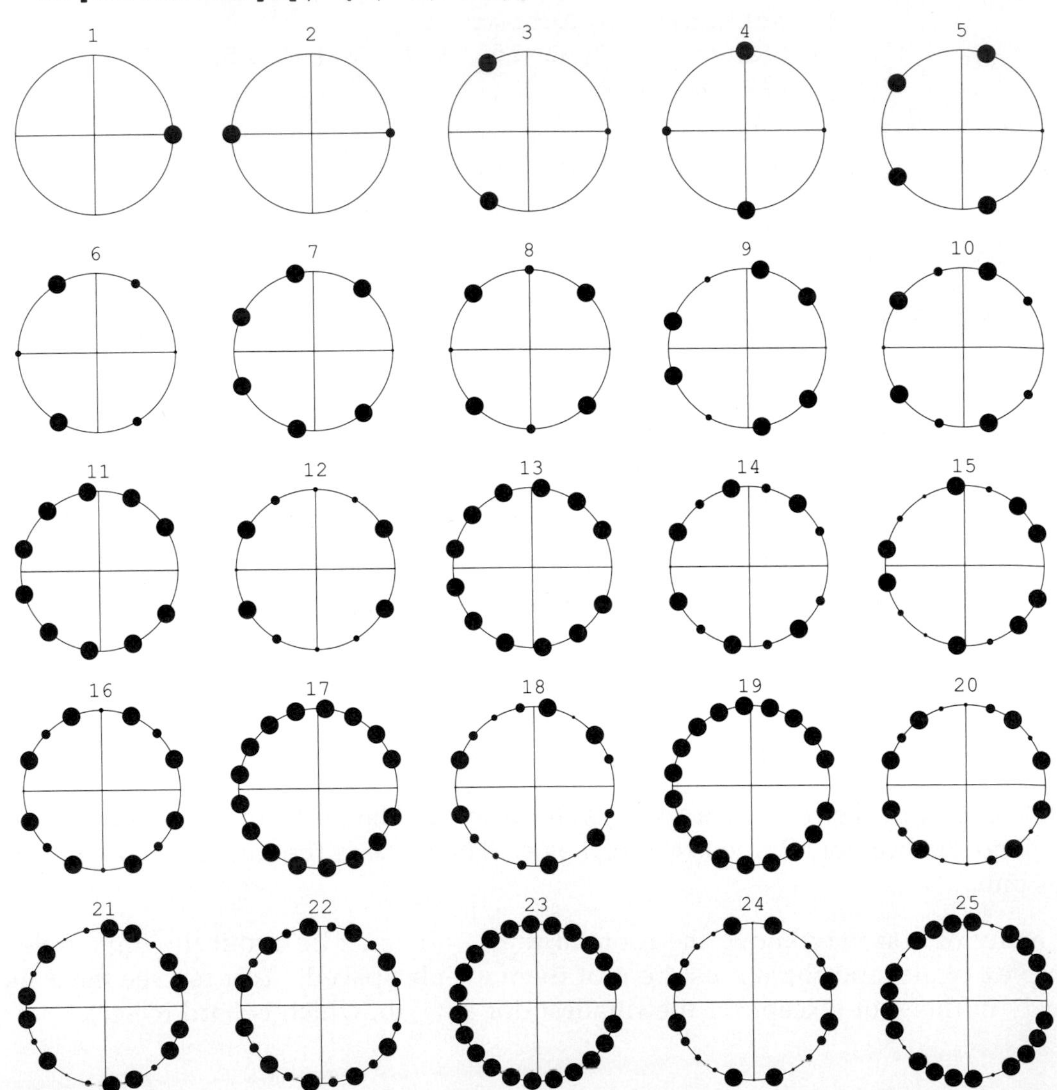

If you stare at these for long enough, you start to notice that there seem to be patterns that repeat themselves. These patterns seem to show up in plots for numbers that have the same divisors. I am, therefore, prepared to make the following claim:

"For each divisor of a number n, there is a certain polynomial that will show up in the factored form of **x^n - 1**."

Jerry: So, if I take 4, which has divisors 1, 2, and 4, together with 8 which has divisors 1, 2, 4, and 8, you are claiming that the factored forms of **x^4 - 1** and **x^8 - 1** will have the same terms, except for one, which belongs to the divisor 8 in **x^8 - 1**.

Theo: Yes, that's what I claim.

```
justFactorList[x^4 - 1]
```

$\{-1 + x,\ 1 + x,\ 1 + x^2\}$

```
justFactorList[x^8 - 1]
```

$\{-1 + x,\ 1 + x,\ 1 + x^2,\ 1 + x^4\}$

Jerry: It seems that the factor belonging to the divisor 8 must be **1 + x^4**. By making a table of many different **n**'s and comparing them, we could figure out which factor belongs to which divisor.

Theo: I wrote a *Mathematica* function to do this automatically, and showed it to Dan. He said we had missed the point, and should try writing fractions next to the dots in our dot plots. Oh, well. According to him, we should number the dots counterclockwise starting at zero on the far right. Then we should divide each of these numbers by n and reduce the fraction to lowest terms.

Jerry: That sounds like something we could do by hand. I think I'll do that while you write the function to do it automatically, which I'm sure you will.

Theo: Of course! Here's a new version of **sizedPlot**:

```
sizedFractionPlot[n_] :=
    Show[
        Graphics[{
            Circle[{0, 0}, 1],
            Line[{{-1, 0}, {1, 0}}],
            Line[{{0, -1}, {0, 1}}],
            Map[makePoints,
                Transpose[{justFactorList[x^n - 1],
                                Divisors[n] 0.1/n}]
            ],

            Table[
                Text[
                    InputForm[m/n],
                    1.4 {Cos[2Pi m/n], Sin[2Pi m/n]},
                    {0, 0}
                ],
                {m, 0, n - 1}
            ]
        }],
        AspectRatio -> Automatic,
        PlotRange -> {{-1.5, 1.5}, {-1.5, 1.5}},
        PlotLabel -> n
    ]
```

We can make a plot for **x^12 - 1**:

```
sizedFractionPlot[12];
```

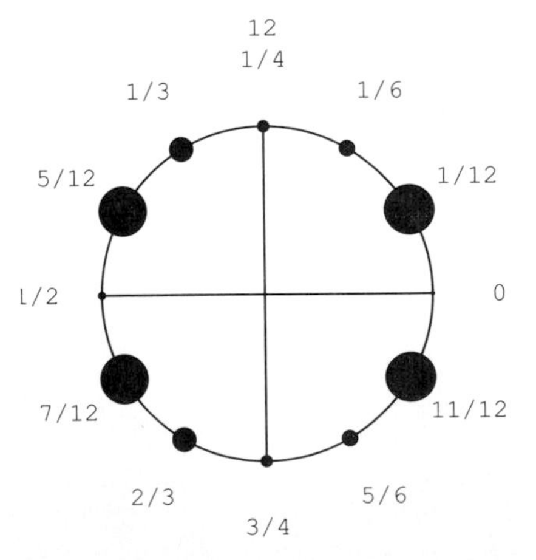

Jerry: The denominators seem to be connected with the size of the dots. All the 12ths have dots of the same size, as do the 6ths.

Theo: It's clear that dots of the same size correspond to fractions with the same denominator. The size of the dots, in turn, represents which term in the factored form the factor belongs to. Therefore, fractions with the same denominator must belong to the same factor. This could be a very useful thing to know.

Jerry: We can calculate what any one of these dots (roots) is, just by using the formula for a circle in the complex plane. For example, the dot corresponding to 1/6 is at a radius of one and an angle of 1/6 of a full circle from the x axis. Its value is:

```
Cos[1/6 2Pi] + I Sin[1/6 2Pi]
```

$$\frac{1}{2} + \frac{I}{2} \text{ Sqrt[3]}$$

Theo: Let's make a function to tabulate all the roots for a given n, and sort them according to what their denominator is:

```
rootTable[n_] :=
    Sort[Table[
        {
            Denominator[m/n],
            Cos[2Pi m/n] + I Sin[2Pi m/n]
        },
        {m, 0, n - 1}
    ]]
```

Here's what happens when we apply it to 12:

```
rootTable[12]//TableForm
```

1	1
2	-1
3	$-(\frac{1}{2}) + \frac{-I}{2}$ Sqrt[3]
3	$-(\frac{1}{2}) + \frac{I}{2}$ Sqrt[3]
4	-I
4	I
6	$\frac{1}{2} + \frac{-I}{2}$ Sqrt[3]
6	$\frac{1}{2} + \frac{I}{2}$ Sqrt[3]
12	$\frac{-I}{2} - \frac{\text{Sqrt[3]}}{2}$
12	$\frac{I}{2} - \frac{\text{Sqrt[3]}}{2}$
12	$\frac{-I}{2} + \frac{\text{Sqrt[3]}}{2}$
12	$\frac{I}{2} + \frac{\text{Sqrt[3]}}{2}$

This is a very powerful table. If we want to know what one of the terms in the factored form of **x^n - 1** is, all we have to do is to multiply together linear terms corresponding to the values in the table above which belong to a given denominator. Here's a function that picks out the terms corresponding to a given denominator:

```
pickRoot[m_, n_] :=
    Last[Transpose[
        Select[rootTable[n], (#[[1]] == m)&]]]
```

Let's use it to find the roots corresponding to 6ths:

```
pickRoot[6, 12]
```

```
  1   -I            1   I
{- + -- Sqrt[3],  - + - Sqrt[3]}
  2   2             2   2
```

We can modify this function to produce the actual polynomial:

```
pickRoot[m_, n_] :=
    Apply[Times, Map[
                (x - #)&,
                Last[Transpose[Select[
                    rootTable[n],
                    (#[[1]] == m)&
                ]]]
    ]]
```

Let's use it:

```
pickRoot[6, 12]
```

```
     1   -I                      1   I
(-(- + -- Sqrt[3]) + x)  (-(- + - Sqrt[3]) + x)
     2   2                       2   2
```

If we expand this out, we should see something familiar:

```
Expand[%]
```

```
         2
1 - x + x
```

Jerry: Sure enough, that's one of the terms in the factored form of **x^12 - 1**!

Theo: It's a simple extension to make this function factor **x^n - 1** completely. All we have to do is to multiply together all these terms. Here's a function to do this:

```
newFactor[n_] :=
    Apply[Times,
        Map[Expand[pickRoot[#, n]]&, Divisors[n]]]
```

We can factor **`x^12 - 1`** in two different ways now:

```
Factor[x^12 - 1]

(-1 + x) (1 + x) (1 + x^2) (1 - x + x^2) (1 + x + x^2)
  (1 - x^2 + x^4)

newFactor[12]

(-1 + x) (1 + x) (1 + x^2) (1 - x + x^2) (1 + x + x^2)
  (1 - x^2 + x^4)
```

Jerry: Amazing! Factoring with no **`Factor`** in sight! And just as fast as the original!

Theo: Well, I sort of doubt it. Also, our version relies on being able to reduce to rational numbers the **`Sin`** and **`Cos`** values that show up. This is not generally an easy thing to do. Let's try a different number:

```
newFactor[5]

(-1 + x) (x^4 - x^3 Cos[2 Pi/5] - x^3 Cos[4 Pi/5] + x^2 Cos[2 Pi/5] Cos[4 Pi/5] - x^3 Cos[6 Pi/5] + x^2 Cos[2 Pi/5] Cos[6 Pi/5] +
  x^2 Cos[4 Pi/5] Cos[6 Pi/5] - x Cos[2 Pi/5] Cos[4 Pi/5] Cos[6 Pi/5] - x^3 Cos[8 Pi/5] + x^2 Cos[2 Pi/5] Cos[8 Pi/5] + x^2 Cos[4 Pi/5] Cos[8 Pi/5] -
  x Cos[2 Pi/5] Cos[4 Pi/5] Cos[8 Pi/5] + x^2 Cos[6 Pi/5] Cos[8 Pi/5] - x Cos[2 Pi/5] Cos[6 Pi/5] Cos[8 Pi/5] - x Cos[4 Pi/5] Cos[6 Pi/5] Cos[8 Pi/5] +
  Cos[2 Pi/5] Cos[4 Pi/5] Cos[6 Pi/5] Cos[8 Pi/5] + -I x^3 Sin[2 Pi/5] + I x^2 Cos[4 Pi/5] Sin[2 Pi/5] + I x^2 Cos[6 Pi/5] Sin[2 Pi/5] +
  -I x Cos[4 Pi/5] Cos[6 Pi/5] Sin[2 Pi/5] + I x^2 Cos[8 Pi/5] Sin[2 Pi/5] + -I x Cos[4 Pi/5] Cos[8 Pi/5] Sin[2 Pi/5] +
  -I x Cos[6 Pi/5] Cos[8 Pi/5] Sin[2 Pi/5] + I Cos[4 Pi/5] Cos[6 Pi/5] Cos[8 Pi/5] Sin[2 Pi/5] + -I x^3 Sin[4 Pi/5] + I x^2 Cos[2 Pi/5] Sin[4 Pi/5] +
  I x^2 Cos[6 Pi/5] Sin[4 Pi/5] + -I x Cos[2 Pi/5] Cos[6 Pi/5] Sin[4 Pi/5] + I x^2 Cos[8 Pi/5] Sin[4 Pi/5] +
  -I x Cos[2 Pi/5] Cos[8 Pi/5] Sin[4 Pi/5] + -I x Cos[6 Pi/5] Cos[8 Pi/5] Sin[4 Pi/5] + I Cos[2 Pi/5] Cos[6 Pi/5] Cos[8 Pi/5] Sin[4 Pi/5] -
  x^2 Sin[2 Pi/5] Sin[4 Pi/5] + x Cos[6 Pi/5] Sin[2 Pi/5] Sin[4 Pi/5] + x Cos[8 Pi/5] Sin[2 Pi/5] Sin[4 Pi/5] -
  Cos[6 Pi/5] Cos[8 Pi/5] Sin[2 Pi/5] Sin[4 Pi/5] + -I x^3 Sin[6 Pi/5] + I x^2 Cos[2 Pi/5] Sin[6 Pi/5] + I x^2 Cos[4 Pi/5] Sin[6 Pi/5] +
  -I x Cos[2 Pi/5] Cos[4 Pi/5] Sin[6 Pi/5] + I x^2 Cos[8 Pi/5] Sin[6 Pi/5] + -I x Cos[2 Pi/5] Cos[8 Pi/5] Sin[6 Pi/5] +
  -I x Cos[4 Pi/5] Cos[8 Pi/5] Sin[6 Pi/5] + I Cos[2 Pi/5] Cos[4 Pi/5] Cos[8 Pi/5] Sin[6 Pi/5] - x^2 Sin[2 Pi/5] Sin[6 Pi/5] +
  x Cos[4 Pi/5] Sin[2 Pi/5] Sin[6 Pi/5] + x Cos[8 Pi/5] Sin[2 Pi/5] Sin[6 Pi/5] - Cos[4 Pi/5] Cos[8 Pi/5] Sin[2 Pi/5] Sin[6 Pi/5] -
  x^2 Sin[4 Pi/5] Sin[6 Pi/5] + x Cos[2 Pi/5] Sin[4 Pi/5] Sin[6 Pi/5] + x Cos[8 Pi/5] Sin[4 Pi/5] Sin[6 Pi/5] -
  Cos[2 Pi/5] Cos[8 Pi/5] Sin[4 Pi/5] Sin[6 Pi/5] + I x Sin[2 Pi/5] Sin[4 Pi/5] Sin[6 Pi/5] + -I Cos[8 Pi/5] Sin[2 Pi/5] Sin[4 Pi/5] Sin[6 Pi/5] +
  -I x^3 Sin[8 Pi/5] + I x^2 Cos[2 Pi/5] Sin[8 Pi/5] + I x^2 Cos[4 Pi/5] Sin[8 Pi/5] + -I x Cos[2 Pi/5] Cos[4 Pi/5] Sin[8 Pi/5] +
  I x^2 Cos[6 Pi/5] Sin[8 Pi/5] + -I x Cos[2 Pi/5] Cos[6 Pi/5] Sin[8 Pi/5] + -I x Cos[4 Pi/5] Cos[6 Pi/5] Sin[8 Pi/5] +
  I Cos[2 Pi/5] Cos[4 Pi/5] Cos[6 Pi/5] Sin[8 Pi/5] - x^2 Sin[2 Pi/5] Sin[8 Pi/5] + x Cos[4 Pi/5] Sin[2 Pi/5] Sin[8 Pi/5] +
  x Cos[6 Pi/5] Sin[2 Pi/5] Sin[8 Pi/5] - Cos[4 Pi/5] Cos[6 Pi/5] Sin[2 Pi/5] Sin[8 Pi/5] - x^2 Sin[4 Pi/5] Sin[8 Pi/5] +
  x Cos[2 Pi/5] Sin[4 Pi/5] Sin[8 Pi/5] + x Cos[6 Pi/5] Sin[4 Pi/5] Sin[8 Pi/5] - Cos[2 Pi/5] Cos[6 Pi/5] Sin[4 Pi/5] Sin[8 Pi/5] +
  I x Sin[2 Pi/5] Sin[4 Pi/5] Sin[8 Pi/5] + -I Cos[6 Pi/5] Sin[2 Pi/5] Sin[4 Pi/5] Sin[8 Pi/5] - x^2 Sin[6 Pi/5] Sin[8 Pi/5] +
  x Cos[2 Pi/5] Sin[6 Pi/5] Sin[8 Pi/5] + x Cos[4 Pi/5] Sin[6 Pi/5] Sin[8 Pi/5] - Cos[2 Pi/5] Cos[4 Pi/5] Sin[6 Pi/5] Sin[8 Pi/5] +
  I x Sin[2 Pi/5] Sin[6 Pi/5] Sin[8 Pi/5] + I Cos[4 Pi/5] Sin[2 Pi/5] Sin[6 Pi/5] Sin[8 Pi/5] + I x Sin[4 Pi/5] Sin[6 Pi/5] Sin[8 Pi/5] +
  -I Cos[2 Pi/5] Sin[4 Pi/5] Sin[6 Pi/5] Sin[8 Pi/5] + Sin[2 Pi/5] Sin[4 Pi/5] Sin[6 Pi/5] Sin[8 Pi/5])
```

Unfortunately, what happened is that it didn't simplify the **Sin[n Pi/5]** terms, so they all ended up in the answer. This is very bad. Just to see if the answer is right anyway, we can make a numerical approximation with the **N** function, then apply the **Chop** function (which removes numbers that are very close to zero) and the **Rationalize** function (which turns floating point numbers back into rational numbers, provided they are reasonably close to rational):

```
Rationalize[Chop[N[%]]]

                    2    3    4
(-1 + x) (1 + x + x  + x  + x )
```

This looks about right, and certainly less frightening than the thing above.

Jerry: Maybe we should just stick to the **Factor** command. On the other hand, by constructing our own version, we have learned something about factoring. In particular, to factor some polynomials, you might start by looking at circles.

Theo: As I've said before, the way a computer does something like factoring is often very different from the way people would do the same thing. That's why it's often not useful to ask if *Mathematica* can "show you what it's doing". It probably looks at hyperbolic ellipses instead of circles.

This chapter is really about cryptography. We spent a long time exploring the factors of **x^n - 1**, finally arriving at a table that allows us to factor any **x^n - 1** very easily. We did it by finding patterns in seemingly random tables of polynomials, and connecting those patterns to things we knew about n, such as its divisors.

This is very similar to how you decode secret messages. First you try to guess what the message might be, which is like guessing that the divisors of n might be important. Then you try to see if some pattern in the coded message corresponds to some pattern in what you think the message really says. When you see a connection, you may have found the rule that was used to encode the message.

In this case, we were looking at encoded forms of the divisors of n. They were hidden in the complicated polynomials that were the factors of **x^n - 1**. Once we saw the pattern, we were able to construct the rule that was used to generate the polynomials. If we wanted to send secret messages to our friends, we could do it by numbering each letter of the alphabet from 1 to 26 and then making a list of the polynomials corresponding to each letter in our message.

In a sense we haven't learned anything new, since Dan Grayson, who wrote the *Mathematica* **Factor** function, already knew this rule, and in fact used it as part of the internal code for **Factor**. So we were just discovering something that had been put in there deliberately.

But then, that's usually the case when decoding secret messages. Someone else who knew the rule used it to do the encoding. The reason that cryptography is so interesting even to people outside the security service is that sometimes it's not a person who did the encoding.

Cryptography is called science when nature did the encoding. A famous case is the DNA code. Over the eons nature made up a code to match 3-"letter" sequences to the 20 or so different amino acids that Nature strings together into proteins. It was Marshall Nirenberg who first saw how to work out the table, and he got a Nobel prize.

Jerry: That's all very interesting, but what does it really mean? We know, after all, that we can't trust our eyes, even if we try 100 cases. How do we know that your claim about one divisor going with one factor is true past the 50 cases we tried?

Theo: I suppose we can't, unless we can use what we've learned to understand a general rule to show why it should be so.

Fortunately the problem of factoring **`x^n - 1`** is a classic one in number theory, and many people before us have thought about it. We can read about them in the books and papers mentioned at the end of this section. The factors we have been studying are called cyclotomic polynomials, because their roots all lie on a circle in the complex plane. The roots of the factors are known as the roots of unity, since they are all numbers that, when raised to some power, give 1.

Many things are known about cyclotomic polynomials and the roots of unity, including that my claim about one divisor = one factor is true for all n. It is also known that if you choose n large enough, you can get coefficients as large as you like (remember that for n up to 1000 the biggest coefficient we found was 3). It is not known, however, how big an n you have to choose to get a given size of coefficient, only that a large enough n must exist.

The grand table that we printed out in the last section can be generated trivially using polynomial long division. It is therefore possible to factor **`x^n - 1`** very easily, by first constructing this table and then figuring out the divisors of n. Both are much easier than factoring **`x^n - 1`** by some more general method. In fact, *Mathematica* uses exactly this method to factor **`x^n - 1`**, which is why it is so much faster than one might expect it to be. There is even a built-in *Mathematica* function, **`Cyclotomic`**, which can be used to generate automatically the factor that goes with a particular divisor.

Visiting Professor: Dan Grayson

Dan: With each root of **x^n - 1** you guys associated a fraction between 0 and 1 which expresses how far we must go counterclockwise around the unit circle in the complex plane, starting at 1, to get to the root. For **n = 15** the possible fractions are:

```
Table[i / HoldForm[15], {i, 0, 14}]
```

$$\{0, \frac{1}{15}, \frac{2}{15}, \frac{3}{15}, \frac{4}{15}, \frac{5}{15}, \frac{6}{15}, \frac{7}{15}, \frac{8}{15}, \frac{9}{15}, \frac{10}{15}, \frac{11}{15}, \frac{12}{15}, \frac{13}{15}, \frac{14}{15}\}$$

and when we let these fractions reduce to lowest terms like this:

```
Table[i / 15, {i, 0, 14}]
```

$$\{0, \frac{1}{15}, \frac{2}{15}, \frac{1}{5}, \frac{4}{15}, \frac{1}{3}, \frac{2}{5}, \frac{7}{15}, \frac{8}{15}, \frac{3}{5}, \frac{2}{3}, \frac{11}{15}, \frac{4}{5}, \frac{13}{15}, \frac{14}{15}\}$$

we see that not all of them keep 15 as their denominator. We manufacture the polynomial whose roots correspond to the fractions here with denominator 15:

```
pickRoot[15, 15]
```

$$\begin{aligned}&(x - (\mathrm{Cos}[\tfrac{2\ Pi}{15}] + I\ \mathrm{Sin}[\tfrac{2\ Pi}{15}]))\\ &\quad (x - (\mathrm{Cos}[\tfrac{4\ Pi}{15}] + I\ \mathrm{Sin}[\tfrac{4\ Pi}{15}]))\\ &\quad (x - (\mathrm{Cos}[\tfrac{8\ Pi}{15}] + I\ \mathrm{Sin}[\tfrac{8\ Pi}{15}]))\\ &\quad (x - (\mathrm{Cos}[\tfrac{14\ Pi}{15}] + I\ \mathrm{Sin}[\tfrac{14\ Pi}{15}]))\\ &\quad (x - (\mathrm{Cos}[\tfrac{16\ Pi}{15}] + I\ \mathrm{Sin}[\tfrac{16\ Pi}{15}]))\\ &\quad (x - (\mathrm{Cos}[\tfrac{22\ Pi}{15}] + I\ \mathrm{Sin}[\tfrac{22\ Pi}{15}]))\\ &\quad (x - (\mathrm{Cos}[\tfrac{26\ Pi}{15}] + I\ \mathrm{Sin}[\tfrac{26\ Pi}{15}]))\\ &\quad (x - (\mathrm{Cos}[\tfrac{28\ Pi}{15}] + I\ \mathrm{Sin}[\tfrac{28\ Pi}{15}]))\end{aligned}$$

The general theory says that when this polynomial is expanded, and like powers of the variable **x** are collected together, we get a polynomial whose coefficients are integers and which cannot be factored further as a product of polynomials with rational coefficients. We can get this polynomial more easily using the built-in function **Cyclotomic**:

```
Cyclotomic[15, x]
```

```
         3    4    5    7    8
1 - x + x  - x  + x  - x  + x
```

The proof that this polynomial cannot be factored further is commonly taught to graduate students of mathematics in their first algebra course, but since it's rather intricate I won't explain it now.

But I can explain why this polynomial has integer coefficients. The reason is that there is a way to separate out from the polynomial **x^15 - 1** the part that accounts for the other roots! The other roots correspond to the following fractions:

```
Select[ Table[ i/15, {i,0,14}], Denominator[#]!=15 &]
```

```
    1  1  2  3  2  4
{0, -, -, -, -, -, -}
    5  3  5  5  3  5
```

and their denominators are:

```
Map[Denominator, %]
```

```
{1, 5, 3, 5, 5, 3, 5}
```

The corresponding complex numbers are **1**, the four other roots of **x^5 - 1**, and the two other roots of **x^3 - 1**. Since every root of **x^5 - 1** is also a root of **x^15 - 1**, the polynomial **x^15 - 1** is divisible by **x^5 - 1**. We can compute the quotient:

```
Cancel[ (x^15-1) / (x^5-1) ]
```

```
     5    10
1 + x  + x
```

This polynomial will have as roots those numbers corresponding to fractions whose denominator is 3 or 15, as we removed (by division) the roots corresponding to fractions whose denominators are 1 or 5. We try the same thing again with **x^3 - 1**:

```
Cancel[ (1 + x^5 + x^10) / (x^3 - 1) ]
```

```
         3    4    5    7    8
1 - x + x  - x  + x  - x  + x
-----------------------------
           -1 + x
```

The function **Cancel** has managed to remove from the numerator and denominator of that fraction the part they share, namely, the part whose roots correspond to the fractions **1/3** and **2/3**. Thus the numerator is now that polynomial whose 8 roots correspond to the fractions with denominator 15:

```
Select[ Table[i/15, {i,0,14}], Denominator[#]==15 &]
```

```
  1   2   4   7   8   11  13  14
{--, --, --, --, --, --, --, --}
 15  15  15  15  15  15  15  15
```

The astute reader who suspects that **Cancel** might enlist the services of **Factor** will be reassured to learn that it doesn't, and that **Cancel** always manages to reduce the fractions to lowest terms in such a way that the numerator and denominator do not have any common factors with rational coefficients, nor do they share any complex roots. This is what allows me to assert that the polynomial we constructed above, with roots just those corresponding to fractions with denominator 15, is a polynomial with rational coefficients.

The miracle that imparts to **Cancel** its speedy and thorough behavior is the same miracle that allows one to reduce common fractions to lowest terms without factoring integers. Schoolchildren are taught to bring fractions to lowest terms by factoring the numerator and denominator, and then cancelling common factors.

Theo: That's the way they tried to get me to do it when I was in school!

Dan: Well, it's not a very good way to do it, and as an algorithm it would be totally useless in a program like *Mathematica*, because of the notorious difficulty of the problem of factoring numbers with fifty digits or more. What's used instead is an ancient algorithm of Euclid that speedily locates the largest factor that the numerator and denominator share. Let's illustrate this with an example:

```
 78583477505990094517195124787814569951571 1 /
115750583895356383671434271794080859878698 7
```

$$\frac{882377}{1299709}$$

The cancellation done above happens in 0.03 seconds on my computer, but running **FactorInteger** on numbers of this size is quite likely to be fruitless, or at least it will take a very long time.

The ancient algorithm of Euclid is easy to describe. Start with a pair of positive integers (the numerator and denominator of the fraction to be reduced). Subtract the smaller from the larger, and use the result to replace the larger number. In this way we get a new pair of positive integers, one of which has become smaller. We repeat this until one member of the pair is 0, and then the other number is the result.

Let's illustrate this with some smaller numbers:

```
x = 970679
y = 1030721
While[ True,
    Print[{x,y}];
    If[ x==0 || y==0, Break[]];
    If[x<y, y-=x, x-=y]]
x+y
```

```
{970679, 1030721}
{970679, 60042}
{910637, 60042}
{850595, 60042}
{790553, 60042}
{730511, 60042}
{670469, 60042}
{610427, 60042}
{550385, 60042}
{490343, 60042}
{430301, 60042}
{370259, 60042}
{310217, 60042}
{250175, 60042}
{190133, 60042}
{130091, 60042}
{70049, 60042}
{10007, 60042}
{10007, 50035}
{10007, 40028}
{10007, 30021}
{10007, 20014}
{10007, 10007}
{0, 10007}
```

```
10007
```

Running the algorithm, as we did above, has told us that 10007 is the common factor that must be cancelled in order to reduce the fraction 970679/1030721 to lowest terms.

```
970679/10007
```

```
97
```

```
1030721/10007
```

```
103
```

Indeed, the system knows this, too:

```
970679/1030721
```

$$\frac{97}{103}$$

Theo: This would go a lot faster if instead of subtracting 60042 fifteen times, you would divide and keep the remainder.

Dan: Yes, it would. So we could rephrase the algorithm in this way: repeatedly subtract some multiple of the smaller number from the larger until one of the two numbers becomes zero. It is this way of phrasing the algorithm that allows one to generalize it in such a way that it works for polynomials.

Functions That We Defined

Theo: In this section we defined the function **justFactorList**, which is like the built-in function **FactorList** except that it returns just a list of the factors of a polynomial, ignoring the number of times that each factor appears in the polynomial:

```
justFactorList[expr_] :=
    First[Transpose[FactorList[expr]]]
```

We also defined a function, **sizedPlot**, that plots the roots of the factors of **x^n - 1** with differently sized dots:

```
makePoints[{eqn_, size_}] :=
    Block[
        {t},
        t = x /. N[Solve[eqn == 0, x]];
        {
            PointSize[size],
            Map[Point, Transpose[{Re[t], Im[t]}]]
        }
    ]
```

```
sizedPlot[n_] :=
    Show[
        Graphics[{
            Circle[{0, 0}, 1],
            Line[{{-1, 0}, {1, 0}}],
            Line[{{0, -1}, {0, 1}}],
            Map[makePoints,
                Transpose[{justFactorList[x^n - 1],
                          Divisors[n] 0.1/n}]
            ]
        }],
        AspectRatio -> Automatic,
        PlotRange -> {{-1.15, 1.15}, {-1.15, 1.15}},
        PlotLabel -> n
    ]
```

A variant also plots fractions next to each of the dots:

```
sizedFractionPlot[n_] :=
    Show[
        Graphics[{
            Circle[{0, 0}, 1],
            Line[{{-1, 0}, {1, 0}}],
            Line[{{0, -1}, {0, 1}}],
            Map[makePoints,
                Transpose[{justFactorList[x^n - 1],
                          Divisors[n] 0.1/n}]
            ],

            Table[
                Text[
                    InputForm[m/n],
                    1.4 {Cos[2Pi m/n], Sin[2Pi m/n]},
                    {0, 0}
                ],
                {m, 0, n - 1}
            ]
        }],
        AspectRatio -> Automatic,
        PlotRange -> {{-1.5, 1.5}, {-1.5, 1.5}},
        PlotLabel -> n
    ]
```

We defined three more functions, culminating in a function that works like **Factor**, at least for certain cases of **x^n - 1**.

```
rootTable[n_] :=
    Sort[Table[
        {
            Denominator[m/n],
            Cos[2Pi m/n] + I Sin[2Pi m/n]
        },
        {m, 0, n - 1}
    ]]
```

```
pickRoot[m_, n_] :=
    Apply[Times, Map[
            (x - #)&,
            Last[Transpose[Select[
                rootTable[n],
                (#[[1]] == m)&
            ]]]
    ]]
```

```
newFactor[n_] :=
    Apply[Times,
        Map[Expand[pickRoot[#, n]]&, Divisors[n]]]
```

References

1) Richard K. Guy, *The Strong Law of Small Numbers*, The American Mathematical Monthly, Volume 95, Number 8, October 1988, p697.

Chapter Eighteen
The Square Root of Beethoven, and Other Really Big Numbers

Jerry and Theo experience an excess of digits.

Dialog–First Day

Jerry: One of the first characteristics of *Mathematica* that got me interested was its ability to generate exact answers to problems like:

```
1000!
```

```
40238726007709377354370243392300398571937486421071463254379991042993851239862902059204420848696940480047998861019719605\
  863166687299480855890132382966994459099742450408707375991882362772718873251977950595099527612087497546249704360141827\
  809464649629105639388743788648733711918104582578364784997701247663288983595573543251318532395846307555740911426241747\
  434934755342864657661166779739666882029120737914385371958824980812686783837455973174613608537953452422158659320192809\
  087829730843139284440328123155861103697680135730421616874760967587134831202547858932076716913244842623613141250878020\
  800026168315102734182797770478463586817016436502415369139828126481021309276124489635992870511496497541990934222156683\
  257208082133318611681155361583654698404670897560290095053761647584772842188967964624494516076535340819890138544248798\
  495995331910172335555660213945039973628075013783761530712776192684903435262520001588853514733161170210396817592151090\
  778801939317811419454525722386554146106289218796022383897147608850627686296714667469756291123408243920816015378088989\
  396451826324367161676217916890977991190375403127462228998800519544441428201218736174599264295658174662830295557029902\
  432415318161721046583203678690611726015878352075151628422554026517048330422614397428693306169089796848259012545832716\
  822645806652676995865268227280707578139185817888965220816434834482599326604336766017699961283186078838615027946595513\
  115655203609398818061213855860030143569452722420634463179746059468257310379008402443243846565724501440282188525247093\
  519062092902313649327349756551395872055965422874977401141334696271542284586237738753823048386568897646192738381490014\
  076731044664025989949022222176590433990188601856652648506179970235619389701786004081188972991831102117122984590164192\
  106888438712185564612496079872290851929681937238864261483965738229112312502418664935314397013742853192664987533721894\
  069428143411852015801412334482801505139969429015348307764456909907315243327828826986460278986432113908350621709500259\
  738986355427719674282224875758676575234422020757363056949882508796892816275384886339690995982628095612145099487170124\
  451646126037902930912088908694202851064018215439945715680594187274899809425474217358240106367740459574178516082923013\
  535808184009699637252423056085590370062427124341690900415369010593398383577793941097002775347200000000000000000000000\
  000000000000000000000000000000000000000000000000000000000000000000000000000000000000000000000000000000000000000000000\
  000000000000000000000000000000000000000000000000000000000000000000000000000000000000000000000000000000000000000000000\
  000000000000000000000000000000000000000000000000000000000000000000000000000000000000000000000000000000000000
```

I admit that this is a trivial reason, but it is the truth.

Theo: Why do you say trivial? When I first saw numbers like the one above, I was convinced that there must be a bug in the program, because I couldn't believe there would be *so many* zeros at the end.

Jerry: I had the same reaction. We must have lacked experience with exact answers to factorials.

Theo: Right. The standard 69! (the largest you can do on almost all pocket calculators) has many zeros at the end, but you would never guess that from the answer you get from the pocket calculator. In fact, it's not hard to figure out how many zeros there should be.

Jerry: Let's see. `10!` has in its factors a `10`, a `5`, and a `2`, so I would expect two zeros:

```
10!
3628800
```

On the other hand, `25!` has the `10`, the `5`, and the `2` from 10!, a `5` each from `15` and `20`, and two `5`'s from `25.` There are lots of `2`'s to combine with these fives, so I am expecting a total of `6` zeros:

```
25!
15511210043330985984000000
```

Theo: So, as we go to higher and higher factorials, every factor of `10`, or every pair of factors `5` and `2` gives us a new zero at the end. We can never get rid of any zeros, since each factorial contains all the factors of all lower factorials. No wonder there are a lot of zeros!

Jerry: Let's ask *Mathematica* how many 5's and 2's there are in `1000!`.

Theo: The function is called `FactorInteger`. Let's try it on `1000!`:

```
FactorInteger[1000!]
{{2, 994}, {3, 498}, {5, 249}, {7, 164}, {11, 98}, {13, 81}, {17, 61}, {19, 54},
  {23, 44}, {29, 35}, {31, 33}, {37, 27}, {41, 24}, {43, 23}, {47, 21}, {53, 18},
  {59, 16}, {61, 16}, {67, 14}, {71, 14}, {73, 13}, {79, 12}, {83, 12}, {89, 11},
  {97, 10}, {101, 9}, {103, 9}, {107, 9}, {109, 9}, {113, 8}, {127, 7}, {131, 7},
  {137, 7}, {139, 7}, {149, 6}, {151, 6}, {157, 6}, {163, 6}, {167, 5}, {173, 5},
  {179, 5}, {181, 5}, {191, 5}, {193, 5}, {197, 5}, {199, 5}, {211, 4}, {223, 4},
  {227, 4}, {229, 4}, {233, 4}, {239, 4}, {241, 4}, {251, 3}, {257, 3}, {263, 3},
  {269, 3}, {271, 3}, {277, 3}, {281, 3}, {283, 3}, {293, 3}, {307, 3}, {311, 3},
  {313, 3}, {317, 3}, {331, 3}, {337, 2}, {347, 2}, {349, 2}, {353, 2}, {359, 2},
  {367, 2}, {373, 2}, {379, 2}, {383, 2}, {389, 2}, {397, 2}, {401, 2}, {409, 2},
  {419, 2}, {421, 2}, {431, 2}, {433, 2}, {439, 2}, {443, 2}, {449, 2}, {457, 2},
  {461, 2}, {463, 2}, {467, 2}, {479, 2}, {487, 2}, {491, 2}, {499, 2}, {503, 1},
  {509, 1}, {521, 1}, {523, 1}, {541, 1}, {547, 1}, {557, 1}, {563, 1}, {569, 1},
  {571, 1}, {577, 1}, {587, 1}, {593, 1}, {599, 1}, {601, 1}, {607, 1}, {613, 1},
  {617, 1}, {619, 1}, {631, 1}, {641, 1}, {643, 1}, {647, 1}, {653, 1}, {659, 1},
  {661, 1}, {673, 1}, {677, 1}, {683, 1}, {691, 1}, {701, 1}, {709, 1}, {719, 1},
  {727, 1}, {733, 1}, {739, 1}, {743, 1}, {751, 1}, {757, 1}, {761, 1}, {769, 1},
  {773, 1}, {787, 1}, {797, 1}, {809, 1}, {811, 1}, {821, 1}, {823, 1}, {827, 1},
  {829, 1}, {839, 1}, {853, 1}, {857, 1}, {859, 1}, {863, 1}, {877, 1}, {881, 1},
  {883, 1}, {887, 1}, {907, 1}, {911, 1}, {919, 1}, {929, 1}, {937, 1}, {941, 1},
  {947, 1}, {953, 1}, {967, 1}, {971, 1}, {977, 1}, {983, 1}, {991, 1}, {997, 1}}
```

The first number in each pair is the factor, and the second number is how many of them there are. There are **994** factors of **2**, and **249** factors of **5** (this includes the number of **10**'s, since **FactorInteger** gives us prime factors).

Jerry: So, there should be **249** zeros at the end of **1000!**. There are (I counted).

Theo: The number of times a factor appears in a given factorial seems to go down in an interesting way as the size of the factor goes up. Let's make a plot where the x-axis is the factor, and the y-axis is the number of times that factor occurs. Fortunately, the function **ListPlot** does exactly what we want when applied to a list like the one above. Here's the plot:

```
ListPlot[FactorInteger[1000!]];
```

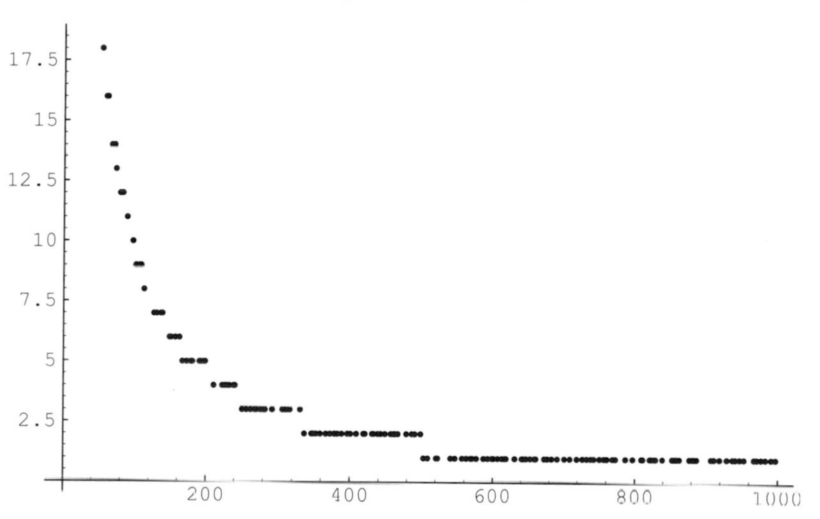

Jerry: This graph says, you get a lot of small factors and not very many larger factors. It looks sort of logarithmic, or hyperbolic.

Theo: You can also see holes in the horizontal lines that represent non-prime numbers.

Jerry: There seem to be a lot of prime numbers, since the lines are quite solid. Can we make a plot of prime numbers?

Theo: Well, we could make a plot of the **n**th prime number vs. **n**. We can use the **Prime** function. **Prime[10]** gives the **10**th prime number:

```
Prime[10]
29
```

Here's a plot of the first 100 primes.

```
ListPlot[Table[{Prime[n], n}, {n, 1, 100}],
        AspectRatio -> Automatic];
```

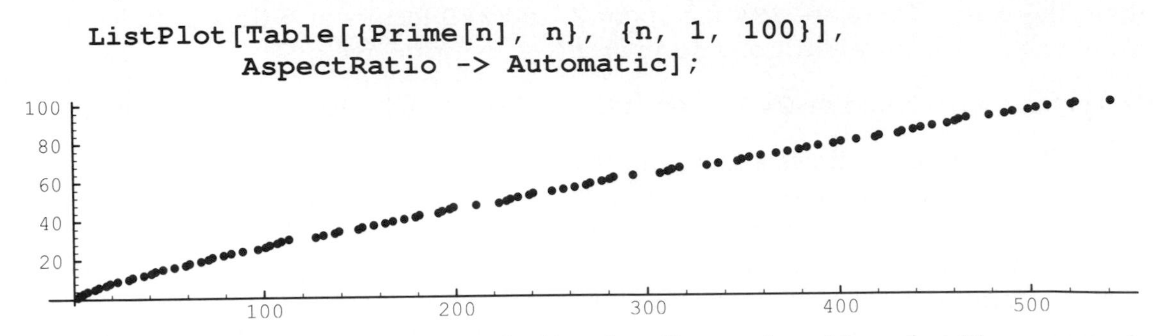

Jerry: This seems to be a nearly straight line, but I'm confused by what it's supposed to mean. Can we try a smaller number of points, and maybe label the axes in a more sensible way?

Theo: Sure, no problem. I'll label the horizontal axis with the prime numbers, and the vertical axis with integers. Let's go up to the `10`th prime number:

```
ListPlot[
    Table[{Prime[n], n}, {n, 1, 10}],
    Ticks ->
        {
            Table[
                If[PrimeQ[i],
                    {i, i, {0.0, 0.015}},
                    {i, "",{0.0, 0.005}}
                ],
                {i, 1, Prime[10]}
            ],
            Table[i, {i, 1, 10}]
        },
    AspectRatio -> Automatic,
    PlotRange -> {{0, Prime[10]}, {0, 10}}
];
```

Jerry: Excellent! I've never seen such a properly labeled graph. To find the 4th prime number, I look up 4 on the vertical axis, find the dot at that height, and look down at the label, 7 in this case. Now the earlier graph makes sense.

Theo: Let's go back to the simpler plot:

```
ListPlot[Table[{Prime[n], n}, {n, 1, 100}],
    AspectRatio -> 0.2];
```

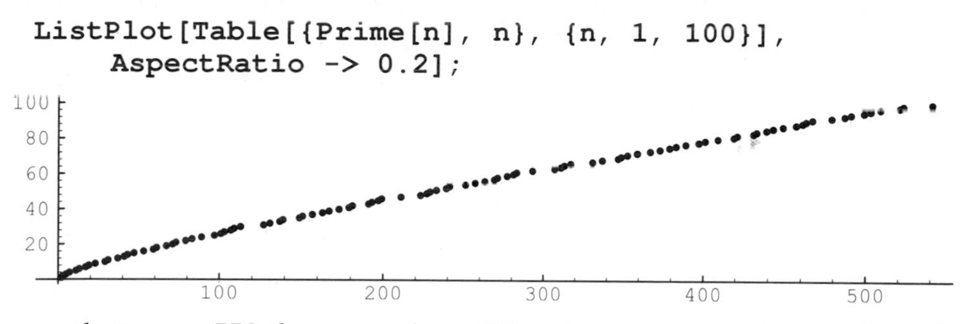

This says that up to 550 there are about 100 primes, or one prime in five. Let's extend the plot to the 1000th prime.

Jerry: If the ratio continues as is, we would expect the 1000th prime to be about 5000.

```
ListPlot[Table[{Prime[n], n}, {n, 1, 1000}],
    AspectRatio -> 0.2];
```

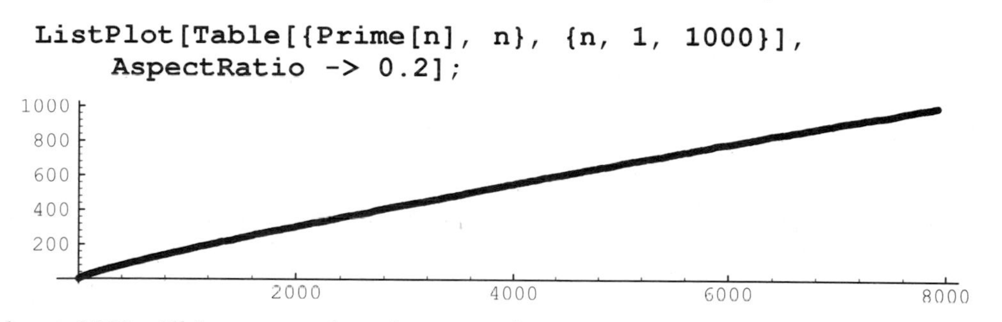

It's about 8000. This means that there are fewer primes, on average, between 100 and 1000 than between 1 and 100. The curve is flattening, something like a log curve.

Theo: Let's see what happens if we go really far out. We can plot every 1000th prime out to the 1,000,000th prime:

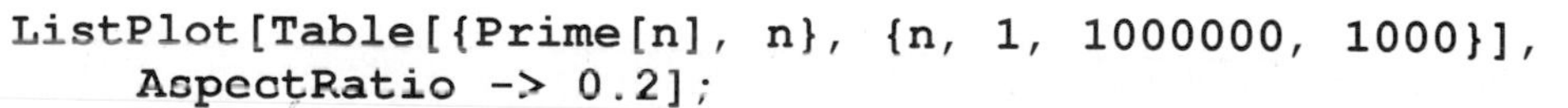

```
ListPlot[Table[{Prime[n], n}, {n, 1, 1000000, 1000}],
    AspectRatio -> 0.2];
```

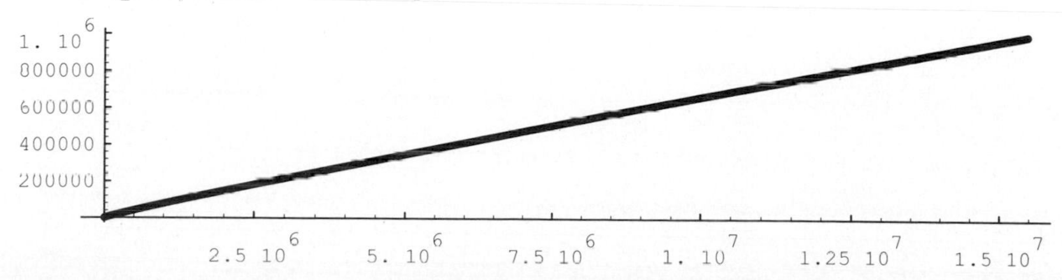

Jerry: It looks like the ratio is now about:

```
1.5 10^7 / 10^6
15.
```

Or one number in 15 is prime. Clearly the primes are getting more and more scarce. One might think that eventually they die out altogether, but of course we know that's not true. I notice *Mathematica* is able to give us prime numbers very quickly. In this last plot we went up to the 1,000,000th prime, and it didn't take very long. Is there a limit on the largest prime that can be produced quickly?

Theo: *Mathematica* uses a very clever mechanism to produce the first few prime numbers. Calculating anything up to about the 105,000,000th prime is almost instantaneous. Beyond that it gets slower.

```
Prime[105000000]
2145390523
```

Out here the ratio is:

```
N[Prime[105000000] / 105000000]
20.4323
```

Or one prime in twenty.

Our visiting professor can explain all this. It's really quite fascinating, and many people have thought about it. We have a chapter (see Chapter 16) in which you can hear a sound very much related to this curve.

Dialog–Second Day

Theo: As soon as people learn that *Mathematica* can calculate numerical functions to many decimal places, they usually want to see lots of digits of π. In case you're one of these people, I'll put in a few thousand here:

```
N[Pi, 5000]
```

```
3.14159265358979323846264338327950288419716939937510582097494459230781640628620899862803482534211706798214808651328230664\
4709384460955058223172535940812848111745028410270193852110555964462294895493038196442881097566593344612847564823378678\
3165271201909145648566923460348610454326648213393607260249141273724587006606315588174881520920962829254091715364367892\
5903600113305305488204665213841469519415116094330572703657595919530921861173819326117931051185480744623799627495673518\
8575272489122793818301194912983367336244065664308602139494639522473719070217986094370277053921717629317675238467481846\
7669405132000568127145263560827785771342757789609173637178721468440901224953430146549585371050792279689258923542019956\
1121290219608640344181598136297747713099605187072113499999983729780499510597317328160963185950244594553469083026425223\
0825334468503526193118817101000313783875288658753320838142061717766914730359825349042875546873115956286388235378759375\
1957781857780532171226806613001927876611195909216420198938095257201065485863278865936153381827968230301952035301852968\
9957736225994138912497217752834791315155748572424541506959508295331168617278558890750983817546374649393192550604009277\
0167113900984882401285836160356370766010471018194295559619894676783744944825537977472684710404753464620804668425906949\
1293313677028989152104752162056966024058038150193511253382430035587640247496473263914199272604269922796782354781636009\
3417216412199245863150302861829745557067498385054945885869269956909272107975093029553211653449872027559602364806654991\
1988183479775356636980742654252786255181841757467289097777279380008164706001614524919217321721477235014144197356854816\
1361157352552133475741849468438523323907394143334547762416862518983569485562099219222184272550254256887671790494601653\
4668049886272327917860857843838279679766814541009538837863609506800642251252051173929848960841284886269456042419652850\
2221066118630674427862203919494504712371378696095636437191728746776465757396241389086583264599581339047802759009946576\
4078951269468398352595709825822620522489407726719478268482601476990902640136394437455305068203496252451749399651431429\
8091906592509372216964615157098583874105978859597729754989301617539284681382686838689427741559918559252459539594310499\
7252468084598727364469584865383673622262609912460805124388439045124413654976278079771569143599770012961608944169486855\
5848406353422072225828488648158456028506016842739452267467678895252138522549954666727823986456596116354886230577456498\
0355936345681743241125150760694794510965960940252288797108931456691368672287489405601015033086179286809208747609178249\
3858900971490967598526136554978189312978482168299894872265880485756401427047755513237964145152374623436454285844479526\
5867821051141354735739523113427166102135969536231442952484937187110145765403590279934403742007310578539062198387447808\
4784896833214457138687519435064302184531910484810053706146806749192781911979399520614196634287544406437451237181921799\
9839101591956181467514269123974894090718649423196156794520809514655022523160388193014209376213785595663893778708303906\
9792077346722182562599661501421503068038447734549202605414665925201497442850732518666002132434088190710486331734649651\
4539057962685610055081066587969981635747363840525714591028970641401109712062804390397595156771577004203378699360072305\
5876317635942187312514712053292819182618612586732157919841484882916447060957527069572209175671167229109816909152801735\
0671274858322287183520935396572512108357915136988209144421006751033467110314126711136990865851639831501970165151168517\
1437657618351556508849099898599823873455283316355076479185358932261854896321329330898570642046752590709154814165498594\
6163718027098199430992448895757128289059232332609729971208443357326548938239119325974636673058360414281388303203824903\
7589852437441702913276561809377344403070746921120191302033038019762110110044929321516084244485963766983895228684783123\
5526582131449576857262433441893039686426243410773226978028073189154411010446823252716201052652272111660396665573092547\
1105578537634668206531098965269186205647693125705863566201855810072936065987648611791045334885034611365768675324944166\
8039626579787718556084552965412665408530614344431858676975145661406800700237877659134401712749470420562230538994561314\
0711270004078547332699390814546646458807972708266830634328587856983052358089330657574067954571637752542021149557615814\
0025012622859413021647155097925923099079654737612551765675135751782966645477917450112996148903046399471329621073404375\
1895735961458901938971311179042978285647503203198691514028708085990480109412147221317947647772622414254854540332157185\
3061422881375850430633217518297986622371721591607716692547487389866549494501146540628433663937900397692656721463853067\
3609657120918076383271664162748888007869256029022847210403172118608204190004229661711963779213375751149595015660496318\
6294726547364252308177036751590673502350728354056704038674351362222477158915049530984448933309634087807693259939780541\
9341447377441842631
```

Jerry: That's nice. Now, what can we do with these digits? They look pretty random.

Theo: Aha, that sounds like a controversial conjecture you're making there. Saying that these digits are random has many serious consequences.

For example, imagine that each pair of digits stands for a letter (as in 01 = a, 02 = b, ... 26 = z). If an infinite sequence of digits is random, then any arbitrary sequence of digits we choose will most likely exist somewhere in the list of digits. For example, the Encyclopedia Britanica would be encoded somewhere in the decimal expansion of π.

Jerry: Yes, but where? It's the "somewhere" that seems to be the problem here. Can you calculate approximately how far you would have to look to find the whole Encyclopedia, or even a single sentence from it?

Theo: Hmmm. Let's think about how we might calculate this number. Clearly, we should write out the digits in base 30 (the 26 letters plus a few punctuation marks). That way we don't have to worry about how the encoding should work. Now, we have to agree on what it means to say "how far we have to go". I suggest that we take that to mean the number of digits we have to look at to have a 50/50 chance of finding the Encyclopedia.

Jerry: Let's try finding a single digit (letter) first. If we look at only one digit, we have a 1/30 chance of its being our digit.

Theo: Or, a 29/30 chance of its *not* being our digit. If we then pick a second digit, there is also a 29/30 chance of it not being ours. The combined probability of both of them not being our digit is **(29/30) (29/30)**, and so the probability of either one (or both) of the two digits being our digit is:

```
1 - 29^2 / 30^2
```

$$\frac{59}{900}$$

Jerry: If we picked more digits, say **n** digits, the probability of finding ours would be:

```
1 - (29/30)^n
```

$$1 - \left(\frac{29}{30}\right)^n$$

Theo: We want to know what value of **n** to use to make this probability at least 1/2. Let's set it equal to 1/2 and solve for **n**:

```
Solve[1 - (29/30)^n == 1/2, n]

Solve::ifun:
   Warning: inverse functions are being used by Solve, so some
     solutions may not be found.
```

$$\{\{n \to \frac{\text{Log}[\frac{1}{2}]}{\text{Log}[\frac{29}{30}]}\}\}$$

A numerical approximation is:

```
N[%]

{{n -> 20.4459}}
```

Jerry: This says that if we pick 21 digits, we will have about a 50/50 chance of getting the one we were looking for. Can we test this calculation by picking a bunch of lists of

21 digits, and looking for a particular value in each list? If our calculation is correct, about half of the lists will have our digit in them.

Theo: Sure. The function **RealDigits** can be use to get digits of **Pi** to any base. Let's store up a few thousand (decimal) digits of **Pi**:

```
nPi = N[Pi, 1000];
```

We can convert these into a list of digits in base 30 using the **RealDigits** function. **RealDigits** returns a list of two objects, the first of which is the list of digits itself (the second object is the exponent by which these digits are shifted, but we don't care about that):

```
digits = First[RealDigits[nPi, 30]];
```

To get a list of 21 digits taken from this long list, we can use **Take**. For example, to get the 100th through 120th digits, we use this command:

```
Take[digits, {100, 120}]

{24, 26, 18, 12, 14, 10, 25, 6, 17, 2, 1, 12, 16, 1, 11, 17,
  2, 0, 7, 27, 18}
```

Jerry: Let's decide that the digit we are looking for is 8 (my granddaughter Jessica just had her eighth birthday). How do we ask *Mathematica* to tell us if 8 is in this list?

Theo: Using the **FreeQ** function. The first argument to **FreeQ** is an expression to be searched, and the second argument is a pattern to look for, 8 in our case:

```
FreeQ[Take[digits, {100, 120}], 8]

True
```

This says that there is *no* 8 in the list. Let's make a table looking for 8 in 30 successive blocks of 21 digits:

```
Table[FreeQ[Take[digits, {21 n, 21 n + 20}], 8], {n, 1, 30}]

{True, False, False, True, True, True, True, True, True,
  False, False, True, False, True, True, False, False, True,
  False, False, False, True, False, False, True, False, True,
  True, True, False}
```

Jerry: So, out of 30 tests, 14 of them contained our digit, which is pretty close to 1/2. All right, now I believe it.

Theo: If we want to look for more than one letter at a time, say **m**-letter sequences, things get a bit trickier. Let's imagine writing out the digits in base **30^m** instead of in base 30. The same analysis as before applies, so to find an **m**-digit sequence we need to solve the equation:

```
Solve[1 - ((30^m - 1) / 30^m)^n == 1/2, n]
Solve::ifun:
   Warning: inverse functions are being used by Solve, so some
     solutions may not be found.

                 1
             Log[-]
                 2
{{n -> ------------------}}
                        m
                 -1 + 30
             Log[--------]
                     m
                   30
```

We can make a function out of this that will tell us the number of digits we need to find any **m**-digit sequence:

```
howFar[m_] := Log[1/2] / Log[(30^m - 1) / 30^m]
```

Jerry: Let's test it on the answer we already know, **m = 1**:

```
N[howFar[1]]
20.4459
```

Theo: Looks the same. How about for 2-digit sequences:

```
N[howFar[2]]
623.486
```

Jerry: Big jump! Luckily, your analysis is not easily tested, since we would need a lot of digits to do so.

Theo: I'm actually quite sure that my analysis is not correct, because it forces the sequence to start at fixed positions from the beginning. Perhaps our visiting professor can sort it out. Anyway, let's try a really long sequence, like 100 letters:

```
N[howFar[100]]
                                         1
Power::infy: Infinite expression -- encountered.
                                         0.

ComplexInfinity
```

To get this answer, we are going to have to do the numerical approximation to more decimal places:

```
N[howFar[100], 150]
```

```
3.6 10^147
```

So, to have a 50/50 chance of finding a 100-letter sequence, we would have to compute 10^{147} digits. This is larger than either the number of particles in the universe, or the number of minimal time units in the age of the universe (where by "minimal time unit" we mean something like the time it takes light to travel the diameter of a proton).

Jerry: You're saying that even if we had a computer the size of the universe that had existed since the beginning of time, we could not find even a 100-letter sentence in the digits of π. I guess there's not much chance of finding the whole Encyclopedia anytime soon.

Theo: Sad but true.

Now, I'd like to make a plot of the digits, so we can see if they really look random. Let's plot the first 100 digits

```
ListPlot[First[RealDigits[N[Pi, 100]]],
    PlotJoined -> True,
    PlotRange -> All,
    AspectRatio -> 0.2];
```

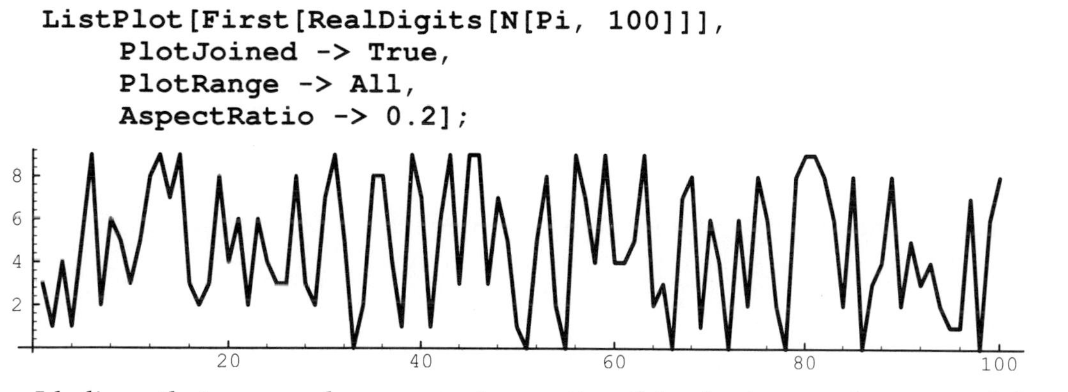

Jerry: I believe that your only purpose in creating this plot is to make a sound from it.

Theo: Well, now that you mention it, it would be nice to hear what these digits sound like. The ear is quite good at picking up any pattern in a waveform. We can define a function to play the digits of a number, like this:

```
playNumber[n_, length_] := Block[{digits},

    digits = First[RealDigits[N[n, length]]];

    ListPlot[Take[digits, 100], PlotJoined -> True,
        AspectRatio -> 0.2, PlotRange -> {0, 9},
        PlotLabel -> InputForm[n],
        Epilog -> SampledSoundList[digits / 10., 7418]
    ]
]
```

The first argument is the number to play, and the second argument is the number of digits to play. Every 7418 digits will give one second of sound.

```
playNumber[Pi, 7418];
```

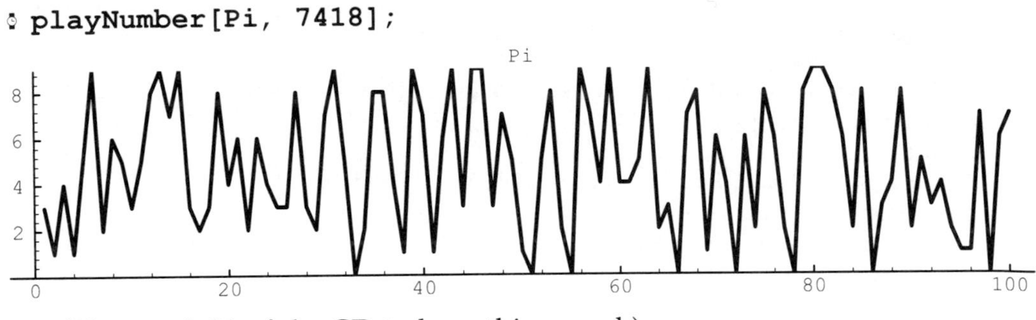

(Play track 31 of the CD to hear this sound.)

Jerry: That sounds pretty random. What we need is some non-random digits to compare this to. Let's try the decimal expansion of 1/97:

```
N[1/97, 110]

0.0103092783505154639175257731958762886597938144329896907216494\
  8453608247422680412371134020618556701030927835051 5
```

This may look random, but it starts to repeat towards the end. The sequence 010392... is both at the beginning and near the end. If we listed more digits, the whole sequence would repeat.

Here is what the first 100 digits of 1/97 looks and sounds like:

```
playNumber[1/97, 100];
```

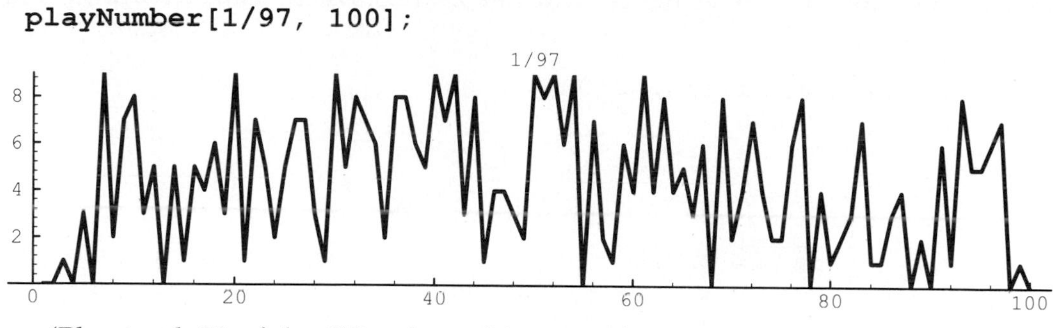

(Play track 32 of the CD to hear this sound.)

Jerry: This is too short to hear. We have to play more digits.

Theo: OK, let's play one second's worth (7418 digits). Note that **`playNumber`** always plots only the first 100, because otherwise the picture would get too crowded.

```
playNumber[1/97, 7418];
```

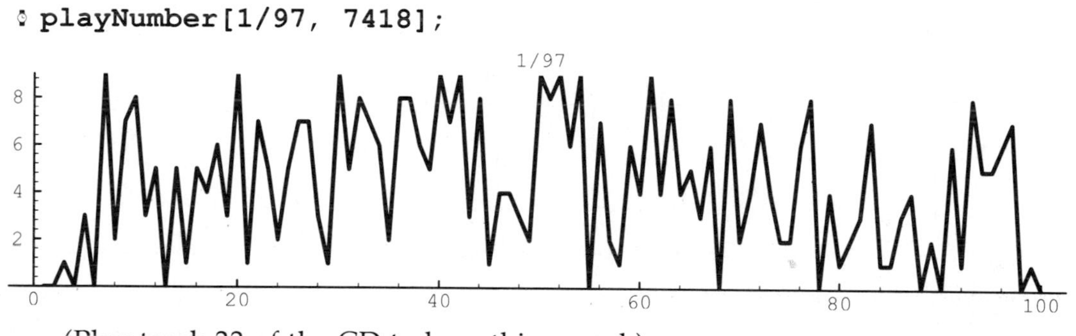

(Play track 33 of the CD to hear this sound.)

Jerry: It certainly does not sound like noise, even though superficially the graph looks just about as random as **`Pi`**. It sounds sort of like a buzzer. I'd like to hear more wonderful fraction sounds! Let's do all of them up to 1/100:

Theo: Here they are. To hear them, double-click on the picture of the first one. This will start them playing in sequence, with the graphs synchronized with the sounds.

```
Do[
    playNumber[1/n, 2400],
    {n, 5, 100}
]
```

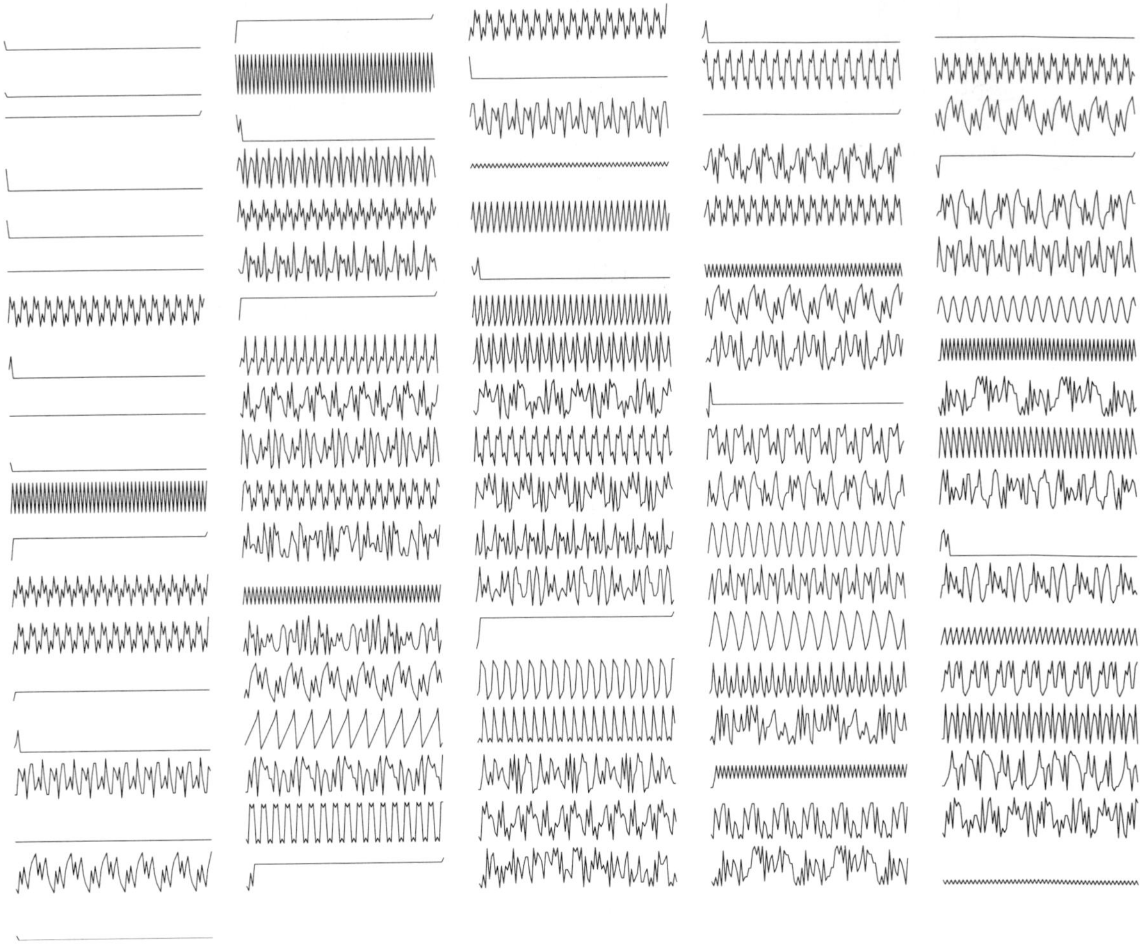

(Play track 34 of the CD to hear these sounds.)

Jerry: Now I know how Beethoven felt when he had finished the 9th Symphony, or maybe John Cage or Paul Koonce....

Theo: It does have a certain catchiness to it. But don't give up your day job yet.

Jerry: Let's try to understand where these different sounds come from. Some of the fractions have a high pitch, others have a low pitch. Some are loud, others are soft.

Some just make a click. These things must be related to features of the decimal expansions.

Theo: Some of the expansions have a long cycle of digits, like 1/97, which starts repeating after 96 digits. Others have a shorter cycle, like 1/11, which alternates between two digits. Some have a cycle of one (the same digit repeated over and over), like 1/3.

Jerry: By clicking the 1/97 sound, I notice it's quite deep, while the 1/11 has a high pitch, and 1/3 just makes a click. Length of the cycle is related to the pitch: long cycles give low pitch.

Theo: This makes sense, since low pitch comes from low frequency, or long cycle, waveforms. Expansions that don't alternate between two or more digits don't make any sound (other than a click), because, after the first few digits, the waveform is flat.

Jerry: Volume must be related to which numbers occur in the expansion. A sequence that alternates between 0 and 9 will be louder than a sequence that alternates between 3 and 4. One that alternates between 3 and 3 would be silent, as you said.

Just for comparison, let's try an irrational number, such as the square root of two.

```
playNumber[Sqrt[2], 7418];
```

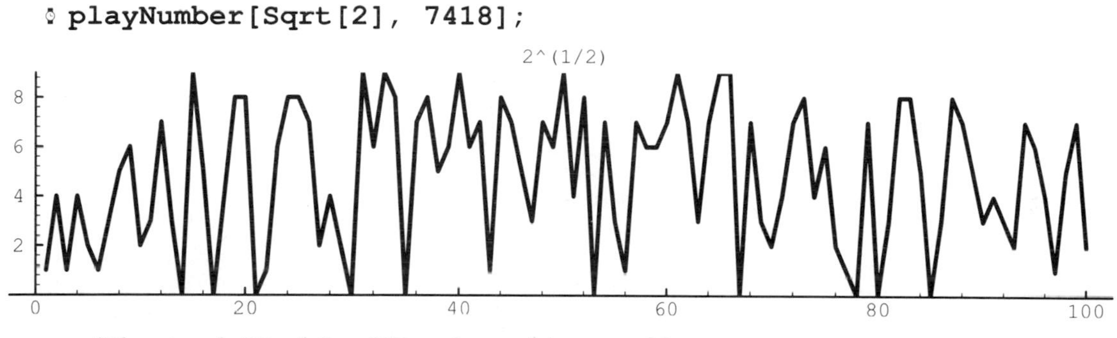

(Play track 35 of the CD to hear this sound.)

Theo: Sounds just like `Pi`: noise. I would be surprised if any irrational number sounded like anything other than noise.

Jerry: So would I.

Dialog–Third Day

Jerry: I'd like to try the strengths of *Mathematica*'s factoring ability. It had no trouble at all factoring **1000!**, a very big number. Since this number has lots of small factors (and no factors above 1000), I would expect it to be easy to factor. I assume *Mathematica* starts by trying to divide by 2, and works its way up from there. A number like 1000! and its divisors are divisible by most numbers, so it's easy to cut it down to size.

Suppose we try to factor a number made out of two large primes. Then *Mathematica* won't be able to find any small factors, and will have to search a long time to find the first factor. Let's try the 1000th and 1001st prime numbers:

```
Prime[1000] Prime[1001]
62773913
FactorInteger[%]
{{7919, 1}, {7927, 1}}
```

Theo: This was no problem (and took almost no time), since the 1000th prime is a small number.

Jerry: OK, try the 5000th prime:

```
Prime[5000] Prime[5001]
2363418209
FactorInteger[%]
{{48611, 1}, {48619, 1}}
```

Theo: Still small numbers, not much time.

Jerry: OK, OK, the millionth!

```
Prime[1000000] Prime[1000001]
239812014798221
FactorInteger[%]
{{15485863, 1}, {15485867, 1}}
```

Theo: Finally, an answer we had to wait a minute or two for (when running on a NeXT computer).

Jerry: Let's try adding 2 to this product, and see if that makes it easier:

```
Prime[1000000] Prime[1000001] + 2
239812014798223
FactorInteger[%]
{{7, 1}, {239, 1}, {143342507351, 1}}
```

Theo: This didn't take very long at all to factor, because *Mathematica* was able to divide by a couple of small factors, and then use a very fast test to check that the remaining factor was prime.

Jerry: This large factor is much larger than the largest prime that the **`Prime`** function is able to return quickly. You're saying that *Mathematica* can check that a number is or is not prime much faster than it can actually factor it. Is there a limit on the size of numbers that can quickly be tested for primality?

Theo: Not really, although past a few hundred digits it gets slow. You can perform this test directly using the function **`PrimeQ`**, which returns **`True`** or **`False`** depending on whether its argument is prime or not.

```
PrimeQ[96]
False
```

```
PrimeQ[97]
True
```

Jerry: Let's use **`PrimeQ`** to investigate some of the classic prime numbers, such as the Fermat numbers. Fermat conjectured that all numbers of the form **`2^(2^n) + 1`** were prime. Here's one:

```
PrimeQ[2^(2^4) + 1]
True
```

Theo: Looks good so far. Let's print a table of the first few. (To save space, I'm going to print out an approximate value for the numbers; of course the calculations internal- ly are done with full precision):

```
Do[Print[{N[2^(2^n)+1], PrimeQ[2^(2^n)+1]}], {n,0,8}]
{3., True}
{5., True}
{17., True}
{257., True}
{65537., True}
             9
{4.29497 10 , False}
             19
{1.84467 10  , False}
             38
{3.40282 10  , False}
             77
{1.15792 10  , False}
```

Jerry: So, he was right up to the fifth one, and wrong after that. I have read that it is currently believed that *only* the first five are prime.

Theo: So, perhaps Mr. Fermat would be happier if everyone would just stop calling them Fermat numbers, since it must be at least a little embarassing to be correct in only 5 out of an infinite number of cases.

Jerry: What happens if we use 3 instead of 2?

```
Do[Print[{N[3^(3^n)+1], PrimeQ[3^(3^n) + 1]}], {n,0,8}]
```

```
{4., False}
{28., False}
{19684., False}
           12
{7.6256 10   , False}
            38
{4.43426 10   , False}
            115
{8.71896 10    , False}
                     347
{6.62818605424187 10    , False}
                      1043
{2.911951061431853 10     , False}
                      3130
{2.469176958933363 10     , False}
```

I guess they won't call them Glynn primes. It seems to be an excellent way of generating non-primes.

How about Mersenne numbers, $2^{prime} - 1$?

Theo: Let's print a table of the first few:

```
Do[Print[{Prime[n], 2^Prime[n]-1, PrimeQ[2^Prime[n]-1]}],
     {n, 1, 18}]
```

```
{2, 3, True}
{3, 7, True}
{5, 31, True}
{7, 127, True}
{11, 2047, False}
{13, 8191, True}
{17, 131071, True}
{19, 524287, True}
{23, 8388607, False}
{29, 536870911, False}
{31, 2147483647, True}
{37, 137438953471, False}
{41, 2199023255551, False}
{43, 8796093022207, False}
{47, 140737488355327, False}
{53, 9007199254740991, False}
{59, 576460752303423487, False}
{61, 2305843009213693951, True}
```

It looks like Mr. Mersenne has better luck. In fact, quite a few Mersenne "primes" *are* prime. I'm sure our visiting professor for this chapter, Dan Grayson, will enlighten us some more on this subject.

Dialog–Fourth Day

Theo: I recently received the Beethoven's Ninth Symphony disk in the CD Companion series from The Voyager Company (Santa Monica, 213-451-1383). It's a HyperCard stack that allows us to read commentary text and play associated passages from the Symphony under complete control, using a CD-ROM drive. Everyone should get one of these (both the CD-ROM drive and the Voyager disk).

Jerry: I'm glad to hear that you've found a good CD-ROM. We've seen so many poor ones while preparing to do our own. Is this one related to *Mathematica*?

Theo: Not especially, but since *Mathematica* can generate and analyze sounds and music, all music is related to *Mathematica*.

Jerry: Hmm.... Moving right along....

Above, we were able to hear sounds from the digits of `Pi` (which sounded like noise), and the digits of fractions (which sounded like tones). Can we do the reverse, and get digits from the sound of Beethoven's Ninth?

Theo: Sure, and then we can take the square root of those digits, and see what *that* looks and sounds like.

Jerry: The *Third* Symphony, of course.

Theo: No, no, I mean the square root of the *digits* of the Ninth Symphony. To get the Third Symphony you have to take every third note from the Ninth--something completely different.

Jerry: In order to take the square root of the digits of the Ninth, we first have to get those digits. How do you propose to do that?

Theo: Well, as anyone who owns a CD player should know, music is numbers. *[Longwinded philosophical controversy edited out at this point].* Sound as we hear it is waves of air pressure bumping up against our ears. At any given time, you can describe the air pressure at your ear with a number. If you list a whole bunch of these numbers in a row, you've got the Ninth. For example, an ordinary CD lists the pressure it would like to have at your ear for each 1/44100th of a second starting at the beginning of the piece. (Actually it lists two numbers at each time, one for each ear. That's called stereo.)

Jerry: What kind of numbers? You mean 27, or 38.5, or -0.036, or what?

Theo: The numbers on a CD are usually thought of as being integers in the range 0 to 65535 (which is 2^{16}), but you can think of them any way you like. They represent inches of displacement of your speaker cones, with the exact scaling factor being supplied by your volume control.

Jerry: How do we capture these numbers from a CD?

Theo: We could try to read the numbers directly using fancy driver software, but for the time being it's easier to play the CD through speakers, and record the output with a MacRecorder digitizer connected to a Macintosh. Here is what you get when you copy from MacRecorder and paste into *Mathematica*:

Sound Cell

(Play track 36 of the CD to hear this sound.)

To hear this sound, double-click on the small speaker-like icon at the top of the cell bracket around this picture (in the electronic edition, of course).

We can use the Convert To InputForm menu command to convert this into a list of *Mathematica* numbers. After a bit of manual editing of the result, you get the following few seconds of the Ninth Symphony written as a list of floating point numbers and assigned to the variable **beethoven**:

```
beethoven = {
0.00781, 0.00000, -0.01562, -0.01562, -0.03125, -0.00781, 0.03125, 0.05469,
0.03906, -0.03906, -0.08594, -0.10156, -0.10156, -0.05469, -0.01562, 0.00781,
0.03906, 0.03125, 0.03906, 0.06250, 0.08594, 0.08594, 0.02344, -0.00781,
-0.00781, -0.04688, -0.05469, -0.03906, -0.03906, -0.03906, -0.07031, -0.04688,
0.02344, 0.03125, 0.00781, -0.04688, -0.08594, -0.08594, -0.10156, -0.08594,
-0.03125, -0.00781, 0.01562, 0.03125, 0.04688, 0.08594, 0.06250, 0.02344,
0.00781, 0.00781, -0.00781, -0.06250, -0.09375, -0.06250, -0.05469, -0.03906,
-0.01562, -0.00781, 0.02344, -0.00781, -0.03125, -0.00781, 0.00781, -0.01562,
-0.07031, -0.08594, -0.04688, -0.02344, 0.00000, 0.04688, 0.07031, 0.09375,
0.10938, 0.14062, 0.14844, 0.14062, 0.11719, 0.01562, -0.07031, -0.08594,
-0.10156, -0.09375, -0.03125, -0.00781, 0.00000, 0.00781, 0.03906, 0.05469,
0.01562, -0.03906, -0.09375, -0.16406, -0.15625, -0.10156, -0.04688, 0.01562,
0.01562, -0.00781, 0.01562, 0.05469, 0.11719, 0.09375, 0.00000, -0.03906,
-0.11719, -0.15625, -0.11719, -0.09375, -0.03906, -0.03906, -0.02344, 0.07812,
0.10156, 0.10156, 0.07812, -0.04688, -0.12500, -0.17188, -0.21094, -0.16406,
-0.07031, 0.01562, 0.01562, 0.00781, 0.10938, 0.15625, 0.16406, 0.16406,
0.13281, 0.03906, -0.08594, -0.14062, -0.09375, -0.03906, 0.02344, 0.07812,
0.07812, 0.07812, 0.08594, 0.03125, -0.06250, -0.10156, -0.08594, -0.12500,
-0.10938, -0.01562, 0.00000, -0.00781, 0.02344, 0.03125, 0.06250, 0.08594,
0.10156, 0.09375, 0.01562, 0.00781, 0.00781, -0.08594, -0.07812, -0.05469,
```

The numbers go on for many, many more pages even at this small font size.

Jerry: Some of these numbers look familiar. 0.06250 must be 5/80, and 0.03125 must be 5/160. What's going on here?

Theo: These fractions are otherwise known as 1/16 and 1/32. MacRecorder recorded each number as an integer between 0 and 255. When *Mathematica* converted them into floating point numbers, it divided each one by 256. Therefore, every number in this list is a rational number with divisor 256. (In case you're interested, we've gone from 16 bit samples at a rate of 44100 per second to 8 bit samples at a rate of 7418 per second.)

Jerry: Good, now I get it.

Theo: We can find out how many numbers we have by asking for the length of the list named **beethoven**:

```
Length[beethoven]
28480
```

We need to convert all the numbers in this list into one big one. Let's do that by pretending that each number in the list is one digit in the big number.

Jerry: How can floating point numbers like the ones in this list be digits? Digits are always integers.

Theo: Good point. Since MacRecorder originally recorded these numbers as integers between 0 and 255, let's convert them back into that form. If we take the numbers, add to each one the value of the smallest one, and then multiply by 255 over the range between smallest and largest, that will scale all of them to lie in this range. **Round** will round each digit to the nearest integer:

```
min = Min[beethoven];
max = Max[beethoven];

beethovenDigits =
    Round[(beethoven - min) 255 / (max - min)];
```

We can check that it worked by asking for the maximum and minimum again:

```
{Min[beethovenDigits], Max[beethovenDigits]}
{0, 255}
```

Jerry: OK, so now we have a list of integers between 0 and 255. Sounds like these are the digits of a number in base 256.

Theo: Right. Now for the exciting part! In the **Functions That We Defined** section below I made up a function that takes a list of digits and converts them into a single number. I want to get a decimal less than 1, so this function starts by putting the first digit just to the right of the decimal point, and continues from there. The **256** means to do it in base 256.

```
beethovenNumber =
      inverseOfDigits[beethovenDigits, 256];
```

Jerry: Whew! That took a *long* time. We could listen to the whole symphony several times over while this was calculating!

Theo: Yes, but now that we have it, we can look at it (with apologies to Kurt Gödel):

```
beethovenNumber
```

Jerry: Do you really expect me to believe that this number, just a bit more than 1/2, is a piece of the glorious Ninth Symphony? I have a hard time adjusting to the notion that something so rich and multifaceted could be reduced to a fraction.

Theo: Well now, I wouldn't be so quick to conclude that it's a fraction (rational). It might be irrational. Many great artists are....

Jerry: Let's think about this. The ratio of two integers always has a decimal expansion that starts repeating one or more digits, eventually. Sometimes the repeated digit is zero, as in 1/8 = 0.12500000..., which is usually written as 0.125 without the zeros. Others end in some other repeated digit, like 191/1500 = 0.1273333333.... (1/8 could also be written as 0.1249999..., if you like, so there's no real difference between repeated zeros and other repeated digits.) Still others end in a cycle of several digits, like 1/7 = 0.142857142857142857....

Getting back to The Beethoven Number, we could say that after we have written the decimal for the whole symphony, we just stop writing. If we mean to imply by this that all subsequent digits are zero (as we do when we write 0.125), then the symphony is a rational number.

Theo: I prefer to say that the number continues forever, representing the sounds in the hall after the symphony is recorded, and in the ruins of that hall when it crumbles to dust. Ignoring certain cosmological questions such as how, if, and when (?) time comes to an end, this means it's an irrational number.

Of course, as with all irrational numbers, for purely practical reasons, we can't calculate *all* its digits. If we wanted to get a few more digits, we could find out where the symphony was recorded, take our MacRecorder there, and record for as long as we like.

Jerry: This seems a bit tortured, but not in conflict with my more reasonable view. Given a choice of having Beethoven rational or irrational, I would prefer rational. All the irrational numbers we've listened to (such as π and the square root of two) sound like noise, whereas all the rational numbers we've listened to sound like tones. Beethoven shares more characteristics with the rational numbers than the irrational ones.

Theo: Even my continuation scheme might result in a rational number, if you accept the Hindu view that time runs in cycles, the largest of which, if I remember correctly, is 88 million years. The number would repeat after:

```
7418 digits/second *
3600 second/hour *
24 hour/day *
365 day/year *
88000000 year
20586196224000000000 digits
```

This repeating Beethoven decimal could be represented by a rational fraction with no more than this number of digits in the numerator and denominator (and potentially fewer, if we're lucky).

Jerry: Well, enough metaphysics. Can we, just to be sure, play **beethovenNumber** back, as we did the numbers earlier in this Chapter?

Theo: Sure, and we can use a slight variation of the function we used before (see the **Functions That We Defined** section below). The command below says to play

beethovenNumber in base 256, which corresponds to how we made up the number in the first place:

```
playFloatingPointNumber[beethovenNumber, 256];
```

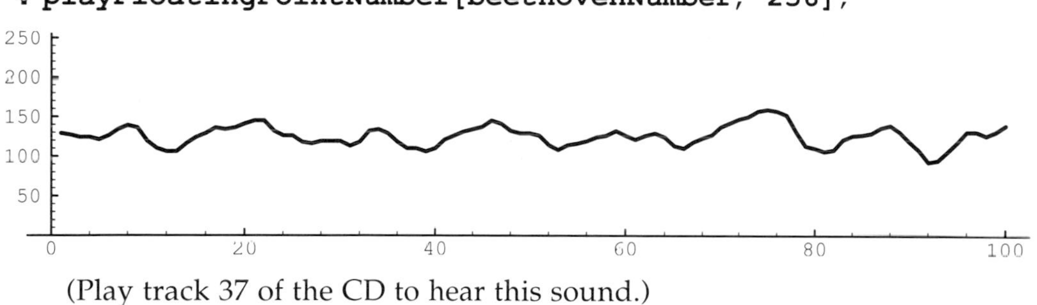

(Play track 37 of the CD to hear this sound.)

Jerry: Yes, it sounds just like it did before. Remarkable.

Theo: Now, let's take the square root! I'll print the whole thing out, so we can all marvel at the mystery of it:

```
squareRootOfBeethoven = Sqrt[beethovenNumber]
```

This number, when played as a sound, sounds like this:

```
playFloatingPointNumber[squareRootOfBeethoven, 256];
```

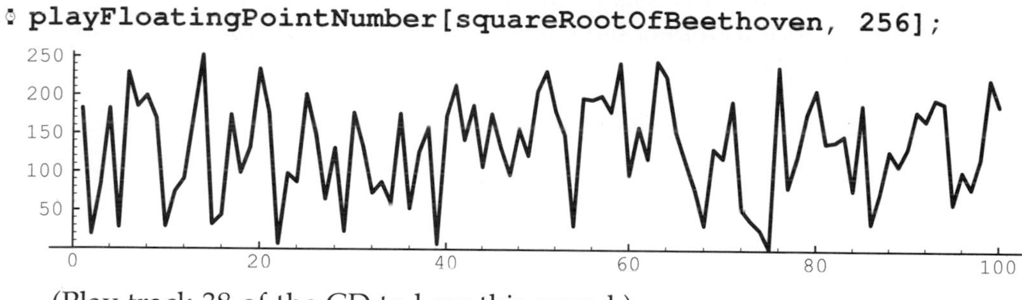

(Play track 38 of the CD to hear this sound.)

Jerry: It's just random noise! Just another irrational number. Oh, what have we done!

Theo: What we have here is a remarkable number. You could show this number to anyone in the world, and not a single one of them would ever, ever, figure out what it was. It's better encoded than any secret code. No digit in this number is related in any way to any of the digits in the original number, except by referring to every other digit.

This is a random number in every regard except that when you square it you get Beethoven's Ninth Symphony.

Visiting Professor: Dan Grayson

The Frequency Plot for the Prime Factors of 1000!.

Dan: Let's look at the plot of factors from earlier in this chapter:

```
ListPlot[FactorInteger[1000!]];
```

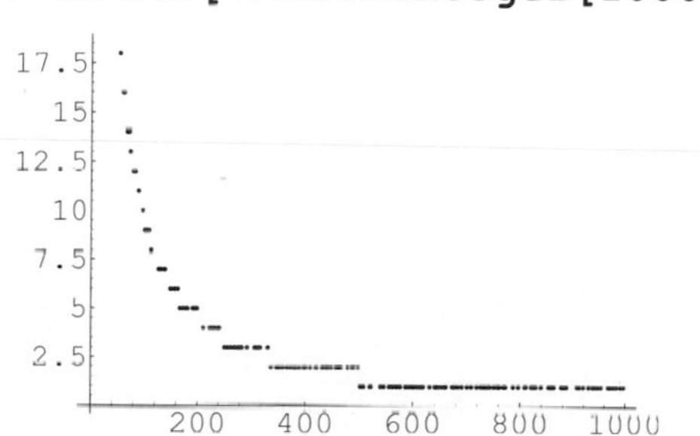

Think of a prime number like 37 and how many times it will appear as a factor in 1000!. The number 1000! is constructed by multiplying together the numbers 1,2,3,..., 1000, and some of those numbers have 37 as a factor, namely: 37, 74, 111, and so on. Here is the complete list:

```
Table[ 37 n, {n, 1000/37} ]
```

```
{37, 74, 111, 148, 185, 222, 259, 296, 333, 370, 407,
  444, 481, 518, 555, 592, 629, 666, 703, 740, 777, 814,
  851, 888, 925, 962, 999}
```

So how many factors of 37 are there in 1000! ? We can compute it this way:

```
Length[%]
```

```
27
```

or this way:

```
Floor[1000/37]
```

```
27
```

Let's try plotting the numbers computed this way for the various primes up to 1000. By trial and error we find that there are 168 primes less than 1000:

```
Prime[168]
```

```
997
```

```
Prime[169]
```

```
1009
```

and that tells us how far to go in the following plot.

```
ListPlot[Table[
     {Prime[i], Floor[1000/Prime[i]]},
     {i,168}
]];
```

17.5
15
12.5
10
7.5
5
2.5
200 400 600 800 1000

This is the same as our original plot. Now if we remove the **Floor** function from the expression being plotted, and run the plot over all values of the horizontal coordinate (instead of just the prime values) we get this picture of the hyperbola y = 1000/x which closely approximates the earlier plots:

```
Plot[ 1000/x, {x, 1, 1000}, PlotRange -> {0,19} ];
```

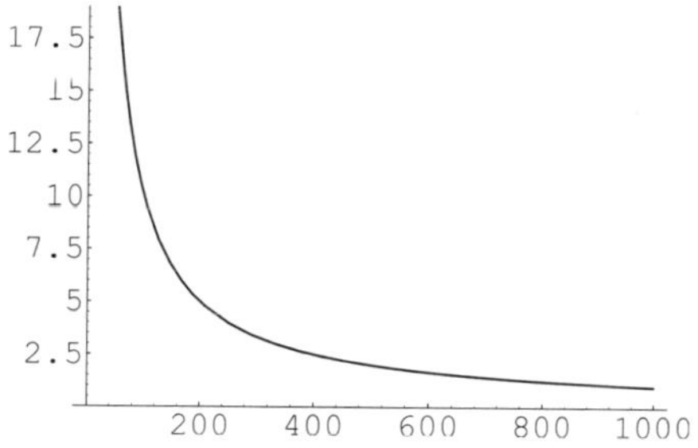

This explains the remark in the text about the curve looking like a hyperbola. It really does!

Actually, the expression **Floor[1000/Prime[i]]** is not a very accurate formula for the number of times **Prime[i]** appears as a factor of 1000!. If the **ListPlot** routine had not discarded the data points corresponding to small primes, then we would have seen some discrepancy between our first two plots. For example, the multiples of 31 less than 1000 are

```
Table[ 31 i, {i, 1000/31} ]
{31, 62, 93, 124, 155, 186, 217, 248, 279, 310, 341,
  372, 403, 434, 465, 496, 527, 558, 589, 620, 651, 682,
  713, 744, 775, 806, 837, 868, 899, 930, 961, 992}
```

Each one of these numbers contains a factor of 31, but one of them contains two!

```
FactorInteger[961]
{{31, 2}}
```

Since the square of 31 is less than 1000, we get an extra factor that we didn't account for before. The correct formula for the number of times a prime occurs in 1000! is:

```
Sum[ Floor[1000/Prime[i]^j], {j, 1, Infinity} ]
```

■ The Scarcity (or Abundance) of Large Primes

Dan: Consider a big prime.

```
Prime[888888]
13651973
N[ Prime[888888]/888888 ]
15.3585
```

This tells us that between 2 and 13651973 are 888888 primes, so in that range, only about one number in 15.3 is prime, on average. Let's call the number 13651973 the "extent" of our search for primes, and call the number 15.3585 the "scarcity" of primes in our search range. Now let's plot the scarcity of primes versus the extent of the search:

```
ListPlot[ Table[ {Prime[n], Prime[n]/n} ,
      {n, 1, 1000000, Round[1000000/130]} ]];
```

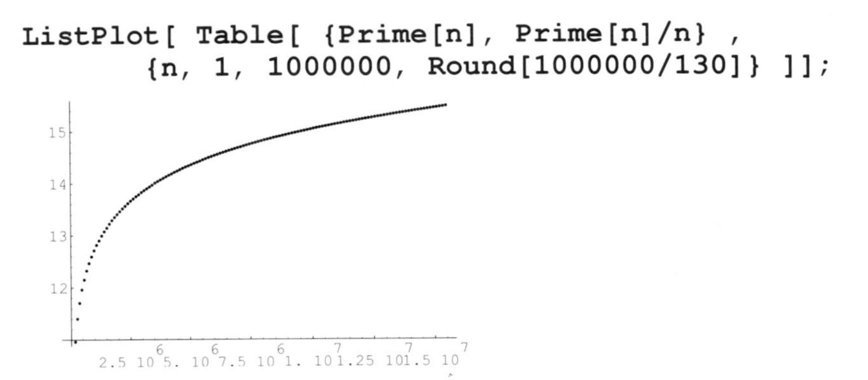

This looks logarithmic, so let's try re-plotting the curve on "log paper", and simultaneously save the data for later use:

```
ListPlot[
   data = Table[ N[{Log[Prime[n]], Prime[n]/n}],
   {n, 1, 1000000, Round[1000000/130]} ]
   ];
```

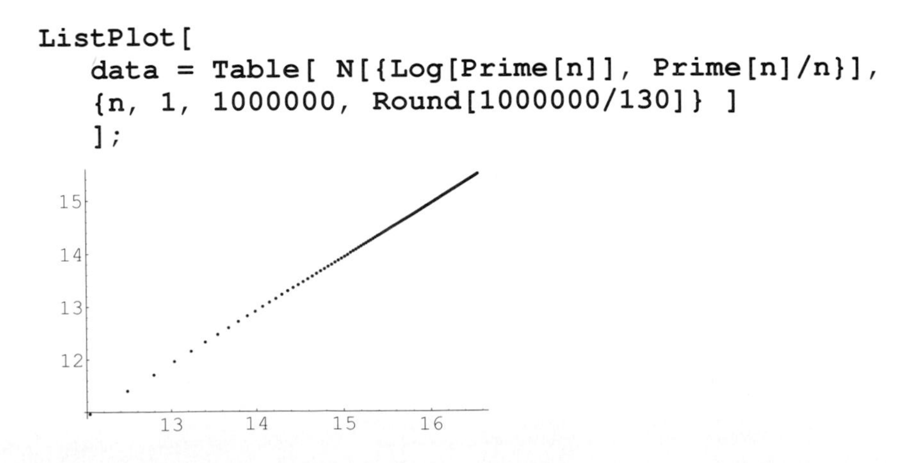

This looks very much like a straight line, so our original guess about logarithmic behavior was correct. Let's find the equation for the straight line in the plot above using the function **Fit**.

```
Fit[ data, {1,x}, x ]
0.463979 + 0.901216 x
```

I don't see anything special about the coefficients in that fitted line, so let's try the same experiment again, but this time go much further.

```
ListPlot[
   data2 = Table[ N[{Log[Prime[n]], Prime[n]/n}],
   {n, 1, 100000000, Round[100000000/130]} ]
   ];
```

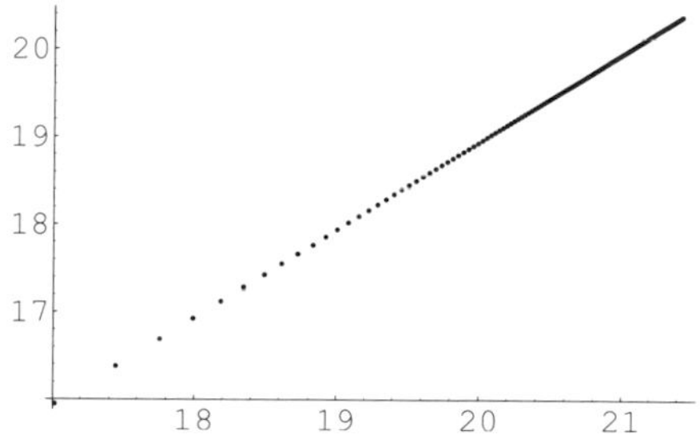

It still looks like a straight line. Let's fit it to an equation, as before.

```
Fit[ data2, {1,x}, x ]
0.774271 + 0.910422 x
```

It's still a straight line. I find this mysterious, but I notice that the slope of the line has increased a bit. Instead of plotting **Prime[n]/n** against **Log[Prime[n]]**, let's plot just the ratio of the two numbers against **n**.

```
ListPlot[  Table[
    {n,  Prime[n]/n  /  Log[Prime[n]]},
    {n,  1,  100000000,  Round[100000000/130]}
   ]];
```

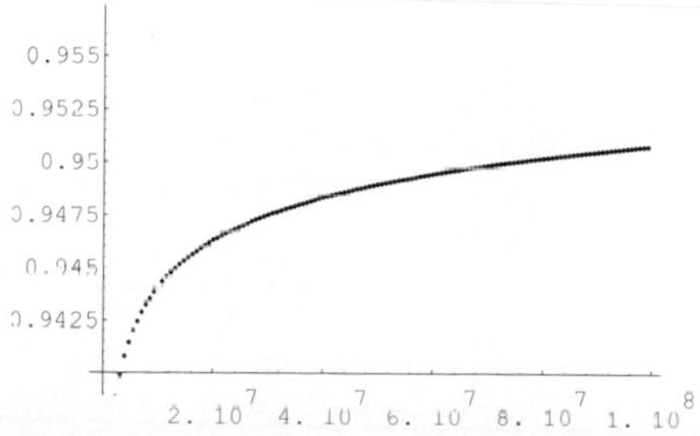

There is a famous theorem called the Prime Number Theorem, which asserts that the limiting value for the function we just plotted is **1**. But we can see from the graph that the convergence is very slow, for with **n** equals 10^8 we have gotten only to **.95** or so.

The prime number theorem confirms the observation in the body of this chapter that the large primes are rarer than small primes. Another way of phrasing it is to say that the scarcity of primes near a large number **x** is near to **Log[x]**, and that this approximation is better and better for larger and larger values of **x**.

Yet another way of phrasing the theorem is to say that the number of primes less than a large number **x** is about equal to **x/Log[x]**.

This is one approximation, but there is a function called the logarithmic integral that gives a much better approximation. Here is an example.

```
Prime[88888888]
1800562657
```

Now we know that the number of primes less than or equal to 1800562657 is 88888888. Let's try out the **LogIntegral** function on this prime:

```
Round[LogIntegral[1800562657.]]
88890430
```

So, the logarithmic integral is nearly the inverse function to **Prime[x]**. (I wanted to put an exclamation point at the end of the last sentence, but people might have thought it was a factorial function, so I didn't. Here it is: !)

■ Decimal Expansions of Simple Fractions

Dan: I'd like to explain a little bit about decimal expansions of fractions, namely, how soon do the digits begin to repeat?

Here is an example that repeats after 96 digits:

```
N[1/97, 200]
0.010309278350515463917525773195876288659793814432989690707\
  21649484536082474226804123711340206185567010309278350515\
  46391752577319587628865979381443298969072164948453608247\
  4226804123711340206185567010309278
```

and here is an example that repeats after 6 digits:

```
N[1/7, 50]
0.142857142857142857142857142857142857142857142857
```

Let's display how long division (technically known as schoolboy arithmetic) would have led to this latter answer.

```
           .142857
      -----------
 7 )   1.000000
         7
       ---
         30
         28
         --
          20
          14
          --
           60
           56
           --
            40
            35
            --
             50
             49
             --
              10
               7
              --
               3
```

Keep in mind that in long division, what we do at each stage is append a zero to the current remainder (thereby multiplying it by 10) and then divide (in this case by 7), putting the quotient up on top and putting the remainder underneath. We will ignore the quotient digits, which we know anyway to be 1 4 2 8 5 7 1 4 2 8 5 7 ... and instead concentrate on the remainders, which, including the initial remainder of 1, form the sequence 1 3 2 6 4 5 1 3 2 6 4 5 ... . The value of the current remainder at any stage determines the complete behavior of the rest of the sequence. Thus, if the sequence of remainders repeats itself, then the sequence of quotient digits repeats itself with the same period.

Here is the way to use *Mathematica* to get the sequence of remainder digits. Start with the initial remainder of 1, which is also the numerator of the fraction 1/7.

```
1
1
```

Multiply that by 10:

```
10 %
10
```

Take the remainder upon division by 7 to get the next remainder.

```
Mod[%,7]
3
```

We can automate this with the **NestList** function as follows.

```
NestList[Mod[10 #, 7]&, 1, 50]
{1, 3, 2, 6, 4, 5, 1, 3, 2, 6, 4, 5, 1, 3, 2, 6, 4, 5, 1,
  3, 2, 6, 4, 5, 1, 3, 2, 6, 4, 5, 1, 3, 2, 6, 4, 5, 1,
  3, 2, 6, 4, 5, 1, 3, 2, 6, 4, 5, 1, 3, 2}
```

Let's call this sequence the *remainder sequence* for **1/7**. Notice that there are no zeroes in this sequence. The reason is that if we would ever get a zero remainder, then all subsequent remainders would be zero, and our decimal expansion would terminate in an infinite sequence of zeroes. This happens only for fractions whose denominators are divisors of a power of 10, such as 1/4=.2500000000000000... .

Theo: What power of 10 is 4 a divisor of?

Jerry: 100/4 = 25.

Theo: So, are we supposed to believe that only numbers that have only factors of 2 and 5 in them give terminating decimals? It seems hard to believe.

Dan: But it's true. The reason is that if some fraction **x/y** in lowest terms is represented by a terminating decimal like **.00008388608000000000**... then that means multiplying **x** by some power of **10** yields a number like **x/y* 10000000000 = 8388608.0000000**... which is an integer. We deduce that **x/y** is equal to **8388608/10000000000**. If we reduce the fraction **8388608/ 10000000000** to lowest terms by entering it into Matematica, we get:

```
8388608/10000000000
  4096
--------
48828125
```

which tells us (by thinking about how one reduces a fraction to lowest terms) that **y** is **48828125**, and that it is a divisor of **10000000000**.

Another thing about the remainder sequence for 1/7 is that no entry in it is bigger than 6. The reason is that every member of the sequence is the remainder resulting from a division of some number by 7, and thus is smaller than 7.

It follows from these two observations that for a number n, the digits in the decimal expansion of 1/n must begin to repeat after at most n-1 digits, for there are only n-1 different values the remainder can assume, namely, 1,2,3,4,5,...,n-1.

With a bit of programming we can make a table which we can use to compare how the length of the repetition cycle compares to n-1 for the fractions 1/1, 1/2, 1/3, ..., 1/99.

```
repcyc = Table[
    {
        n,
        Block[{firstseen, i=0, r=1},
            While[ !IntegerQ[firstseen[r]],
                firstseen[r] = i;
                r = Mod[10 r, n];
                i++;
            ];
            i - firstseen[r]
        ]
    },
    {n,2,100}
]
```

```
{{2, 1}, {3, 1}, {4, 1}, {5, 1}, {6, 1}, {7, 6}, {8, 1},
  {9, 1}, {10, 1}, {11, 2}, {12, 1}, {13, 6}, {14, 6},
  {15, 1}, {16, 1}, {17, 16}, {18, 1}, {19, 18}, {20, 1},
  {21, 6}, {22, 2}, {23, 22}, {24, 1}, {25, 1}, {26, 6},
  {27, 3}, {28, 6}, {29, 28}, {30, 1}, {31, 15}, {32, 1},
  {33, 2}, {34, 16}, {35, 6}, {36, 1}, {37, 3}, {38, 18},
  {39, 6}, {40, 1}, {41, 5}, {42, 6}, {43, 21}, {44, 2},
  {45, 1}, {46, 22}, {47, 46}, {48, 1}, {49, 42},
  {50, 1}, {51, 16}, {52, 6}, {53, 13}, {54, 3}, {55, 2},
  {56, 6}, {57, 18}, {58, 28}, {59, 58}, {60, 1},
  {61, 60}, {62, 15}, {63, 6}, {64, 1}, {65, 6}, {66, 2},
  {67, 33}, {68, 16}, {69, 22}, {70, 6}, {71, 35},
  {72, 1}, {73, 8}, {74, 3}, {75, 1}, {76, 18}, {77, 6},
  {78, 6}, {79, 13}, {80, 1}, {81, 9}, {82, 5}, {83, 41},
  {84, 6}, {85, 16}, {86, 21}, {87, 28}, {88, 2},
  {89, 44}, {90, 1}, {91, 6}, {92, 22}, {93, 15},
  {94, 46}, {95, 18}, {96, 1}, {97, 96}, {98, 42},
  {99, 2}, {100, 1}}
```

The entry **{65,7}** in the table above indicates that 1/64 has a repeating pattern 6 digits long. We can check this:

```
N[1/65, 40]
0.0153846153846153846153846153846153846153
```

Let's select those entries where 1/n has a repeating pattern n-1 digits long.

```
longones = Select[repcyc, (#[[1]]-1 == #[[2]])&]
{{2, 1}, {7, 6}, {17, 16}, {19, 18}, {23, 22}, {29, 28},
  {47, 46}, {59, 58}, {61, 60}, {97, 96}}
```

For example, 1/29 has repeats after 28 digits:

```
N[1/29, 40]
0.03448275862068965517241379310344827586207
```

Notice that all of the numbers **n** whose repetition cycle has length **n-1** are prime:

```
Map[PrimeQ[#[[1]]]&, longones]
{True, True, True, True, True, True, True, True, True,
  True}
```

Theo: So couldn't some sort of test for primality of numbers be developed from this?

Dan: Yes, and it's extremely astute of you to suggest that, for *Mathematica* contains a primality tester which incorporates your suggestion. Let's launch into a more detailed discussion about that.

The observation that any number **n** whose repetition cycle has length **n-1** is prime can be verified theoretically. The verification involves the following two statements.

- The length of the repetition cycle for **1/n** is a divisor of **EulerPhi[n]**.
- The number **EulerPhi[n]** is equal to **n-1** only when **n** is a prime.

The Euler Phi function used here is built into *Mathematica*: given a number **n**, it computes the number of fractions in lowest terms whose denominator is **n** and whose numerical value is between 0 and 1. This definition makes the second statement above clear. The first statement is a bit harder, so we will just take it for granted.

We illustrate the definition **EulerPhi** with n=27 by constructing a table of all the fractions between 0 and 1 which have a chance of having 27 as denominator when reduced to lowest terms, but we use **HoldForm** to prevent the fractions from getting reduced.

```
Table[ HoldForm[#/27]&[i], {i, 0, 27} ]
```

$$\{\frac{0}{27}, \frac{1}{27}, \frac{2}{27}, \frac{3}{27}, \frac{4}{27}, \frac{5}{27}, \frac{6}{27}, \frac{7}{27}, \frac{8}{27}, \frac{9}{27}, \frac{10}{27}, \frac{11}{27}, \frac{12}{27}, \frac{13}{27}, \frac{14}{27}, \frac{15}{27}, \frac{16}{27}, \frac{17}{27}, \frac{18}{27}, \frac{19}{27}, \frac{20}{27}, \frac{21}{27}, \frac{22}{27}, \frac{23}{27}, \frac{24}{27}, \frac{25}{27}, \frac{26}{27}, \frac{27}{27}\}$$

Now we let the fractions reduce to lowest terms:

```
Table[i/27, {i, 0, 27}]
```

$$\{0, \frac{1}{27}, \frac{2}{27}, \frac{1}{9}, \frac{4}{27}, \frac{5}{27}, \frac{2}{9}, \frac{7}{27}, \frac{8}{27}, \frac{1}{3}, \frac{10}{27}, \frac{11}{27}, \frac{4}{9}, \frac{13}{27}, \frac{14}{27}, \frac{5}{9}, \frac{16}{27}, \frac{17}{27}, \frac{2}{3}, \frac{19}{27}, \frac{20}{27}, \frac{7}{9}, \frac{22}{27}, \frac{23}{27}, \frac{8}{9}, \frac{25}{27}, \frac{26}{27}, 1\}$$

and now we select those fractions whose denominator is 27:

```
Select[%, Denominator[#]==27 & ]
```

$$\{\frac{1}{27}, \frac{2}{27}, \frac{4}{27}, \frac{5}{27}, \frac{7}{27}, \frac{8}{27}, \frac{10}{27}, \frac{11}{27}, \frac{13}{27}, \frac{14}{27}, \frac{16}{27}, \frac{17}{27}, \frac{19}{27}, \frac{20}{27}, \frac{22}{27}, \frac{23}{27}, \frac{25}{27}, \frac{26}{27}\}$$

and then count the number of fractions we have:

```
Length[%]
18
```

So, the value of the Euler Phi function for n=27 should be 18. We check the built-in version to see whether it agrees:

```
EulerPhi[27]
18
```

and it does.

Let's check the assertion I made above about the length of the repetition cycle for **1/27** being a divisor of **EulerPhi[27]**.

```
N[1/27,36]
0.037037037037037037037037037037037037
```

The length of the cycle is 3 in this case, and that certainly is a divisor of 18. It is also evidently not a divisor of n-1, which in this case is 26. If 27 had been a prime, then EulerPhi[27] would have been 26, and 26 would have to be divisible by 3. But 26 isn't divisible by 3, so 27 couldn't have been a prime. This reasoning may seem a bit pedantic when applied to such a small number as 27, but suppose we have a number like **n=36113011195088134404 91** at hand and we are wondering whether it is prime or not. It turns out that one can quickly compute whether or not the length of the repetition cycle for n is a divisor of n-1 using the **PowerMod** function.

```
PowerMod[10, 36113011195088134404 90,
             36113011195088134404 91]
```

```
22348536669061476222 34
```

In the example above, we computed the **36113011195088134404 90**th remainder in the remainder sequence for **1/36113011195088134404 90** and got **22348536669061476222 34**. If the length of the cycle had been a divisor of n-1, then the (n-1)st remainder would have cycled back to 1, but it didn't; so we conclude that **3611301119508813440 49** can't possibly be prime.

Here is another example, with **n=30812642497573142641141141**.

```
PowerMod[10, 30812642497573142641141140,
             30812642497573142641141141]
```

```
1
```

Here we observe that the (n-1)st remainder actually *is* equal to 1, so the length of the repetition cycle is a divisor of n-1, and thus it might well be the case that the number **30812642497573142641141141** is a prime number. In order to conclude that n is a prime number, we would have to convince ourselves that the number 1 does not occur at any earlier point in the remainder sequence. If we don't know the prime factorization of n-1 this turns out to be quite hard to do, practically as hard as computing all the members of the remainder sequence!

Mathematica contains a built-in function called **PrimeQ** which incorporates a fancy version of considerations just discussed, and tries to tell you whether a number is prime or not.

```
PrimeQ[36113011195088134404 91]
```

```
False
```

The answer **False** here indicates that the number **36113011195088134404 91** is definitely not prime.

```
PrimeQ[30812642497573142641141141]
```

```
True
```

The answer **True** here indicates that **30812642497573142641141141** might well be prime, and probably is.

The details concerning all these things are usually taught in those college courses with names like *Modern Algebra*, where the theory of groups is presented.

■ Mersenne Primes

Dan: The Mersenne primes are those prime numbers of the form **`2^p-1`**. Such a number can be prime only if **`p`** is also prime, for the number **`2^(m*n)-1`** is always divisible by the number **`2^m-1`**. Here is an example of that:

```
2^(7*11) - 1
151115727451828646838271
2^7 - 1
127
(2^(7*11) - 1) / (2^7 - 1)
1189887617730934227073
```

Since **`2^(m*n)-1`** equals **`(2^m)^n-1`**, its divisibility by **`(2^m)-1`** follows from divisibility of the polynomial **`x^n-1`** by **`x-1`** upon substitution of **`2^m`** for **`x`**. One divides **`x^n-1`** by **`x-1`** and gets the quotient **`x^(n-1)+x^(n-1)+...+x^2+x+1`** with no remainder. Here is an example of that:

```
Cancel[ (x^11-1) / (x-1) ]
```

$$1 + x + x^2 + x^3 + x^4 + x^5 + x^6 + x^7 + x^8 + x^9 + x^{10}$$

Here is the list of numbers **`p`** for which it is known (as of October 1, 1990) that **`2^p-1`** is prime:

> 2, 3, 5, 7, 13, 17, 19, 31, 61, 89, 107, 127, 521, 607, 1279, 2203, 2281, 3217, 4253, 4423, 9689, 9941, 11213, 19937, 21071, 23209, 44497, 86243, 110503, 132049, 216091.

The list above is rather short, so it appears that Mersenne primes are not particularly common. Their popularity arises from the fact that there is a relatively fast algorithm that determines whether a number of the form **`2^p-1`** is prime; there are no other algorithms which yield the primality of such large numbers. Thus a popular pastime for people with access to fast computers is to look ahead for the next Mersenne prime, for when you find it, you become the discoverer of the largest known prime, and articles about you are published in newspapers.

To give an idea of the size of these numbers, we display 500 digits of the biggest one:

```
N[2^216091-1,500]
```

```
7.46093103064661343687339579400511489540228754084977328800\
   51133049777936627252709687806643956351409557300083644 9\
   41548827574272300629992209408195687757874506481981443 9\
   46824524867402922449581677424837048648328299865631531 6\
   35212123406637340315061577534318140942529858971588283 8\
   80868134463451237239916853909168796333674502254039533 2\
   66955915753912516109458086972343184814561433194919942 4\
   23248129480825283814784335269926552925800359826862374 4\
   06711419642986578165168295620281522325379423774986019 5\
   335965009575 10^65049
```

Notice that this Mersenne prime has 65050 digits!

I understand that many readers of this book want to get into the newspapers, so here is the algorithm for checking whether a number of the form **2^p-1** is prime, encoded as a *Mathematica* function:

```
MersenneExponentQ[p_] :=
    PrimeQ[p] ~And~
        (
            (p==2)
        ~Or~
            Block[
                {
                    N=2^p-1,
                    r=4,
                    s=2^((p+1)/2)
                },
                Do[ r = Mod[r^2-2,N], {p-2} ];
                r==0
            ]
        )
```

A reference for the correctness of this algorithm is **[1]**.

Theo: I think this is a very pretty algorithm. I particularly like the use of **~And~** and **~Or~** instead of **&&** and **||**, which mean the same thing.

Dan: Here we use the algorithm to determine which exponents between 1 and 70 give Mersenne primes:

```
Select[ Table[i, {i, 70}], MersenneExponentQ ]
{2, 3, 5, 7, 13, 17, 19, 31, 61}
```

The corresponding Mersenne primes are:

```
2^%-1

{3, 7, 31, 127, 8191, 131071, 524287, 2147483647,
  2305843009213693951}
```

Finally, let's check a rather large one which appeared on our list above:

```
Mersenne[2281]//Timing

{276.317 Second, True}
```

That test was done on a regular-speed workstation (1990 model).

Functions That We Defined

The function **playNumber** plays **length** digits of the number **n** as a decimal expansion. The decimal digits of the number are played as an amplitude waveform at a sampling rate of 7418 digits per second.

```
playNumber[n_, length_] := Block[{digits},

    digits = First[RealDigits[N[n, length]]];

    ListPlot[Take[digits, 100], PlotJoined -> True,
        AspectRatio -> 0.2, PlotRange -> {0, 9},
        PlotLabel -> InputForm[n],
        Epilog -> SampledSoundList[digits / 10., 7418]
    ]
]
```

The function **playFloatingPointNumber** plays all the digits of the number **n** in base **base**, at a sampling rate of 7418 digits per second.

```
playFloatingPointNumber[n_, base_] := Block[{digits},

    digits = First[RealDigits[n, base]];

    ListPlot[Take[digits, 100], PlotJoined -> True,
        AspectRatio -> 0.2,
        PlotRange -> {0, base - 1},
        Epilog ->
            SampledSoundList[digits / N[base], 7418]
    ]
]
```

The function **inverseOfDigits** takes a list of integers and converts it into a floating point number less than 1. The numbers in the list are interpreted as digits in base **base**, with the first one being just to the right of the decimal point.

```
inverseOfDigits[digits_, base_] := Block[
    {length},
    length = Length[digits];
    Fold[(base #1 + #2)&, 0, digits] /
        N[base^length, Round[N[length Log[10, base]]]]
]
```

References

1) D. H. Lehmer, *Journal of the London Mathematical Society*, volume 10, 1935, p162-165.

Chapter Nineteen
Complex Maps, or How to Make a Tangled Mess

Learning from others, Jerry and Theo make a better tangled mess.

Dialog

Jerry: One thing that the reader may not be aware of is that Theo is a Macintosh/NeXT person.

Theo: And proud of it.

Jerry: A friend of mine, Greg Smith, can go on for hours about the differences between Macintosh people and DOS people. Another friend of mine, Martin Lapidus, the author of "*f(z)*", "The Intelligent Calculator", "Fields and Operators", and who-knows-what-else **[1]**, is the archetypal DOS user. Marty's program *f(z)* is, in my opinion, a sensational way of experimenting with the geometry of transformations of complex numbers. I tried for months to get Theo to look at it; but since it's a DOS program, he found one excuse after another to avoid it. Finally we looked at *f(z)* together, and this inspired Theo to use *Mathematica* to do complex transformation plots "the right way", as he put it. In effect, Marty and *f(z)* triggered Theo and *Mathematica* to bring the ideas to more robust, general (and possibly slower-running) form.

Theo: My main objection to the DOS environment is the hopeless inconsistencies among different programs. Although *f(z)* is a very nice program, I don't have a clue how to use it, nor do I intend to learn. If I did learn how to use it, I would not be able to use that knowledge for anything other than using *f(z)*. Since I am unlikely to want to use *f(z)* more than once a year, I am sure to forget most things by the time I want to use it next. This is a general problem with DOS programs, almost without exception.

Jerry: What you accurately describe is also, in a perverse way, a source of richness. Many DOS programs are written by "lone wolves" who don't seem to spend much time worrying about how anyone other than themselves will use their programs.

Theo: The Macintosh developer community has encouraged and sometimes aggressively enforced the notion that people writing applications for the Macintosh are not writing their own programs, but are cooperating in the writing of a larger whole. It's very important to Macintosh users and developers that each program fit together with all the other programs in the world. As a result, what I learn in one program is probably useful knowledge in other programs as well. Studies show that the average Macintosh user uses more than twice as many different applications as the average DOS user.

Jerry: It's exactly this rational, organized, team-playing view of the world that is fundamentally distasteful to some DOS developers. I don't believe they get up one morning and think this through, it's just something in their bones.

Theo: Fortunately, users are free to choose which set of developers they want to base their lives and livelihoods on. The Macintosh program *ComplexPaint* by Mark Parmet **[2]** does many of the things that *f(z)* does. I was able to use all its features from the very start, because it works basically the same way as many other programs, particularly "MacPaint"-like programs, that I am familiar with.

Jerry: I love *ComplexPaint*, although *f(z)* has qualities that are missing in *ComplexPaint*. Both will be used widely, and it will be interesting to see what Marty's Macintosh version of *f(z)* looks like.

Theo: In *Mathematica* we can't do the kinds of interactive manipulation possible in *f(z)* or *ComplexPaint*. What we're going to do is not going to be nearly as much fun. In exchange we will be able to explore while leaving a record of what we did. Because Notebooks like this one are complete documents, we can use them to explain to other people what we did. What follows is not as flashy as *f(z)*, but it's considerably more powerful. The small amount of learning about *Mathematica* syntax that people need to do is the kind of learning that they can use later for other things.

Jerry: Now, back to complex maps in *Mathematica*. Complex numbers are usually expressed in the form `a + b I`, with `a` and `b` being real numbers. So, `2 + 3 I`, and `-4`, and `5 I`, and `1.732 - 3.14 I` are all complex numbers. We can associate the complex number `a + b I` with the ordered pair `{a, b}`, and we can plot a point corresponding to this ordered pair. So, we can translate any complex number into a point on the (x, y) plane, and likewise we can translate any point on the plane into a complex number. By convention, most people use the horizontal axis as the `a`-axis, and the vertical axis as the `b`-axis.

Theo: We can combine three steps into one; first, convert a point on the plane into a complex number; second, apply some mathematical function to this complex number, resulting in another complex number; third, convert this complex number back into a point on the plane. The combined operation is a mapping that takes every point on the plane and moves it somewhere else (or sometimes leaves it in the same place). We can imagine laying down a grid of points or lines on the plane, and then seeing what the image of the grid looks like after we've done the mapping. I've written a function, **ComplexMapPlot**, that lets you make complex map plots starting from any arbitrary set of initial points, lines, circles, arcs, etc.

Jerry: For a given function we can get very different plots depending on what grid of lines or points we start with. How do we specify in *Mathematica* what we want as our starting grid?

Theo: I've written several functions that provide you with suitable sets of initial lines and/or circles for use with **ComplexMapPlot**. Here is an example:

```
ComplexMapPlot[Sin[z], z,
    RectangularGrid[{{-1, 1}, {-1, 1}}]];
```

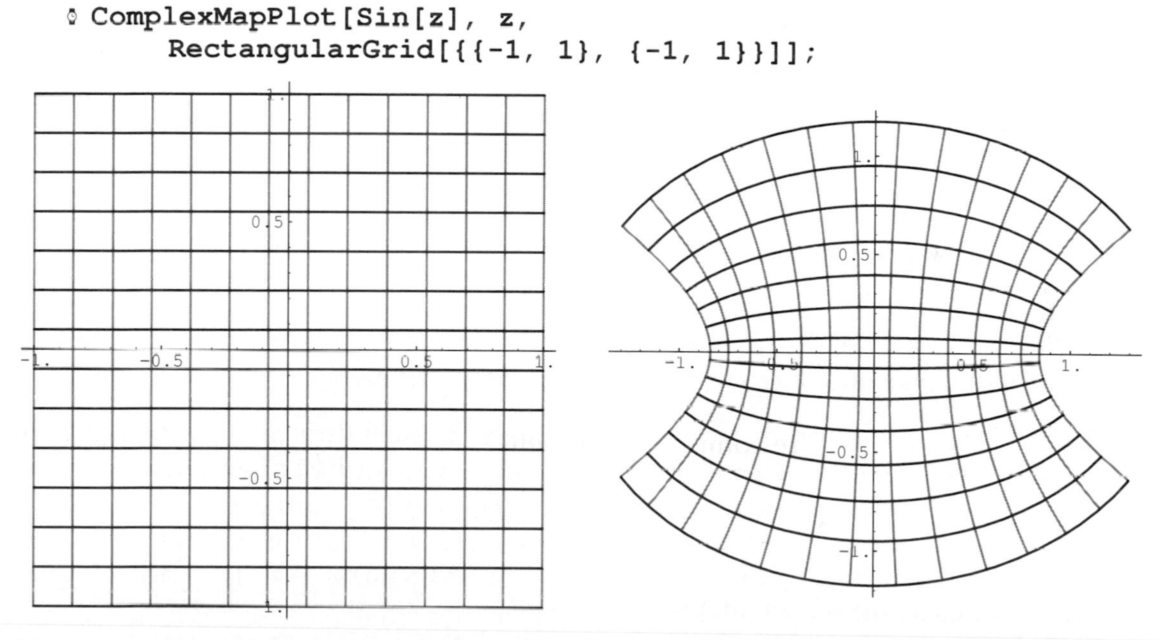

This is a grid of 14x14 lines, over the range -1 to 1 in each direction. The original grid is on the left, and the mapped image is on the right. The horizontal lines are black, while the vertical lines are gray (dotted, on a black-and-white screen).

Jerry: The mapping looks like a sheaf of wheat. Could you explain the arguments of **ComplexMapPlot** and **RectangularGrid**, so I could reap my own golden harvest?

Theo: The first argument to **ComplexMapPlot** is the function you want to plot. Typically it's a function of one variable called **z**. The second argument is the name of the variable you used in your function, typically **z**. The third argument is the set of lines, point, circles, etc., that you want to have mapped:

```
ComplexMapPlot[function, variable, preimage]
```

Usually you would use one of the following functions to generate the **preimage** argument automatically. The arguments to these functions are as follows:

RectangularGrid[{{a1, a2}, {b1, b2}}]
Horizontal and vertical lines over the range **a1** to **a2** in the horizontal direction and **b1** to **b2** in the vertical direction. The horizontal lines are black while the vertical lines are gray.

HorizontalLines and **VerticalLines**
These take the same arguments as **RectangularGrid**, but produce only one set of lines each.

PolarGrid[{a, b}, r]
Concentric circles and radial lines with center at **{a, b}** and maximum radius of **r**. The circles are black while the radial lines are gray.

PolarGrid[{a, b}, {r1, r2}]
Donut of concentric circles and radial lines with center at **{a, b}**, inner radius of **r1**, outer radius of **r2**.

PolarCircles and **PolarLines**
These take the same kinds of arguments as **PolarGrid**, but produce only concentric circles or radial lines, respectively.

Each of these functions produces 14 lines or circles in each direction, by default. You can change this using the **PlotPoints** option. If you add **PlotPoints -> n** to any of these functions, you will get **n** lines or circles in each direction. With **RectangularGrid** you can use **PlotPoints -> {n, m}** to get **n** horizontal and **m** vertical lines. With **PolarGrid** you can use **PlotPoints -> {n, m}** to get **n** concentric circles and **m** radial lines.

Jerry: I would like to see the function **1/z** with each of the different pre-images: rectangular grid, horizontal lines, vertical lines, concentric circles and radial lines, concentric circles, and radial lines.

Theo: OK, no problem. Here are the rectangular ones:

```
ComplexMapPlot[1/z, z,
    RectangularGrid[{{-2, 2}, {-2, 2}}]];
```

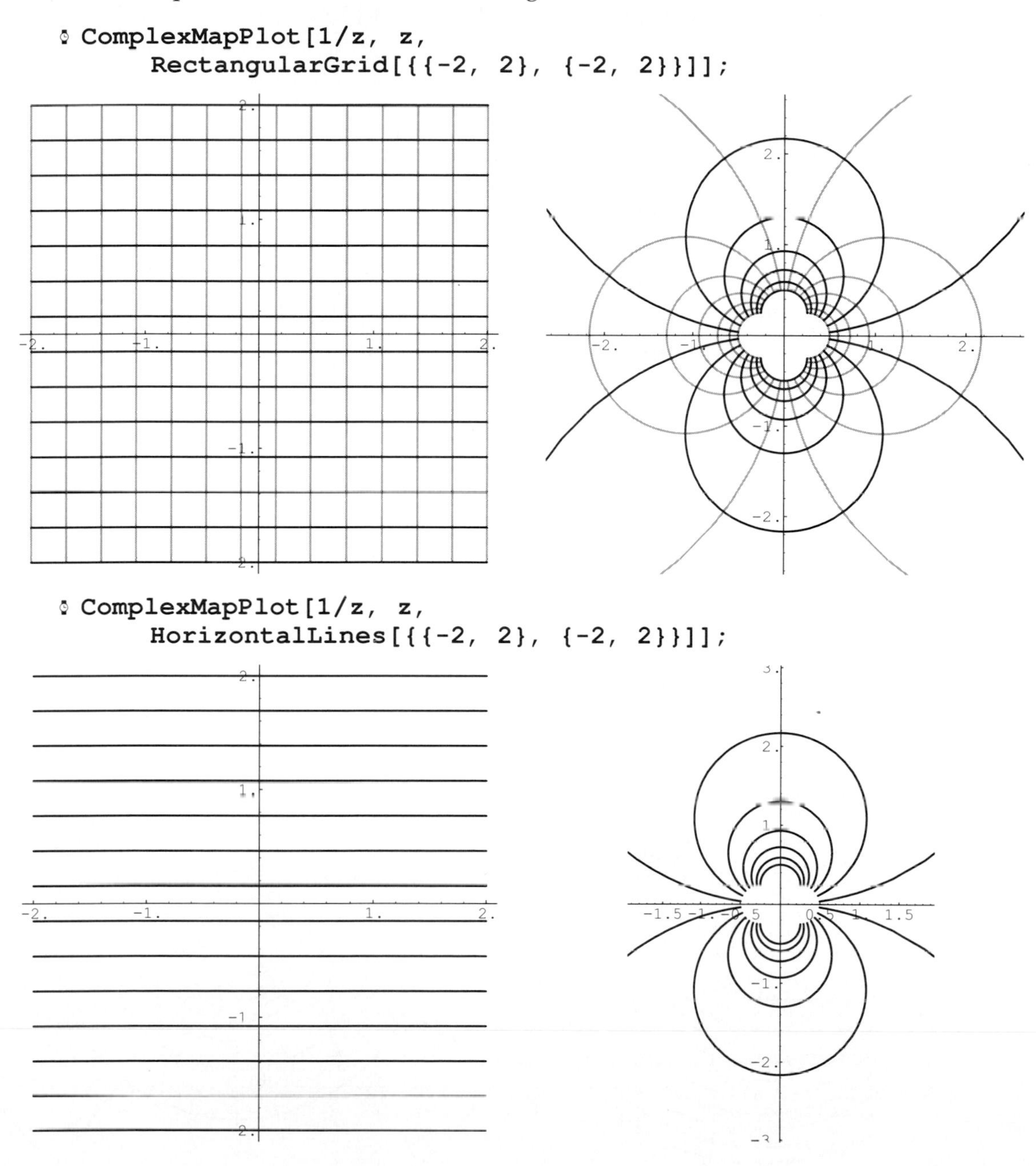


```
ComplexMapPlot[1/z, z,
    HorizontalLines[{{-2, 2}, {-2, 2}}]];
```

```
ComplexMapPlot[1/z, z,
    VerticalLines[{{-2, 2}, {-2, 2}}]];
```

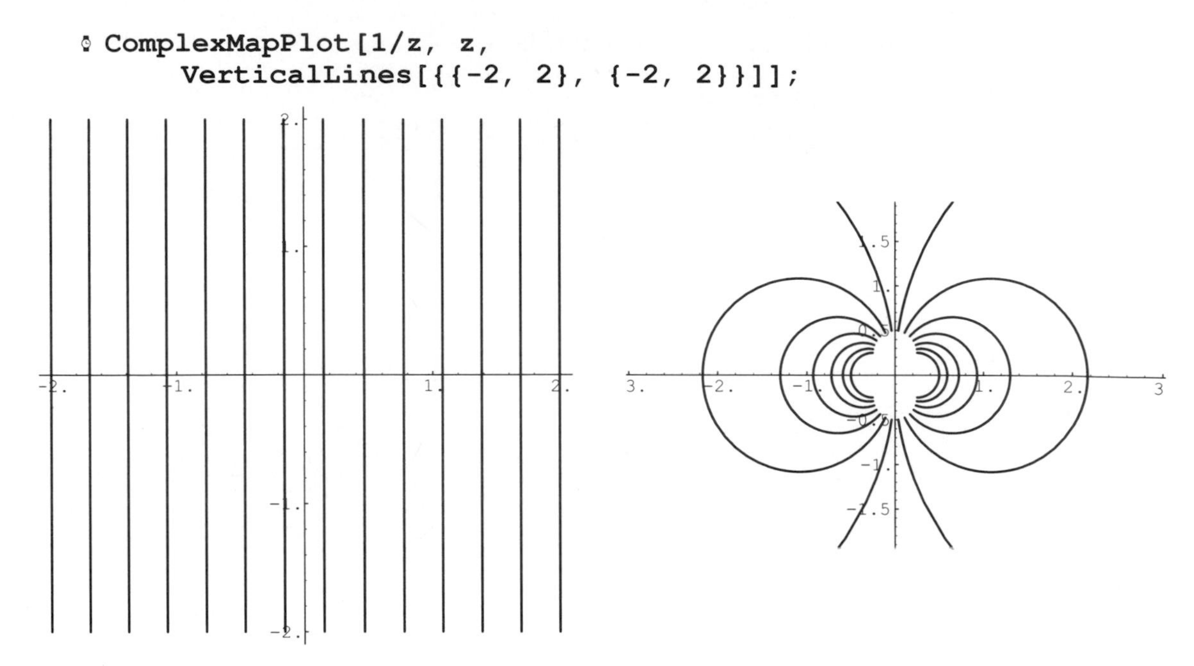

Here are the polar ones (note that I've used a range of radii from 0.1 to 2, because there is an undefined point at 1/0):

```
ComplexMapPlot[1/z, z,
    PolarGrid[{0, 0}, {0.1, 2}]];
```

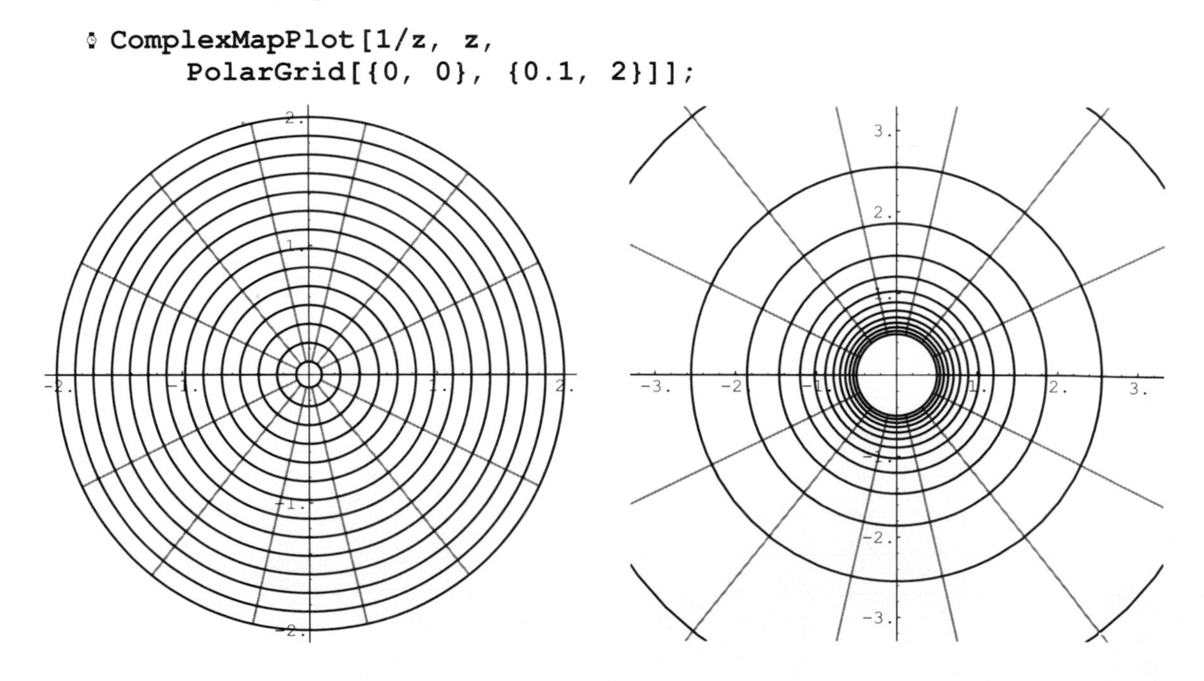

```
ComplexMapPlot[1/z, z,
     PolarCircles[{0, 0}, {0.1, 2}]];
```

```
ComplexMapPlot[1/z, z,
     PolarLines[{0, 0}, {0.1, 2}]];
```

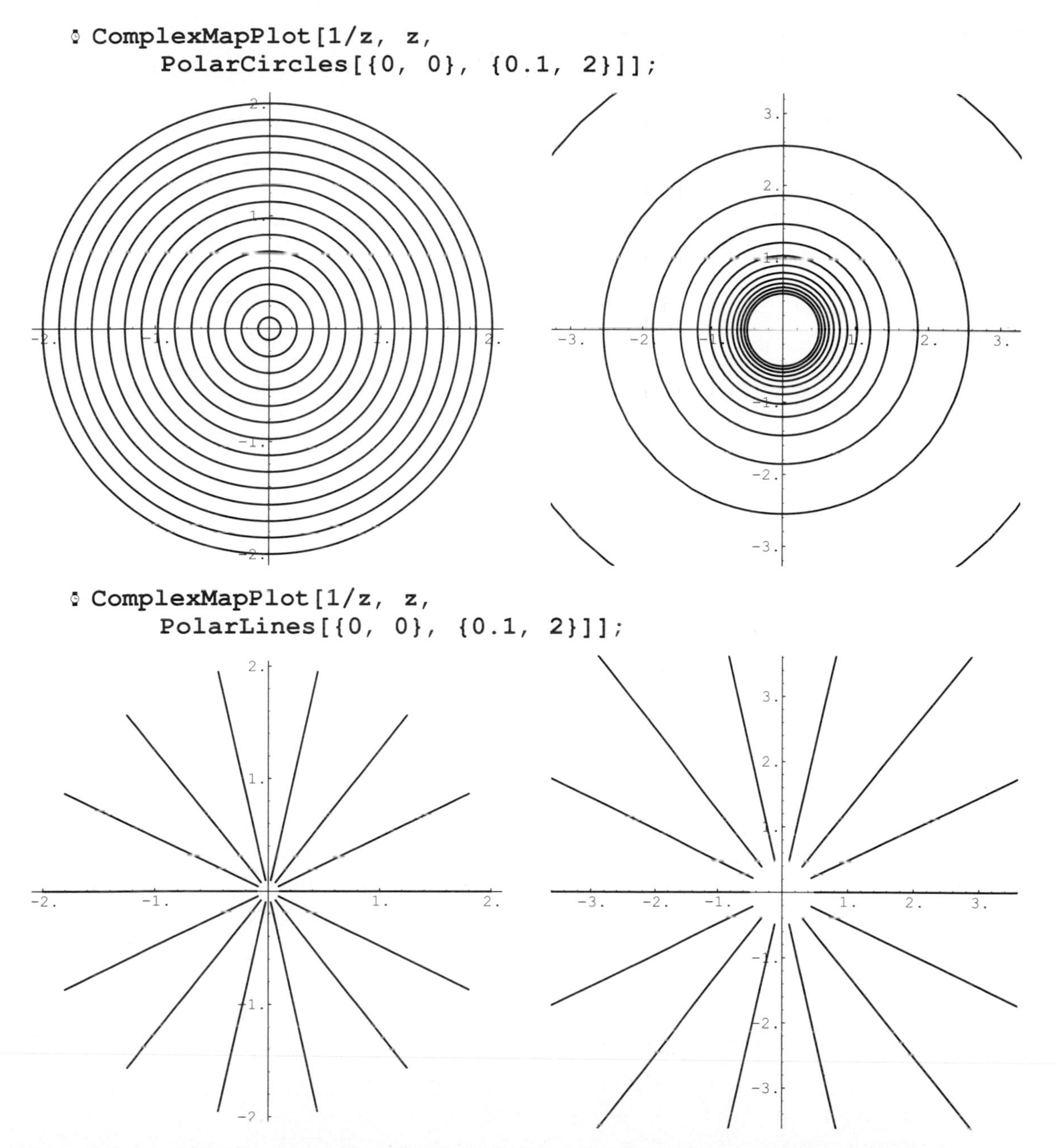

Theo: It looks like horizontal and vertical lines get turned into circles, but radial lines stay as lines and concentric circles stay as circles, though possibly not in the same order. As we'll see later we can use animation to see how the circles in the input and output are related to each other.

Jerry: There is a transformation, **`z + 1/z`**, which when applied to a circle with center on the a-axis and passing through the point (-1, 0) produces an attractive shape called the *Joukowski airfoil.*

```
ComplexMapPlot[z + 1/z, z, {Circle[{0.25, 0}, 1.25]}];
```

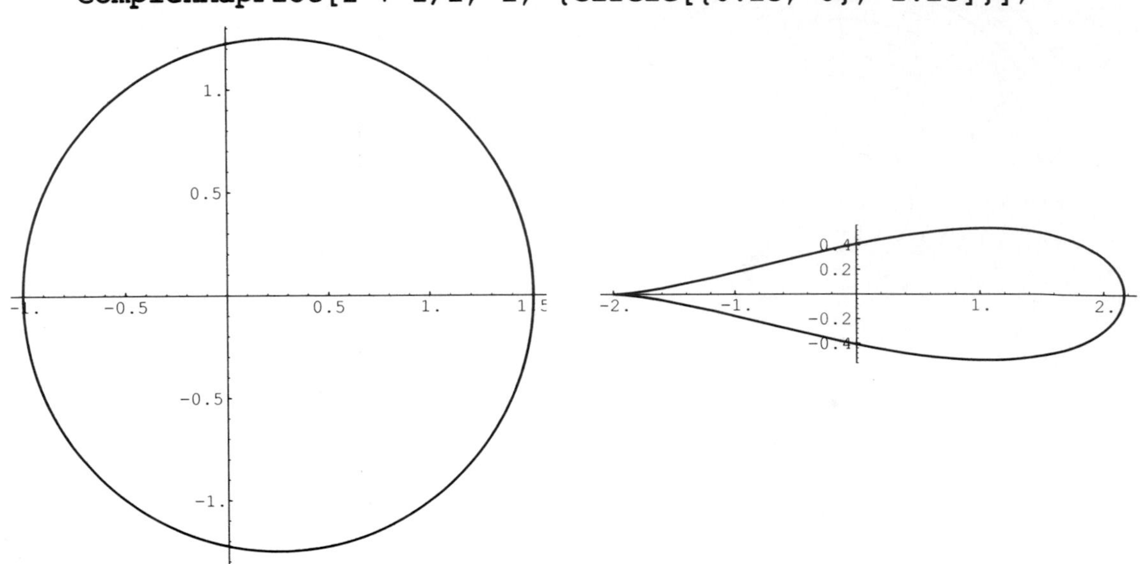

The above example is from *Complex Variables and Applications* **[3]**. I'd like to see what this airfoil looks like when the circle has its center at different points on the horizontal axis, but still passes through (-1, 0). Let's make an animation!

Theo: OK. Let's also replace the **`Circle`** graphics object in the above example with a **`Disk`** object. **`Disk`** is the same as **`Circle`** except that it draws a filled-in circle instead of the outline of a circle. Note that using filled shapes (like **`Disk`** and **`Polygon`**) will not work with functions that cause the inside of the filled area to end up outside in the transformed shape. The fill will be drawn incorrectly. Fortunately this is not a problem with this example.

```
Do[
    ComplexMapPlot[
        z + 1/z, z, {Disk[{n, 0}, n + 1]},
        PlotRange -> {{-2, 3.5}, {-2, 2}}
    ],
    {n, 0, 1, 0.1}
]
```

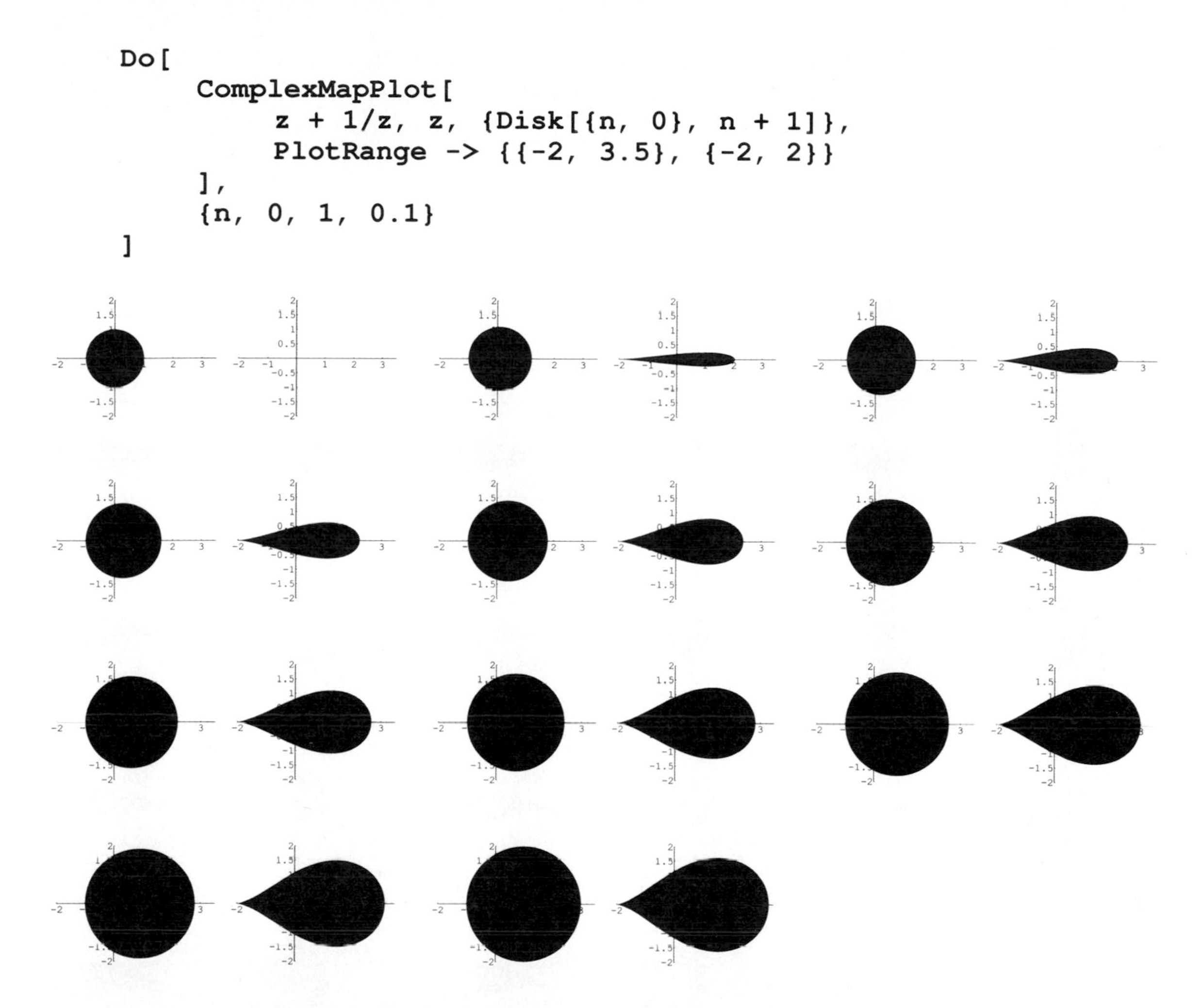

Nice, but none of these really look like useful airfoils. If it's going to fly, it's got to be flatter on the bottom side than on the top!

Jerry: Let's try moving the circle up or down off the horizontal axis. Maybe that will make it come out less symmetrical.

Theo: We can make another animation, this time changing the vertical coordinate of the circle while leaving the horizontal coordinate fixed at 0.4:

```
Do[
    ComplexMapPlot[
        z + 1/z, z, {Disk[{0.3, n}, 0.3 + 1]},
        PlotRange -> {{-2, 3.5}, {-2, 2}}
    ],
    {n, -0.3, 0.3, 0.05}
]
```

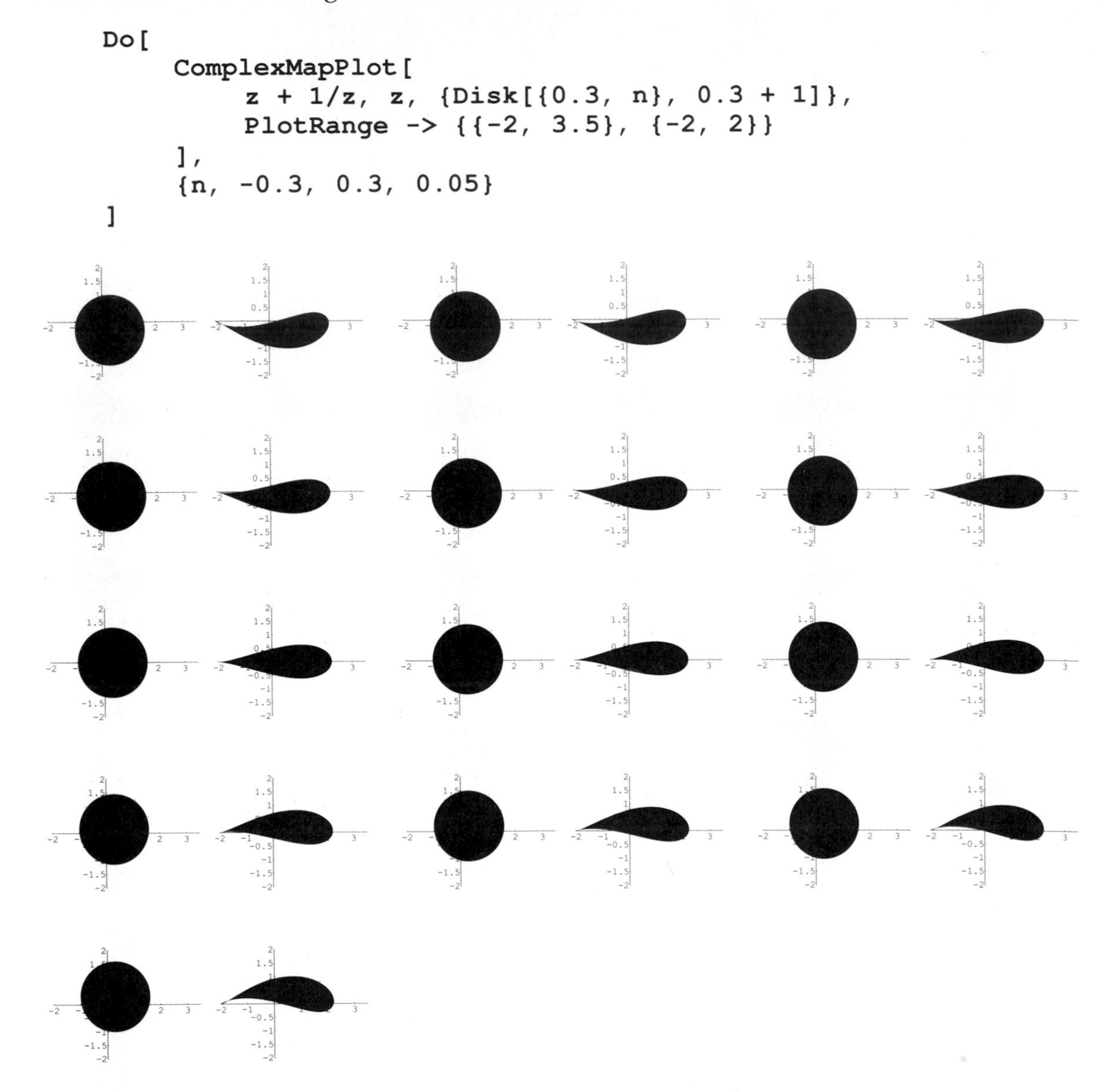

Who needs wind tunnels! Let's move on to some other functions.

Jerry: There are interesting shapes I call "cuspy" circles (other people, for example Philip J. Davis in *The Schwarz Function and its Applications* **[4]**, have more dignified names for them), which can be generated nicely by **ComplexMapPlot**. Let's try **z + z^6/6**, and start with a single unit circle centered at (0, 0).

```
ComplexMapPlot[z + z^6/6, z, {Circle[{0, 0}, 1]}];
```

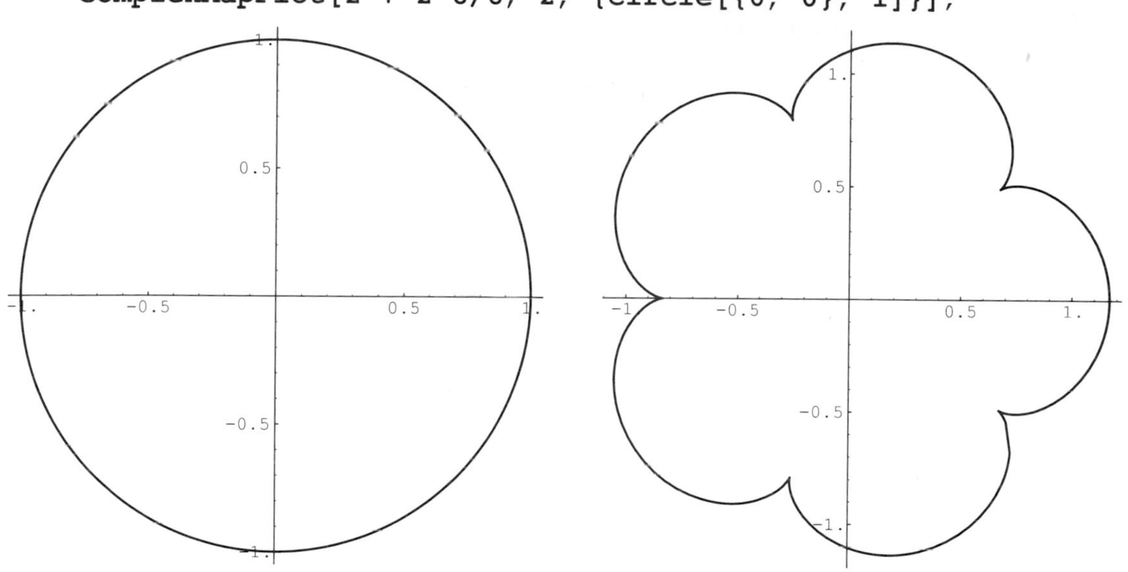

I can imagine hundreds of possibilities here, such as change the plus to minus, try a different exponent, use an exponent different from the number we divide by, or make the function $z + z^3/3! + z^5/5!$.

Theo: I'd like to see what happens if we use circles of different radii. Since we can give **ComplexMapPlot** an arbitrary list of graphics objects, we can do something clever. We can make a single thick circle of radius 1, and several thinner circles with different radii.

The graphics primitive, **Thickness**, specifies the thickness of any following lines or circles, so our pre-image consists of a thickness specification, followed by a single circle, followed by another thickness, followed by a **PolarCircles** function, which generates seven equally-spaced circles with radii inside and outside the first circle.

```
ComplexMapPlot[z + z^6/6, z,
    {
        Thickness[0.008],
        Circle[{0, 0}, 1],
        Thickness[0.004],
        PolarCircles[{0, 0}, {0.8, 1.2}, PlotPoints -> 7]
    },
    IncludePreimage -> False,
    Axes -> None];
```

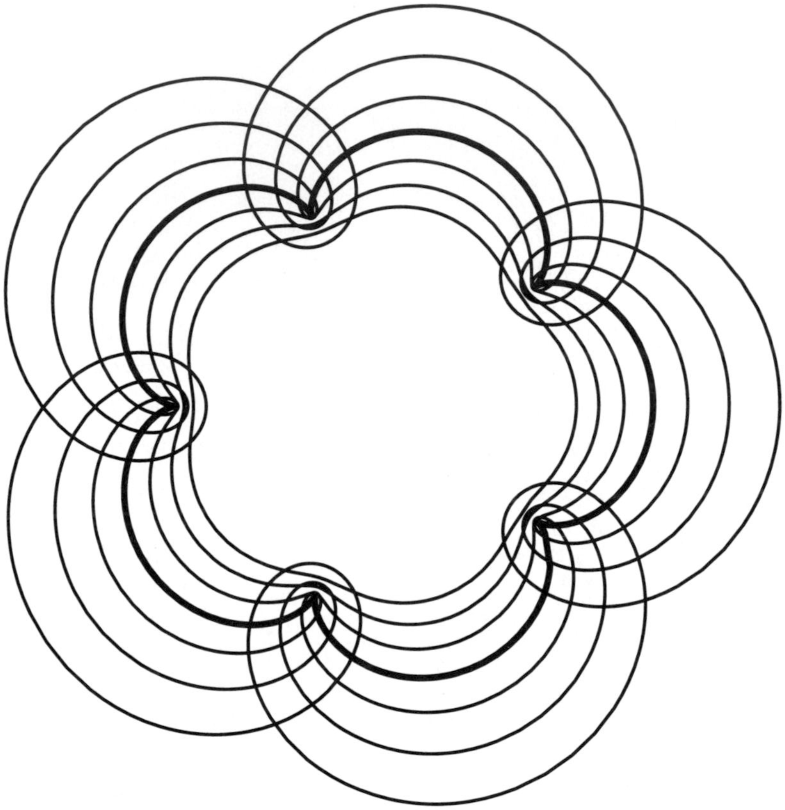

Jerry: This is my favorite graphic in the book so far. I especially like the thick, ropey effect. This should be on the cover!

Theo: (Well, I guess the readers will know if it made it or not.) I think we can do even better. One problem is that we can't know for sure which of these loopy things came from which of the original circles. I'd like to add some `GrayLevel` graphics primitives (which specify the darkness with which to draw the following circles), to make the outside circles lighter than the inside ones. I'd also like to have more circles over a wider range of radii.

```
◊ ComplexMapPlot[z + z^6/6, z,
      {
           Thickness[0.002],
           GrayLevel[0.5],
           PolarCircles[{0, 0}, {1.05, 1.4},
                          PlotPoints -> 8],
           Thickness[0.005],
           GrayLevel[0],
           Circle[{0, 0}, 1],
           Thickness[0.002],
           PolarCircles[{0, 0}, {0.35, 0.95},
                          PlotPoints -> 14]
      },
      IncludePreimage -> False,
      Axes -> None];
```

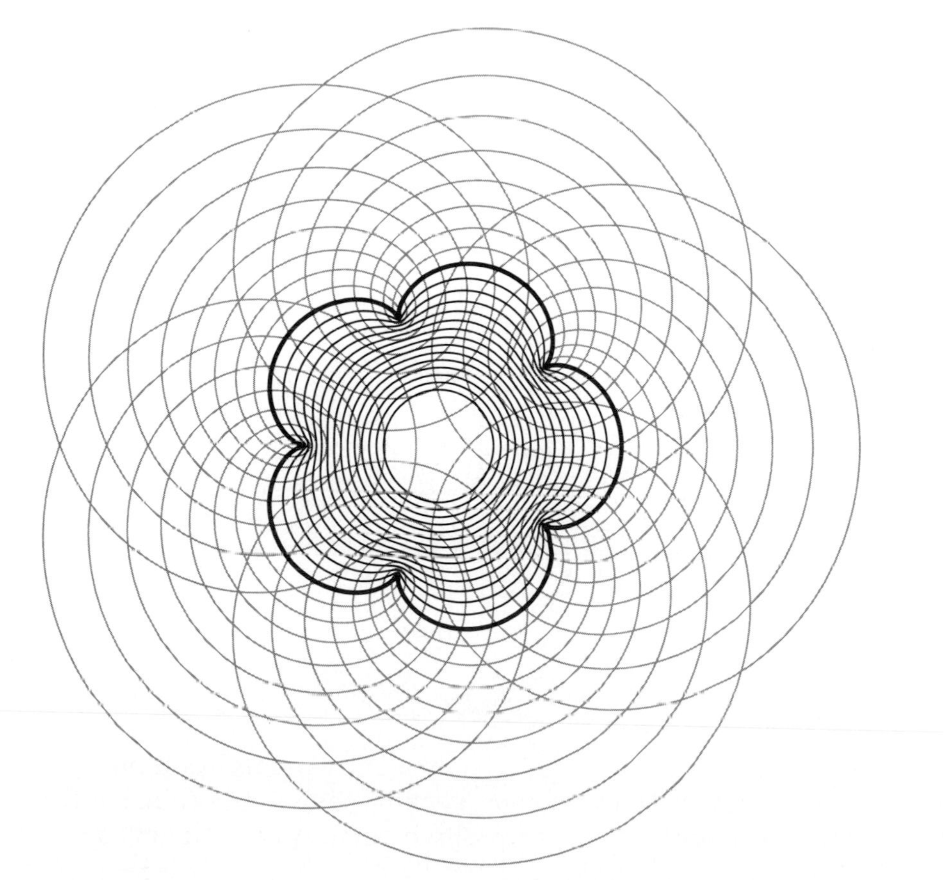

Jerry: Good try.

Theo: Well, *I* like it.

Jerry: Let's try a variation of this function, using a negative exponent:

```
ComplexMapPlot[z + z^-6/6, z, {Circle[{0, 0}, 1]}];
```

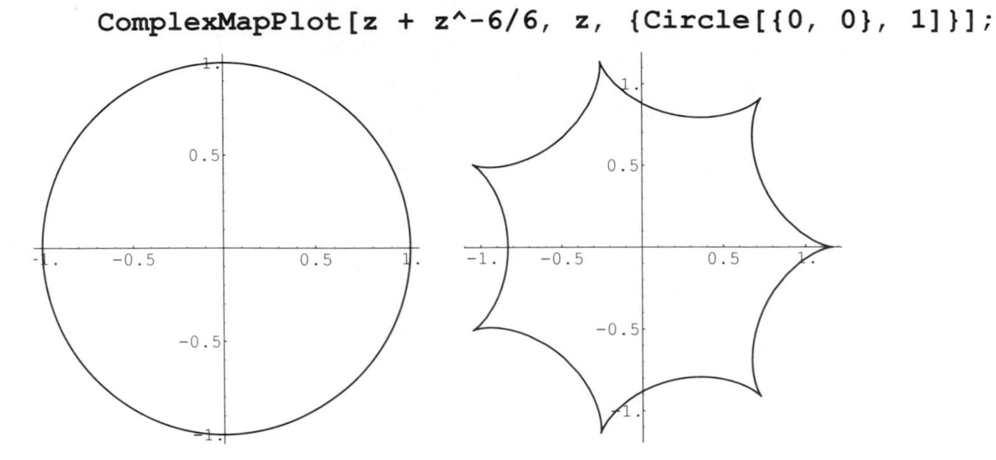

Theo: That looks like it might fit together with the positive exponent version, making a nice combined plot. Fortunately it is possible to use **ComplexMapPlot** to plot several different transformation functions on top of each other, starting from the same pre-image. To do that, we give **ComplexMapPlot** a list of functions as its first argument:

```
ComplexMapPlot[{z + z^6/6, z - z^-6/6}, z,
    {Circle[{0, 0}, 1]}];
```

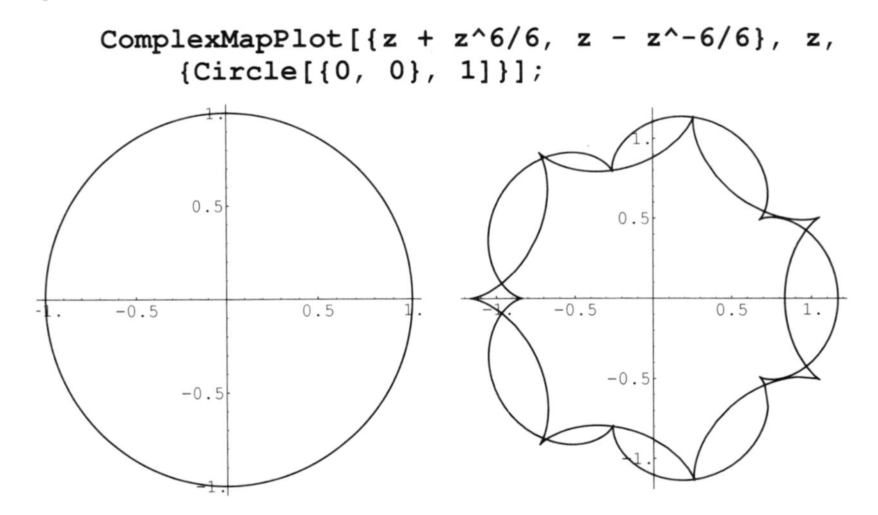

Jerry: I didn't notice before that the negative exponent version has more spikes than the positive exponent version. The negative version has one more spike than the absolute value of the exponent, while the positive version has one fewer.

Theo: So, to calculate the number of spikes, you can subtract 1 from the value of the exponent, and then take the absolute value. It sounds like the number of spikes is equal to the exponent after taking a derivative. That's probably interesting.

Jerry: This picture might look nicer if the number of spikes were the same. Let's adjust one of the exponents:

```
ComplexMapPlot[{z + z^8/8, z - z^-6/6}, z,
    {Circle[{0, 0}, 1]}];
```

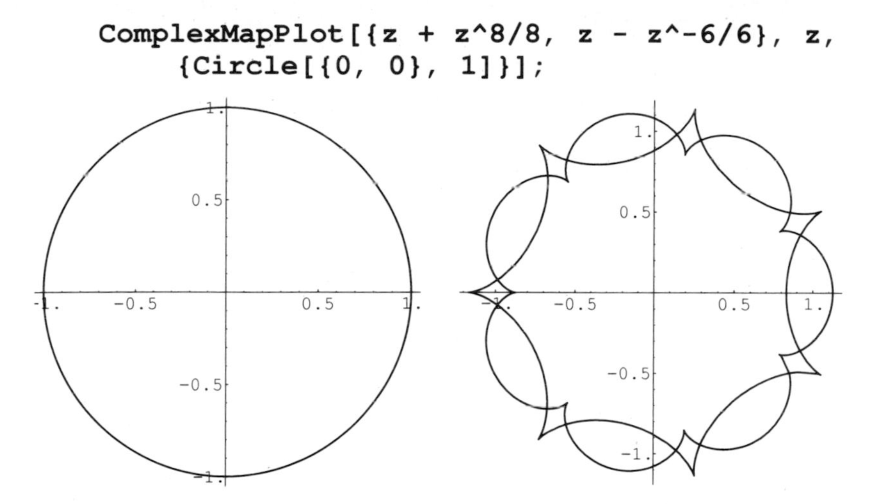

Theo: I think if we made the outwards-spiky circle a bit smaller, we could get the spikes to just touch each other. I don't know by how much to make it smaller, but I'm going to try combining two numbers that seem to be important in these plots and multiply by the ratio 6/8:

```
ComplexMapPlot[{z + z^8/8, (6/8) (z - z^-6/6)}, z,
    {Circle[{0, 0}, 1]}];
```

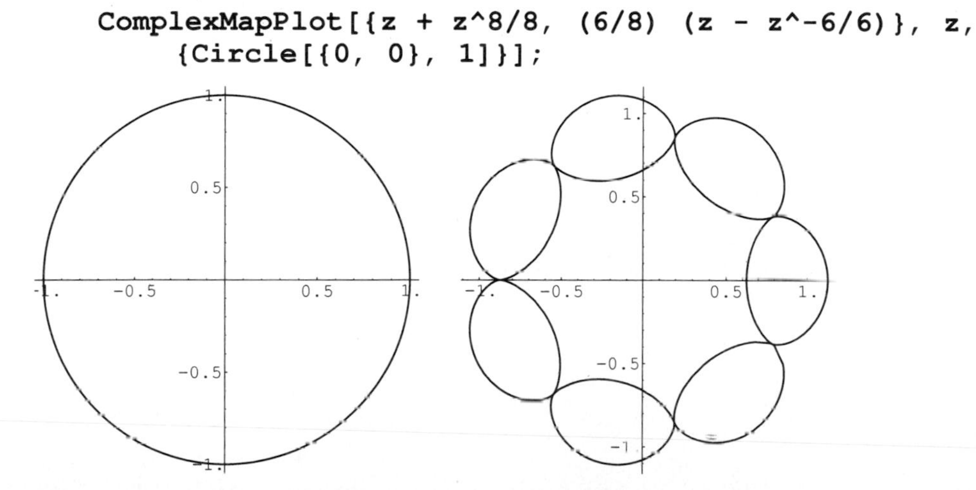

It looks like a circle sausage. I wonder how they make it.

Jerry: I'm getting a bit overwhelmed by all the possibilities before me. Where to start? Let's try **z + z^n/6** as the transformation, with **n** going from **3** to **9**, so that we get seven versions in one plot.

Theo: Rather than write out all seven functions by hand, we can use the **Table** command to generate the list automatically:

```
Table[z + z^n/6, {n, 3, 9}]
```

$$\{z + \frac{z^3}{6}, z + \frac{z^4}{6}, z + \frac{z^5}{6}, z + \frac{z^6}{6}, z + \frac{z^7}{6}, z + \frac{z^8}{6}, z + \frac{z^9}{6}\}$$

We can use the **Table** command right inside the **ComplexMapPlot** command:

```
ComplexMapPlot[
    Table[z + z^n/6, {n, 3, 9}], z,
    {Circle[{0, 0}, 1]}];
```

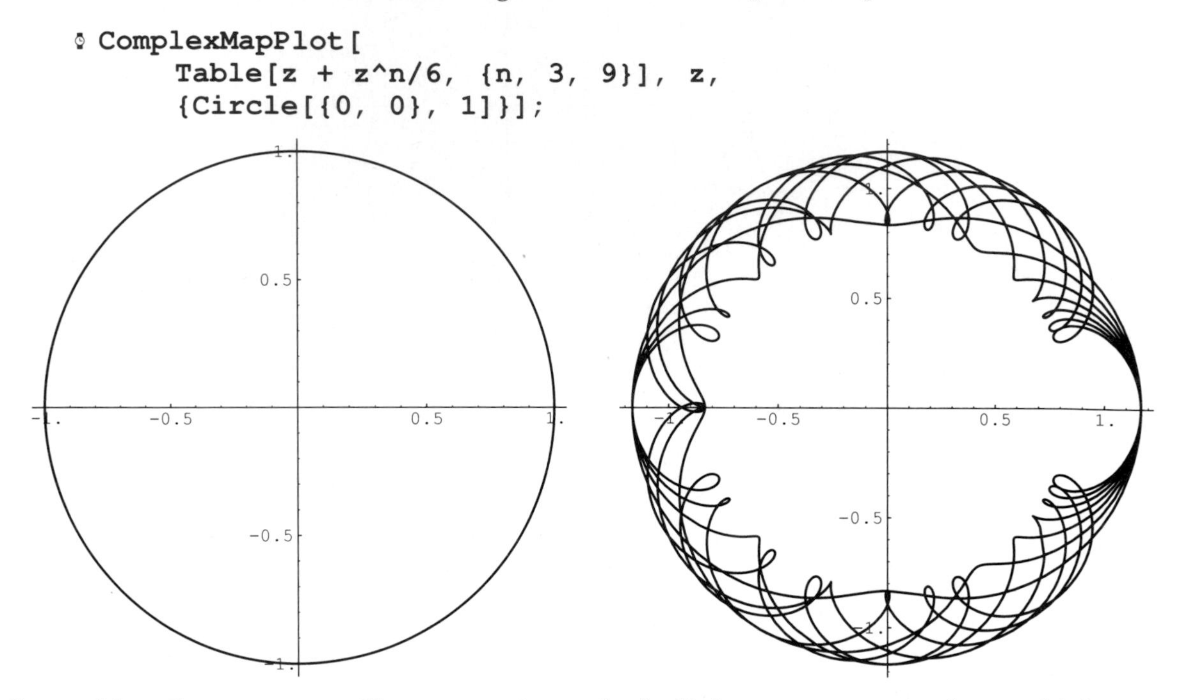

Jerry: Now I suppose you'll want to tie up the half the computers in the world for the next four hours doing a hundred superimposed functions.

Theo: Sure! I don't like this arbitrary dividing by 6, though. Let's take out the 6, and let the exponent run from 1 to 20 (despite what you think, I am able to exhibit restraint every now and then).

```
ComplexMapPlot[
     Table[
          z + z^n, {n, 1, 20}
     ], z,
     {Circle[{0, 0}, 1]},
     PlotPoints -> 100];
```

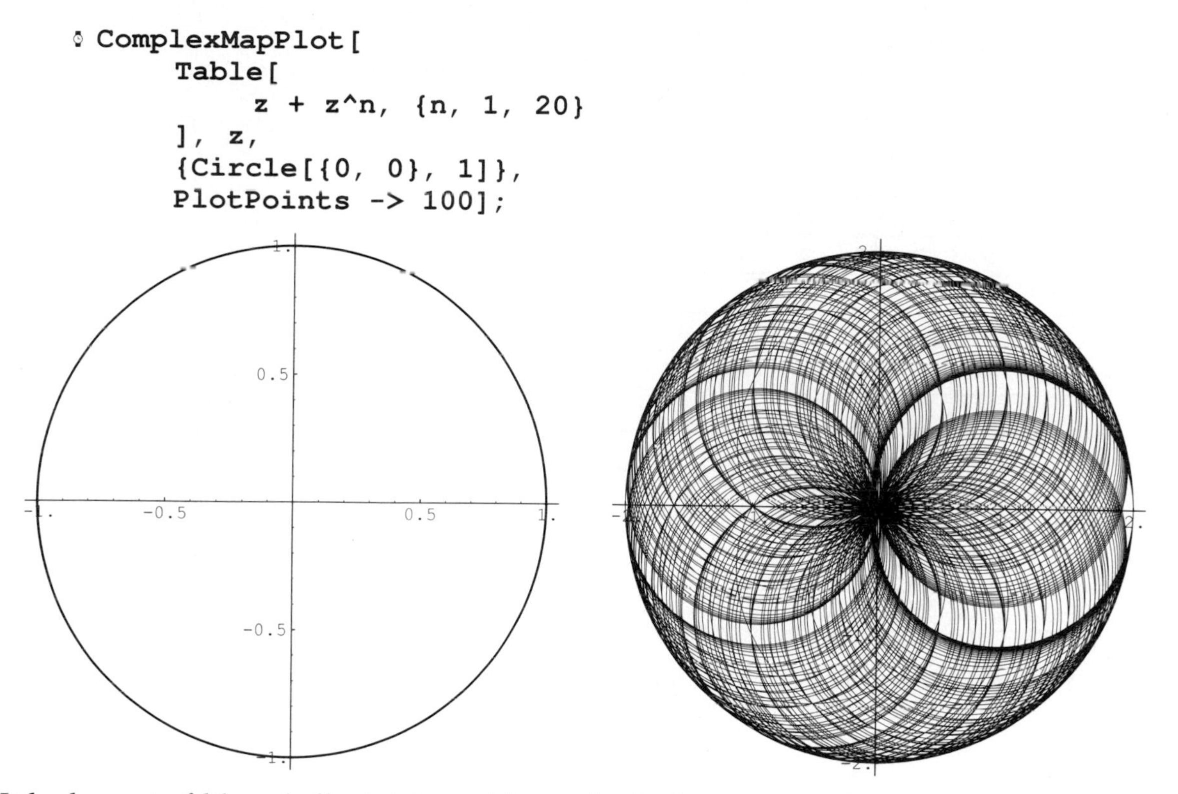

It looks sort of like a ball of string with smaller balls inside. I am particularly struck by the circles that form on the right side.

Jerry: Well, this is all very interesting, but there are a few standard, classic functions that everyone wants to look at. Can we make plots of **`Log[z]`**, **`E^z`**, **`Sin[z]`**, **`z^2`**, **`E^(1/z)`**, and **`Sqrt[z]`**?

Theo: OK, here they are. They are in pairs, first with a rectangular grid then with a polar grid. Since several of these functions go funny at the point (0, 0), I've made the polar grid start at a larger radius in those cases.

```
ComplexMapPlot[Log[z], z,
    RectangularGrid[{{-2, 2}, {-2, 2}}]];
```

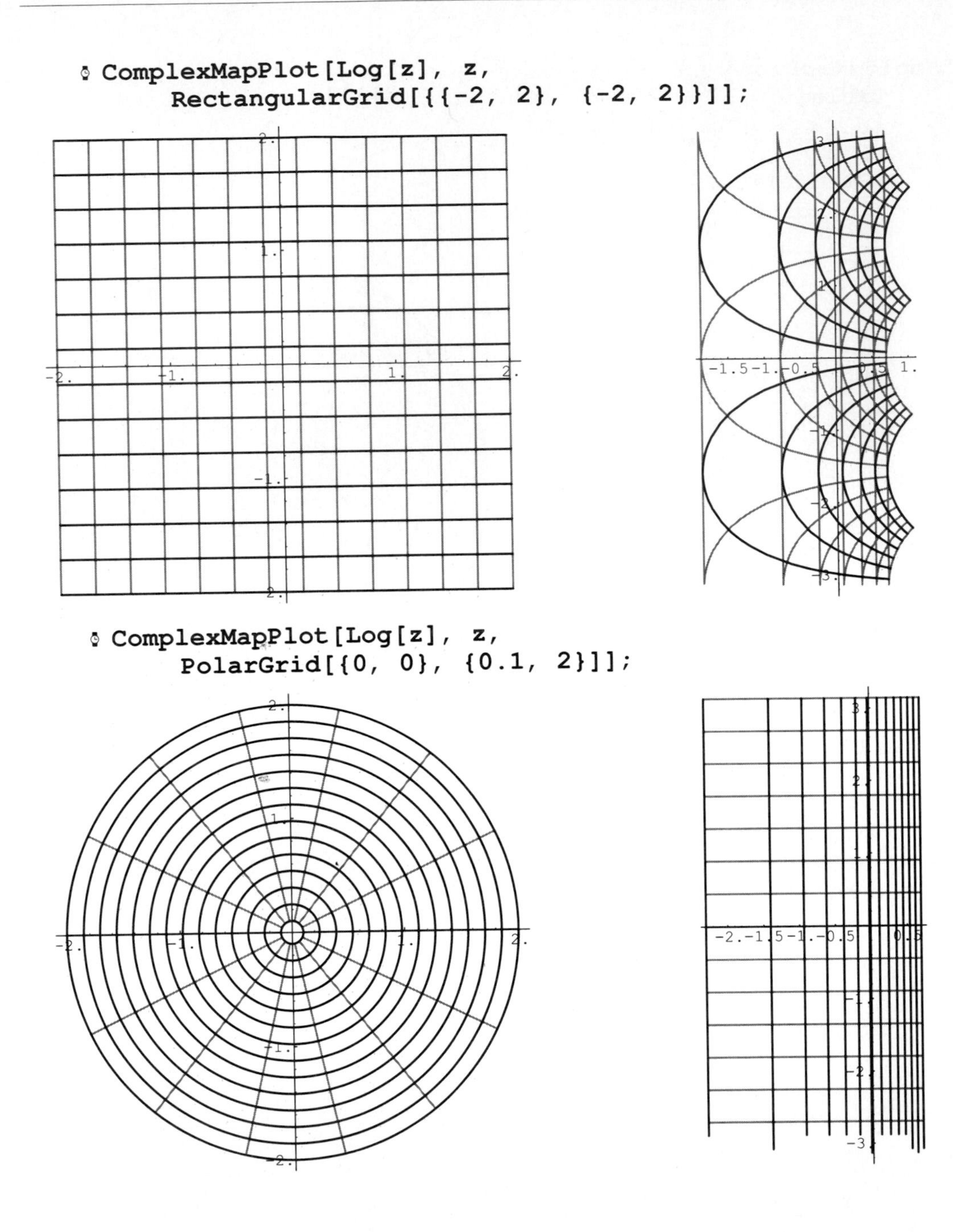

```
ComplexMapPlot[Log[z], z,
    PolarGrid[{0, 0}, {0.1, 2}]];
```

```
ComplexMapPlot[E^z, z,
    RectangularGrid[{{-2, 2}, {-2, 2}}]];
```

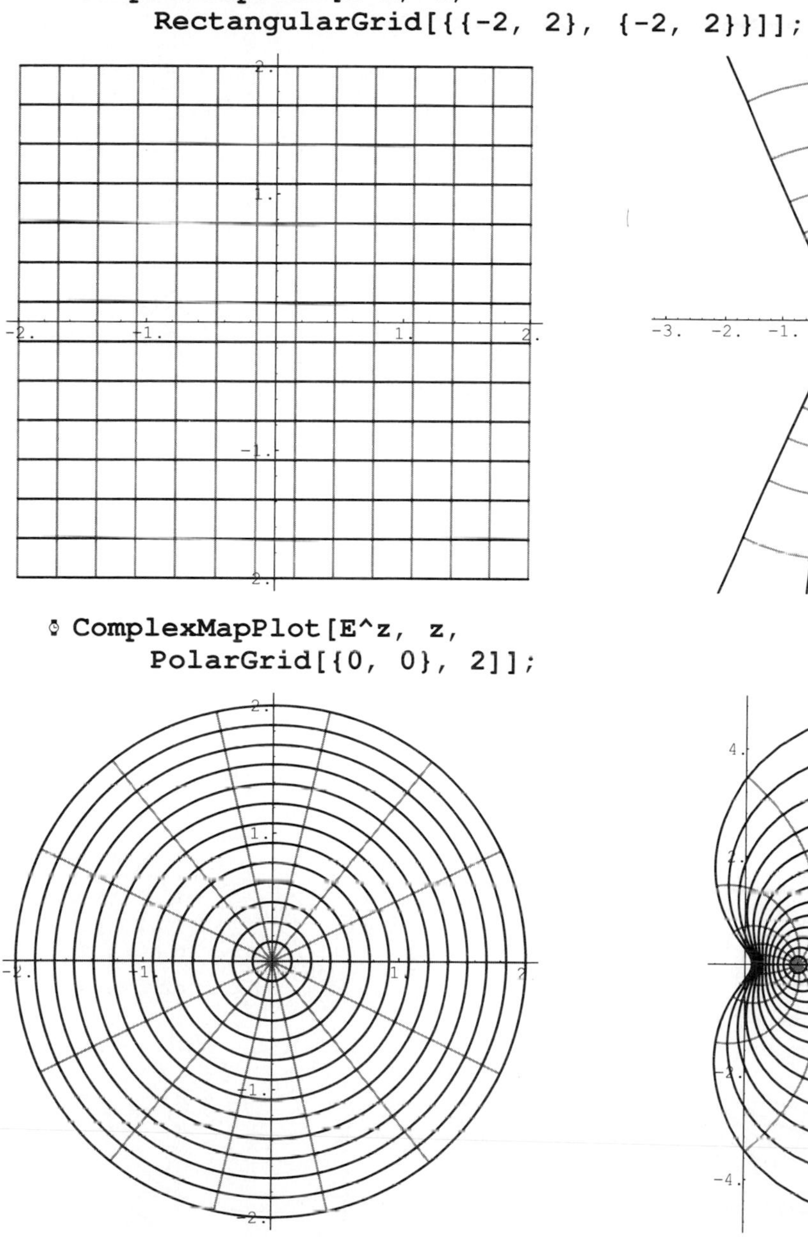

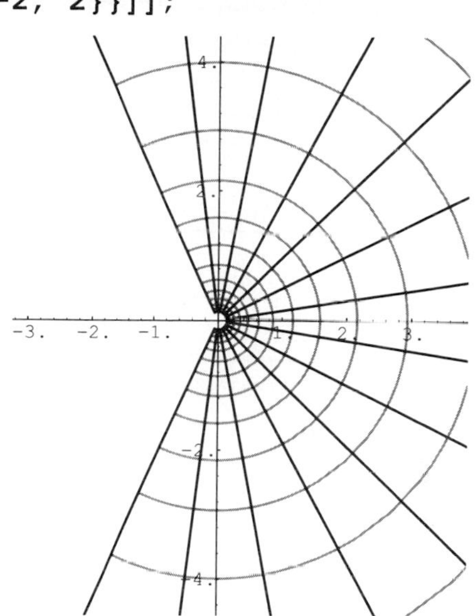

```
ComplexMapPlot[E^z, z,
    PolarGrid[{0, 0}, 2]];
```

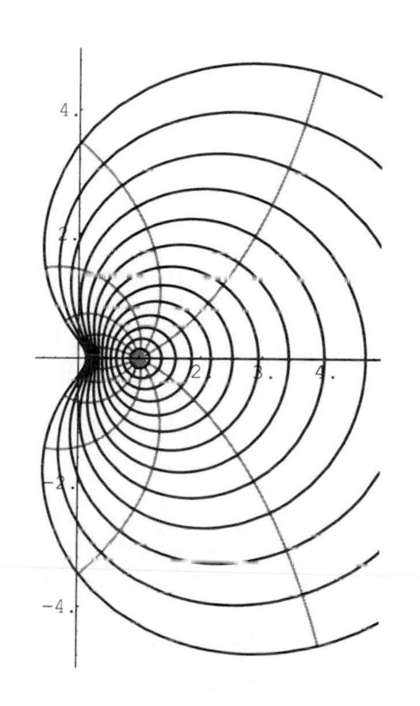

```
ComplexMapPlot[Sin[z], z,
     RectangularGrid[{{-2, 2}, {-2, 2}}],
     PlotRange -> All];
```

```
ComplexMapPlot[Sin[z], z,
     PolarGrid[{0, 0}, 2]];
```

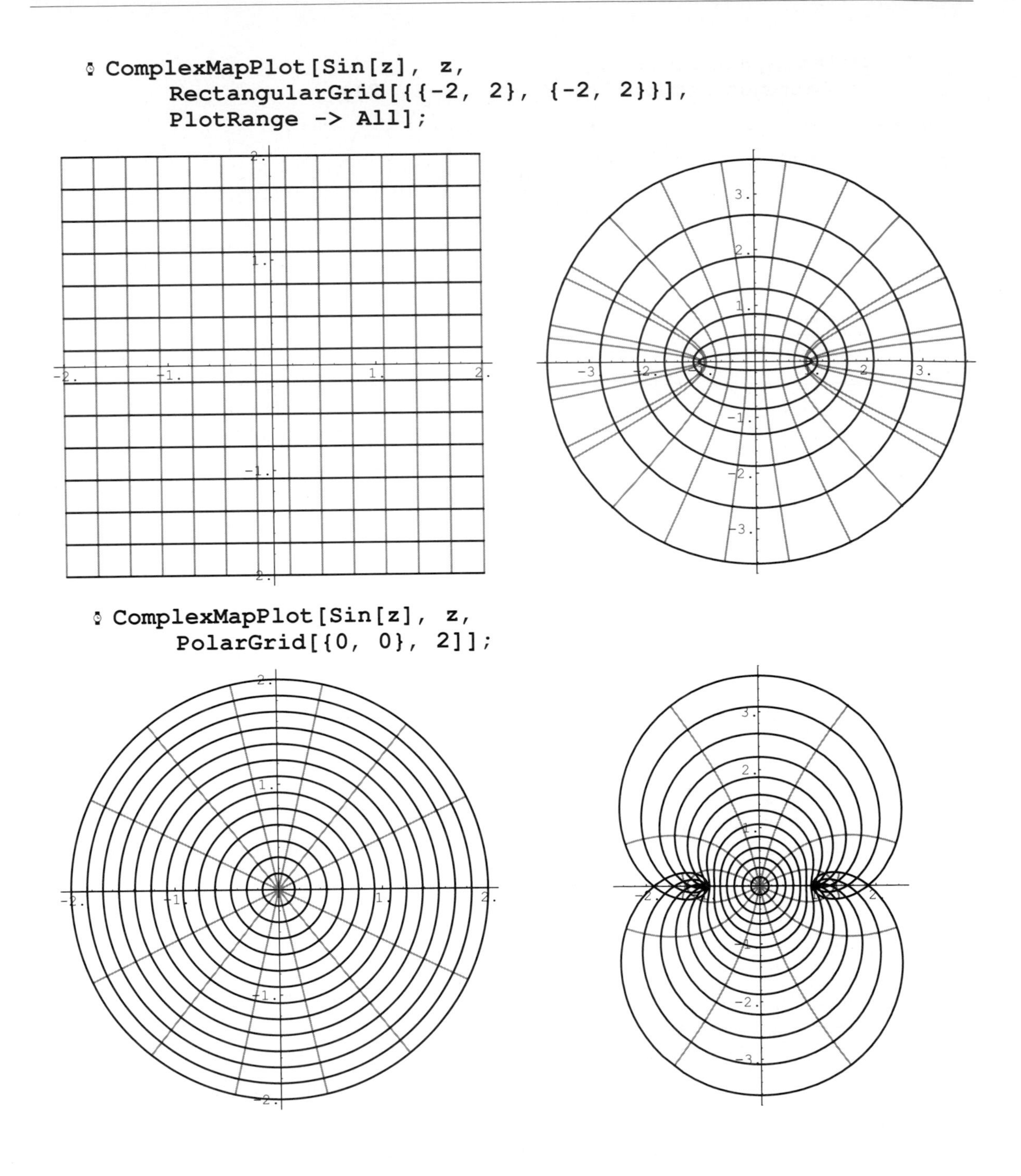

```
ComplexMapPlot[z^2, z,
     RectangularGrid[{{-2, 2}, {-2, 2}}],
     PlotRange -> All];
```

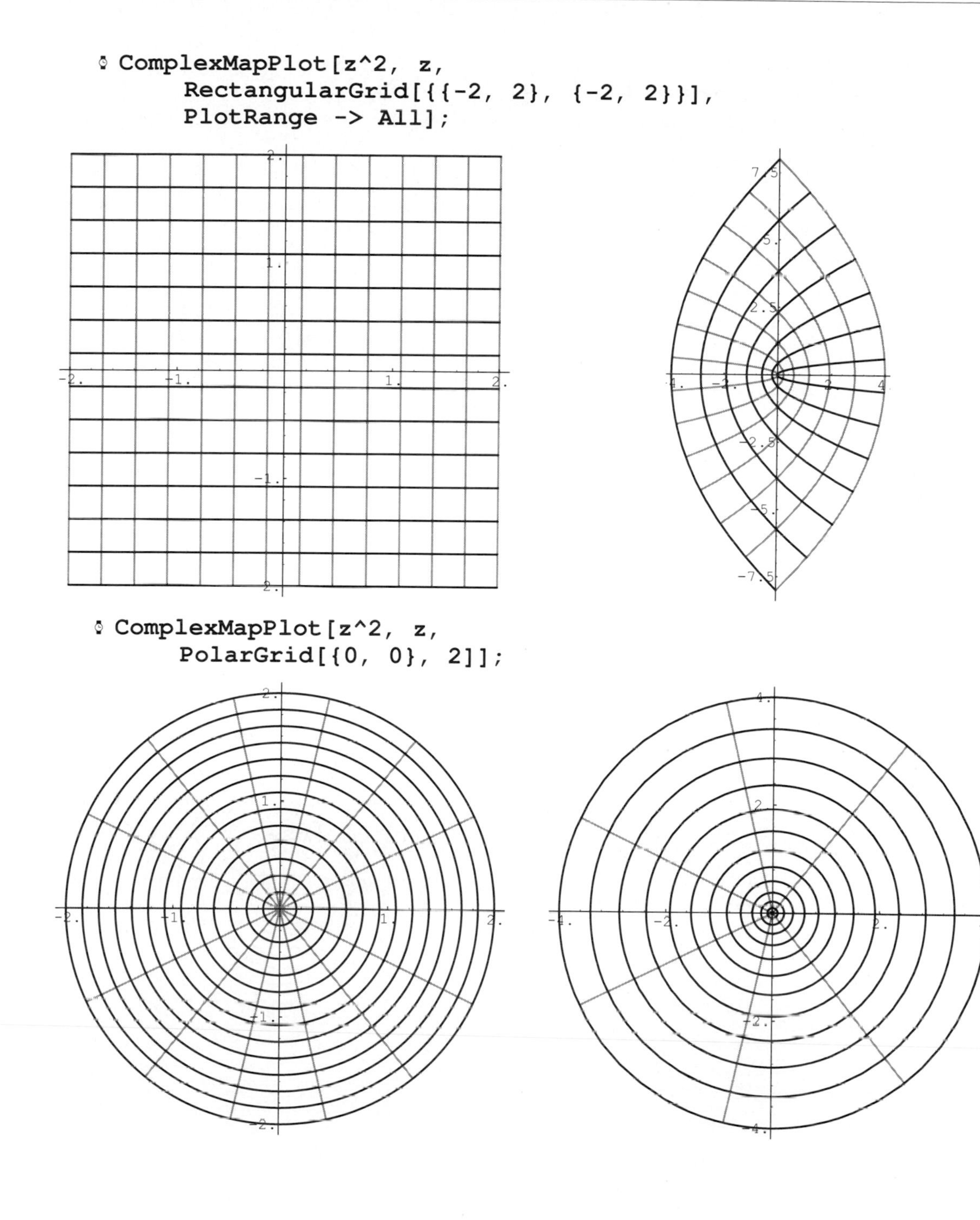

```
ComplexMapPlot[z^2, z,
     PolarGrid[{0, 0}, 2]];
```

```
ComplexMapPlot[E^(1/z), z,
    RectangularGrid[{{-2, 2}, {-2, 2}}]];
```

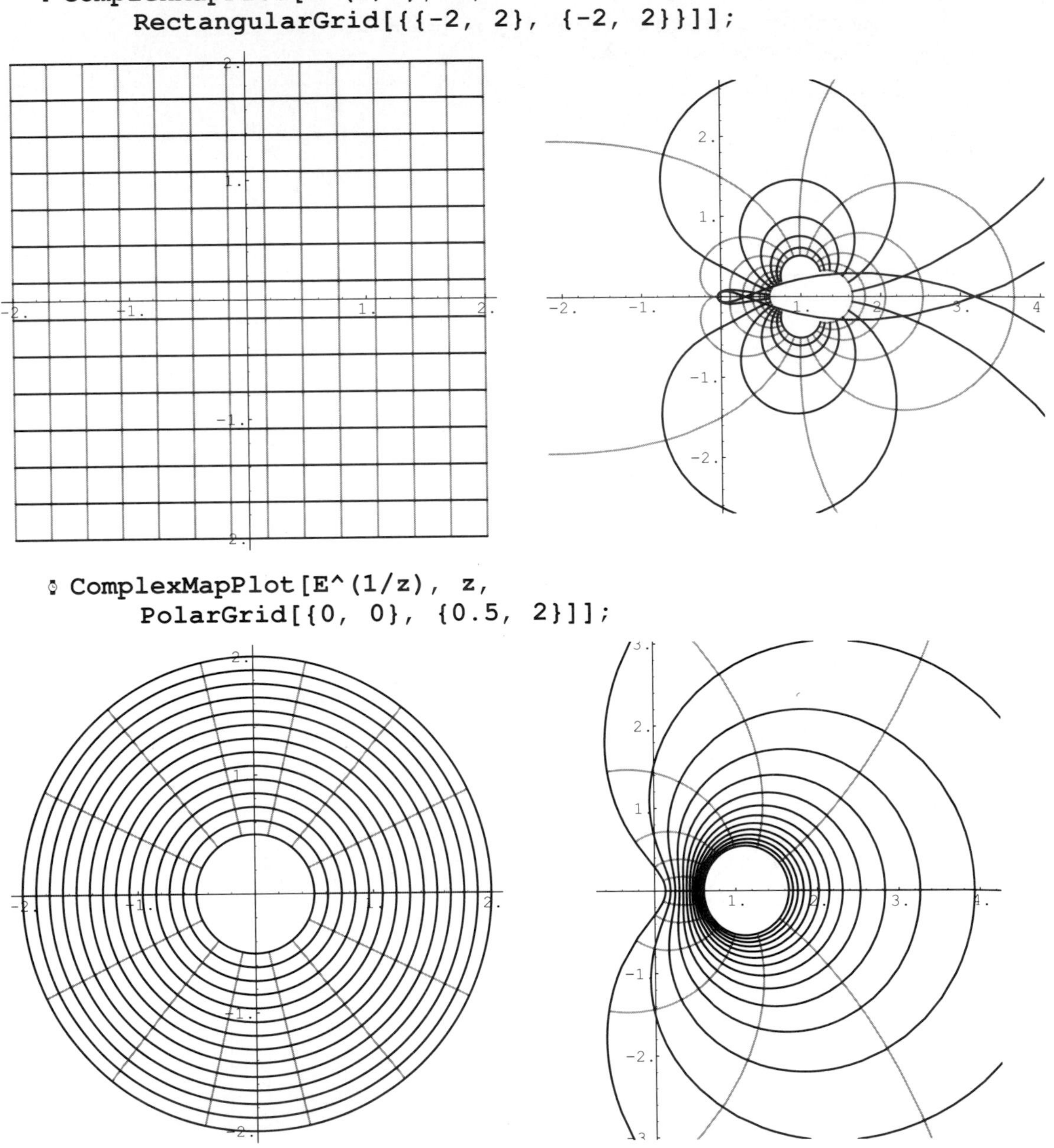

```
ComplexMapPlot[E^(1/z), z,
    PolarGrid[{0, 0}, {0.5, 2}]];
```

```
ComplexMapPlot[Sqrt[z], z,
    RectangularGrid[{{-2, 2}, {-2, 2}}]];
```

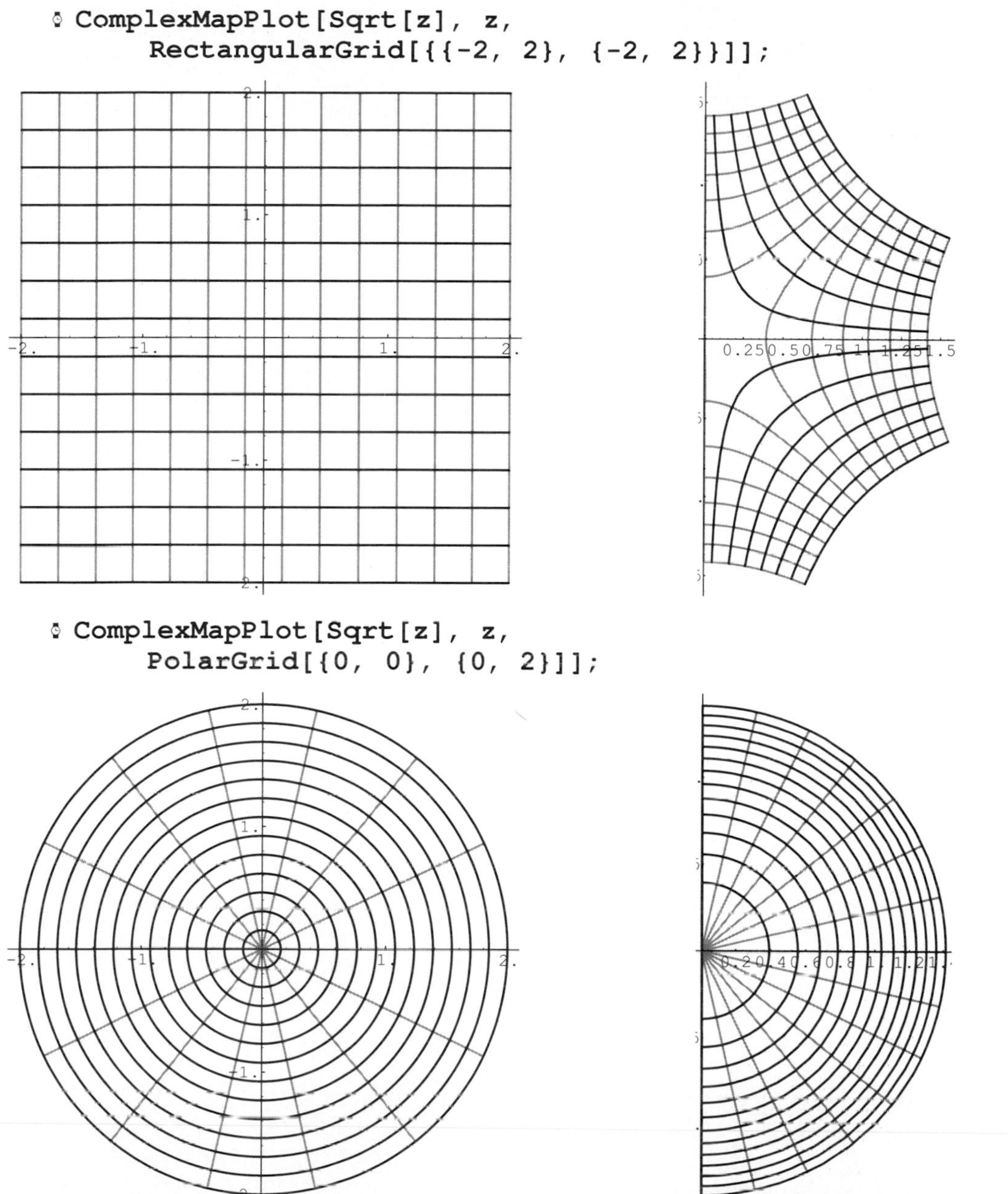

```
ComplexMapPlot[Sqrt[z], z,
    PolarGrid[{0, 0}, {0, 2}]];
```

Jerry: These are all very nice.

Many interesting plots result from using a polar grid with a center *not* at (0, 0). How can we do that?

Theo: Easy, just change the argument to **PolarGrid**, for example like this (which puts the center at the point (0.1, 0.2)):

```
ComplexMapPlot[E^(1/z), z,
     PolarGrid[{0.1, 0.2}, {0.5, 2}],
     PlotRange -> All];
```

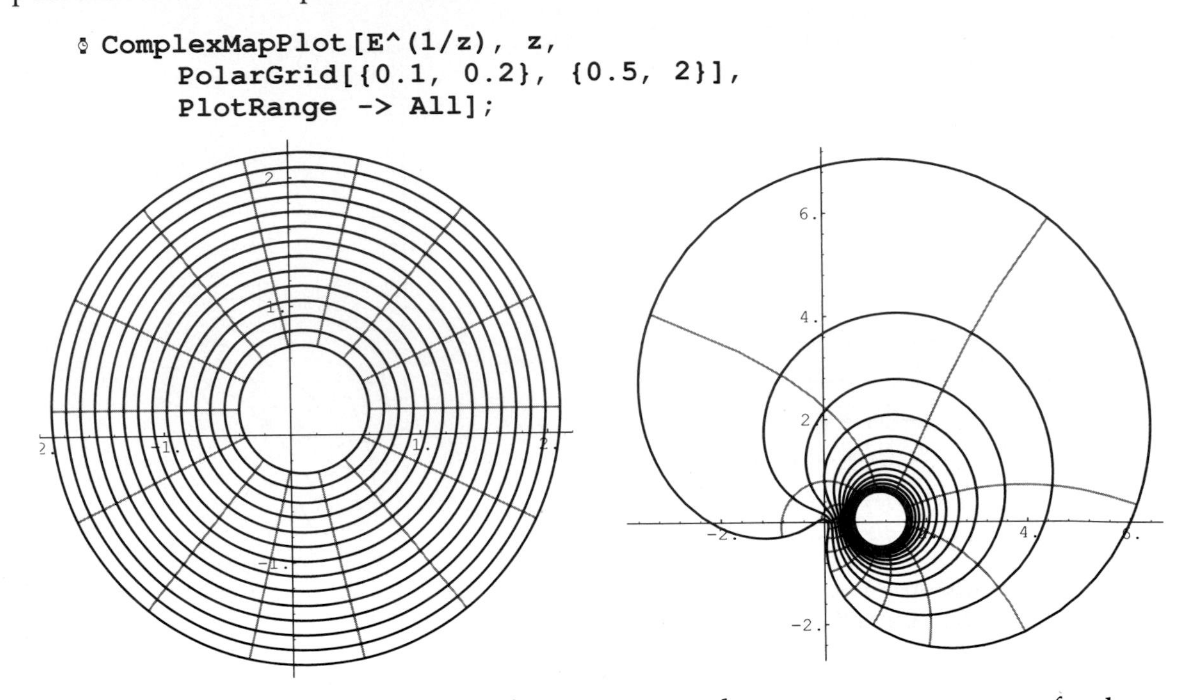

Let's make this into an animation where we move the center over a range of values, following a line from 0 to 0.3 + 0.3I;

```
Do[
    ComplexMapPlot[E^(1/z), z,
        PolarGrid[{center, center}, {0.5, 2}],
        PlotRange -> {{-10, 10}, {-10, 10}}],
    {center, 0, 0.3, 0.015}
]
```

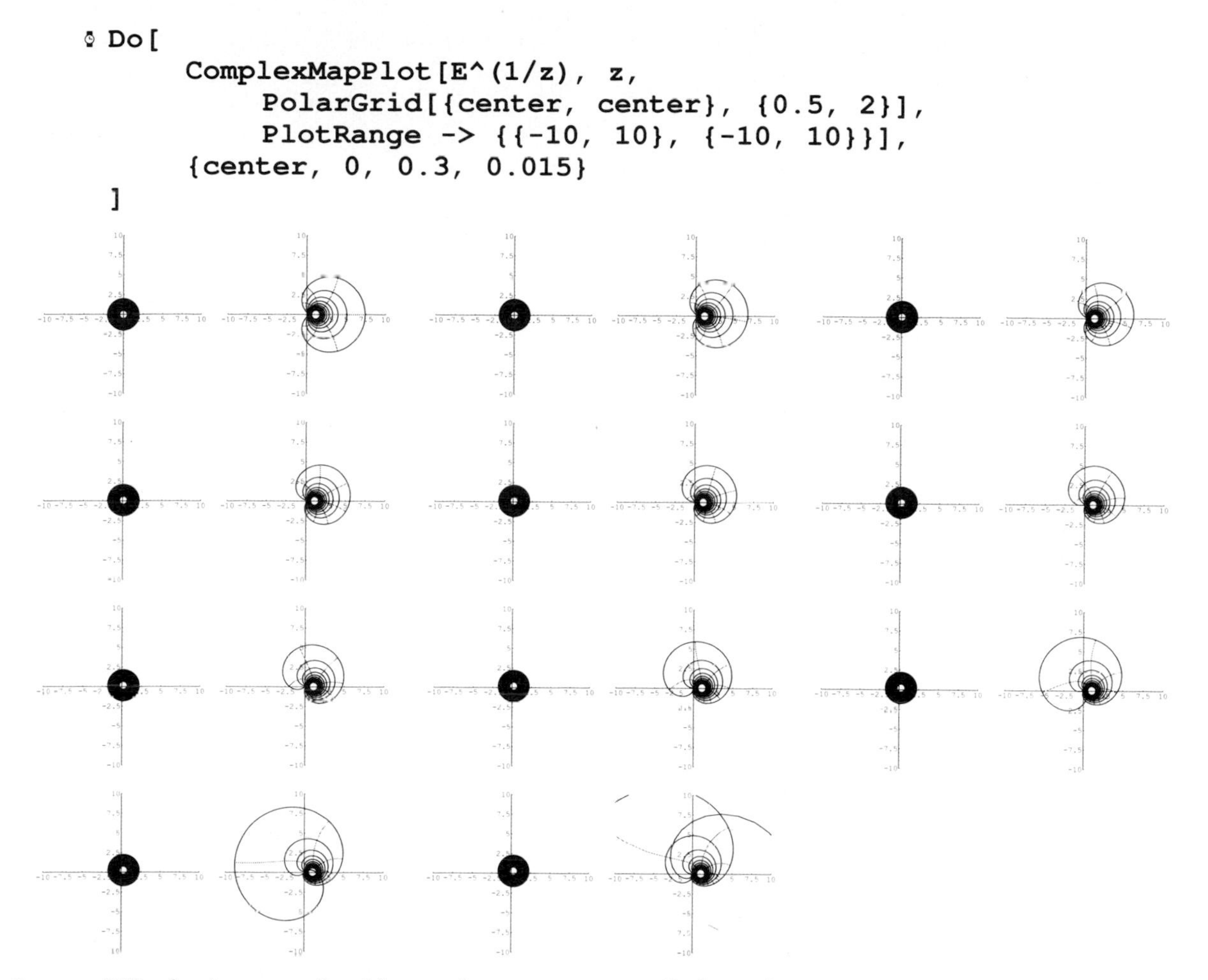

Jerry: OK, that's enough. If people want to see all the other amazing plots you can get this way, they can do it themselves.

References

1) Lapidus, Martin, *f(z)*, published by Lascaux Graphics, 3220 Steuben Ave., The Bronx, New York, NY 10467, 212-654-7429.
2) Parmet, Mark, *ComplexPaint*, Ithaca, NY.
3) Churchill, Brown, and Verhey, *Complex Variables and Applications*, McGraw-Hill, New York, 1974.
4) Philip J. Davis, *The Schwarz Function and its Applications*, The Mathematical Association of America, 1974.

Chapter Twenty
The Gasket Chapter

Jerry and Theo develop an affinity for gaskets.

Dialog

Theo: My friend Doug Stein suggested that we take a look at iterated affine maps **[1]**.

Jerry: Hold it right there. If you're going to use language like that, I'm going to have to put my foot down. Before we go any farther, I want you to explain what he meant by those last three words. Let's start with "map". We've seen the *Mathematica* command **`Map`** many times, but I have the feeling that Doug is using the word in a different way.

Theo: Yes, he is. By "map" Doug means you start with a geometric object and apply some geometrical transformation to it. For example, the operation "rotate by 90 degrees" is a "map" that can be applied to any figure. Another map might be: "Stretch by a factor of two and then move two units to the left.".

Jerry: That's the normal mathematical definition of a map. It's nice to know that Doug isn't completely off the wall. Normally one thinks of such a mapping being done by applying a function, for example multiplication by a matrix, to each point in the object being transformed.

Come to think of it, this might explain why the *Mathematica* **`Map`** function has its name, a fact that has always puzzled me. **`Map`** takes a function and applies it to every element in a list. If you think of the function as the mapping operation, and the list as the object being transformed, it makes perfect sense. To map an object, you apply the operation, or function, to each point, or element, in the object, which is exactly what **`Map`** does.

Theo: You're absolutely right. I never thought of it like that before, but it's undoubtedly the reason for the name.

Jerry: Moving right along, what does Doug mean by "affine"? It's one of those words like diophantine that would sound more at home in a Shakespearean play than in a math book.

Theo: Sadly the definition of affine is just about as dull as the definition of diophantine. It's a just a transformation that you get by multiplying each point by a 2x2 matrix and then adding a translation.

Jerry: So, if a point in our geometric object is represented by the vector p, an affine transformation is defined by two objects, a matrix m, and a translation vector c. The new point, $p' = m \, . \, p + c$. Where "$m \, . \, p$" means dot product. How do we translate that into *Mathematica*?

Theo: In *Mathematica*, matrices and vectors are represented by lists. For example, a point in the plane can be represented by a vector (x, y), or by the *Mathematica* list:

```
{x, y}
{x, y}
```

A 2x2 matrix is a list of lists. For example:

```
{{a, b}, {c, d}}
{{a, b}, {c, d}}
```

Jerry: This doesn't look like a matrix to me! I realize that we can't expect *Mathematica* to let us enter a matrix in a matrix-like way, but can we at least get it to print out in a rectangle, so we can tell which element is which?

Theo: Yes, the function `MatrixForm` will print any list of lists like a matrix:

```
MatrixForm[{{a, b}, {c, d}}]
a  b
c  d
```

Jerry: So, the first element of the first sublist is the top left element of the matrix. The first sublist list, `{a, b}`, is the first row of the matrix, and the second sublist is the second row.

Theo: Right. The dot product function is just "`.`", so we can do a whole affine transformation like this:

```
{{a, b}, {c, d}} . {x, y} + {t1, t2}
{t1 + a x + b y, t2 + c x + d y}
```

Jerry: Since we got a list with two element in it, we can see that this operation takes a point, `{x, y}`, and gives back another point. It can be used to transform, or map, any geometric object by applying it to each point in the object. Let's make up a specific example:

```
m = {{0.5, 0}, {0, 0.5}}
{{0.5, 0}, {0, 0.5}}
```

We can apply this matrix to the vector (x, y):

```
m . {x, y}
{0.5 x, 0.5 y}
```

This matrix multiplies both coordinates by 1/2. Can we see what it does to a specific point?

Theo: Sure, we can plot a point, say (1, 2), and the transformed point.

```
ShowPoints[
        Point[{2, 3}],
        Point[m . {2, 3}]
];
```

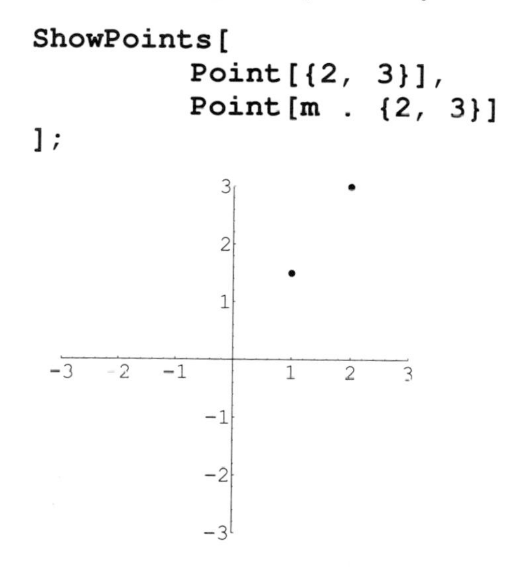

Jerry: Now I'd like to plot what happens if we apply the transformation several times in a row, because I assume that's what Doug meant by iterated.

```
ShowPoints[
    Point[{2, 3}],
    Point[m . {2, 3}],
    Point[m . m . {2, 3}],
    Point[m . m . m . {2, 3}]
];
```

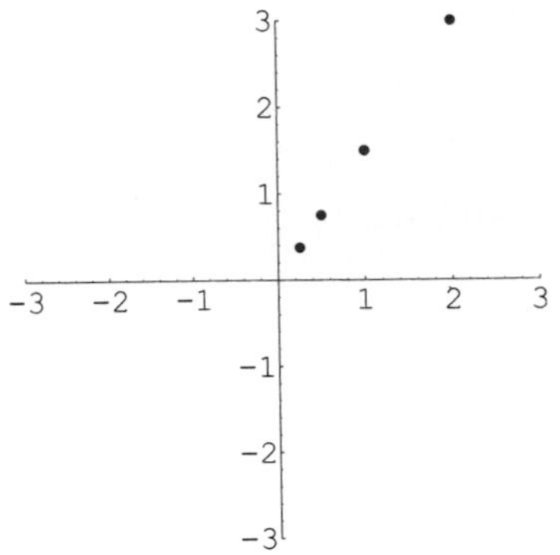

So the matrix **m** is a device for shrinking things toward the origin along straight lines. What happens if we add a translation to our transformation?

```
t = {0.2, 0.2}
{0.2, 0.2}
```

We can plot a few points:

```
ShowPoints[
    Point[{2, 3}],
    Point[m . {2, 3} + t],
    Point[m . (m . {2, 3} + t) + t],
    Point[m . (m . (m . {2, 3} + t) + t) + t]
];
```

Theo: We need to see several more points before we can tell what is happening. Rather than type in lots of **Point** commands, let's generate them automatically. We can write a function that takes one point and transforms it:

```
doTransformation[Point[{x_, y_}]] :=
    Point[m . {x, y} + t]
```

If we use this function on a point, we get another point:

```
doTransformation[Point[{2, 3}]]
```

```
Point[{1.2, 1.7}]
```

We can use the function **NestList** to apply **doTransformation** repeatedly to a starting point and return a list of all the points along the way:

```
NestList[doTransformation, Point[{2, 3}], 10]
```

```
{Point[{2, 3}], Point[{1.2, 1.7}], Point[{0.8, 1.05}],
  Point[{0.6, 0.725}], Point[{0.5, 0.5625}],
  Point[{0.45, 0.48125}], Point[{0.425, 0.440625}],
  Point[{0.4125, 0.420312}], Point[{0.40625, 0.410156}],
  Point[{0.403125, 0.405078}],
  Point[{0.401562, 0.402539}]}
```

Now we can plot these points:

```
ShowPoints[
    NestList[doTransformation, Point[{2, 3}], 10]
];
```

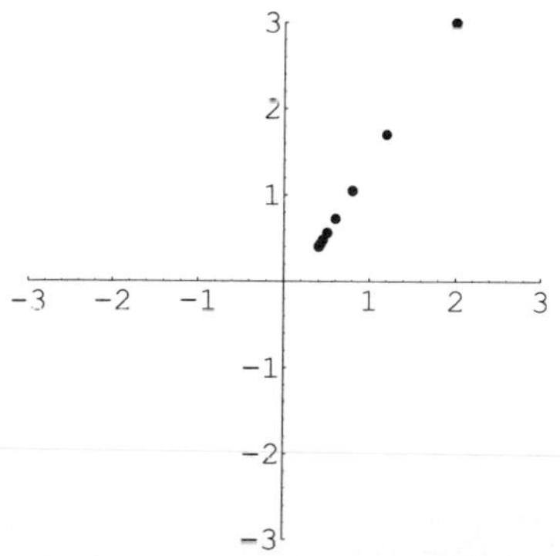

Jerry: It looks like it's converging on a different point. I can think of all sorts of things to try! What happens if we pick a different starting point? What about different matrices? Suppose the object we start with is not a point, but is a triangle, or a house, or something? I suppose you're going to want to define a whole elaborate system of functions to let us do these experiments?

Theo: Of course! In the file `AffineMaps.m` (see the appendix by the same name) I've define a function **`IteratedMapping`** that does what we want. It's a relatively complicated function, but it's very powerful and flexible. It's actually very similar to the functions we used in the complex mapping chapter (see Chapter 19), both in the way it's used and in the way it's implemented.

Jerry: How does it work?

Theo: `IteratedMapping` takes three arguments. The first argument is a list of one or more transformations. The second argument is a number that specifies how many times to apply the transformations. The third argument is a list of two-dimensional graphics objects to be transformed.

The transformations are specified using the function **`AffineMap`**. **`AffineMap`** is mainly a place-holder kind of function. It doesn't actually do anything other than indicate that you are talking about an affine transformation. The first argument to it is a matrix, and the second argument is a translation vector. For example, here is a transformation that does absolutely nothing:

```
AffineMap[{{1, 0}, {0, 1}}, {0, 0}]
AffineMap[{{1, 0}, {0, 1}}, {0, 0}]
```

The list of graphics objects to be transformed can contain any of the standard *Mathematica* graphics objects, such as **`Line`**, **`Point`**, **`Circle`**, or **`Polygon`**. Here is an example:

```
myObject = {
    Polygon[{{0, 1}, {1, 1}, {1, 0}, {0, 1}}]
}
{Polygon[{{0, 1}, {1, 1}, {1, 0}, {0, 1}}]}
```

This is a list containing a single filled triangle. We can look at it like this:

```
Show[Graphics[myObject],
    Axes -> Automatic, AspectRatio -> Automatic];
```

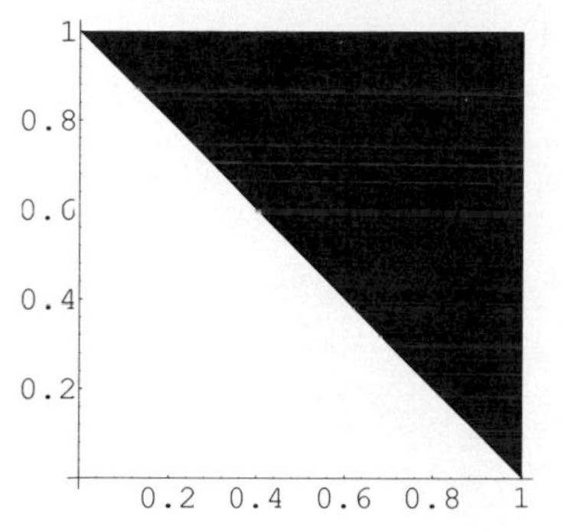

Jerry: I'll propose an action on this triangle and you adjust the function to carry out this action. First, shrink the sides of the triangle by 1/2 and then move the resulting object 1/2 unit to the right.

Theo: OK:

```
IteratedMapping[
    {
        AffineMap[{{0.5, 0}, {0, 0.5}}, {0.5, 0}]
    },
    1,
    myObject,
    PlotRange -> {{0, 1}, {0, 1}}
];
```

Jerry: Fine. Now shrink the original triangle by one half, and move the resulting object up 1/2 unit.

Theo: OK:

```
IteratedMapping[
    {
        AffineMap[{{0.5, 0}, {0, 0.5}}, {0, 0.5}]
    },
    1,
    myObject,
    PlotRange -> {{0, 1}, {0, 1}}
];
```

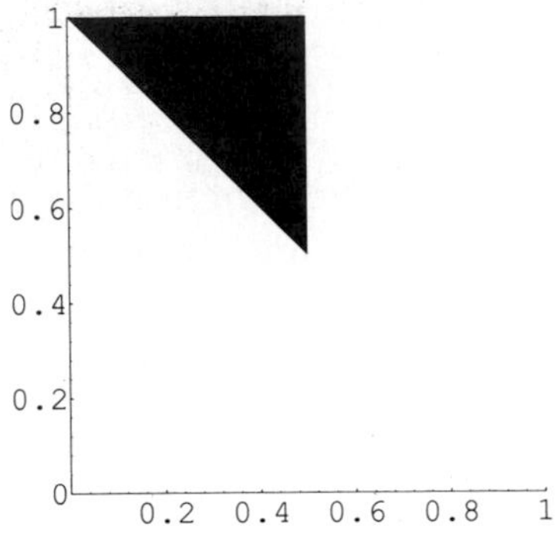

Theo: We can get **IteratedMapping** to do both these transformations at once, by putting both of them in the list:

```
IteratedMapping[
    {
        AffineMap[{{0.5, 0}, {0, 0.5}}, {0.5, 0}],
        AffineMap[{{0.5, 0}, {0, 0.5}}, {0, 0.5}]
    },
    1,
    myObject,
    PlotRange -> {{0, 1}, {0, 1}}
];
```

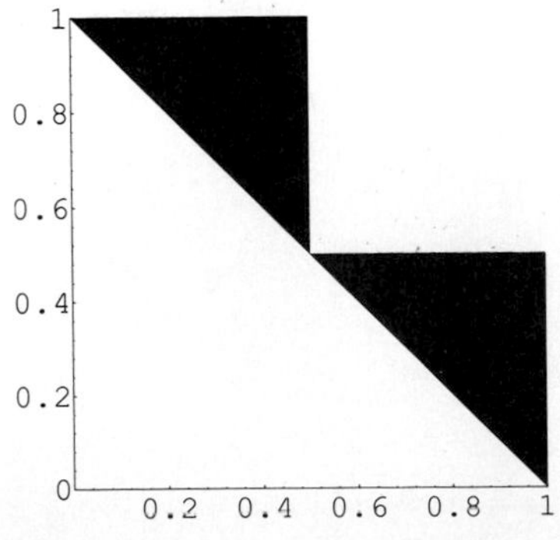

Let's add a third transformation:

```
IteratedMapping[
    {
        AffineMap[{{0.5, 0}, {0, 0.5}}, {0.5, 0  }],
        AffineMap[{{0.5, 0}, {0, 0.5}}, {0,   0.5}],
        AffineMap[{{0.5, 0}, {0, 0.5}}, {0.5, 0.5}]
    },
    1,
    myObject
];
```

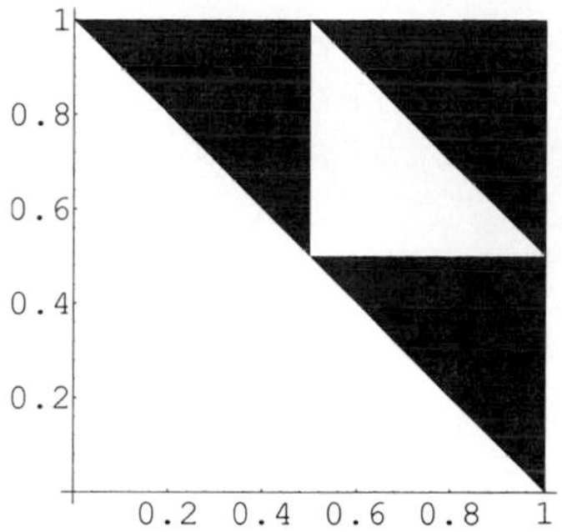

Jerry: We got three smaller copies of the object, in different locations. Now, what happens if we tell **IteratedMapping** to do more than one iteration?

Theo: It does the first iteration and gets a list of, in this case, three new objects. Then it combines these three objects into a single new object and applies the transformation again. This results in three copies of the combined object, or nine copies of the original object. Let's see what it looks like:

```
IteratedMapping[
    [
        AffineMap[{{0.5, 0}, {0, 0.5}}, {0.5, 0  }],
        AffineMap[{{0.5, 0}, {0, 0.5}}, {0,   0.5}],
        AffineMap[{{0.5, 0}, {0, 0.5}}, {0.5, 0.5}]
    },
    2,
    myObject
];
```

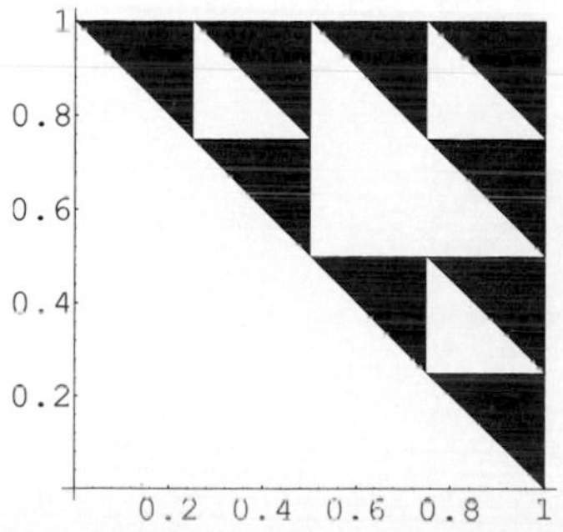

Jerry: Oooh, I can just imagine what this will look like if we iterate some more. Can we make an animation where we increase the number of iterations up to, say, 8 times?

Theo: Sure. Let's start at zero iterations, which means just show the original image. (Double-click the picture to see the animation.)

```
Do[
        IteratedMapping[
            {
                AffineMap[{{0.5, 0}, {0, 0.5}}, {0.5, 0  }],
                AffineMap[{{0.5, 0}, {0, 0.5}}, {0,   0.5}],
                AffineMap[{{0.5, 0}, {0, 0.5}}, {0.5, 0.5}]
            },
            n,
            myObject
        ],
        {n, 0, 8}
]
```

(The printed version includes fewer steps, for space reasons.)

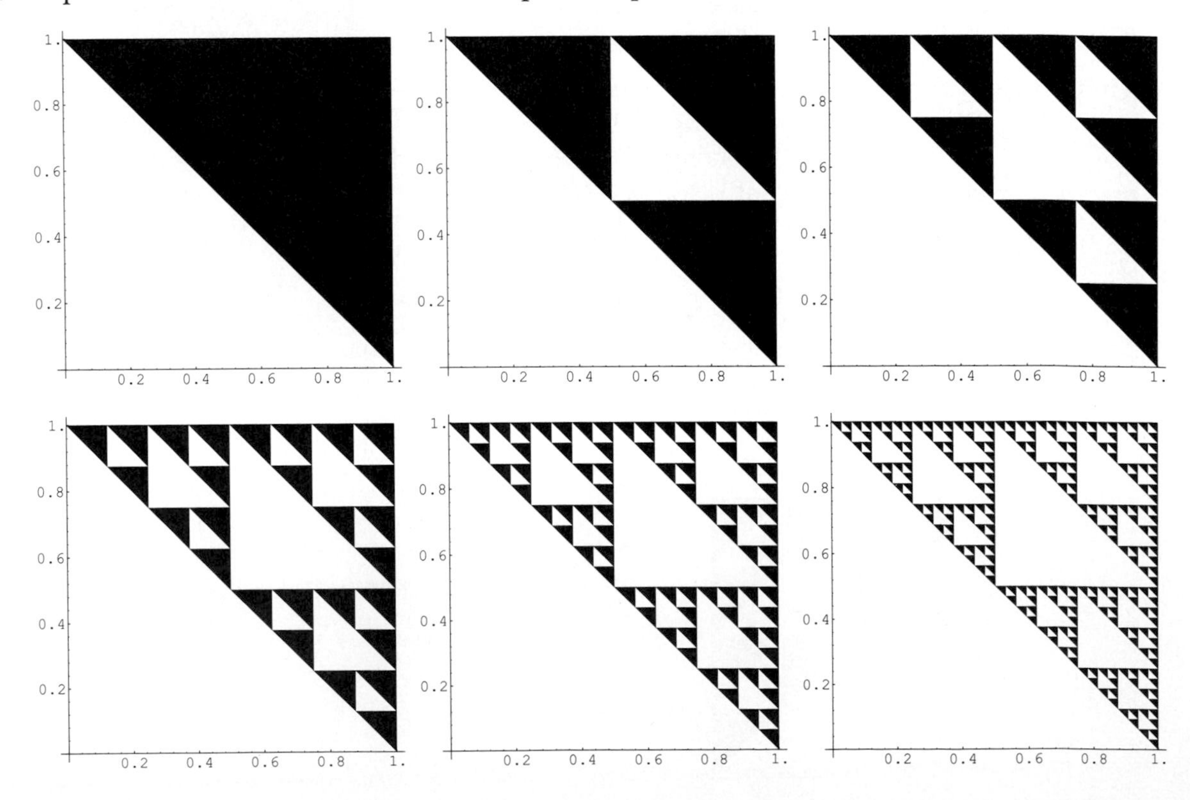

Jerry: Fabulous! Magical! To be able to take those small steps that we took and then say do all of that many, many times and to see it almost immediately is sensational!

We started with a filled triangle--pick another object and repeat the process.

Theo: I happen to have a house lying around from when Don Cohen wanted to see things being transformed. I drew the house in MacDraw and pasted it into *Mathematica*. Here it is:

We can transform the house into *Mathematica* graphics functions (like the **`Polygon`** function we used above to draw a triangle) using the Convert To InputForm command in the Graph menu. Here is the command that you get by converting the house: (I've manually added the "**`house =`** " so we can use the **`house`** variable later. All but the first line is in a very small font, because there's a lot of it, and it's not very interesting.)

```
Show[Graphics[house = {
Thickness[0.007752],
GrayLevel[0.49999],
Polygon[{{0.228682, 0.236434}, {0.825581, 0.236434}, {0.825581, 0.686047}, {0.228682, 0.686047}, {0.228682, 0.236434}}],
GrayLevel[0.00000],
Line[{{0.228682, 0.236434}, {0.825581, 0.236434}}], Line[{{0.825581, 0.236434}, {0.825581, 0.686047}}],
Line[{{0.825581, 0.686047}, {0.228682, 0.686047}}], Line[{{0.228682, 0.686047}, {0.228682, 0.236434}}],
GrayLevel[1.00000],
Polygon[{{0.275194, 0.500000}, {0.406977, 0.500000}, {0.406977, 0.639535}, {0.275194, 0.639535}, {0.275194, 0.500000}}],
GrayLevel[0.00000],
Line[{{0.275194, 0.500000}, {0.406977, 0.500000}}], Line[{{0.406977, 0.500000}, {0.406977, 0.639535}}],
Line[{{0.406977, 0.639535}, {0.275194, 0.639535}}], Line[{{0.275194, 0.639535}, {0.275194, 0.500000}}],
GrayLevel[1.00000],
Polygon[{{0.461240, 0.500000}, {0.593023, 0.500000}, {0.593023, 0.639535}, {0.461240, 0.639535}, {0.461240, 0.500000}}],
GrayLevel[0.00000],
Line[{{0.461240, 0.500000}, {0.593023, 0.500000}}], Line[{{0.593023, 0.500000}, {0.593023, 0.639535}}],
Line[{{0.593023, 0.639535}, {0.461240, 0.639535}}], Line[{{0.461240, 0.639535}, {0.461240, 0.500000}}],
GrayLevel[1.00000],
Polygon[{{0.647287, 0.500000}, {0.779070, 0.500000}, {0.779070, 0.639535}, {0.647287, 0.639535},{0.647287, 0.500000}}],
GrayLevel[0.00000],
Line[{{0.647287, 0.500000}, {0.779070, 0.500000}}], Line[{{0.779070, 0.500000}, {0.779070, 0.639535}}],
Line[{{0.779070, 0.639535}, {0.647287, 0.639535}}], Line[{{0.647287, 0.639535}, {0.647287, 0.500000}}],
GrayLevel[1.00000],
Polygon[{{0.468992, 0.236434}, {0.569767, 0.236434}, {0.569767, 0.414729}, {0.468992, 0.414729}, {0.468992, 0.236434}}],
GrayLevel[0.00000],
Line[{{0.468992, 0.236434}, {0.569767, 0.236434}}], Line[{{0.569767, 0.236434}, {0.569767, 0.414729}}],
```

```
Line[{{0.569767, 0.414729}, {0.468992, 0.414729}}], Line[{{0.468992, 0.414729}, {0.468992, 0.236434}}],
GrayLevel[1.00000],
Polygon[{{0.275194, 0.298450}, {0.406977, 0.298450}, {0.406977, 0.430233}, {0.275194, 0.430233}, {0.275194, 0.298450}}],
GrayLevel[0.00000],
Line[{{0.275194, 0.298450}, {0.406977, 0.298450}}], Line[{{0.406977, 0.298450}, {0.406977, 0.430233}}],
Line[{{0.406977, 0.430233}, {0.275194, 0.430233}}], Line[{{0.275194, 0.430233}, {0.275194, 0.298450}}],
GrayLevel[1.00000],
Polygon[{{0.647287, 0.298450}, {0.779070, 0.298450}, {0.779070, 0.430233}, {0.647287, 0.430233}, {0.647287, 0.298450}}],
GrayLevel[0.00000],
Line[{{0.647287, 0.298450}, {0.779070, 0.298450}}], Line[{{0.779070, 0.298450}, {0.779070, 0.430233}}],
Line[{{0.779070, 0.430233}, {0.647287, 0.430233}}], Line[{{0.647287, 0.430233}, {0.647287, 0.298450}}],
GrayLevel[0.87500],
Polygon[{{0.228682, 0.686047}, {0.523256, 0.980620}, {0.817830, 0.686047}, {0.228682, 0.686047}, {0.228682, 0.686047}}],
GrayLevel[0.00000],
Line[{{0.228682, 0.686047}, {0.523256, 0.980620}}], Line[{{0.523256, 0.980620}, {0.817830, 0.686047}}],
Line[{{0.817830, 0.686047}, {0.228682, 0.686047}}], Line[{{0.228682, 0.686047}, {0.228682, 0.686047}}],
GrayLevel[0.24999],
Polygon[{{0.647287, 0.856589}, {0.647287, 0.965116}, {0.724806, 0.965116},
{0.724806, 0.779070}, {0.647287, 0.856589}, {0.647287, 0.856589}}],
GrayLevel[0.00000],
Line[{{0.647287, 0.856589}, {0.647287, 0.965116}}], Line[{{0.647287, 0.965116}, {0.724806, 0.965116}}],
Line[{{0.724806, 0.965116}, {0.724806, 0.779070}}], Line[{{0.724806, 0.779070}, {0.647287, 0.856589}}],
Line[{{0.647287, 0.856589}, {0.647287, 0.856589}}],
GrayLevel[0.68750],
Polygon[{{0.484496, 0.236434}, {0.321705, 0.003876}, {0.445736, 0.003876}, {0.562016, 0.236434}, {0.484496, 0.236434}}],
GrayLevel[0.00000],
Line[{{0.484496, 0.236434}, {0.321705, 0.003876}}], Line[{{0.321705, 0.003876}, {0.445736, 0.003876}}],
Line[{{0.445736, 0.003876}, {0.562016, 0.236434}}], Line[{{0.562016, 0.236434}, {0.484496, 0.236434}}]
}], AspectRatio->1.00000,
PlotRange->{{0.00000, 1.00000}, {0.00000, 1.00000}}];
```

Jerry: Now we'll run exactly the same command as before, except that the variable **house** is there in place of the triangle.

Theo: But we'd better not go all the way to 8 iterations. That's going to be a whole lot of houses to draw. I think it would be better to stop at 5 iterations.

```
Do[
     IteratedMapping[
          {
               AffineMap[{{0.5, 0}, {0, 0.5}}, {0.5, 0  }],
               AffineMap[{{0.5, 0}, {0, 0.5}}, {0,   0.5}],
               AffineMap[{{0.5, 0}, {0, 0.5}}, {0.5, 0.5}]
          },
          n,
          house,
          PlotRange -> {{0, 1}, {0, 1}}
     ],
     {n, 0, 5}
];
```

(The printed version includes fewer steps, for space reasons.)

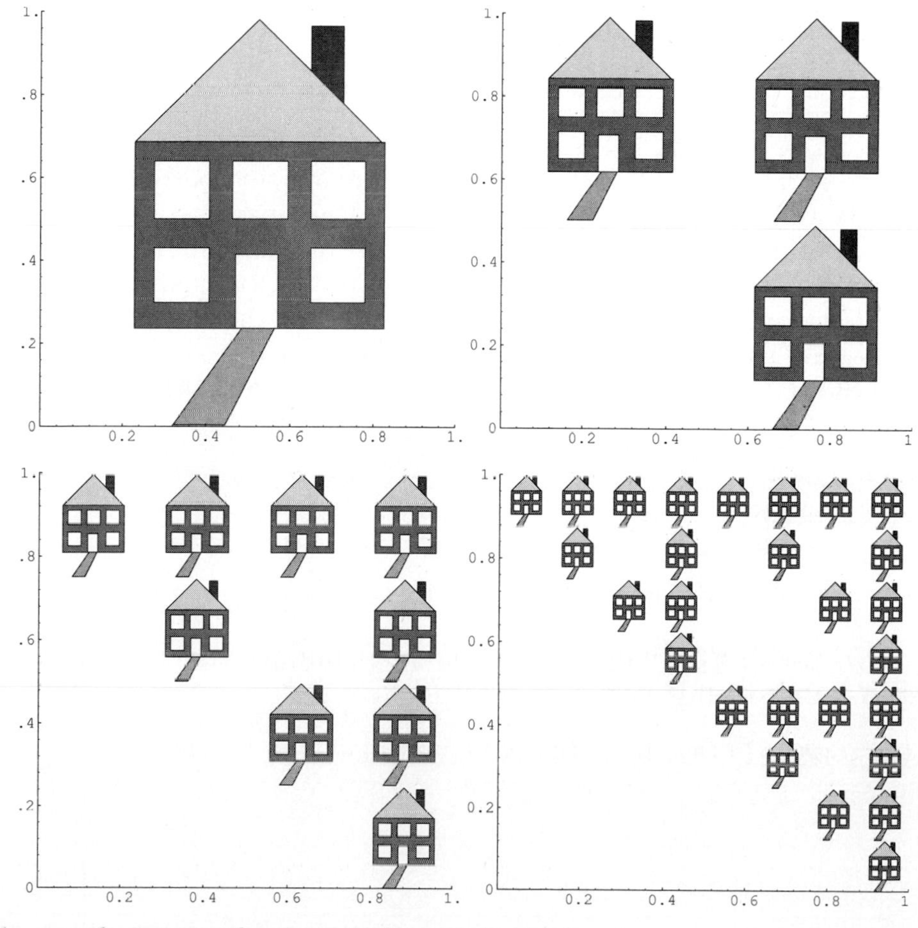

Finally, a solution to the population explosion!

Jerry: Especially if the people in successive generations get smaller and smaller. By the fifth generation these houses look like dots. In fact, they look pretty much the same as the triangle did. Could it be that no matter what object we start with, from a triangle to a whole house, after enough iterations the pictures look the same?

Theo: Could be! I'm having a hard time understanding what is really going on here. It looks as if each triangle is being split up into three smaller triangles at each new iteration. I don't see why the global transformations we are applying should cause each tiny triangle to split up into three triangles occupying the same area.

Jerry: Perhaps if we made an animation in which we show each step in the iteration being applied *slowly*, instead of all in one step....

Theo: Hmm.... That's an interesting idea. Let's think about what it might mean to apply a transformation *slowly*. By slowly we mean generate intermediate steps that lead up to the full transformation. We want to define a transformation that is, say, *half way between* the identity transformation and the transformation we want.

Jerry: Perhaps we can just use a linear combination of the identity and the desired transformation.

Theo: Let's define a function that does this. Since there are two parts to the transformation, the matrix and the translation vector, we need two parameters to specify how much of each of these parts we want to use. Our function will have as its first argument a matrix, as its second argument a translation vector (just like **AffineMap**), as its third argument a number between zero and one that says how much of the matrix to use, and as its fourth argument a number between zero and one that says how much of the translation vector to use. It will return an **AffineMap** function. Here it is:

```
partAffineMap[matrix_, trans_, xm_, xt_] :=
    AffineMap[
        xm matrix + (1 - xm) {{1, 0}, {0, 1}},
        xt trans
    ]
```

For example, if we use this function on a general map, but use zero for both coefficients, we get back the identity map:

```
partAffineMap[{{a, b}, {c, d}}, {t1, t2}, 0, 0]
AffineMap[{{1, 0}, {0, 1}}, {0, 0}]
```

On the other hand, if we use 1 for both coefficients, we get what we started with:

```
partAffineMap[{{a, b}, {c, d}}, {t1, t2}, 1, 1]
AffineMap[{{a, b}, {c, d}}, {t1, t2}]
```

If we use 1 for the matrix coefficient, but zero for the translation coefficient, we get a map that includes only the matrix component, with no translation:

```
partAffineMap[{{a, b}, {c, d}}, {t1, t2}, 1, 0]
AffineMap[{{a, b}, {c, d}}, {0, 0}]
```

Using this function, we can imagine an animation when we start with an object and slowly apply the three transformations. Since the matrix components of all three transformations are the same, we might as well bring them all on at the same time. Then, we can bring on the translation parts of each of the three transformations separately. The program to do this is a little lengthy (although not very complicated conceptually), so it's down in the Functions That We Defined section. Here is the animation applied to the triangle:

```
doTheNiceAnimation[myObject, 4]
```

(We didn't include any pictures in the printed edition, because you can't really tell what's going on unless the pictures are actually moving.)

Jerry: This animation is wonderful in itself. It will inspire me to think of many other situations that are hard to explain but may become very clear by the same kind of technique.

Theo: Finally I understand how the shapes come to be. It's not that each triangle splits into three; instead it's that the whole figure gets shrunk down and then triplicated.

Jerry: Visually, the area of the objects seems to be getting smaller and would have zero as a limit. When we shrink the object by a factor of two in each linear dimension, we reduce its area by a factor of four. When we then triplicate the object, we increase the area by a factor of three. So the net effect of each iteration is to multiply the area by 3/4. Clearly the limit is zero.

Theo: What about the perimeter? When we shrink the object by a factor of two, the perimeter shrinks by a factor of two as well. Then when we triplicate, the perimeter increases by a factor of three, so the net effect is to multiply the area by 3/2. So, we have the peculiar fact that while the area goes to zero, the perimeter goes to infinity.

Jerry: In some ways this reminds me of the snowflake curve (which is made out an infinite number of ever smaller triangles glued to each other), whose perimeter heads off toward infinity while the area has a finite, non-zero limit.

Theo: I think our object is much more interesting. The snowflake curve may have an infinite perimeter, but then it has some *area* about which this perimeter is. Our curve has an infinite perimeter, but all that perimeter encloses *no area*.

Jerry: Good point! *[sic]*.

Functions That We Defined

```
ShowPoints[dots__] :=
     Show[
          Graphics[{
               PointSize[0.02],
               dots
          }],
          Axes -> Automatic,
          AxesOrigin -> {0, 0},
          AspectRatio -> Automatic,
          PlotRange -> {{-3, 3}, {-3, 3}}
     ]
```

```
partAffineMap[matrix_, trans_, xm_, xt_] :=
     AffineMap[
          xm matrix + (1 - xm) {{1, 0}, {0, 1}},
          xt trans
     ]
```

```
ourMaps[xm_, xt1_, xt2_, xt3_] :=
{
     partAffineMap[{{0.5, 0}, {0, 0.5}}, {0.5, 0  },
                         xm, xt1],
     partAffineMap[{{0.5, 0}, {0, 0.5}}, {0,   0.5},
                         xm, xt2],
     partAffineMap[{{0.5, 0}, {0, 0.5}}, {0.5, 0.5},
                         xm, xt3]
}
```

```
fullMap[full_] :=
        First[IteratedMapping[
            ourMaps[1, 1, 1, 1],
            full,
            myObject,
            DisplayFunction -> Identity
        ]]
```

```
partMap[start_, xm_, xt1_, xt2_, xt3_, howMany_] :=
    IteratedMapping[
        Take[
            ourMaps[xm, xt1, xt2, xt3],
            howMany
        ],
        1,
        start,
        Axes -> Automatic,
        AxesOrigin -> {0, 0},
        PlotRange -> {{0, 1}, {0, 1}}
    ]
```

```
doTheNiceAnimation[start_, iterations_] :=
    Do[
        theFullMap = fullMap[full];
        Do[
            partMap[theFullMap, n, 0, 0, 0, 1],
            {n, 0, 1, 0.1}
        ];
        Do[
            partMap[theFullMap, 1, n, 0, 0, 2],
            {n, 0, 1, 0.1}
        ];
        Do[
            partMap[theFullMap, 1, 1, n, 0, 3],
            {n, 0, 1, 0.1}
        ];
        Do[
            partMap[theFullMap, 1, 1, 1, n, 3],
            {n, 0, 1, 0.1}
        ],
        {full, 0, iterations}
    ]
```

References

1) Michael Barnsley, *Fractals Everywhere*, Academic Press, Boston, 1988.

Chapter Twenty-One
Mathematica for Young People Ages 7, Yes 7, and Up

Some math you can do with children who are obviously too young.

■ Dialog–First Day

Jerry: Our title for this chapter was inspired by Don Cohen, my partner in The Math Program for the past 15 years. His exciting and charming book, *Calculus By and For Young People Ages 7, Yes 7, and Up* **[1]**, has shown all of us the wonderful possibilities that exist if you mix young people and mathematics in the right environment. (See the review of Don's book by the Morrisons in *Scientific American*, December 1988.)

Theo: I was delighted by the number of things in his book that I didn't know. I guess that's what the "and Up" part is for.

Jerry: Don and I have been hearing about *Mathematica* for three and a half years, as you know, and using it with our students for three years. Our students have ranged from five-year-olds through adults.

Although few college, community college, or secondary school teachers doubt that *Mathematica* could be useful, almost no elementary school teachers have experience with *Mathematica*. We may be the only math teachers in the world who have teaching experience with young children and *Mathematica*. I would not teach for five minutes without having either *Mathematica* or Derive running in the room.

I have taught older students and teachers, but in my opinion the greatest potential for progress is with the younger students.

Theo: Although I'm surprised to hear this from someone who has actually tried it, in general I would have said the same thing. Children who grow up with personal computers take them for granted. The idea of a whole generation of students growing up taking programs like *Mathematica* for granted is very exciting.

So, what can you do with obviously-too-young students?

Jerry: Recently I began Tali, a second grader, with the following sequence (she had previously worked a little with *Mathematica*). I sat her in front of *Mathematica* and said "I'll show you a way to do five times twelve". I typed:

```
5 * 12 == 5 * 10 + 5 * 2
True
```

Then I said, "you try it on four times thirteen and let me know if you have any trouble" and walked away. She worked for the next half hour or so. A few times she called for help. I would walk by and look every few minutes and suggest a new problem, or smile and say "fine".

Theo: This is an interesting use of ==, and one that I would not have recommended myself. Unfortunately in more complicated cases there is no guarantee that *Mathematica* will return either **True** or **False**: It may just return the equation unevaluated if it can't see right away whether it's true or false. For example, **x^2 + 2 x + 1** is the same as **(x + 1)^2**, and yet:

```
x^2 + 2 x + 1 == (x + 1)^2
              2          2
1 + 2 x + x  ==  (1 + x)
```

Jerry: Interesting, we've run into exactly the same situation, and I explained to my students that what they've written is quite correct, but not understood by *Mathematica* the way they mean it. I suggest two ways to fix it:

```
x^2 + 2 x + 1 == Expand[(x + 1)^2]
True

Factor[x^2 + 2 x + 1] == (x + 1)^2
True
```

Theo: Another thing you might suggest is to wrap the **Simplify** function around the equation. This will tend to make a much larger set of equations reduce to either **True** or at least to a simpler form if the two sides are not, in fact, equal to each other:

```
Simplify[x^2 + 2 x + 1 == (x + 1)^2]
True
```

There are always going to be some equations that are true, but that *Mathematica* will never be able confirm as such. (A function that would prove any given equation true or false is an example of a non-computable function. If you could devise a function to evaluate the truth of any statement by a purely mechanical procedure, you would be able to prove or disprove any theorem you liked. Aside from getting you Nobel prizes on demand, this would violate several theorems, among them the famous one by Gödel.)

Jerry: Another starting point with students is equations. I remember starting Carolyn, a fourth grader, with

```
Solve[2 x + 1 == 7, x]
{{x -> 3}}
```

My explanation was that **x** represents any number, and that *Mathematica* has found that if **x** is three, then the equation is true, because two times three plus one is seven. I then showed her

```
Solve[5 x + 3 == 13, x]
{{x -> 2}}
```

and talked her though it. I showed her how to use the "Copy Input From Above" command (Command-L on the Macintosh and NeXT versions, or Alt-L on the MS-DOS version) to save on typing, and showed her how to use the mouse to change one or more numbers in the equation. She absorbed this with a few minutes practice and some reassurance from me, and then was free to do her own problems.

Theo: So what did she learn from this?

Jerry: She learned to work at a serious computer program (not a game) without worrying about it. Initially this is the most important thing. She also began to anticipate the solution and to see cases where the solution was not a positive number and cases where the solution was not an integer. She got used to working independently on math even though surprising new situations kept popping up. Later she worked on quadratic equations and learned about the number of roots, and how to predict the roots of new equations.

A good way to get a feeling for the possibilities is to look at files created by students.

Theo: I'm glad to know that your students keep Notebooks of what they have been doing. I've always thought it important that *Mathematica* actively *encourages* people to keep records of what they are doing, instead of just having one night stands with *Mathematica*.

Jerry: I suggest that students use the "Copy Input from Above" command instead of editing away "errors". This way a rather complete record is kept. Let's have a look at an excerpt from Carolyn's Notebook:

```
Expand[(x+2)(x+3)]

6 + 5 x + x^2
Expand[(x+5)(x+3)]

15 + 8 x + x^2
Expand[(x+5)(x+4)]==20+9 x+x^2

True

Expand[(x+6)(x+4)]==24+10 x+x^2

True

Expand[(x+6)(x+8)]==48+14 x+x^2

True

Expand[(x+6)(x+8)(x+4)]==192+18 x+x^2

192 + 104 x + 18 x^2 + x^3 == 192 + 18 x + x^2
Expand[(x+6)(x+8)(x+4)]==192+18 x+x^3

192 + 104 x + 18 x^2 + x^3 == 192 + 18 x + x^3
Expand[(x+4)(48+14 x+x^2)]

192 + 104 x + 18 x^2 + x^3
```

Theo: Looking at this small section of the Notebook (the whole file is about 20 pages long, so far), one can see many interesting things. Perhaps we should talk our way through it.

Jerry: In the first two lines, Carolyn is letting the machine produce the expansion:

```
Expand[(x+2)(x+3)]

6 + 5 x + x^2

Expand[(x+5)(x+3)]

15 + 8 x + x^2
```

In the next three, she is ready to guess at the solution, and is able to do them correctly:

```
Expand[(x+5)(x+4)]==20+9 x+x^2

True

Expand[(x+6)(x+4)]==24+10 x+x^2

True
```

```
Expand[(x+6)(x+8)]==48+14 x+x^2
True
```

Theo: I am pleased to see that she chose to write out her guesses in order of increasing powers of x. This is the opposite of what all algebra text books use, but is the way preferred by symbolic algebra systems (for good reasons, I might add).

Jerry: Carolyn also works on Derive, where the same polynomials would be written out in the "usual" form, and where only one equal sign is needed. Many schools try to create a uniform environment, so the students will not be "confused". My attitude is that we should arrange experiences for the students that reflect what happens in the real world. Young students adjust easily to whatever cockeyed system we set up.

In the next example, we discover that Carolyn is not perfect! Imaginatively, she expands three factors, and guesses wrong. *Mathematica* does not say **False**, because, after all, the equation is true for several different values of **x** (like zero, for example). Instead it says "I don't know yet if this is true or false".

```
Expand[(x+6)(x+8)(x+4)]==192+18 x+x^2
```

$$192 + 104\ x + 18\ x^2 + x^3 == 192 + 18\ x + x^2$$

```
Expand[(x+6)(x+8)(x+4)]==192+18 x+x^3
```

$$192 + 104\ x + 18\ x^2 + x^3 == 192 + 18\ x + x^3$$

Theo: From the fact that she guesses wrong twice on the same problem, I deduce that she did not notice that *Mathematica* gave her the correct answer the first time (on the left hand side of the result). Presumably she interprets anything other than **True** as being false.

Jerry: Yes, and I think sometimes the kids ignore the correct answer on the left side because they want to make another guess. Other times they simply forget that the answer is there.

Theo: Sort of gives a whole new perspective on cheating.

In the last step, we see that Carolyn has starting asking *Mathematica* again, presumably because she has decided she doesn't know enough to guess:

```
Expand[(x+4)(48+14 x+x^2)]
```

$$192 + 104\ x + 18\ x^2 + x^3$$

Jerry: But notice she has combined an earlier answer into this unsolved problem. Later in this session she began guessing these problems correctly.

Dialog–Second Day

Jerry: Let me show you some of Anand's work. He was a fourth grader at the time, and said something to me about changing miles/hour into other units. I think the speed limit was changing from 55 to 65 miles/hour at the time, and it was on his mind.

Theo: That's a pretty radical idea there: A student who actually has a mathematical question on his mind.

Jerry: This happens much more with younger students, and also in the summer. When school closes, the kids come in to The Math Program acting three time smarter. After school opens again in the fall, they act really stupid when they first come in. Fortunately this "school-effect" wears off after a few minutes. Maybe they're just tired.

Theo: I remember that from my years in The Math Program. But, back to speed limits.

Jerry: I showed Anand how to load the units package that comes with *Mathematica*, and gave him one example of how to use the **Convert** command:

```
Needs["Miscellaneous`Units`"];

Convert[55 Mile/Hour,Meter/Second]

24.5872 Meter
-------------
   Second
```

He started on his own from here:

```
Convert[65 Mile/Hour,Meter/Second]

29.0576 Meter
-------------
   Second

Convert[75 Mile/Hour,Meter/Second]

33.528 Meter
------------
   Second
```

Theo: At this point he clearly become interested in large numbers. Personally I've always wondered how many millimeters you could go in a millennium.

```
Convert[75 Mile/Hour,Millimeter/Millennia]
Convert::incomp:
                                  75 Mile     Millimeter
     Incompatible units in ------- and ----------.
                                   Hour       Millennia

75 Mile
-------
 Hour

Convert[75 Mile/Hour,Millimeter/Hour]
Convert::incomp:
                                  75 Mile     Millimeter
     Incompatible units in ------- and ----------.
                                   Hour          Hour

75 Mile
-------
 Hour
```

Unfortunately for Anand, **Millimeter** is not a recognized unit. *Mathematica* handles only the basic units like **Meter**, **Foot**, **Pony**, **Hogshead**, **AvoirdupoisOunce**, **WineBottle**, **Jeroboam**, etc. Millimeter is a compound unit, and if *Mathematica* tried to list all of those too, there would be far too many.

He also used the plural form of Millennium, while *Mathematica* puts all units in the singular form. An additional complication is that in the Units package distributed with the version he was using, Stephen had misspelled "Millennium" with one "n". In case you're interested, we can do the calculation here:

```
Convert[75 Mile/Hour, Meter/Millenium]

              12
1.05734 10      Meter
---------------------
    Millenium
```

Multiplying by 1000 gives us the answer in millimeters.

Jerry: So, my poor students have to learn to misspell certain words to use this program successfully. A heavy burden! Is it true that all computer people spell poorly, or just the ones I know?

Anand continued with speed conversions, clearly demonstrating that he should not be allowed in a car for a while:

```
Convert[85 Mile/Hour,Meter/Second]

37.9984 Meter
-------------
   Second
```

```
Convert[95 Mile/Hour,Meter/Second]
```

$$\frac{42.4688\ Meter}{Second}$$

```
Convert[105 Mile/Hour,Meter/Second]
```

$$\frac{46.9392\ Meter}{Second}$$

Here he switches to minutes:

```
Convert[65 Mile/Hour,Meter/Minute]
```

$$\frac{1743.46\ Meter}{Minute}$$

Temperatures:

```
Convert[10 Fahrenheit,Centigrade]
5.55556 Centigrade
```

Theo: I bet this was in the winter.

Jerry: Probably! Now his age in minutes:

```
Convert[10 Year,Minute]
```

$5.256\ 10^{6}$ Minute

Many other topics on many other days followed. About 18 pages later, we found the following examples:

```
1 day*24 hours/day
24 hours
```

```
7 day*24 hours/day
168 hours
```

```
7 day/week*24 hours/day
```

$$\frac{168\ hours}{week}$$

```
365 day/year*3600 seconds/day
```

$$\frac{1314000\ seconds}{year}$$

```
365 day/year*86400 seconds/day
```

$$\frac{31536000\ seconds}{year}$$

```
36500 day/century*86400 seconds/day
3153600000 seconds
------------------
     century
36500 day/century*1440 min/day
52560000 min
------------
  century
```

Theo: He didn't seem interested in using the built-in `Units.m` package anymore, preferring to do it himself.

Jerry: Some teachers and parents worry that students will not carry through the details of their work in math, if they have access to programs as powerful as *Mathematica*. I do not find this at all. In this case, Anand wanted to carry through a calculation (which I think he's done before), and *Mathematica* was willing to make a modest contribution.

Theo: Where did he get the 86400 from?

Jerry: I'm not sure, but sometimes the kids bring hand calculators with them to *Mathematica*. Odd behavior, but I see it. He may have calculated the number with the calculator, or on paper (or he might just know it).

■ Dialog–Third Day

Jerry: This is a complete record of Michael's work, which was done over a period of four sessions. I made a few suggestions, but Michael, a fifth grader, works very well independently, so I mostly smiled and said "fine". I think we can all learn a lot about how people learn and think by reading this "stream of consciousness" record.

Theo: Let's give this one without comment. Don't just glance at what follows: To get anything out of it, you have to look at and try to understand why he did each new thing.

```
7
7
2x+3x+4x
9 x
2p+4p+3p
9 p
2a+3a+4a
9 a
```

```
2k+3k+4k==9k
True
6m+4m+7m==17m
True
6a+6a+6a==18a
True
4h+5h+1h==10h
True
7a+7a+4a==18a
True
8l+8l+4l==20l
True
2n+9n+8n==19
19 n == 19
2n+9n+8n==19n
True
9n+9n+9n-6n==21n
True
4a+5a+6a-9a*5n==25
15 a - 45 a n == 25
4a+5a+6a-9a*5a==25
            2
15 a - 45 a  == 25
4a+5a+6a-9a*5a==15a-45a^2
True
a*a*a*a
 4
a
a*a*a*a*a*a*a*a*a
 9
a
a*a*a*a*a*a*a*a*a*a*a*a*a*a*a==15
 15
a   == 15
a*a*a*a*a*a*a*a*a*a*a*a*a*a*a==15
 15
a   == 15
a*a*a*a*a*a*a*a*a*a*a*a*a*a*a
 15
a
a*a*a*a*a*a*a*a
 8
a
```

```
p*p*p*p==4
 4
p  == 4
p*p*p*p==p^4
True
k*k*k*k*k*k*k*k*k*k*k*k*k*k*k*k*k*k*k*k*k*k*k==k^23
True
m*m*m*m*m*m*m*m*m*m*m*m*m*m*m*m*m*m*m*m*m*m*m*m*m==m^25
True
a*b*c*a
 2
a  b c
a*b*c*a
 2
a  b c
a*b*a*b*a*b
 3  3
a  b
a*a*a*b*b*b
 3  3
a  b
a*b*a
 2
a  b
2*7+3==17
True
8*8+9==73
True
9*9+20==101
True
10*9+50==140
True
12*12+16==160
True
9*3+80==107
True
12*12+12==156
True
4*2+145==153
True
9*9+70==151
True
```

```
8*8+56==120
True
14*14+14==210
True
15*15+15==240
True
16*16+16==272
True
17*17+17==306
True
18*18+18==332
False
18*18+18==342
True
19*19+19==380
True
20*20+20==420
True
21*21+21==263
False
463
463
21*21+21==463
False
15*15+21==246
True
23*23+23==552
True
21*21+21==462
True
22*22+22==506
True
24*24+24==590
False
24*24+24==592
False
24*24+24==593
False
```

`25*25+25==635`

False

`a*a*a`

a^3

`a*b*a*b*a*a`

$a^4\ b^2$

`a*a*a*a*a*r*y`

$a^5\ r\ y$

`a*n*u*i*a*a*a*a*a*a`

$a^7\ i\ n\ u$

`a*b*c*d*e`

a b c d e

`l*m*n*o*p*q*r*s*t*u*v*w*x*y*z`

l m n o p q r s t u v w x y z

`h*g*f*p*q*n`

f g h n p q

`e*o*s*k*f*o*o*o*o*r*k*l`

$e\ f\ k^2\ l\ o^5\ r\ s$

`a*b*c*a*b*c`

$a^2\ b^2\ c^2$

`e*r*g*h*u*i`

e g h i r u

`a*b*a==a^2b`

Truc

`a*a*b*a*b*a==a2^b`

$a^4\ b^2 == a2^b$

`a*a*b*a*b*a==a^2b`

$a^4\ b^2 == a^2\ b$

`a*a`

a^2

`a*a*a`

a^3

`a*a*a*a`

a^4

```
a*a*a*a*a==a5
 5
a  == a5
a*a*a*a*a==a^5
True
a*a*a*a*a*a*a*a*a*a*a*a*a*a*a*a*a==a^17
True
Factor[x^2-81]
(-9 + x) (9 + x)
Factor[x^2-10x+21]
(-7 + x) (-3 + x)
Factor[x^2-8x+15]
(-5 + x) (-3 + x)

Factor[x^2-11x+30]
(-6 + x) (-5 + x)

Factor[x^2-11x+28]
(-7 + x) (-4 + x)
Factor[x^2-10x+21]
(-7 + x) (-3 + x)
Factor[x^2-11x+18]
(-9 + x) (-2 + x)
Factor[x^2-8x+16]
        2
(-4 + x)
Factor[x^2-12x+8]
              2
8 - 12 x + x
Expand[(x-2)(x+6)]
               2
-12 + 4 x + x
Factor[x^2-8x+12]
(-6 + x) (-2 + x)
Factor[x^2-23x+132]
(-12 + x) (-11 + x)
```

Jerry: I've handed data like this to a variety of teachers, and been surprised at the lack of interest shown. I think records like these are far more valuable than, for example, homework or test results, in understanding what the student has done and learned. It guides me in my attempt to teach the student.

I tell a student to sit down and write some stuff. If I've just spent some time trying to teach her something, and it doesn't show up, I know I'm in trouble. If she regurgitates what we've done earlier, or doesn't write anything, it's bad. If she starts to *extend* what I showed her, that's when I know things are working.

You can see the experimentation and stretching going on in the record above: This student is using his natural curiosity and talent to learn more about math.

I believe that this stream-of-consciousness activity is appropriate for all students, independent of age or topic. This is not just something for fifth graders.

■ References

1) Don Cohen, *Calculus By and For Young People (ages 7, yes 7, and up)*, ISBN #0-9621674-1-X.

In selected bookstores and science museums, and available from Don Cohen, 809 Stratford Dr, Champaign, IL, 61821, for $14.50 including postage and handling.

Chapter Twenty-Two
High School Math

In which spaceship Mathematica lands in your local schoolyard.

Jerry: Some high school students and teachers I've worked with have had difficulties getting *Mathematica* to do everyday high school math. If we can show examples of how to do these things, I'm sure they'd begin to see why *Mathematica* is worth getting to know.

Theo: *Mathematica* can do high school math, of course. It's just not obvious how, at first. By showing specific examples we can let people get right to work. They can learn the rest of what they need to know as it comes up, when they need it.

Jerry: In the following sections let's talk about some of the things we think high school math consists of. We won't cover every topic in complete detail, but each one will have enough to get people started.

■ Cartesian Plots

Jerry: Show us how to do a simple graph, like y = 2x - 3, with x going from -5 to 5.

Theo: To do two-dimensional plots, you use the **Plot** command, like this:

```
Plot[2 x - 3, {x, -5, 5}];
```

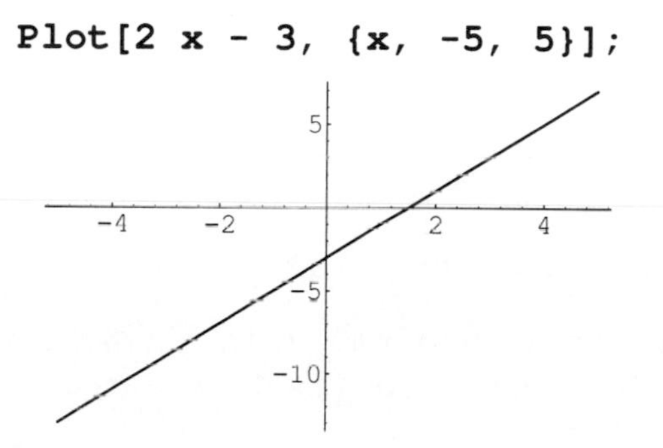

Jerry: I notice it wasn't necessary to include the "y = " in the **`Plot`** command. Do I have the option of entering "y = 2x - 3", or "f(x) = 2x - 3" in the **`Plot`** command?

Theo: No. The **`Plot`** command requires an expression as its first argument. Whatever values that expression has for different values of **`x`** get plotted. In other words, **`Plot`** is interested only in the right hand side of your equations. (Actually, it *would* work, but it's not a good idea to do it that way. It works only by chance, and has the undesirable consequence of assigning a value to the variable **`y`**.)

Jerry: From the equation, this graph has a slope of two, which should look steeper than a graph with a slope of one. This graph doesn't look steep enough.

Theo: *Mathematica* automatically picks the scales in the x and y directions to make the plot fit into a golden rectangle. People making simple plots that don't have a lot of very big or very small numbers often like to have the same scaling in both directions; then slopes come out looking "right". You can do that by adding the option **`AspectRatio -> Automatic`** to your **`Plot`** command:

```
Plot[2x - 3, {x, -5, 5}, AspectRatio -> Automatic];
```

I think you can also see why most of the time it's better to scale the plot to fit into a reasonably shaped rectangle: you can get some extremely thin rectangles using the automatic aspect ratio option.

Jerry: Suppose I want to plot several graphs on the same axes?

Theo: Simple. In *Mathematica* you use curly brackets to make lists. To plot several graphs at the same time, you give **`Plot`** a list of expressions as its first argument:

```
Plot[{2x - 3, -x + 4}, {x, -5, 5},
        AspectRatio -> Automatic];
```

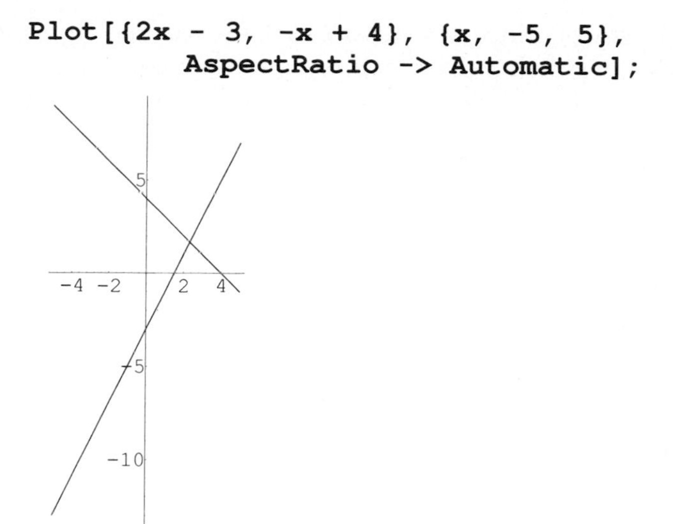

Jerry: How do I locate the intersection?

Theo: Hmm.... Well, you can use the interactive coordinate display feature. Click the mouse when the pointer is anywhere in the graph. Then, holding down the Command key (on the Macintosh or NeXT) or the Alt key (on MS-DOS), move the mouse pointer over the graph. You will see the (x, y) coordinates you are pointing to displayed at the bottom of the window. If you point to the intersection, you will see the coordinates, approximately, of the point where the graphs cross. I get about (2.32, 1.67). Of course, if you really want to know, you use the **Solve** function:

```
Solve[{y == 2x - 3, y == -x + 4}, {x, y}]
```

$$\{\{x \to \frac{7}{3},\ y \to \frac{5}{3}\}\}$$

Jerry: How do I get those two numbers as decimals, so I can compare them with the ones you found with the mouse?

Theo: Anytime you want a decimal instead of an exact answer, use the **N** command:

```
N[Solve[{y == 2x - 3, y == -x + 4}, {x, y}]]
{{x -> 2.33333, y -> 1.66667}}
```

Jerry: So far we've graphed only simple linear expressions. Can we also find intersections of fancier polynomials? Let's pick a cubic expression and a quadratic expression:

```
Plot[{(x + 2)(x - 0)(x - 2), x^2}, {x, -5, 5}];
```

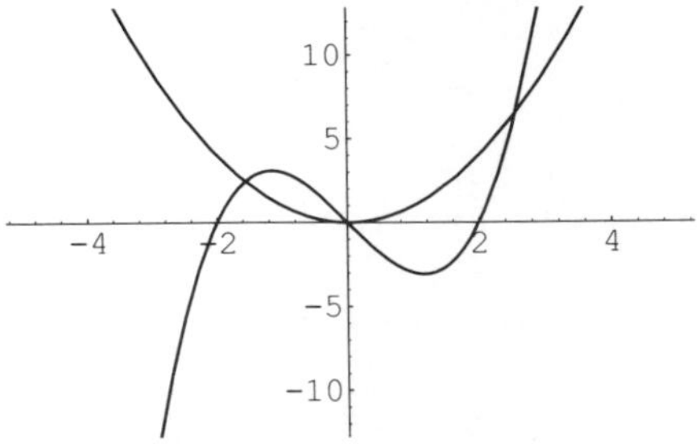

Theo: Just as before, we can use **Solve** to find these three points of intersection.

```
TableForm[Solve[
     {y == (x + 2)(x - 0)(x - 2), y == x^2}, {x, y}]]
```

```
y -> 0                          x -> 0

      18 + 2 Sqrt[17]                1 + Sqrt[17]
y -> ---------------            x -> ------------
             4                            2

      18 - 2 Sqrt[17]                1 - Sqrt[17]
y -> ---------------            x -> ------------
             4                            2
```

Jerry: Very nice! We see the three points of intersection on the graph, and **Solve** gives their coordinates. (**TableForm** arranges the answer in columns, but doesn't change it.)

Theo: Here are some more plots to try:

```
Plot[Abs[x], {x, -3, 3}]

Plot[Abs[Sin[x]], {x, -2Pi, 2Pi}]

Plot[Sin[Abs[x]], {x, -2Pi, 2Pi}]

Plot[{Floor[x], Round[x], Ceiling[x]}, {x, 0, 8}]

Plot[Sec[x], {x, -2Pi, 2Pi}]

Plot[Abs[x] / x, {x, -3, 3}]

Plot[Sin[x] / x, {x, -3Pi, 3Pi}]

Plot[(x^2 - 1) / (2 x + 4), {x, -8, 8}]

Plot[8 / (4 + x^2), {x, -4, 4}]

Plot[4 x / (x^2 + 1), {x, -3, 3}]
```

■ Polar Plots

Jerry: What function can I use to get polar graphs instead of the Cartesian graphs we've seen so far?

Theo: **`PolarPlot`**, of course. The **`PolarPlot`** function is not, however, a built-in *Mathematica* function. It is defined in an external package that is included as part of the standard *Mathematica* program. Before you can use **`PolarPlot`** you have to execute the following command to load this package:

```
Needs["Graphics`Graphics`"]
```

Theo: Then you can use **`PolarPlot`** as if it were built-in. For example:

```
PolarPlot[Sin[4 theta], {theta, 0, 2Pi}];
```

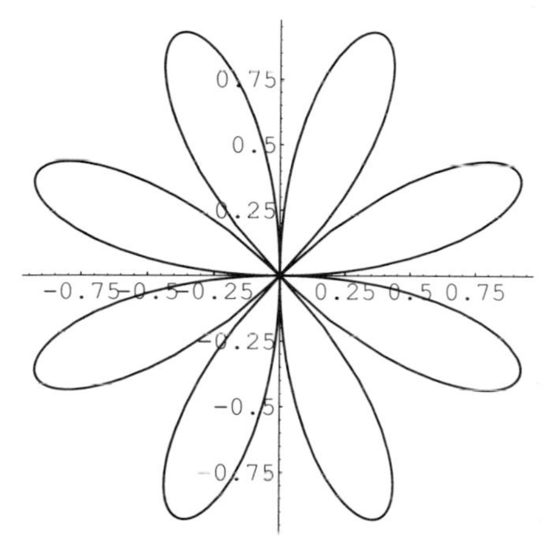

Jerry: Can we plot several expressions at once, as we can with **`Plot`**? I assume we can just give it a list of expressions, as we did using **`Plot`**:

```
PolarPlot[{Sin[4 theta], theta/4}, {theta, 0, 2Pi}];
```

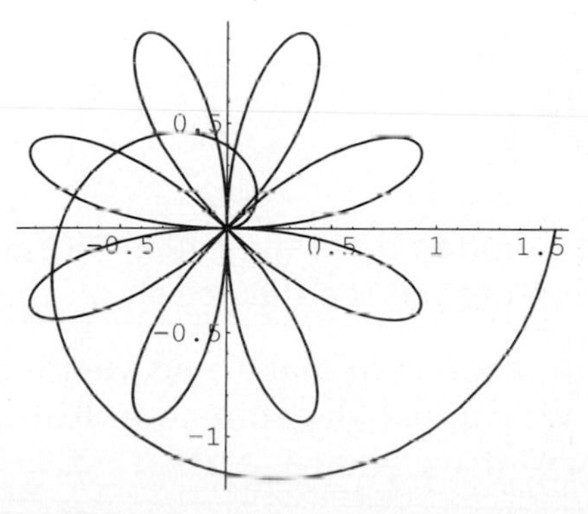

Theo: Polar graphs always seem to be a lot of fun. The nicer ones usually need a `PlotPoints -> 100` or `200` option to make them come out looking right.

```
PolarPlot[
    2 + Sin[8 theta] + Sin[64 theta]/4,
    {theta, 0, 2Pi},
    PlotPoints -> 200];
```

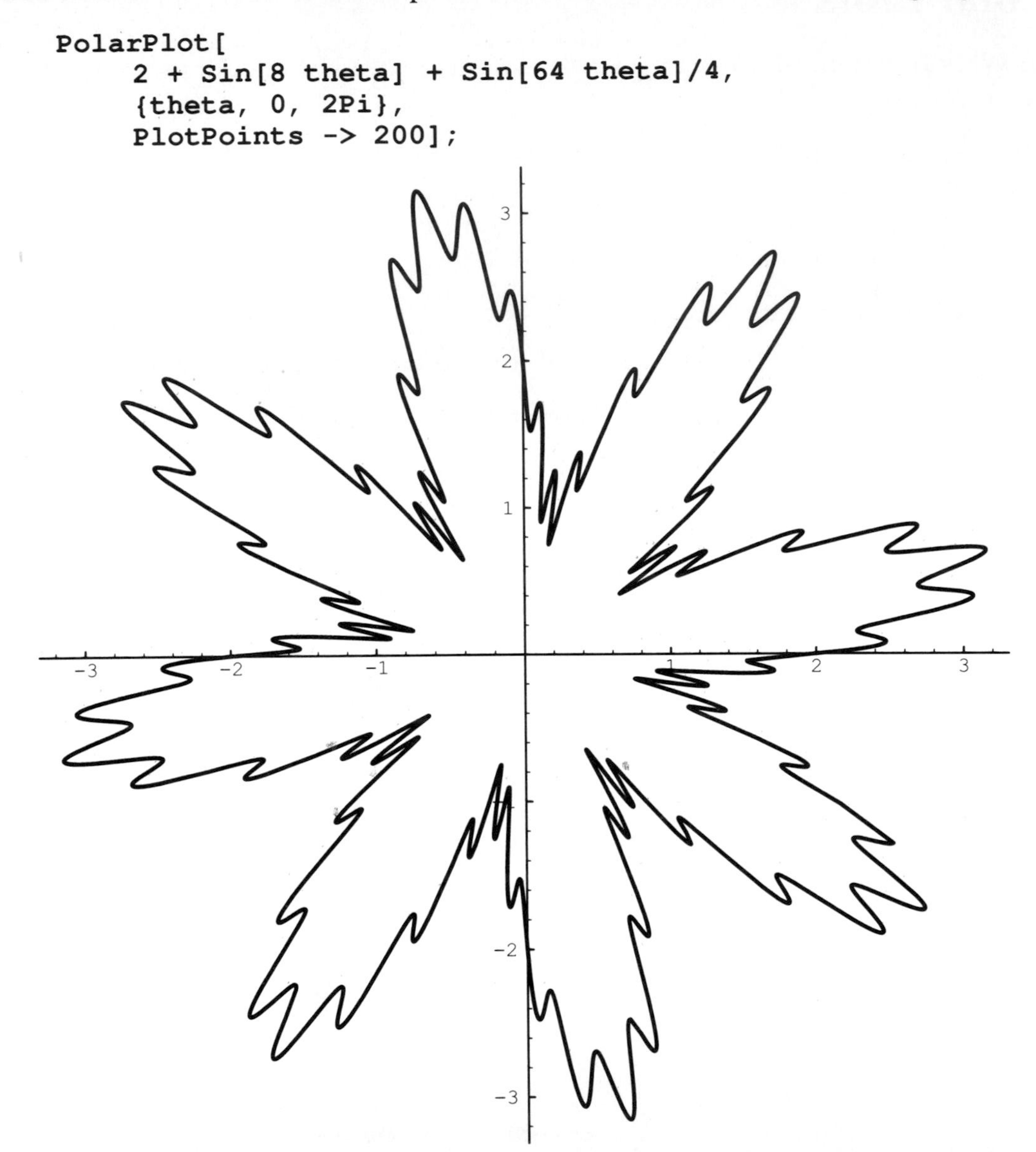

Jerry: I would like to see `Cos[11/12 theta]` plotted a bit at a time. Normally, *Mathematica* plots everything at once without letting me see the action on the way.

Theo: Sure, no problem: That's what animation is for. We make a sequence of plots with a larger range for theta each time. Then we can use the animation feature to see the plot being drawn, and even being "undrawn" if we want.

```
Do[PolarPlot[
      Cos[11/12 theta], {theta, 0, n Pi},
      PlotPoints -> 100,
      PlotRange -> {{-1, 1}, {-1, 1}}],
  {n, 1, 24, 1}]
```

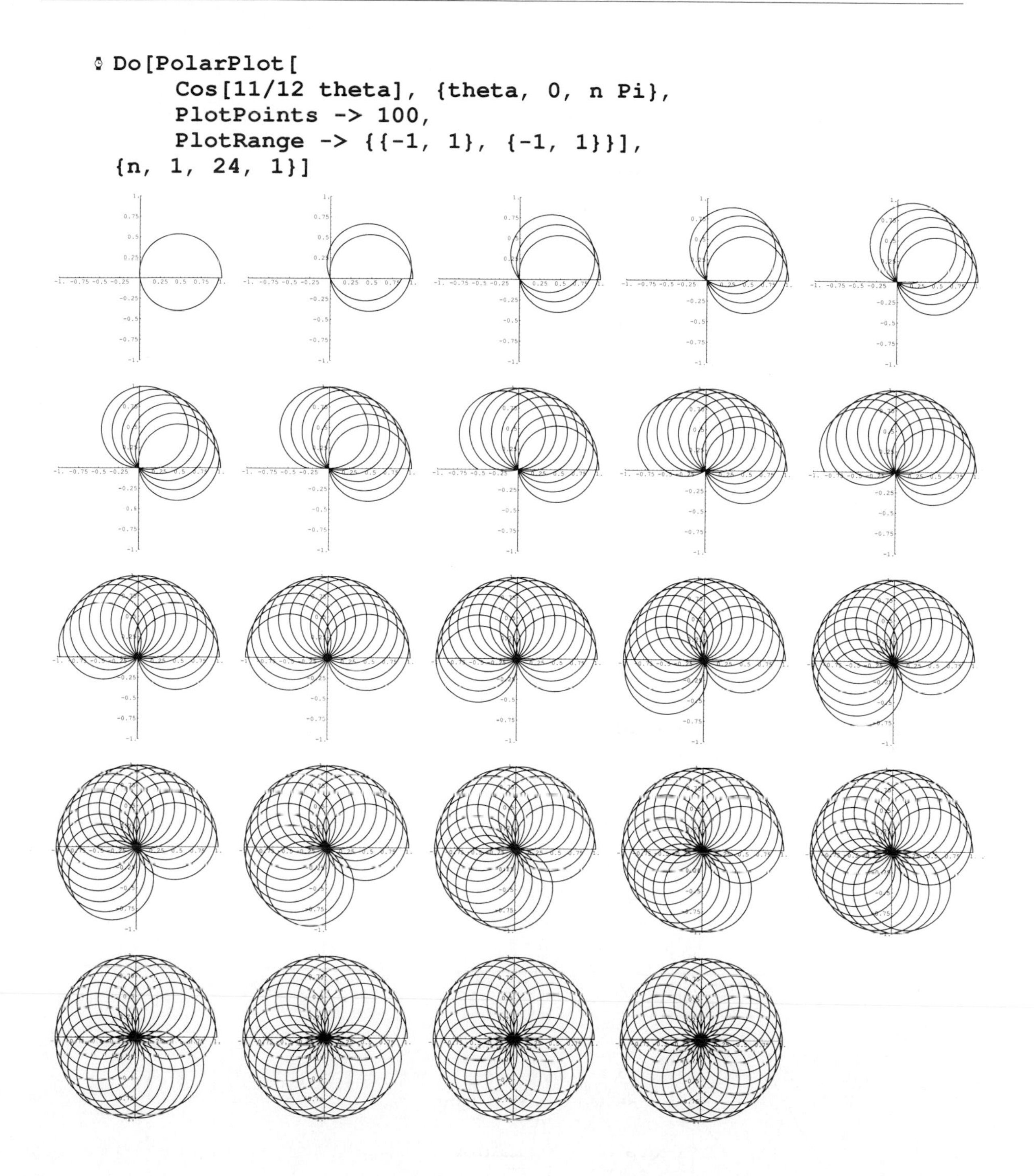

Parametric Plots

Jerry: What about parametric graphs?

Theo: **`ParametricPlot`**.

Jerry: Let me guess: It takes a list of two functions, the x and y coordinates as a function of some parameter, say t:

```
ParametricPlot[{Cos[3 t], Sin[5 t]}, {t, 0, 2Pi}];
```

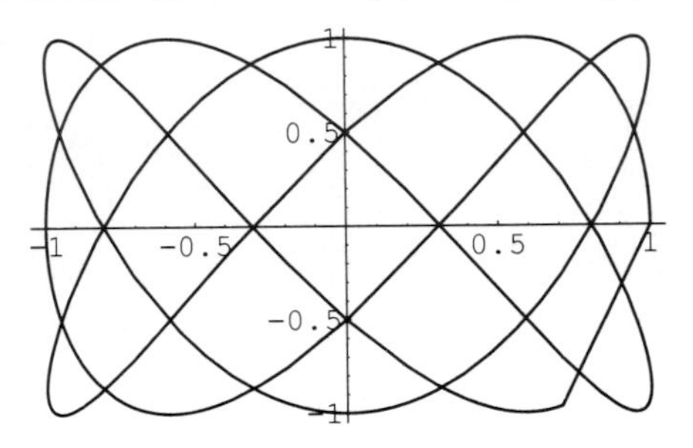

Theo: I suppose you're going want that with the same scale in both directions. I also notice that one corner is chopped off. We can adjust these two things using the following command. (**`AspectRatio -> Automatic`** we saw above, and **`PlotPoints -> 40`** tells *Mathematica* to be more thorough about smoothing out the curve.)

```
ParametricPlot[
    {Cos[3 t], Sin[5 t]}, {t, 0, 2Pi},
    AspectRatio -> Automatic,
    PlotPoints -> 40];
```

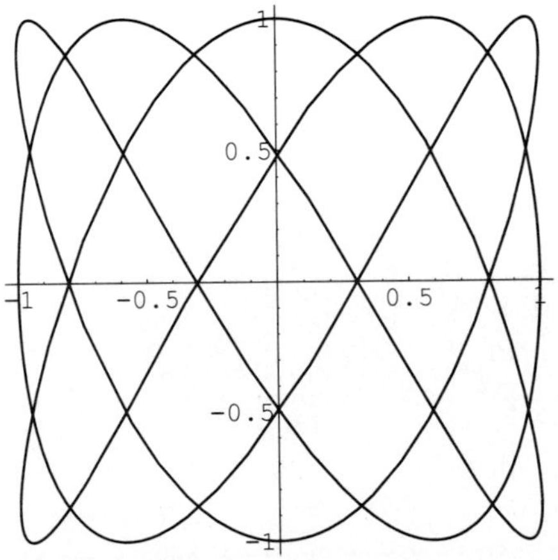

Jerry: Unfortunately, most people are introduced to parametric plots like these for the first time in a calculus class, and so this isn't considered "high school math". To change that idea, let's do a few graphs like lines, parabolas and circles.

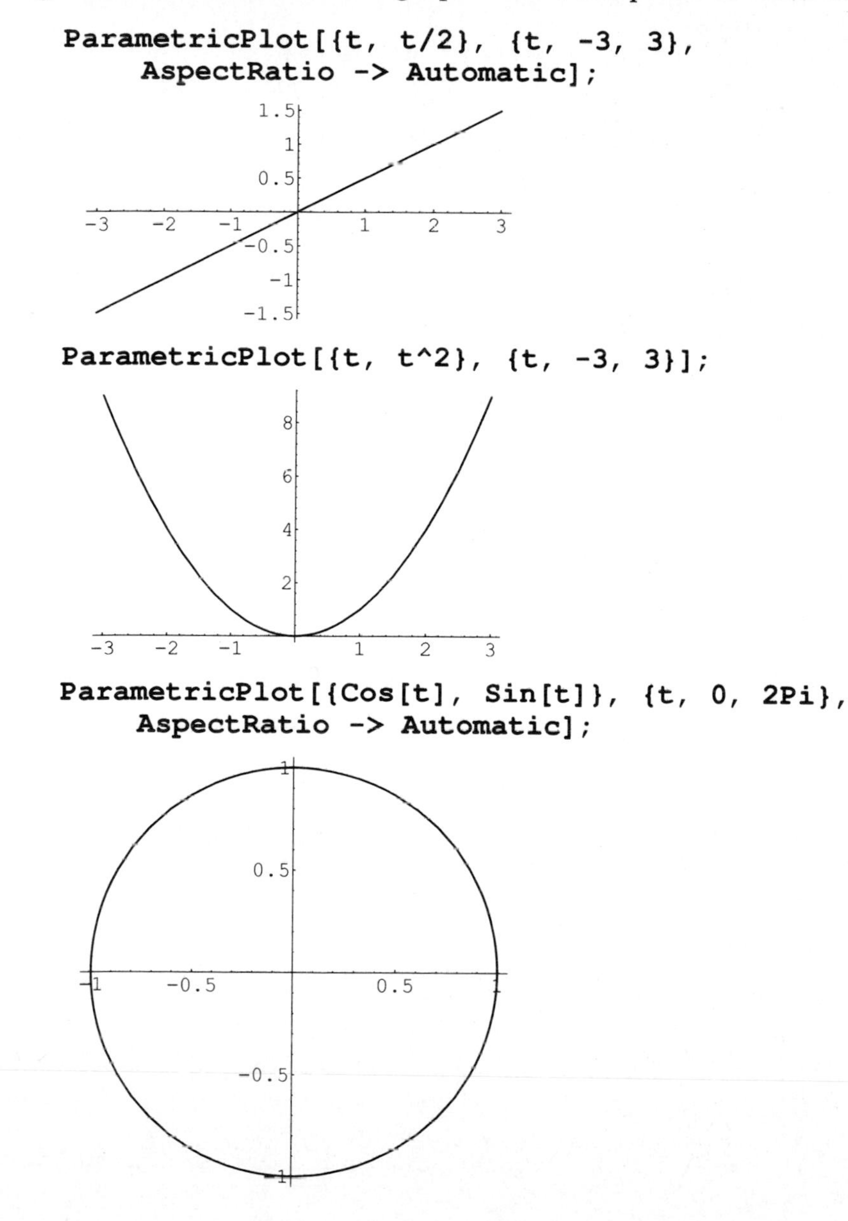

```
ParametricPlot[{t, t/2}, {t, -3, 3},
    AspectRatio -> Automatic];
```



```
ParametricPlot[{t, t^2}, {t, -3, 3}];
```



```
ParametricPlot[{Cos[t], Sin[t]}, {t, 0, 2Pi},
    AspectRatio -> Automatic];
```

Theo: **ParametricPlot** can take a list as its first argument, to plot several things on the same set of axes. Since the normal argument to **ParametricPlot** is already a list of two things, to get multiple plots we give it a list of lists, like this:

```
ParametricPlot[
    {{t, t/2}, {t, t^2}, {Cos[t], Sin[t]}},
    {t, -2Pi, 2Pi},
    AspectRatio -> Automatic,
    PlotRange -> {{-2, 2}, {-1.5, 1.5}}];
```

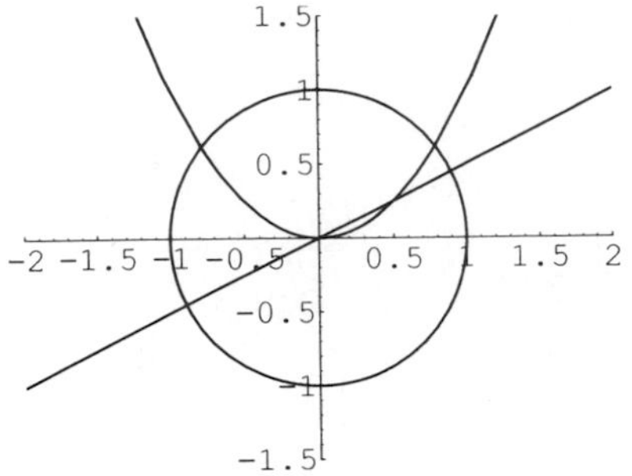

Theo: A nice advantage of parametric plots over Cartesian ones is the ease of plotting non-functions, like circles. To make a plot of a circle in Cartesian coordinates you have to plot two separate solutions to a quadratic equation, and they usually don't join up right in the graph.

Jerry: Text books usually put the most interesting problems way in the back where no one ever sees them. An advantage of *Mathematica* is that it lets you explore these "hard" problems. For example, we found the following example in the Finney/Thomas Calculus book (Addison-Wesley, 1989, p672):

```
ParametricPlot[
    {Cos[t] + 5 Cos[3 t], 6 Cos[t] - 5 Sin[3 t]},
    {t, 0, 2Pi}, AspectRatio -> Automatic];
```

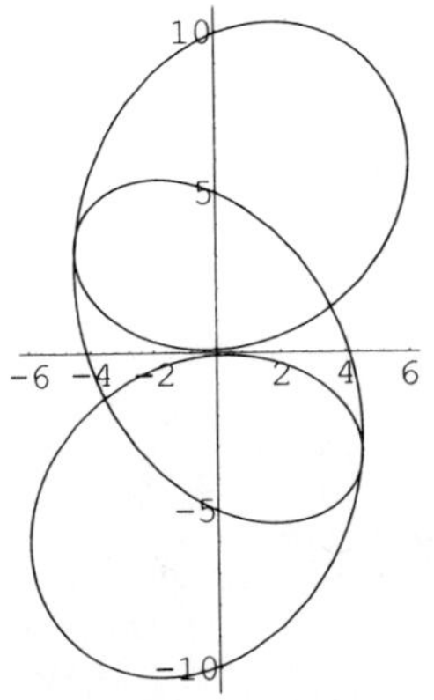

Theo: This is a beautiful curve that I've never seen before.

Another advantage of *Mathematica* is that it's easy to start with an example and then modify it. For example, I wonder what happens if we change the 3's in this plot into different numbers. Let's make an animation with that number going from 1 to 6.

```
Do[
    ParametricPlot[
        {Cos[t] + 5 Cos[n t], 6 Cos[t] - 5 Sin[n t]},
        {t, 0, 2Pi}, AspectRatio -> Automatic,
        PlotPoints -> 50,
        PlotRange -> {{-7, 7}, {-12, 12}}],
    {n, 1, 6, 1}
]
```

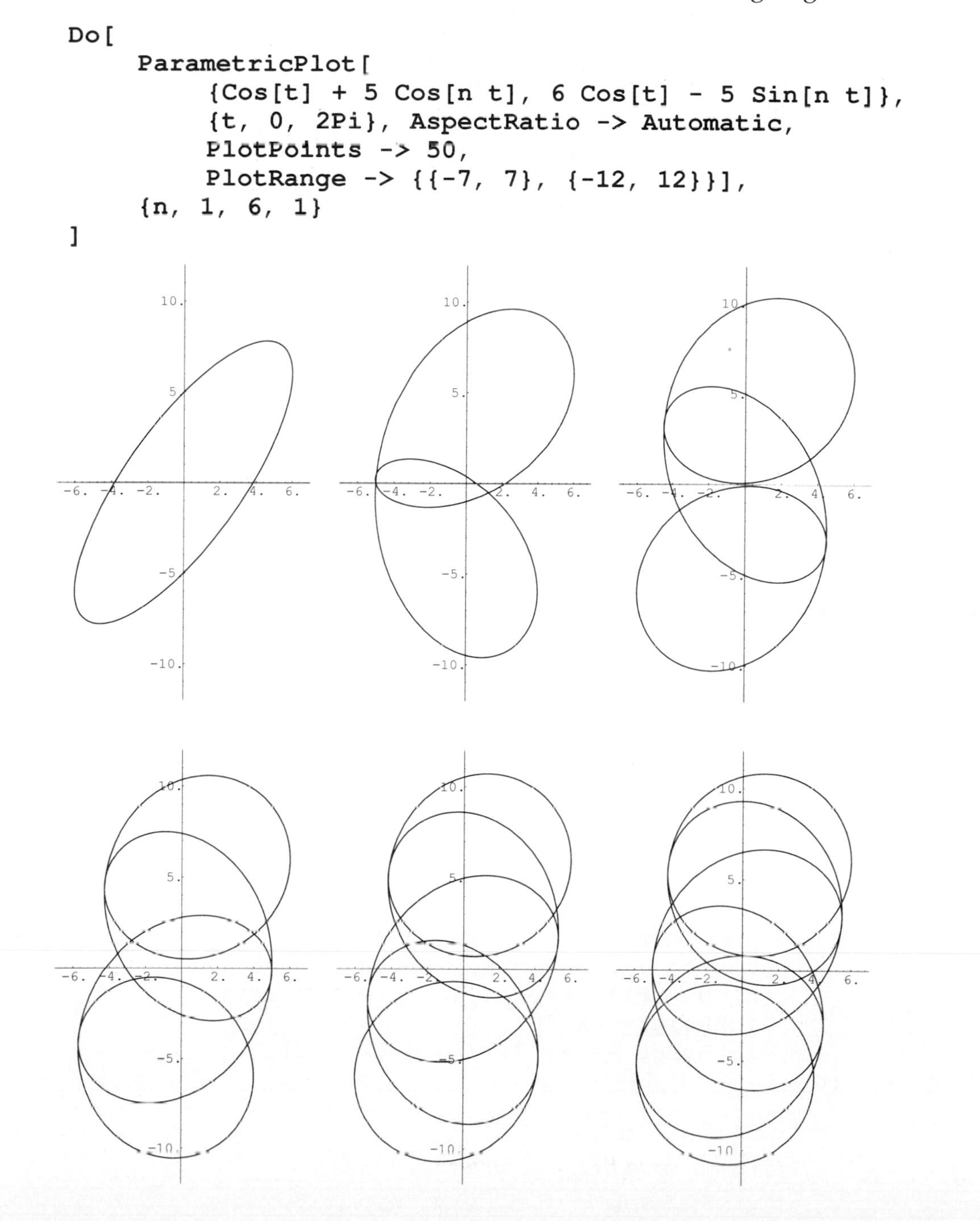

It's fascinating how the loops of each of these curves *just touch* each other. It works no matter what value of n you choose, as we can see from these examples. This is something the reader of Finney/Thomas is likely to miss.

Jerry: Here are some more examples for people to try:

```
ParametricPlot[{Cos[t]^3, Sin[t]^3}, {t, 0, 2Pi},
    AspectRatio -> Automatic]

ParametricPlot[{2 Cos[t] + Cos[2 t],
                    2 Sin[t] - Sin[2 t]},
          {t, 0, 2Pi}, AspectRatio -> Automatic]

Do[ParametricPlot[{n Cos[t] + Cos[n t],
                      n Sin[t] - Sin[n t]},
          {t, 0, 2Pi}, AspectRatio -> Automatic],
     {n, 2, 20}]

ParametricPlot[{9 Cos[t] - Cos[9 t],
                    9 Sin[t] - Sin[9 t]},
          {t, 0, 2Pi}, AspectRatio -> Automatic]

ParametricPlot[{
          {9 Cos[t] - Cos[9 t], 9 Sin[t] - Sin[9 t]},
          {7 Cos[t] + Cos[7 t], 7 Sin[t] - Sin[7 t]}},
     {t, 0, 2Pi}, AspectRatio -> Automatic]

ParametricPlot[{8 Cos[t] + 2 Cos[4 t],
                    8 Sin[t] - 2 Sin[4 t]},
          {t, 0, 2Pi}, AspectRatio -> Automatic]

ParametricPlot[{8 Cos[t] + 2 Cos[4 t],
                    8 Sin[t] - 2 Sin[4 t]},
          {t, 0, 2Pi}, AspectRatio -> Automatic]

ParametricPlot[{Cos[t] + 5 Cos[3 t],
                    6 Cos[t] - 5 Sin[3 t]},
          {t, 0, 2Pi}, AspectRatio -> Automatic]

Do[
     ParametricPlot[
          {Cos[t] + 5 Cos[n t], 6 Cos[t] - 5 Sin[n t]},
          {t, 0, 2Pi}, AspectRatio -> Automatic,
          PlotPoints -> 50,
          PlotRange -> {{-7, 7}, {-12, 12}}],
     {n, 3, 4, 0.1}
]
```

Plotting Implicit Equations

Jerry: People are always asking if *Mathematica* can plot implicit equations. One thing we can always do is try to solve for one variable, y, and then plot.

Theo: This doesn't always work. Sometimes you can't solve the equation; even if you can, the graphs of the separate solutions don't always meet up nicely.

Jerry: Our visiting professor Dan Grayson suggested using **`ContourPlot`** as an implicit equation plotter.

Theo: This sounds shocking. How did Dan say to do that?

Jerry: He said to subtract the right hand side of the equation from the left hand side, and then do a contour plot of the resulting expression. He also said to restrict the plot range (z value) to go from 0 to 0.

Theo: OK, let's try it on the circle **`x^2 + y^2 == 1`**. We rewrite this as **`x^2 + y^2 - 1`**, and do a contour plot of it. The option **`Contours -> {0}`** tells **`ContourPlot`** to draw only one contour at a height of zero. The **`ContourShading`** option makes **`ContourPlot`** draw only contour lines, not shade the spaces between, and the **`ContourSmoothing`** option makes the contour lines come out smoother.

```
ContourPlot[x^2 + y^2 - 1, {x, -2, 2}, {y, -2, 2},
    Contours -> {0}, ContourShading -> False,
    ContourSmoothing -> 4];
```

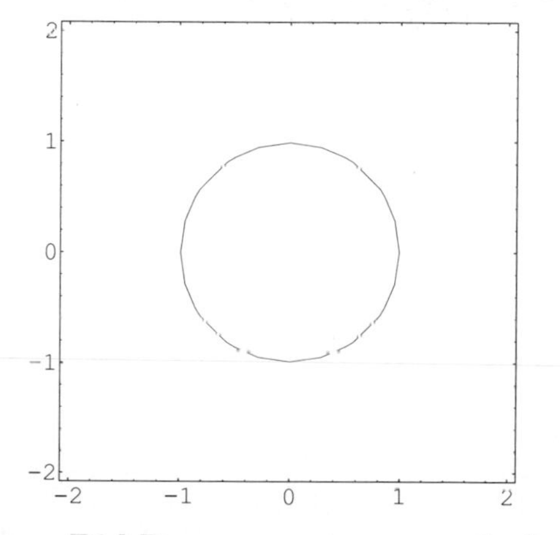

Amazing. Did Dan suggest any particular equations to plot?

Jerry: He said x^3 - x^2 = y^2 - y would be an interesting plot:

```
ContourPlot[(x^3 - x^2) - (y^2 - y),
    {x, -2, 2}, {y, -2, 2},
    Contours -> {0}, ContourShading -> False,
    ContourSmoothing -> 4];
```

Theo: It looks like it was drawn by a 6-year-old with a crayon who knows how to plot implicit equations. We can simulate a somewhat older child by increasing the number of sample points, using the **PlotPoints** option (same as for **Plot** and **Plot3D**). In the case of **ContourPlot**, increasing the number of sample points does not directly change the number of lines you see in the plot; it just makes **ContourPlot** start refining its lines from a finer grid (the default is **PlotPoints -> 15**).

```
ContourPlot[(x^3 - x^2) - (y^2 - y),
    {x, -2, 2}, {y, -2, 2},
    Contours -> {0},
    PlotPoints -> 30, ContourShading -> False,
    ContourSmoothing -> 4];
```

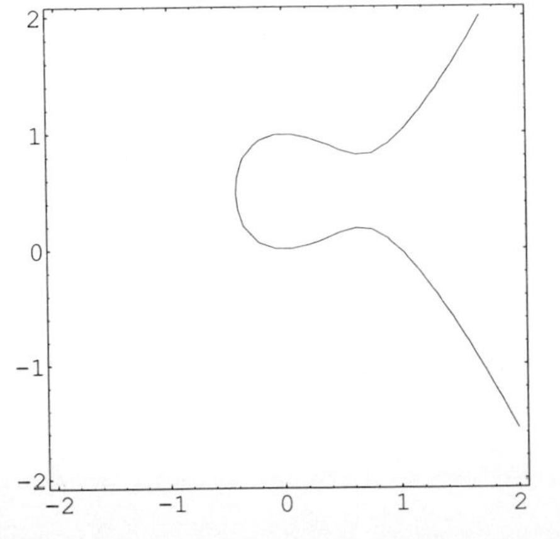

Jerry: Is it correct to think of this as a slice out of a three-dimensional graph of z = (x^3 - x^2) - (y^2 - y) at the level z = 0?

Theo: Yes. **ContourPlot** always plots horizontal slices out of three-dimensional graphs. Normally, it plots about 10 slices at equally spaced heights, giving the standard contour plots people are used to. At Dan's suggestion, we told it to plot only a single contour at zero, which is a clever way of telling it to plot only where the equation is true (i.e. where the left hand side minus the right hand side is zero).

Jerry: Can we make a function to hide the details of using **ContourPlot**?

Theo: The following function lets you plot implicit equations automatically:

```
ImplicitPlot[lhs_ == rhs_,
                  xRange_, yRange_, options___] :=
    ContourPlot[lhs - rhs, xRange, yRange,
        options,
        Contours -> {0},
        PlotPoints -> 30,
        ContourShading -> False,
        ContourSmoothing -> 4]
```

You can enter your equation without having to subtract the left hand side from the right hand side. Here is an example (note that you have to use *Mathematica*'s double equal sign, just as you do when using **Solve** to solve equations):

```
ImplicitPlot[x^3 - x == y^4 - y^2,
                  {x, -3, 3}, {y, -3, 3}];
```

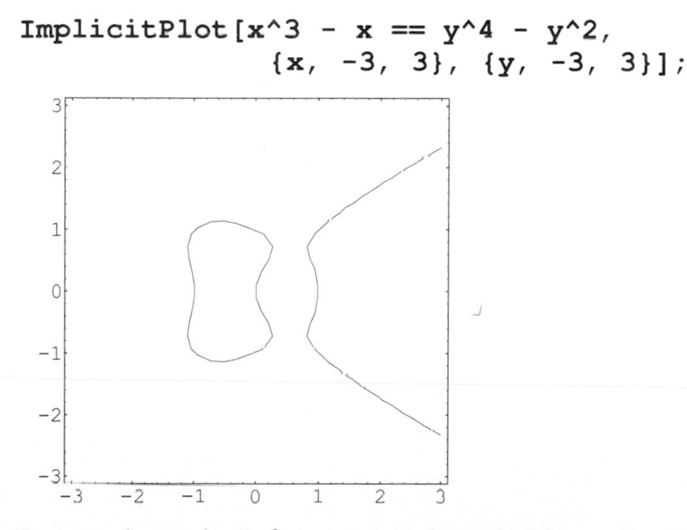

Jerry: As usual we don't have to understand how the function **ImplicitPlot** is defined; we can just use it like any other function.

Can we do an animation now? I would like to see what happens if we add a small constant to one side of the equation Dan suggested.

Theo: You can make an animation out of any *Mathematica* plotting command, including those you define yourself. We'll use **n** as the variable we vary during the animation, like this:

```
Do[
      ImplicitPlot[
          x^3 - x^2 == y^2 - y + n,
          {x, -1, 2}, {y, -1, 2}
      ],
      {n, 0, 0.3, 0.03}
]
```

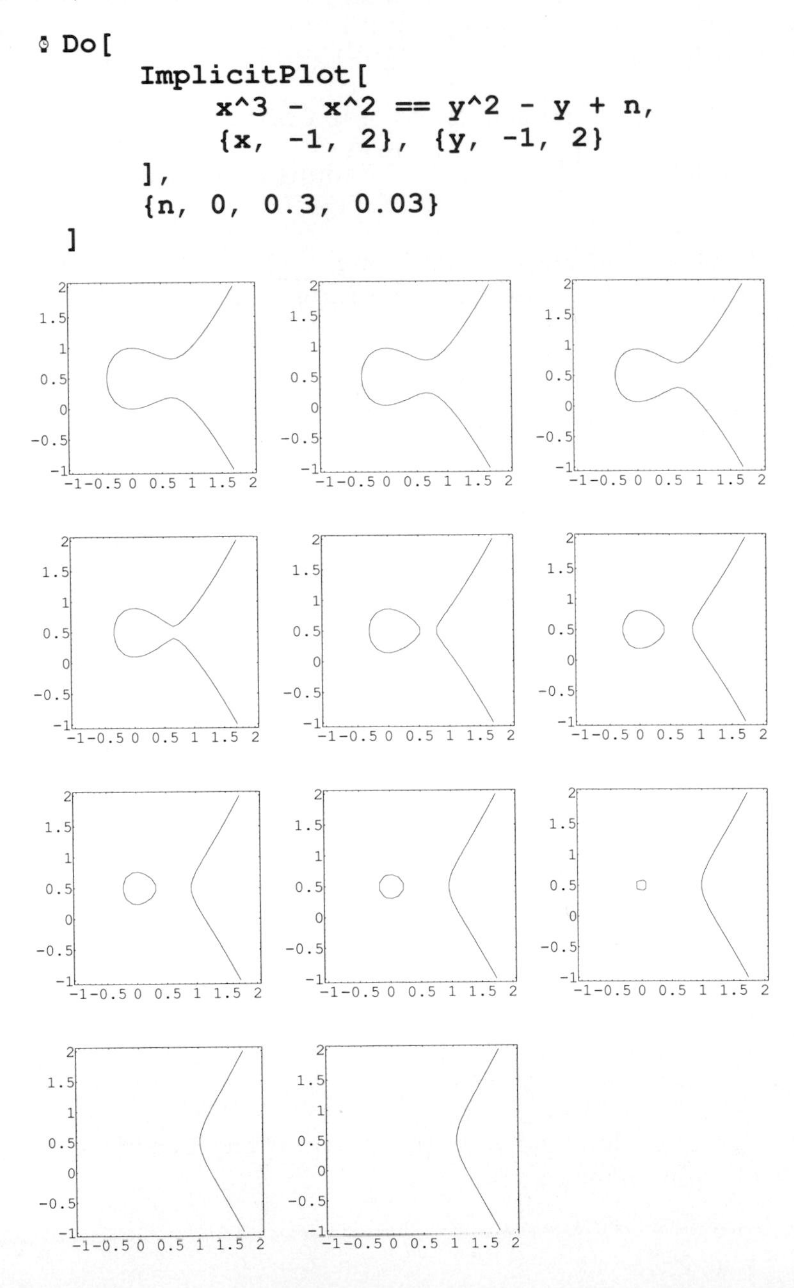

Jerry: Here are some more examples people can try:

```
ImplicitPlot[x^(4/3) + y^(1/3) == 1,
        {x, -2, 2}, {y, -2, 2}]
ImplicitPlot[Sin[2 x] + Cos[3 y] == 1,
        {x, -2Pi, 2Pi}, {y, -2Pi, 2Pi}]
ImplicitPlot[x^2 + x y + y^2 == 1,
        {x, -2Pi, 2Pi}, {y, -2Pi, 2Pi}]
ImplicitPlot[x^3 + x y + y^2 == 1,
        {x, -2Pi, 2Pi}, {y, -2Pi, 2Pi}]
```

■ Plotting Points and Fitting Curves

Theo: *Mathematica* provides several ways to plot points. The simplest is to use the **ListPlot** function, which plots a list of points:

```
ListPlot[{{-2, -3}, {1, 4}}];
```


Jerry: How do we make the dots bigger?

Theo: Using the **PlotStyle** option:

```
ListPlot[{{-2, -3}, {1, 4}},
    PlotStyle -> PointSize[0.04]];
```

Jerry: Excellent. Now, will *Mathematica* give me a linear expression whose graph goes through these two points?

Theo: We can use either **`Solve`** to solve two linear equations of the form **`a x + b == y`**, or **`Fit`** to do a least-squares fit:

```
Solve[{-2 a + b == -3, a + b == 4}, {a, b}]
```

$$\{\{a \to \frac{7}{3},\ b \to \frac{5}{3}\}\}$$

```
f = Fit[{{-2, -3}, {1, 4}}, {1, x}, x]
1.66667 + 2.33333 x
```

Theo: **`Fit`** starts with a sum of terms, each of which is an arbitrary constant multiplied by some function of x. Its job is to find values for all the arbitrary constants that make the sum follow the list of points as closely as possible. You have to give **`Fit`** the list of points to fit to (first argument), a list of the functions of x you want in the sum (second argument), and the name of the variable you want to use as your x variable (last argument). You don't have to include the arbitrary constants anywhere; **`Fit`** just makes them up. For example, if you have **`{1, x}`** as your second argument, it means to fit to an expression of the form **`a + b x`**. (**`Fit`** always gives an approximate answer.)

Jerry: And to graph this on top of the points?

```
Plot[f, {x, -3, 2},
    Epilog -> {PointSize[0.04],
              Point /@ {{-2, -3}, {1, 4}}}];
```

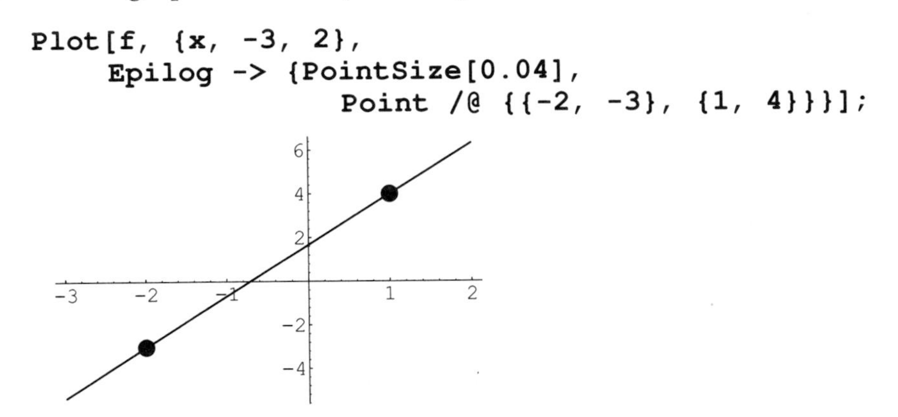

Now I'd like to do the same thing with three points, and fit a line and a parabola through these points.

Theo: First we'll assign the list of points you want to a variable:

```
thePoints = {{-1, 1}, {2, 3}, {4, -1}};
```

Now we'll fit a line and a parabola to these points:

```
theLine = Fit[thePoints, {1, x}, x]
1.52632 - 0.315789 x
```

```
theParabola = Fit[thePoints, {1, x, x^2}, x]
                             2
2.73333 + 1.2 x - 0.533333 x
```

We can plot these two curves together with the points:

```
Plot[{theLine, theParabola}, {x, -2, 5},
    Epilog -> {PointSize[0.04], Point /@ thePoints}];
```

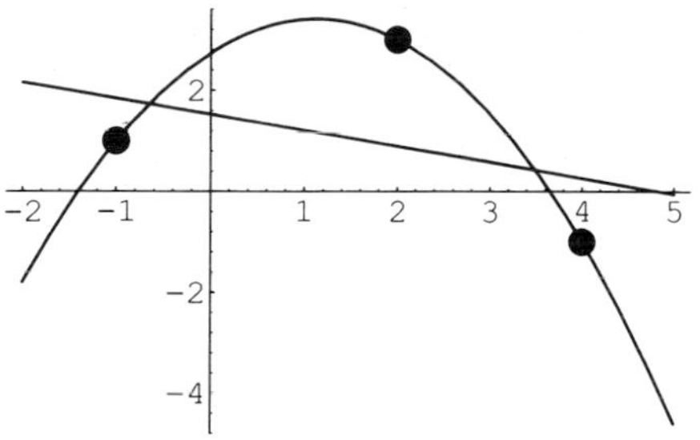

Jerry: Very nice.

Theo: We can also fit curves to lots of points. One way to get lots of points is to use the mouse point-selecting feature. Click the mouse on a graph as you did above to find the intersection (you will see a selection box around the graph). Hold down the Command key (on Macintosh and NeXT) or the Alt key (on MS-DOS), and click the mouse on the graph. You can either drag the mouse while holding down the button, or click individual points. After you've selected some points, use the Copy command (Edit menu) to copy the points onto the clipboard. Then you can paste them into a text cell, such as the following: (note that I typed the **`thePoints =`** part myself)

```
thePoints = {{0.0731502, 2.45363}, {0.091966, 2.16947},
{0.110782, 1.8853}, {0.148413, 1.60114},
{0.186045, 1.31697}, {0.242492, 1.03281},
{0.298939, 0.748644}, {0.374202, 0.606562},
{0.449465, 0.464479}, {0.524728, 0.393438},
{0.599991, 0.393438}, {0.656439, 0.677603},
{0.69407, 0.961767}, {0.712886, 1.24593},
{0.769333, 1.5301}, {0.844596, 1.67218},
{0.919859, 1.8853}, {0.957491, 2.16947}};
```

Theo: We can look at these points using **ListPlot**. We assign the result of the **ListPlot** command to a variable, **plotOfThePoints**, so that we can later combine this plot with another one.

```
plotOfThePoints = ListPlot[thePoints];
```

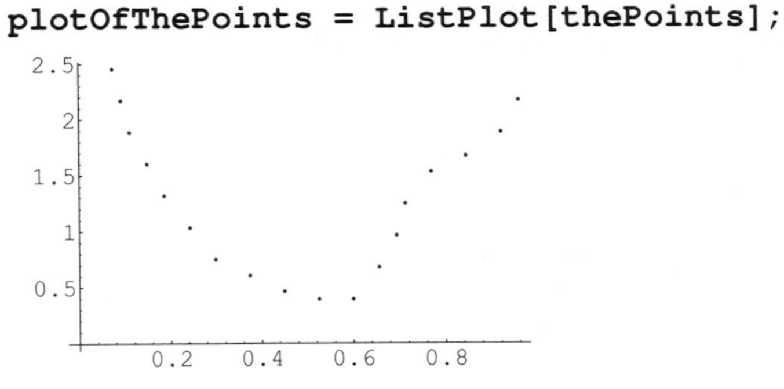

Theo: Now we can fit a function to the points. Let's try a quartic expression; the allowable terms are **1**, **x**, **x^2**, **x^3**, and **x^4**:

```
fitFunction = Fit[thePoints, {1, x, x^2, x^3, x^4}, x]
```

$$3.05802 - 10.2872\ x + 3.83777\ x^2 + 18.8237\ x^3 - 13.2533\ x^4$$

Jerry: Can we see a plot of this together with the points?

Theo: We can first make a plot of the fit expression, and then combine it with the plot of the points, like this:

```
plotOfTheFitFunction = Plot[fitFunction, {x, 0, 1}];
```

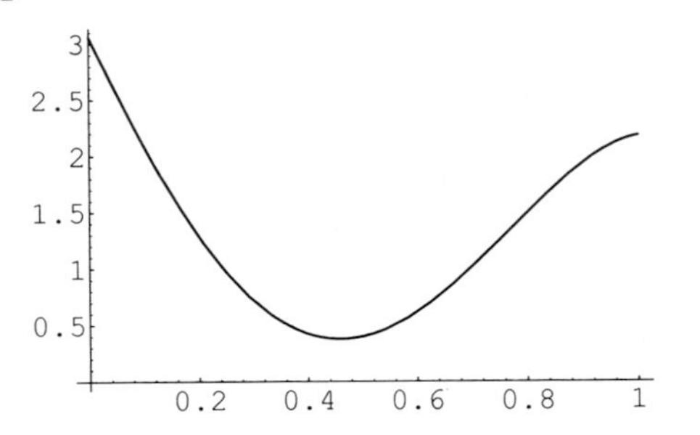

Theo: The **Show** function can be used to combine several plots together. Our two plots are named **plotOfThePoints**, and **plotOfTheFitFunction**, so we use these two names in the **Show** function:

```
Show[plotOfThePoints, plotOfTheFitFunction];
```

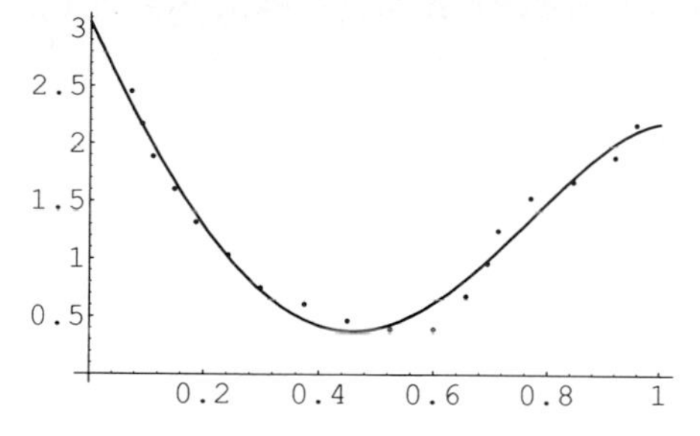

Jerry: This is the quartic curve of best fit?

Theo: Least-squares best fit, to be precise.

Jerry: Which means that the sum of the squares of the vertical distances from the points to the curve is minimized. If we need a better fit, we could try a higher order polynomial.

■ Plotting Roots of Equations

Theo: In another chapter (see Chapter 17) we developed a function for plotting roots of polynomials. Here is a simplified version of it that works for more different kinds of equations:

```
RootPlot[expr_, x_] :=
    Show[
        Graphics[{
            PointSize[0.04],
            Map[
                Point[{Re[#], Im[#]}]&,
                 x /. N[Solve[expr == 0, x]]
            ]
        }],
        AspectRatio -> Automatic,
        Axes -> Automatic,
        AxesLabel -> {"Re", "Im"}
    ]
```

Jerry: I assume we don't have to understand how this function works to use it. It looks as if we should give it two arguments -- a polynomial and then a variable name. Let's try one:

```
RootPlot[x^4 - x^2 + x + 1, x];
```


Let's make up a polynomial with known roots, and see if **RootPlot** plots them in the right place:

```
f = Expand[(x - 1) (x - 2) (x - 3) (x - 4)]
```

$24 - 50\ x + 35\ x^2 - 10\ x^3 + x^4$

```
RootPlot[f, x];
```

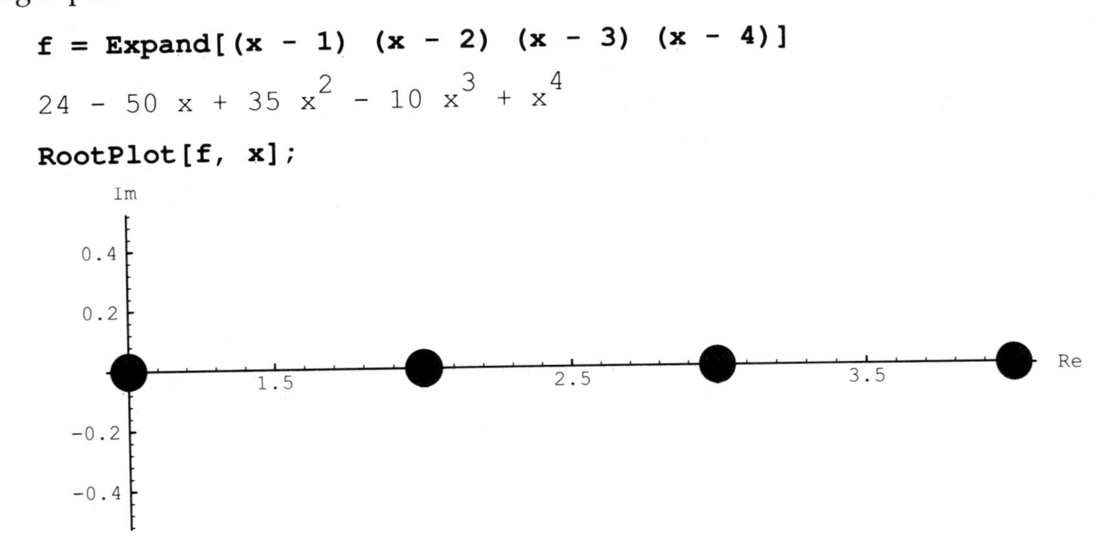

Theo: Let's look at a graph of the expression as well:

```
Plot[f, {x, -1, 5}];
```

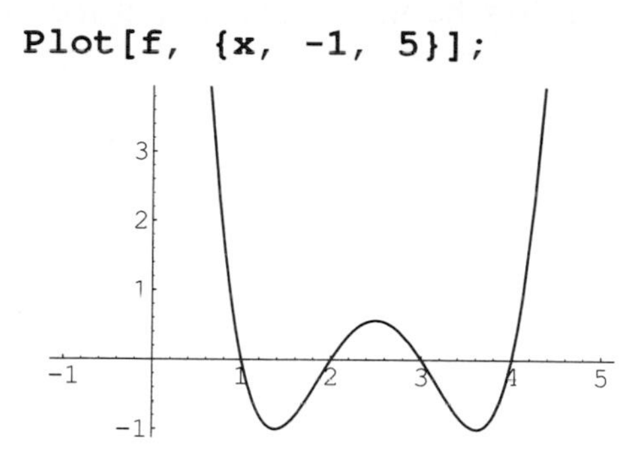

Jerry: It's easy to see that the dots in the `RootPlot` are in the same places as the intersections of the curve and the x-axis. Very nice.

■ Defining and Using Functions

Theo: What do high school math teachers mean by "defining functions"? Is it anything like what the phrase means in *Mathematica*?

Jerry: Probably. How do we define a function in *Mathematica* and work with it?

Theo: We use _ and :=. The underbar character is used on the left hand side to indicate the arguments of the function, and the := is used instead of =. For example, here is a very simple function, which in conventional notation would be defined as f(x) = x^2:

```
f[x_] := x^2
```

Jerry: And I suppose g(x) = sin(x) would be:

```
g[x_] := Sin[x]
```

Theo: A function of 2 variables is defined like this:

```
h[x_, y_] := Sqrt[x^2 + y^2]
```

Jerry: And to evaluate a function, I assume I can just use it as I would a built-in function, so f(3) would be:

```
f[3]
9
```

Theo: Very good. There are a number of different kinds of functions you can define in *Mathematica*. We explain these in more detail in Chapters 1 and 2, but for now we'll work with the examples above.

Of course the names of functions don't have to be just a single letter. Often one uses function names that are whole words or even several words, although one can't put spaces between the words.

Jerry: Can I take these functions and plot and differentiate and multiply by seven and solve and all the rest?

Theo: Sure, let's do all these things to the **g** function we just defined:

```
Plot[g[x], {x, 0, 2Pi}];
```

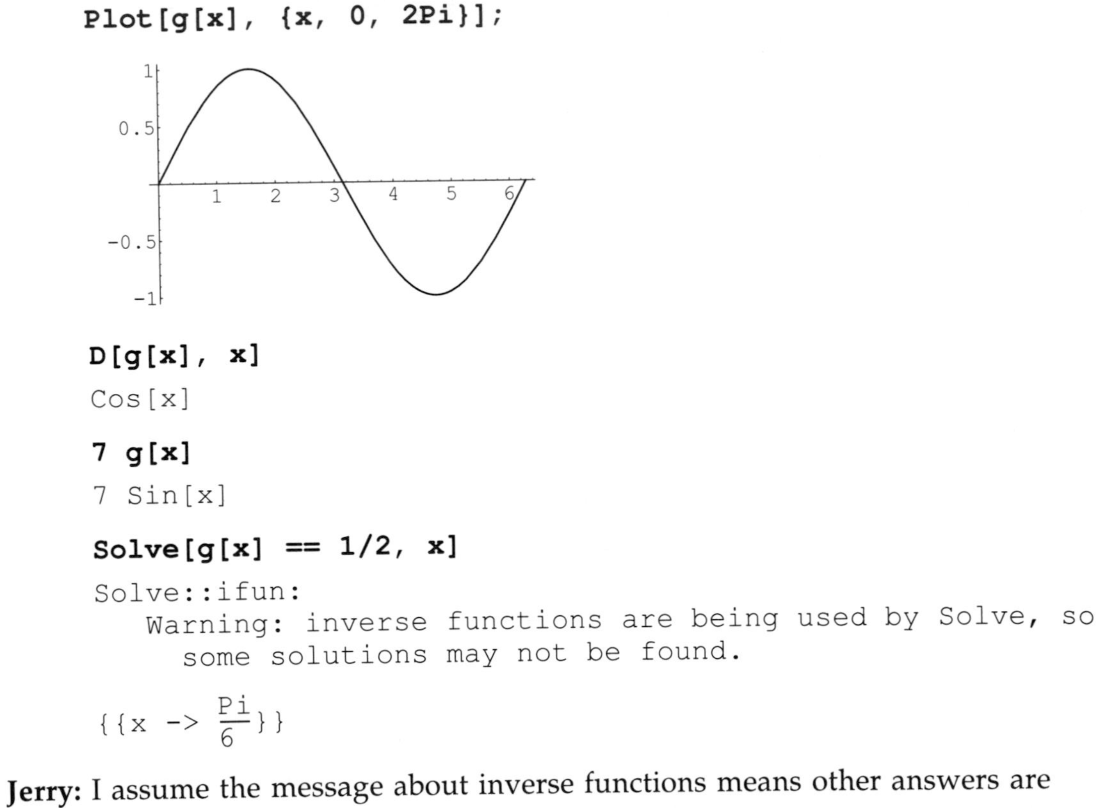


```
D[g[x], x]
```

```
Cos[x]
```

```
7 g[x]
```

```
7 Sin[x]
```

```
Solve[g[x] == 1/2, x]
```

```
Solve::ifun:
   Warning: inverse functions are being used by Solve, so
     some solutions may not be found.

{{x -> Pi/6}}
```

Jerry: I assume the message about inverse functions means other answers are possible?

Theo: Right, *Mathematica* is simply warning us that there may be other answers, in this case an infinite number of them.

Jerry: Teachers like to plot things like f(x), f(x+1) and f(g(x)). Can we do that?

Theo: Yes, and we can even do them all on the same plot. (If you give **`Plot`** a list of expressions as its first argument (indicated by the curly brackets), it plots them all on the same axes.)

```
Plot[{f[x], f[x + 1], f[g[x]]}, {x, -4, 4}];
```

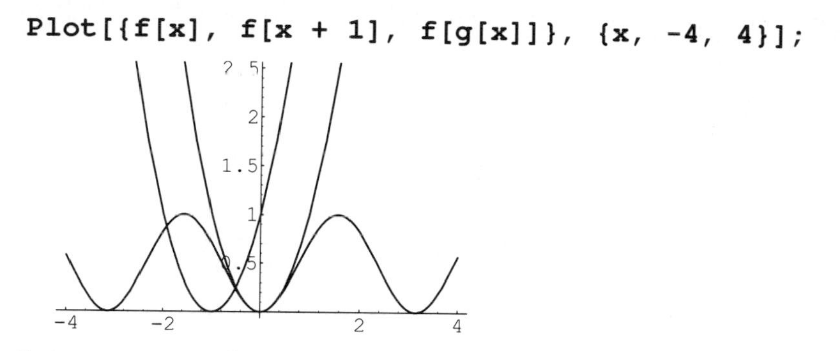

Jerry: Is it easy to evaluate expressions involving several different uses of f(x)?

Theo: Of course:

```
f[a] + f[a+1] + 4 f[5] + f[f[b]]
```

```
            2           2    4
100 + a  + (1 + a)  + b
```

Jerry: How about a table of x, f(x) and g(x), for some values of x?

Theo: Sure. We use the **`Table`** command to make the table, and the **`TableForm`** command to display it in a table-like format. (You can tack a **`//TableForm`** after any command that produces a list (table) of values.)

```
Table[{x, f[x], g[x]}, {x, 0.0, 10.0}]//TableForm
```

```
0.     0.      0.
1.     1.      0.841471
2.     4.      0.909297
3.     9.      0.14112
4.     16.     -0.756802
5.     25.     -0.958924
6.     36.     -0.279415
7.     49.     0.656987
8.     64.     0.989358
9.     81.     0.412118
10.    100.    -0.544021
```

Jerry: Excellent.

Doing Algebra

Theo: What do high school math teachers mean by "doing algebra"? I bet it's completely different from what other people mean.

Jerry: Some day it may be very different, but for now it isn't. Things like 2x + 3x should come out to be 5x, and (a + b)^2 should come out to be a^2 + 2ab + b^2 (and vice-versa).

Theo: Interestingly, *Mathematica* will do the first of these things automatically as soon as you evaluate **`2x + 3x`**, but to do the other two you need to use two functions, **`Expand`** and **`Factor`**. It couldn't really do both of these automatically, since they are opposites of each other. Let's try all three of these things:

```
2x + 3x
5 x
Expand[(a + b)^2]
 2           2
a  + 2 a b + b
Factor[a^2 + 2 a b + b^2]
         2
(a + b)
```

Jerry: Wait a minute, I saw that. You put a space between **`a`** and **`b`** in the last expression. Normally people just write ab.

Theo: That depends on what you mean by "normally". If you are using a language in which "ab" means a times b, you can't at the same time have any variables whose names are more than one letter long. This is a very bad thing, and the alternatives are even worse. If you want to mean multiplication, you've got to use a space.

Jerry: Since I work with lots of young people, I would argue for two modes, in one of which ab means a times b, and in the other it means the variable named "ab".

Theo: Where do these young people get the idea that they can write ab for a times b in the first place? That sounds like a bad habit that should be discouraged right from the start.

Jerry: They learn it from their culture: parents, teachers, aunts, uncles, and me.

Theo: So, you're saying that whatever they learned before, however bad and wrong, we should try to reinforce? I would have a problem with that attitude.

Jerry: If I understand your attitude, you want to introduce only those ideas that you feel are logically correct. You feel that the conflict with the existing culture is none of your concern.

Theo: I don't mind a certain amount of adaptation to reality, and *Mathematica* certainly has a number of logically questionable elements for that reason. But something as fundamental as the difference between a name and an operation should be taught correctly or not at all.

Jerry: Well, I suppose I can't argue, since in *Mathematica* I don't have any choice anyway.

Moving right along: Other situations that secondary school teachers have to deal with include simplification of the following kinds of things:

12!/10!
a/b + c/d
(x^2 - 1)/(x - 1)
(x^2 - 7x + 12)/(x-1)
sin(2x)
sin(pi/2)
sin(pi/8)
sin(a+b)
sqrt(8)
sqrt(x^3y^6)
ln(x^2).

Theo: Well, "simplify" is a difficult word. It's rarely clear which of several possible forms of an expression is "simpler". In *Mathematica* there are three different ways of thinking about and carrying out simplification: automatic, automatic using **`Simplify`**, and manual. Some things just happen automatically and give you what you might expect, as with the following of your examples:

```
12!/10!
132

Sqrt[8]
2 Sqrt[2]

Sin[Pi/4]
Sqrt[2]
-------
   2
```

```
Sqrt[x^3 y^6]

       3  6
Sqrt[x  y ]
```

Jerry: Some people want x sqrt(x) y^3 for the last one, but I assume that would be hard to deliver.

Theo: It could be done, but it would not be easy.

For some of your other examples, just evaluating them will not do anything at all, but the built-in *Mathematica* function **Simplify** will get you the result you expect:

```
(x^2 - 1)/(x - 1)

        2
-1 + x
-------
-1 + x
```

but

```
Simplify[(x^2 - 1)/(x - 1)]

1 + x
```

Theo: For others, even **Simplify** does nothing, because as far as *Mathematica* is concerned the expression is already as simple as it gets. In those cases you have to use a specific function to transform the expression into what you want. For example, to transform a/b + c/d into a single fraction, you have to use the **Together** function:

```
Together[a/b + c/d]

b c + a d
---------
   b d

Apart[(x^3 - 7 x + 12) / (x - 1)]

         6                2
-6 + ------- + x + x
     -1 + x
```

Jerry: This last one looks like the result of polynomial long division.

Theo: It is not at all obvious which of these forms is "simpler", so *Mathematica* does not do these transformations automatically.

Jerry: Many words that we use everyday in the math classroom, such as "simplify", are ill-defined. Somehow students are supposed to "pick up" what they mean. If you look in the back of calculus books at the answers, you quickly learn that "simplify" means different things to different people.

Theo: Two more of your examples can be handled by adding the option **`Trig -> True`** to the **`Factor`** function. This tells *Mathematica* to rearrange trig functions as well as polynomials:

```
Factor[Sin[2 x], Trig -> True]
2 Cos[x] Sin[x]
```

```
Factor[Sin[a + b], Trig -> True]
Sin[a + b]
```

Jerry: Why **`Cos Sin`** instead of **`Sin Cos`**, the order I expected?

Theo: It's alphabetical.

Your last example, ln(x^2), isn't dealt with even in an optional package. We have to write our own rules to deal with the general problem:

```
Unprotect[Log];
Log[x_ y_] := Log[x] + Log[y];
Log[x_^n_] := n Log[x];
Protect[Log];
```

Jerry: Of course, for x = -2 and y = -3, this rule is not true. I suppose we could make the rule more selective, to avoid problems like this.

Theo: Yes, one would have to add conditions to the rules to make them more correct. We won't get into that here. After adding these rules, we can just evaluate the example:

```
Log[x^2]
2 Log[x]
```

Jerry: I notice you use log instead of ln. Is that the natural logarithm?

Theo: **`Log[x]`** means log base *e* of **`x`**. **`Log[b, x]`** means log base **`b`** of **`x`**. Since **`Log`** can be used with any base, whereas ln means specifically natural log, we had to use the general term log.

Jerry: So **`Log[10, 100]`** is 2?

```
Log[10, 100]
2
```

Theo: And finally, sin(pi/8) doesn't transform into anything at all, except a floating point number, if you ask for it with the **`N`** function:

```
N[Sin[Pi/8]]
0.382683
```

Solving Equations

Jerry: Let's start with 2x + 3 = 11, solving for x.

Theo: The **Solve** function solves equations. In *Mathematica* you have to use **==** in equations (= by itself means assignment):

```
Solve[2x + 3 == 11, x]
{{x -> 4}}
```

Jerry: How do we substitute into the original equations the value of 4 for x?

Theo: We use the substitute command, usually abbreviated as "**/.**":

```
2x + 3 /. x -> 4
11
```

Jerry: I propose we try solving the following equations; each one is "harder" than the one before:

$$\begin{aligned} &2x + 3 = 11 \\ &2x + 3 = c \\ &2x + b = c \\ &ax + b = c. \end{aligned}$$

Here we go!

```
Solve[2 x + 3 == 11, x]
{{x -> 4}}

Solve[2 x + 3 == c, x]
          -(3 - c)
{{x -> ---------}}
             2

Solve[2 x + b == c, x]
          -(b - c)
{{x -> ---------}}
             2

Solve[a x + b == c, x]
           b - c
{{x -> -(-----)}}
             a
```

Jerry: Now, how about some quadratic equations?

```
Solve[x^2 - 5 x + 6 == 0, x]
{{x -> 3}, {x -> 2}}
Solve[x^2 - 5 x + 5 == 0, x]
         5 + Sqrt[5]           5 - Sqrt[5]
{{x -> -----------}, {x -> -----------}}
              2                     2
Solve[x^2 - 5 x + 7 == 0, x]
         5 + Sqrt[-3]           5 - Sqrt[-3]
{{x -> ------------}, {x -> ------------}}
              2                      2
```

Jerry: Can I see the results in decimal form?

Theo: In general, anytime you want a decimal result, use the **N** command.

```
N[Solve[x^2 - 5 x + 7 == 0, x]]
{{x -> 2.5 + 0.866025 I}, {x -> 2.5 - 0.866025 I}}
```

Mathematica uses capital `I` for *i*, because of the rule that all built-in *Mathematica* symbols start with a capital letter. (See Chapter 23 for more information about complex numbers.)

Jerry: Suppose I want to substitute the solutions back into the original equation, without having to retype anything (as we had to do before).

Theo: So that we can use the solutions later, I'm going to assign them to a variable named **solutions** (remember that *Mathematica* uses = for assignment):

```
solutions = Solve[x^2 - 5 x + 7 == 0, x]
         5 + Sqrt[-3]           5 - Sqrt[-3]
{{x -> ------------}, {x -> ------------}}
              2                      2
```

Now we can use the `/.` substitution command together with this variable to substitute into the left hand side of the original equation:

```
x^2 - 5 x + 7 /. solutions
                                         2
      5 (5 + Sqrt[-3])   (5 + Sqrt[-3])
{7 - ---------------- + ---------------,
             2                 4
                                         2
      5 (5 - Sqrt[-3])   (5 - Sqrt[-3])
 7 - ---------------- + ---------------}
             2                 4
```

Notice that we got a list of two expressions. That's because there were two solutions to the equation: *Mathematica* gave us two expressions, one for each of the solutions.

Of course, we would expect both of them to be zero. We can use the **Simplify** function to see if these expressions are just zero in disguise (% means the last result):

```
Simplify[%]
{0, 0}
```

Jerry: Suppose I want to have solutions as a simple list of numbers, without those peculiar **->**'s?

Theo: We can substitute **solutions** in the expression "**x**":

```
x /. solutions
 5 + Sqrt[-3]   5 - Sqrt[-3]
{------------, ------------}
      2              2
```

Jerry: Can we solve simultaneous equations, too? How do we ask to solve for two variables at the same time?

Theo: The first argument to **Solve** can be a list of simultaneous equations, and the second argument can be a list of variables to be solved for:

```
Solve[{(x - a)(x - b)(x - c) == y, d x + e == y},
      {x, y}]
```

Theo: This brings up an interesting point: Sometimes an answer is the last thing in the world you want. This answer is so huge we had to reduce it to 2-point type just to fit it in (don't worry, you're not expected to read it, just marvel at the size of it).

Let's try removing the **(x - c)** term, to make things a bit simpler. Also, we don't really want to solve for **y**, we want to eliminate it and then solve for **x**. We can give **Solve** a third argument specifying which variables to eliminate, instead of solve for. With these adjustments, we get a much more reasonable result:

```
Solve[{(x - a)(x - b) == y, d x + e == y}, x, y]
```

$$\{\{x \to \frac{a + b + d + \mathrm{Sqrt}[(-a - b - d)^2 - 4\ (a\ b - e)]}{2}\},$$

$$\{x \to \frac{a + b + d - \mathrm{Sqrt}[(-a - b - d)^2 - 4\ (a\ b - e)]}{2}\}\}$$

Jerry: Can we eliminate **y**, without solving for **x**?

Theo: The **Eliminate** function does exactly this:

```
Eliminate[{(x - a)(x - b) == y, d x + e == y}, y]
```

$$e == a\ b - a\ x - b\ x - d\ x + x^2$$

Jerry: And I remember that in the section on plotting functions, above, we used **Solve** to find the intersection of several graphs, such as this example of a parabola and a straight line:

```
Solve[{x^2 - y == 0, 2 x + y == 5}, {x, y}]
```

```
         -2 - 2 Sqrt[6]        14 + 4 Sqrt[6]
{{x -> --------------, y -> --------------},
              2                     2

         -2 + 2 Sqrt[6]        14 - 4 Sqrt[6]
  {x -> --------------, y -> --------------}}
              2                     2
```

```
N[%]
```

```
{{x -> -3.44949, y -> 11.899},
  {x -> 1.44949, y -> 2.10102}}
```

Mathematica seems to be very good at spitting back these answers, but some people what to know how it did it. Can you ask *Mathematica* how it arrives at these answers?

Theo: Fortunately not. That's sort of like asking a pocket calculator how it calculates square roots. There is a procedure that people used to be taught for calculating square roots. The procedure that the pocket calculator uses is *nothing like* the one that people used to use. It's bizarre, complicated, and involves mathematics that only a few people in Japanese calculator factories understand. Likewise, *Mathematica* uses bizarre, complex techniques for solving equations. The number of people in the world for whom *Mathematica*'s Gröbner-basis algorithm would provide a meaningful explanation of what it did is extremely limited.

Jerry: I think what people are really asking is "How do I use *Mathematica* to teach the things people learn in high school math classes?".

Theo: There is nothing useful that a pocket calculator can tell a high school student about how it calculates square roots. How square roots are calculated is, in general, not something that is of any interest to high school students. They don't have to know, precisely because of pocket calculators. With *Mathematica*, it is no longer important for high school students to know how to solve complicated equations either. The human-labor-oriented algorithms worked out in the past are of little interest anymore.

Jerry: If this is true, then new topics might be introduced into high school math in the space made available by this new technology.

Theo: Absolutely. Why do we know how to build space ships? Simple, it's because most of us *don't* know how to hunt for food.

Jerry: Perhaps a better question yet is, how do we use *Mathematica* to gain insight into mathematics itself?

Theo: Keep reading this book.

Plotting Inequalities

Jerry: In more than a year of showing mathematics and plotting programs like *Mathematica*, Derive, and Omnifarious Plotter, to high school and community college teachers, one of the most common questions I have heard is, "Can it graph inequalities?". My answer has always been *No*. Is this a permanent condition?

Theo: Not for *Mathematica*. Anytime *Mathematica* appears to be unable to do something, the first question should always be whether it's possible to write a Notebook to make it do it. I've written the Notebook `InequalityPlot`, which implements graphing of inequalities in a quite general way. Writing the Notebook took a few hours and a bit of thought, and is not something the beginning user is likely to do. However, since it's included in this book, no one else has to worry about how it works. The following command will load it in, after which the function **`InequalityPlot`** works just like any built-in *Mathematica* function:

```
Needs["InequalityPlot`"]
```

Jerry: So you're saying I can graph an inequality using **`InequalityPlot`** instead of **`Plot`**, but with the same sort of arguments as **`Plot`** uses:

```
InequalityPlot[y <= x^2, {x, -3, 3}];
```

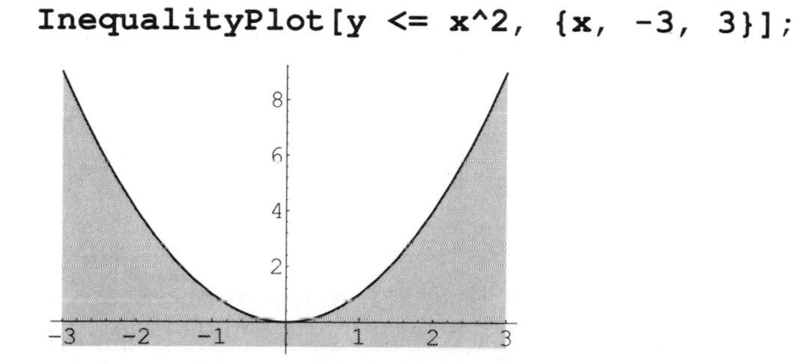

Jerry: Excellent! Does the plot look any different if we use < instead of <=?

```
InequalityPlot[y < x^2, {x, -3, 3}];
```

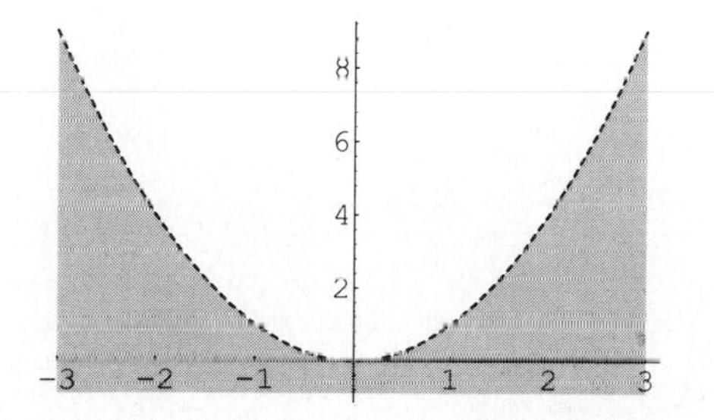

Theo: I decided that if the curve itself is included in the area (<= or >=), then it is drawn as a solid line, but if it is excluded (< or >) then it is drawn as a dashed line. (In the printed book these are easy to tell apart, but people reading this in the electronic version may have a hard time seeing the dashing, due to low screen resolution.)

Jerry: What if a curve is particularly wiggly, for example y <= x sin(1/x)?

```
InequalityPlot[y <= x Sin[1/x], {x, -0.3, 0.3}];
```

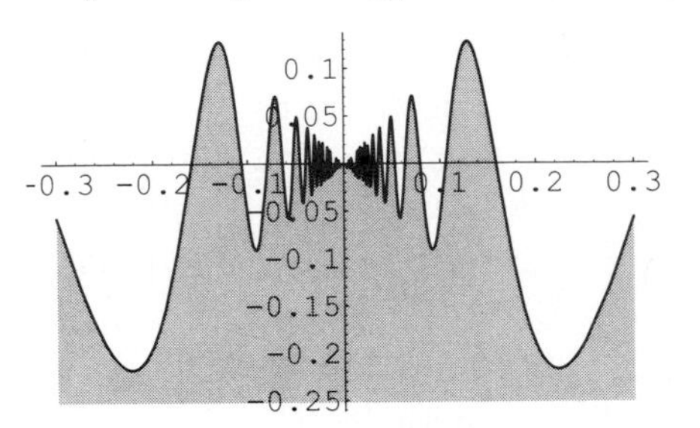

Lovely. What about two curves? Let me try sin(x) <= y <= cos(x):

```
InequalityPlot[Sin[x] <= y <= Cos[x], {x, -2Pi, 2Pi}];
```

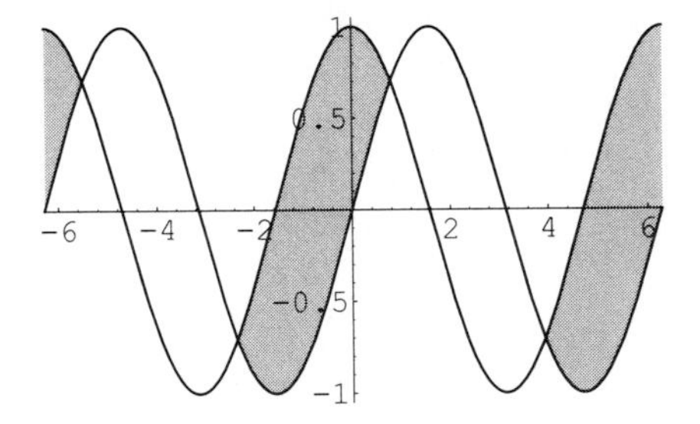

Theo: Fortunately I anticipated that people might want to make plots where y is bracketed between two curves. In fact, **InequalityPlot** is able to deal with all possible combinations of two curves. The following four plots are the four possible combinations of **<=** and **>=** and the curves **4 Sin[4 x]** and **x^2 - 1**:

```
InequalityPlot[4 Sin[4 x] <= y <= x^2 - 1,
               {x, -Pi, Pi}];
```

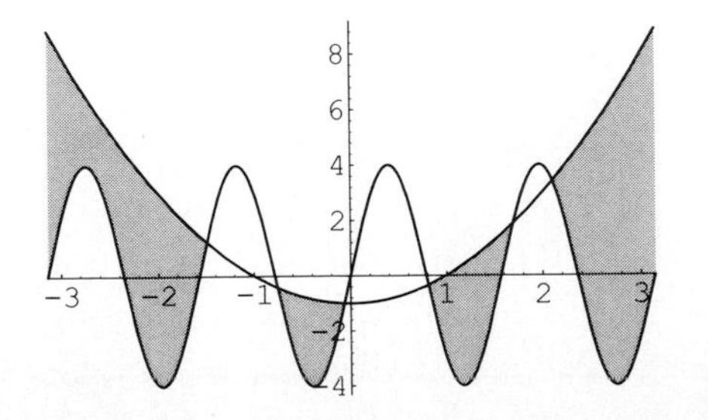

```
InequalityPlot[4 Sin[4 x] >= y >= x^2 - 1,
                   {x, -Pi, Pi}];
```

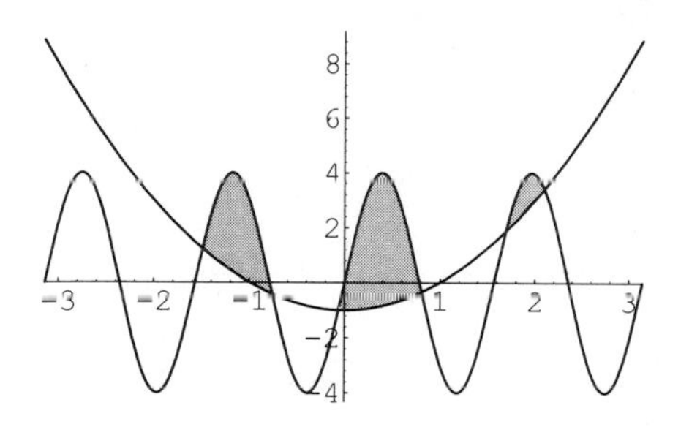


```
InequalityPlot[4 Sin[4 x] >= y <= x^2 - 1,
                   {x, -Pi, Pi}];
```

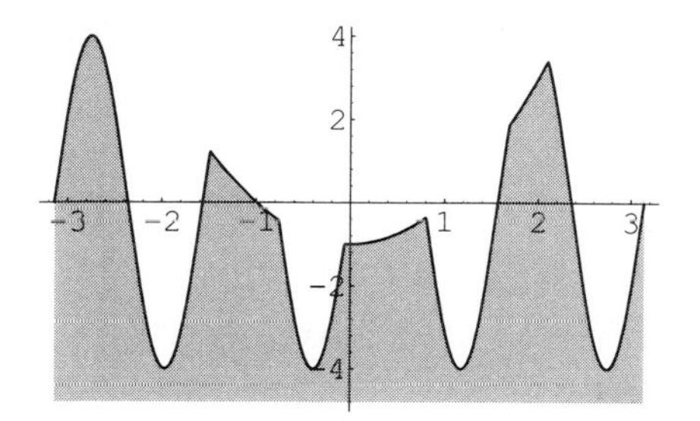


```
InequalityPlot[4 Sin[4 x] <= y >= x^2 - 1,
                   {x, -Pi, Pi}];
```

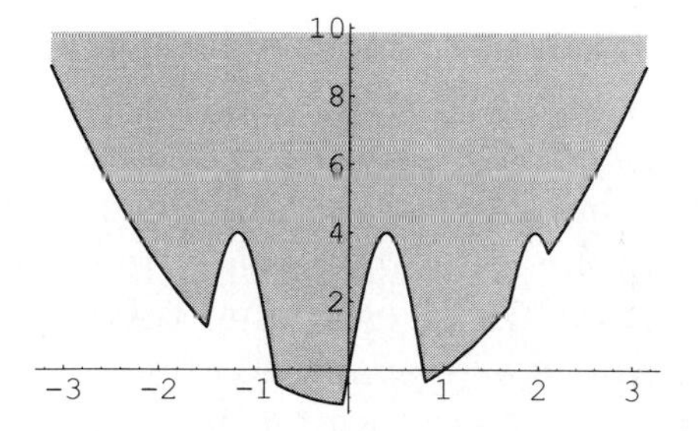

Theo: In the last two cases we are asking for y to be greater or less than both curves at the same time. (In other words, $a < y > b$ means $a < y$ *and* $y > b$.) This is the same as y being greater or less than the maximum or minimum of the two curves.

Jerry: I can think of an example where the solution is empty. Will it handle that case?

```
InequalityPlot[-2 > y > x^2, {x, -3, 3}];
```

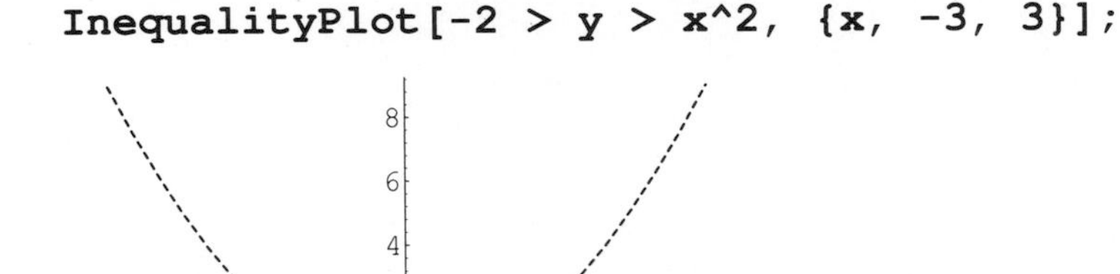

It looks fine.

Someone working with inequalities would probably want to have this function available all the time. If it's so useful, why isn't **`InequalityPlot`** built into *Mathematica*?

Theo: Only a very small fraction of *Mathematica* users ever want to plot inequalities. There are thousands of things like **`InequalityPlot`** that are tremendously useful to a few people, and of no interest to the rest. We can't build in everything for everyone: that's what Notebooks are for.

Jerry: A math teacher's chances of successfully using *Mathematica* will increase as the teacher either learns to program in *Mathematica* or gets to know people like you.

Theo: Or buys the electronic version of this book. A function defined in a Notebook is as good as a built-in function. To make **`InequalityPlot`** available, you could execute the command **`Needs["InequalityPlot`"]`** once each time you start up *Mathematica*. If you use **`InequalityPlot`** all the time, you could put this same command into the file called `init.m` (located in the `Packages` folder in your *Mathematica* directory). The `init.m` file is loaded automatically each time you start up *Mathematica*, so anything you put in there will always be available. (Note: If you use *Mathematica* on a network, or are using someone else's computer, you should probably ask permission before changing the `init.m` file.)

Chapter Twenty-Three
College Math

Theo and Jerry dabble in higher education.

■ Dialog

Theo: There are not many sections in this chapter, because many of the topics we could have put in here have already been discussed in other chapters.

Jerry: Some topics, such as surfaces of revolution, linear programming, convolutions, differential equations, statistics, etc., are not in here either. We haven't done enough yet with these topics to say anything sensible.

Theo: Other people have worked on these topics in connection with *Mathematica*. A good source of information about what people are doing is the *Mathematica Journal*, published quarterly by Addison-Wesley.

■ Complex Numbers

Theo: What do people need to know about complex numbers?

Jerry: How to type them in. What symbol does *Mathematica* use to represent the square root of -1?

Theo: The normal symbol to use is *i*. In *Mathematica* there is a strong rule that all built-in functions, commands, and constants start with a capital letter. So, the *Mathematica* symbol is **I**, as we can see:

```
Sqrt[-1]
I
```

Jerry: And I assume if we solve `x^2 == -9` we should get the usual suspects:

```
Solve[x^2 == -9, x]
{{x -> 3 I}, {x -> -3 I}}
```

Theo: We can add, multiply, divide, and otherwise manipulate complex numbers in a fairly straightforward way. Here are some examples:

```
(3 + 4I) + (7 - 2I)
10 + 2 I

(3 + 4I) (7 - 2I)
29 + 22 I

(3 + 4I) / (7 - 2I)
13    34 I
-- + ----
53    53

(3 + 4I) ^ (7 - 2I)
          7 - 2 I
(3 + 4 I)
```

Jerry: Why didn't *Mathematica* evaluate this last example?

Theo: Probably because it couldn't think of anything sensible to do with it. If we convert these numbers to floating point form using the **N** command (they are currently perfect rational numbers, as you can see from the division example), we might get a different answer:

```
N[(3 + 4I) ^ (7 - 2I)]
-494896. - 65002.5 I
```

Jerry: What about other mathematical functions? Can *Mathematica* evaluate such functions with complex numbers? Let's try a few:

```
Sin[3 + 4I]
Sin[3 + 4 I]
```

Theo: Again, evaluating this numerically will allow *Mathematica* to work out an answer (there is no better *exact* way of writing `Sin[3 + 4I]`):

```
N[Sin[3 + 4I]]
3.85374 - 27.0168 I
```

Jerry: Let me try some more examples. I'll just use **N** all the time:

```
N[ArcSin[2]]
1.5708 - 1.31696 I
N[Log[-1]]
3.14159 I
N[I^I]
0.20788
(1 + I)^16
256
```

People also like to write complex numbers in several different forms. One way of writing complex numbers is in polar coordinates, where a point is described by its distance from the origin and its angle measured from the positive real axis. Can *Mathematica* transform a complex number into this form?

Theo: Yes, but it's not a built-in function. The most natural way to represent a complex number in polar coordinates is as a list containing the magnitude and the phase angle. We can define a function to transform complex numbers into polar coordinates like this:

```
Attributes[ComplexToPolar] = {Listable};
ComplexToPolar[z_] := {Abs[z], Arg[z]};
```

Abs means absolute value, which is interpreted as magnitude for complex numbers; **Arg** means phase angle (an angle between **-Pi** and **Pi** radians). The attribute **Listable** makes this function apply itself automatically to each element of a list.

If we use this function on a real number, we get an angle of zero:

```
ComplexToPolar[2]
{2, 0}
```

Applied to a pure imaginary number, we get a right angle:

```
ComplexToPolar[2 I]
     Pi
{2, --}
     2
```

A mixed complex number gives us an angle in between:

```
ComplexToPolar[2 + 3 I]
{Sqrt[13], ArcTan[2, 3]}
```

As usual, *Mathematica* has refused to make approximations, so we don't see an actual angle. We can use **N** to get this in numerical form:

```
N[ComplexToPolar[2 + 3 I]]
{3.60555, 0.982794}
```

Jerry: Let's see if all this really works. Try solving **x^4 - 1 == 0** and then write the answer in polar form.

Theo: First we solve the equation:

```
Solve[x^4 - 1 == 0, x]
{{x -> 1}, {x -> I}, {x -> -1}, {x -> -I}}
```

Then we make a list of the roots:

```
x /. %
{1, I, -1, -I}
```

Now we can rewrite these numbers in polar form:

```
ComplexToPolar[%]
            Pi                   -Pi
{{1, 0}, {1, --}, {1, Pi}, {1, ---}}
            2                     2
```

Jerry: How about solving **x^6 - 1 == 0**?

```
Solve[x^6 - 1 == 0, x]
               I/3 Pi         (2 I)/3 Pi
{{x -> 1}, {x -> E      }, {x -> E          }, {x -> -1},
        (4 I)/3 Pi         (5 I)/3 Pi
  {x -> E          }, {x -> E          }}
```

I suppose **N** of that will give us more intelligible numbers:

```
N[%]
{{x -> 1.}, {x -> 0.5 + 0.866025 I},
  {x -> -0.5 + 0.866025 I}, {x -> -1.},
  {x -> -0.5 - 0.866025 I}, {x -> 0.5 - 0.866025 I}}
```

We can write these in polar form:

```
ComplexToPolar[x /. %]
{{1., 0}, {1., 1.0472}, {1., 2.0944}, {1., Pi},
  {1., -2.0944}, {1., -1.0472}}
```

Theo: There are several other functions worth mentioning.

`Re` gives the real part of a complex number:

```
Re[2 + 3I]
2
```

`Im` gives the imaginary part:

```
Im[2 + 3I]
3
```

`Conjugate` gives the complex conjugate:

```
Conjugate[2 + 3I]
2 - 3 I
```

`Abs` gives the magnitude:

```
Abs[2 + 3I]
Sqrt[13]
```

`Arg` gives the phase angle in radians (between **`-Pi`** and **`Pi`**):

```
N[Arg[2 + 3I]]
0.982794
```

Jerry: A number of high school and community college math teachers I've worked with have commented that some adjustments are necessary when students are using a symbolic algebra program. A major adjustment is that students get answers to questions that the teachers have often implied have no answers. Some examples are: **`Log[-2]`**, **`Log[10, -3]`**, **`ArcSin[2]`**, and **`Sqrt[-4]`**. Usually the problem is that we are all used to the domain of the real numbers, while *Mathematica* has no hesitations about evaluating these functions in the complex domain. Beware!

Theo: I guess my response is that the whole idea of restricting functions to certain domains is a bit silly. If there is a consistent way to compute the value of a function for complex arguments, or if the value of a normally-real function is complex for certain real arguments, then so be it.

Jerry: In many calculus books, the limit of **`Sqrt[x]`** as **`x`** approaches zero from the left (negative values) is given as "undefined". The limit from the right is given as zero. The explanation is that since the value of **`Sqrt[x]`** is outside the "domain of real numbers" for all negative values of **`x`**, therefore the limit must also be outside.

Theo: This is obviously ridiculous. `Sqrt[x]` when `x` is negative is the same as `I Sqrt[Abs[x]]`. The limit of this as `x` approaches zero is `0 I`, which is exactly, perfectly, `0`, just as it is when you approach from the right. There are many functions that do confusing and "undefined" things at the origin, but `Sqrt[x]` is *not* one of them.

Jerry: People teaching standard calculus courses and using *Mathematica* will have to make a decision: follow the standard practice, or change it to adjust to the new tools. I will enjoy watching the arguments.

Theo: Well, if they've been teaching students that the limit of `Sqrt[x]` as `x` approaches zero depends on the direction, then they have been seriously misinforming them. What are these students going to think when they encounter a function, like `Log`, which really does have complicated behavior at the origin?

■ Calculus

Jerry: Math teachers and students in colleges and universities are very concerned with the teaching and learning of calculus. Can we show some examples of how *Mathematica* can help them?

Theo: Nothing we can say in this chapter will begin to compare with what two math professors and a student, Jerry Uhl, Horacio Porta, and Don Brown, have done with Calculus and *Mathematica*. They are developing a complete university calculus curriculum at the University of Illinois based in large part on *Mathematica* Notebooks. The courseware has been published by Addison-Wesley, in paper and electronic form **[1]**.

Jerry: I agree, but let's offer some suggestions for those who are teaching normal 1990-type calculus classes.

Theo: OK. Somehow when people talk about teaching calculus they don't seem to mean doing integrals and taking derivatives. They seem more interested in drawing boxes under curves, and drawing tangents. While *Mathematica* can do the integrals and derivatives automatically, these other things usually require some work before *Mathematica* can be helpful.

■ Derivatives and Integrals

Jerry: Let's start with what *Mathematica* can do easily, such as taking derivatives, doing integrals, and finding limits. How do we find the derivative of a simple expression like x^2 ?

Theo: One of the few exceptions to the rule that *Mathematica* function names are fully spelled-out words is the **D** function, which takes derivatives. Here is the derivative of **x^2**:

```
D[x^2, x]
2 x
```

The second argument is the variable you want to differentiate with respect to.

Jerry: How about differentiating a list of expressions. Can we give **D** a list as its first argument? Let's try:

```
D[{x, x^2, x^2, x^3, x^4}, x]
               2     3
{1, 2 x, 2 x, 3 x , 4 x }
D[{Sin[x], Cos[x], Tan[x], Cot[x], Sec[x], Csc[x]}, x]
                           2         2
{Cos[x], -Sin[x], Sec[x] , -Csc[x] , Sec[x] Tan[x],
  -(Cot[x] Csc[x])}
D[{a x^2 + b x + c, a x^n + b x^(n-1) + c}, x]
                          -2 + n           -1 + n
{b + 2 a x, b (-1 + n) x        + a n x      }
```

Theo: The first example you typed could be done more easily. The **Table** command can make a list of things, such as terms of the form **x^n**, with **n** going from 1 to 8.

```
D[Table[x^n, {n, 1, 8}], x]
           2     3     4     5     6     7
{1, 2 x, 3 x , 4 x , 5 x , 6 x , 7 x , 8 x }
```

We can use the **TableForm** command to make the output come out as a nice column instead of a horizontal list. Adding "**//TableForm**" to the end of the expression is the same as applying **TableForm** to the whole expression. It just looks a little nicer at the end:

```
D[Table[x^n, {n, 1, 8}], x]//TableForm
```

1
$2\ x$
$3\ x^2$
$4\ x^3$
$5\ x^4$
$6\ x^5$
$7\ x^6$
$8\ x^7$

Jerry: Fine, but I would like to see the function together with its derivative, instead of just the derivatives.

Theo: The easiest way to do that is to put the **D** command inside the **Table** command, instead of the other way around. That way we can make a table of pairs, **x^n** and the derivative of **x^n**:

```
Table[{x^n, D[x^n, x]}, {n, 1, 8}]//TableForm
```

x	1
x^2	$2\ x$
x^3	$3\ x^2$
x^4	$4\ x^3$
x^5	$5\ x^4$
x^6	$6\ x^5$
x^7	$7\ x^6$
x^8	$8\ x^7$

Jerry: Wonderful. Now I want a table of the function and the first, second, third, and fourth derivatives.

Theo: The **D** command can be used in an alternate form, in which we say how many times to differentiate. For example the fourth derivative of **x^10** is:

```
D[x^10, {x, 4}]
```

$5040\ x^6$

Of course the zeroth derivative is just the function itself:

```
D[x^10, {x, 0}]
```

x^{10}

Using this feature we can make a two-dimensional table, again using the **`Table`** command. If you give **`Table`** a list as its second argument and another list as its third argument, it will make a two-dimensional table (matrix).

```
Table[D[x^n, {x, m}], {n, 1, 8}, {m, 0, 5}]//TableForm
```

```
x     1       0        0        0         0

 2
x     2 x     2        0        0         0

 3       2
x     3 x     6 x      6        0         0

 4       3        2
x     4 x     12 x     24 x     24        0

 5       4        3         2
x     5 x     20 x     60 x     120 x     120

 6       5        4          3         2
x     6 x     30 x     120 x     360 x     720 x

 7       6        5          4         3          2
x     7 x     42 x     210 x     840 x     2520 x

 8       7        6          5          4          3
x     8 x     56 x     336 x     1680 x     6720 x
```

Jerry: Great! The numbers coming out are very familiar: 1, 2, 6, 24, 120.

Next I'd like a list of successive derivatives of sin(x). I'll use the same trick as in the above table, this time changing only the number of derivatives, not the function itself.

```
Table[D[Sin[x], {x, m}], {m, 0, 10}]
```

```
{Sin[x], Cos[x], -Sin[x], -Cos[x], Sin[x], Cos[x],
  -Sin[x], -Cos[x], Sin[x], Cos[x], -Sin[x]}
```

Theo: We could do pretty much the same thing with integration instead. The integration function in *Mathematica* is called **`Integrate`**:

```
Integrate[x^2, x]
```

```
 3
x
--
3
```

To make a table like the one above, we might try using the same format for multiple integrals as we did for multiple derivatives:

```
Integrate[x^10, {x, 4}]
```

```
Integrate::limitform:
   Integration limit {x, 4}
     is not of the form {x,xmin,xmax}.

            10
Integrate[x  , {x, 4}]
```

Unfortunately, **Integrate** thinks we are trying to take a definite integral, but got it wrong. We can fix **Integrate** to understand this notation using the following somewhat peculiar definition (you don't need to understand this):

```
Unprotect[Integrate];
Integrate[expr_, {x_, n_}] :=
    Nest[Integrate[#, x]&, expr, n];
Protect[Integrate];
```

Now we can use **Integrate** as we want to:

```
Integrate[x^10, {x, 4}]
```

$$\frac{x^{14}}{24024}$$

We can make the same tables as we did above, just substituting **Integrate** for **D**:

```
Table[Integrate[x^n, {x, m}], {n, 1, 8}, {m, 0, 5}]//
TableForm
```

x	$\frac{x^2}{2}$	$\frac{x^3}{6}$	$\frac{x^4}{24}$	$\frac{x^5}{120}$	$\frac{x^6}{720}$
x^2	$\frac{x^3}{3}$	$\frac{x^4}{12}$	$\frac{x^5}{60}$	$\frac{x^6}{360}$	$\frac{x^7}{2520}$
x^3	$\frac{x^4}{4}$	$\frac{x^5}{20}$	$\frac{x^6}{120}$	$\frac{x^7}{840}$	$\frac{x^8}{6720}$
x^4	$\frac{x^5}{5}$	$\frac{x^6}{30}$	$\frac{x^7}{210}$	$\frac{x^8}{1680}$	$\frac{x^9}{15120}$
x^5	$\frac{x^6}{6}$	$\frac{x^7}{42}$	$\frac{x^8}{336}$	$\frac{x^9}{3024}$	$\frac{x^{10}}{30240}$
x^6	$\frac{x^7}{7}$	$\frac{x^8}{56}$	$\frac{x^9}{504}$	$\frac{x^{10}}{5040}$	$\frac{x^{11}}{55440}$
x^7	$\frac{x^8}{8}$	$\frac{x^9}{72}$	$\frac{x^{10}}{720}$	$\frac{x^{11}}{7920}$	$\frac{x^{12}}{95040}$
x^8	$\frac{x^9}{9}$	$\frac{x^{10}}{90}$	$\frac{x^{11}}{990}$	$\frac{x^{12}}{11880}$	$\frac{x^{13}}{154440}$

```
Table[Integrate[Sin[x], {x, m}], {m, 0, 10}]
```

```
{Sin[x], -Cos[x], -Sin[x], Cos[x], Sin[x], -Cos[x],
  -Sin[x], Cos[x], Sin[x], -Cos[x], -Sin[x]}
```

Theo: Now we've seen integration in its simplest form. Both **D** and **Integrate** are able to handle a lot of different kinds of expressions. Most people will find them satisfactory, although of course there are always going to be things that **Integrate** can't integrate.

Jerry: Maybe now it's time to show off a little bit. Let's try integrating expressions of the form $1/(1 + x^n)$, with n between 1 and 6:

```
Integrate[1/(1 + x^1), x]
```

```
Log[1 + x]
```

```
Integrate[1/(1 + x^2), x]
```

```
ArcTan[x]
```

```
Integrate[1/(1 + x^3), x]
```

$$\frac{\mathrm{ArcTan}[\frac{-1 + 2\ x}{\mathrm{Sqrt}[3]}]}{\mathrm{Sqrt}[3]} + \frac{\mathrm{Log}[1 + x]}{3} - \frac{\mathrm{Log}[1 - x + x^2]}{6}$$

```
Integrate[1/(1 + x^4), x]
```

$$\frac{\mathrm{Sqrt}[2]\ \mathrm{ArcTan}[\frac{-\mathrm{Sqrt}[2] + 2\ x}{\mathrm{Sqrt}[2]}]}{4} + \frac{\mathrm{Sqrt}[2]\ \mathrm{ArcTan}[\frac{\mathrm{Sqrt}[2] + 2\ x}{\mathrm{Sqrt}[2]}]}{4} - \frac{\mathrm{Sqrt}[2]\ \mathrm{Log}[1 - \mathrm{Sqrt}[2]\ x + x^2]}{8} + \frac{\mathrm{Sqrt}[2]\ \mathrm{Log}[1 + \mathrm{Sqrt}[2]\ x + x^2]}{8}$$

```
Integrate[1/(1 + x^5), x]
```

$$\frac{\mathrm{Log}[1 + x]}{5} - \frac{\mathrm{Cos}[\frac{\mathrm{Pi}}{5}]\ \mathrm{Log}[1 + x^2 - 2\ x\ \mathrm{Cos}[\frac{\mathrm{Pi}}{5}]]}{5} - \frac{\mathrm{Cos}[\frac{3\ \mathrm{Pi}}{5}]\ \mathrm{Log}[1 + x^2 - 2\ x\ \mathrm{Cos}[\frac{3\ \mathrm{Pi}}{5}]]}{5} + \frac{2\ \mathrm{ArcTan}[\frac{x - \mathrm{Cos}[\frac{\mathrm{Pi}}{5}]}{\mathrm{Sin}[\frac{\mathrm{Pi}}{5}]}]\ \mathrm{Sin}[\frac{\mathrm{Pi}}{5}]}{5} + \frac{2\ \mathrm{ArcTan}[\frac{x - \mathrm{Cos}[\frac{3\ \mathrm{Pi}}{5}]}{\mathrm{Sin}[\frac{3\ \mathrm{Pi}}{5}]}]\ \mathrm{Sin}[\frac{3\ \mathrm{Pi}}{5}]}{5}$$

```
Integrate[1/(1 + x^6), x]
```

$$\frac{\mathrm{ArcTan}[x]}{3} + \frac{\mathrm{ArcTan}[-\mathrm{Sqrt}[3] + 2\ x]}{6} + \frac{\mathrm{ArcTan}[\mathrm{Sqrt}[3] + 2\ x]}{6} - \frac{\mathrm{Sqrt}[3]\ \mathrm{Log}[1 - \mathrm{Sqrt}[3]\ x + x^2]}{12} + \frac{\mathrm{Sqrt}[3]\ \mathrm{Log}[1 + \mathrm{Sqrt}[3]\ x + x^2]}{12}$$

Jerry: These are probably worth staring at for a while. Let's move on.

■ Areas Between Curves

Jerry: What about definite integrals, such as the integral of **x^2** from 0 to 1, otherwise known as the area between the curve and the x axis?

Theo: To do a definite integral you give **Integrate** a list of three things as its second argument: the variable to be integrated with respect to, the lower limit, and the upper limit. (Notice that this is the same format you use for specifying the domain for **Plot** and **Table** commands, for example.)

```
Integrate[x^2, {x, 0, 1}]
```

$\frac{1}{3}$

Jerry: Now I'd like a list of definite integrals of **x^n**:

```
Table[Integrate[x^n, {x, 0, 1}], {n, 0, 10}]
```

$\{1, \frac{1}{2}, \frac{1}{3}, \frac{1}{4}, \frac{1}{5}, \frac{1}{6}, \frac{1}{7}, \frac{1}{8}, \frac{1}{9}, \frac{1}{10}, \frac{1}{11}\}$

And I'd like plots of **x**, **x^2**, ...

Theo: The **Release** in the following **Plot** command is a little hard to explain. In general, if you want to make a plot of a list of functions you get from a **Table** command, put a **Release** around the **Table**:

```
Plot[Release[Table[x^n, {n, 1, 10}]], {x, 0, 1},
    AspectRatio -> Automatic,
    PlotRange -> {0, 1}];
```

Jerry: Since the definite integrals of **x^n** from 0 to 1 represent the areas below the curves in the picture above, we have a nice match between the list of numbers for the areas and the picture. The first "curve", **y = x**, splits the unit square in half, and indeed its area is 1/2. The next curve, **y = x^2**, encloses 1/3 of the unit area (so the area between those two curves is 1/2 - 1/3), and so on. Leibnitz, one of the people traditionally thought of as an inventor of the calculus, is said to have thought about these facts at about the same time that he was inventing the calculus **[3]**.

Now I'd like to find the area under the curve **y = Sin[x]** between **x = 0** and **x = Pi**. We should look at a graph first:

```
Plot[Sin[x], {x, 0, Pi}];
```

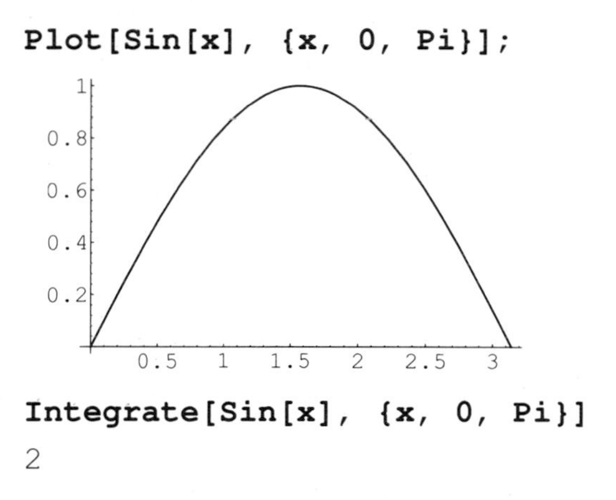


```
Integrate[Sin[x], {x, 0, Pi}]
```

2

This seems provocative! Why should this be 2, instead of 11.6 or anything else? What do you think the area is between **Pi** and **2Pi**? From **0** to **2Pi**? For **0** to **Pi/2**? What about the area between **0** and **Pi/2** for **y = Cos[x]**? Lots of experiments to do here!

Theo: Fortunately, we don't have to do them all, since our readers can try them themselves, just by editing the above input lines in the electronic edition of this book.

■ Limits

Jerry: Students learn that an expression like 2/x is undefined at x = 0. Often it's not completely clear what "undefined" means, and why it happens. There is also some confusion between 2/x and x/2: one becomes undefined but the other doesn't. If students have experience with substituting values for x close to zero and calculating the result, the confusion can go away.

Theo: Let's make a table of values of x, 2/x and x/2 (the function **LabeledTable** is defined in the Functions That We Defined section below. It works just like the **Table** command, but prints out nicely):

```
LabeledTable[{x, 2/x, x/2}, {x, 5, 1, -1}]
```

x	2/x	x/2
5	$\frac{2}{5}$	$\frac{5}{2}$
4	$\frac{1}{2}$	2
3	$\frac{2}{3}$	$\frac{3}{2}$
2	1	1
1	2	$\frac{1}{2}$

Jerry: The results for 2/x and x/2 are quite different, but dramatic differences don't appear until we push it farther. Is it possible to make a table with decimals instead of fractions? Some people relate better to rational numbers, but it's easier to see the magnitude of a decimal number.

Theo: Sure. To convert the rational numbers in the table into decimals, we use the **N** function:

```
LabeledTable[N[{x, 2/x, x/2}], {x, 5, 1, -1}]
```

x	2./x	0.5*x
5.	0.4	2.5
4.	0.5	2.
3.	0.666667	1.5
2.	1.	1.
1.	2.	0.5

Jerry: Now we can move in towards zero! Make the same table with x from 0.1 down to 0.01:

```
LabeledTable[N[{x, 2/x, x/2}],
                    {x, 0.1, 0.01, -0.01}]
```

x	2./x	0.5*x
0.1	20.	0.05
0.09	22.2222	0.045
0.08	25.	0.04
0.07	28.5714	0.035
0.06	33.3333	0.03
0.05	40.	0.025
0.04	50.	0.02
0.03	66.6667	0.015
0.02	100.	0.01
0.01	200.	0.005

This experience keeps us away from any confusion about 2/x and x/2. Clearly, as x gets smaller and smaller, 2/x get larger and larger, while x/2 gets smaller and smaller.

What if we have a ratio where both numerator and denominator are heading for zero? For example, sin(x)/x. A table of x, sin(x), and sin(x)/x from 0.05 down to 0.01 looks like this:

```
LabeledTable[N[{x, Sin[x], Sin[x]/x}],
                           {x, 0.05, 0.01, -0.01}]
```

x	Sin[x]	Sin[x]/x
0.05	0.0499792	0.999583
0.04	0.0399893	0.999733
0.03	0.0299955	0.99985
0.02	0.0199987	0.999933
0.01	0.00999983	0.999983

Now from 0.000005 down to 0.000001:

```
LabeledTable[N[{x, Sin[x], Sin[x]/x}],
                           {x, 0.000005, 0.000001, -0.000001}]
```

x	Sin[x]	Sin[x]/x
$5.\ 10^{-6}$	$5.\ 10^{-6}$	1.
$4.\ 10^{-6}$	$4.\ 10^{-6}$	1.
$3.\ 10^{-6}$	$3.\ 10^{-6}$	1.
$2.\ 10^{-6}$	$2.\ 10^{-6}$	1.
$1.\ 10^{-6}$	$1.\ 10^{-6}$	1.

Wrong answer! It is not true that sin(x)/x is equal to one. Let's do this with more precision.

Theo: You can use a second argument to **N** to specify how many digits of precision to use. We now have to use exact rational numbers in the range specification of the table, because the decimal numbers we used above have less precision than we want in the answer: you can't start with a low precision number and increase its precision just by asking for more.

```
LabeledTable[N[{x, Sin[x], Sin[x]/x}, 17],
                           {x, 5/100000, 1/100000, -1/100000}]
```

x	Sin[x]	Sin[x]/x
0.00005	0.000049999999979166667	0.99999999958333333
0.00004	0.000039999999989333333	0.99999999973333333
0.00003	0.0000299999999955	0.99999999985
0.00002	0.000019999999998666667	0.99999999993333333
0.00001	$9.9999999998333333\ 10^{-6}$	0.99999999998333333

Jerry: We can continue this game, moving closer and closer to zero, appearing to reach 1 for the ratio, increase the precision, and repeating. All of this is to inform our *intuition*. The one thing we can't do is get *all* the way to zero:

```
Sin[0]/0

                                         1
Power::infy: Infinite expression - encountered.
                                         0

Infinity::indet:
   Indeterminate expression 0 ComplexInfinity
     encountered.

Indeterminate
```

Theo: The table shows us, with numbers, that the value very, very close to zero is very, very close to 1. Perhaps a plot would help give us an overall view to reinforce our feeling about what the limit should be:

```
Plot[Sin[x]/x, {x, -8, 8}];
```

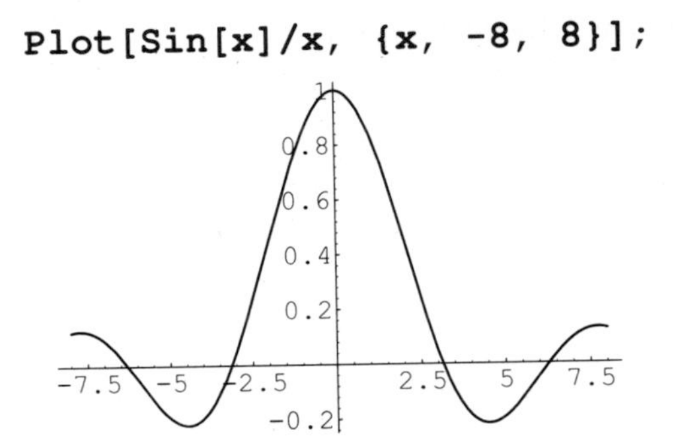

It certainly looks like the value at zero *ought* to be 1, just like all the values close to it. We can use the **Limit** command to ask *Mathematica* what it thinks the limiting value at zero is:

```
Limit[Sin[x]/x, x -> 0]

1
```

Theo: Notice that the result is the exact number 1, not an approximate decimal, as in the table above (you can tell because there is no decimal point in the number).

Jerry: Pleasantly, our intuitive answer and the "official" answer are exactly the same. This will not always be the case.

Here are some other expressions to try. You might find it useful to plot these functions (using **Plot**), to make tables of values (using **LabeledTable**, or **Table**), and to use the **Limit** command to find the limit (if it exists!).

```
Limit[(1+1/x)^x, x -> Infinity]
Limit[x^(1/x), x -> Infinity]
Limit[(E^x - 1)/x, x -> 0]
Limit[(E^(x^2) - 1)/x, x -> 0]
Limit[(2 x^2 - 5 x + 7) / (x^2 + 37 x + 573), x -> Infinity]
Limit[Log[10, x] / Log[E, x], x -> Infinity]
Limit[E^x / x^10, x -> Infinity]
Limit[Sqrt[x] / E^-x^2, x -> 0]
Limit[x^n / n!, n -> Infinity]
Limit[((x + h)^3 - x^3) / h, h -> 0]
Limit[Tan[x] E^x, x -> Pi/4]
Limit[Sin[1/x], x -> 0]
```

Functions That We Defined

Theo: The function **LabeledTable** is just a little utility for printing a table where the first row labels each column.

```
Attributes[LabeledTable] = {HoldFirst};
```

```
LabeledTable[expr_, range_] :=
    TableForm[
        Table[expr, range],
        ColumnLabels -> Map[InputForm, expr],
        ColumnAlignment -> Center
    ]
```

This function turns a complex number into a list of the magnitude and phase angle:

```
ComplexToPolar[z_] := {Abs[z], Arg[z]}
```

This command loads in the trigonometry package:

```
Needs["Algebra`Trigonometry`"]
```

References

1) Brown, Porta, and Uhl, *Calculus&Mathematica*, Addison-Wesley, 1991.
2) Finney, Thomas, *Calculus*, Addison-Wesley, 1989, p676. (Finney/Thomas is not to be confused with the Thomas/Finney. Finney/Thomas is the one with the nice *Mathematica* graphics by Cameron Smith.)
3) C.H. Edwards, Jr, *The Historical Development of the Calculus*, Springer-Verlag, 1979, p237.

Appendices

NicerFourier

By Jerry Keiper, Theodore W. Gray, and Jerry Glynn

This Notebook implements a set of functions similar to the built-in functions **Fourier** and **InverseFourier**, but better suited to lists of real numbers, such as sampled sounds.

These functions are based on a set written by Jerry Keiper, modified by Theodore W. Gray.

Implementation

Preparation

Create the context representing the contents of this Notebook:

```
BeginPackage["NicerFourier`"];
EndPackage[];
```

Load any needed packages:

```
Needs["Graphics`Colors`"];
```

Declare all the globally visible symbols:

```
SinAndCosFourier;
NicerFourier;
FrequencyAmplitudes;
FrequencyAmplitudePhase;
Synthesize;
YValuesOnly;
Resample;
SpectrogramPlot;
```

Begin private context:

```
Begin["NicerFourier`Private`"];
```

Declare all the private symbols:

```
`even;
`odd;
`fcs;
```

■ Definitions

```
even[list_] := Module[
     {dlist = Drop[list, 1]},
     Prepend[(dlist + Reverse[dlist])/2, list[[1]]]
]

odd[list_] := Module[
     {dlist = Drop[list, 1]},
     Prepend[(dlist - Reverse[dlist])/2, 0]
]

fcs[list_] := Fourier[N[even[list] + I odd[list]]]

SinAndCosFourier[list_] := Module[
     {fcslist, clist, slist, length, halfLength},

     length = Length[list];
     halfLength = Floor[(1+length)/2];
     fcslist = fcs[N[list]];

     clist = Chop[even[fcslist]/Sqrt[N[length]]];
     clist = 2 Take[clist, {1, halfLength}];

     slist = Chop[odd[fcslist]/Sqrt[N[length]]];
     slist = -2 Take[slist, {1, halfLength}];

     {
          (First[clist] + First[slist])/2,
          Transpose[{Drop[clist, 1], Drop[slist, 1]}]
     }
]
```

```
NicerFourier[list_, x_] := Module[{sincoslist},
     sincoslist = SinAndCosFourier[list];
     First[sincoslist] +
               Flatten[Last[sincoslist]]
          .
               Flatten[Table[
                    {Cos[k x], Sin[k x]},
                    {k, 1, Length[Last[sincoslist]]}
               ]]
]

FrequencyAmplitudes[list_, sampleRate_] :=
Module[{sincoslist, freq},
     sincoslist = SinAndCosFourier[list];
     numSamples = N[Length[list]];

     Transpose[{
          Table[sampleRate freq / numSamples,
               {freq, 1, Length[Last[sincoslist]]}],
          Map[
               Sqrt[#[[1]]^2 + #[[2]]^2]&,
               Last[sincoslist]
          ]
     }]
]

FrequencyAmplitudePhase[list_, sampleRate_] :=
Module[{sincoslist, freq},
     sincoslist = SinAndCosFourier[list];
     numSamples = N[Length[list]];

     Transpose[{
          Table[sampleRate freq / numSamples,
               {freq, 1, Length[Last[sincoslist]]}],
          Map[
               Sqrt[#[[1]]^2 + #[[2]]^2]&,
               Last[sincoslist]
          ],
          Map[
               ArcTan[#[[2]]/#[[1]]]&,
               Last[sincoslist]
          ]
     }]
]
```

```
Synthesize[list_, var_] := Module[{twoPi},
    twoPi = N[2 Pi];
    Apply[
        Plus,
        Map[
            (#[[2]] Sin[Random[Real, {0, twoPi}] +
                        var twoPi #[[1]]])&,
            list
        ]
    ]
]

YValuesOnly[list_] :=
    Last[Transpose[list]]

Resample[list_] :=
    Transpose[{
        Table[
            N[i 2Pi/Length[list]],
            {i, 0, Length[list]-1}
        ],
        YValuesOnly[list]
    }]

SpectrogramPlot[fc_, options___] :=
    Show[
        Graphics[
            Map[
                Line[{{First[#], 0},
                        {First[#], Last[#]}}]&,
                fc
            ]
        ],
        options,
        Axes -> Automatic
    ]
```

■ Finish

End private context:

```
End[];
```

ComplexMapPlot

by Theodore W. Gray and Jerry Glynn

This Notebook implements a sophisticated form of complex function mapping. The function **ComplexMapPlot** defined in the Implementation section allows you to apply one or more complex functions to any list of graphics objects (lines, circles, points, etc). The graphics objects are interpreted as lying in the complex plane, and **ComplexMapPlot** shows what the objects look like when they are mapped by the given function(s).

Six commands are provided for generating commonly used pre-images (which are lists of graphics objects suitable for using with **ComplexMapPlot**). They are **RectangularGrid**, **HorizontalLines**, **VerticalLines**, **PolarGrid**, **PolarCircles**, and **PolarLines**.

It is possible to use **ComplexMapPlot** on any list of graphics objects, including those returned by **Plot** and other graphics functions.

ComplexMapPlot uses an adaptive sampling algorithm to decide how many times to subdivide each graphical element. For example, a straight line in the input will usually end up as a curved line in the output; **ComplexMapPlot** automatically calculates enough points along the line to make the resulting curve smooth.

Implementation

Preparation

Create the context representing the contents of this Notebook:

```
BeginPackage["ComplexMapPlot`"];
EndPackage[];
```

Load any needed packages:

```
Needs["Graphics`Colors`"];
```

Declare all the globally visible symbols:

```
MapGraphics;
RectangularGrid;
HorizontalLines;
VerticalLines;
PolarGrid;
PolarCircles;
PolarLines;
ComplexMapPlot;

PlotPoints;
IncludePreimage;
```

Begin private context:

```
Begin["ComplexMapPlot`Private`"];
```

Declare all the private symbols:

```
`pairToComplex;
`complexToPair;
`transformList;
`listParameter;
`transformCircle;
`MG0;
```

■ Definitions

□ MapGraphics

```
`parametricPlotOptions;

pairToComplex[{re_, im_}] = re + im I;

complexToPair[z_] = {Re[z], Im[z]};
```

```
transformList[f_, {min:{_, _}, max:{_, _}}] :=
    Module[
        {t, minz, deltaz},
        minz = pairToComplex[min];
        deltaz = pairToComplex[max] - minz;
        ParametricPlot[
            complexToPair[f[minz + t * deltaz]],
            {t, 0.0, 1.0},
            Release[parametricPlotOptions]
        ][[1, 1, 1, 1]]
    ]

transformList[f_, list_] :=
    Module[
        {t},
        ParametricPlot[
            complexToPair[f[pairToComplex[
                listParameter[
                    t,
                    list
                ]
            ]]],
            {t, 0.0, 1.0},
            Release[parametricPlotOptions]
        ][[1, 1, 1, 1]]
    ]

listParameter[t_, {min_, max_}] :=
                    min + t * (max - min);

listParameter[t_, list_] := Module[
    {parameter, length, startIndex, startPoint},
    length = Length[list];
    parameter = t (length - 1);
    startIndex = Floor[parameter] + 1;
    startPoint = list[[startIndex]];
    If[startIndex == length,
        startPoint,
    (*else*)
        startPoint +
            (parameter - startIndex + 1) *
                (list[[startIndex + 1]] - startPoint)
    ]
]
```

```
transformCircle[f_, center_, {rx_, ry_},
                                    {mint_, maxt_}] :=
     Module[
          {t, zcenter},
          zcenter = pairToComplex[center];
          ParametricPlot[
               complexToPair[f[
                    rx Cos[t] + I ry Sin[t] + zcenter
               ]],
               {t, N[mint], N[maxt]},
               Release[parametricPlotOptions]
          ][[1, 1, 1, 1]]
     ]

MapGraphics[Graphics[list_List, opts___],
                                         f_, options___] :=
     (
     parametricPlotOptions =
          Join[{DisplayFunction -> Identity}, options];
     Graphics[ MG0[list, f], opts ]
     )

MG0[d_List, f_] := Map[ MG0[#, f]& , d ]

MG0[Point[d_List], f_] :=
     Point[complexToPair[f[pairToComplex[d]]]]

MG0[Line[d_List], f_] := Line[transformList[f, d]]

MG0[Rectangle[{xmin_, ymin_}, {xmax_, ymax_}], f_] :=
     MG0[Polygon[{
          {xmin,ymin},
          {xmin,ymax},
          {xmax, ymax},
          {xmax, ymin}
     }], f]

MG0[Polygon[d_List]] :=
     MG0[Polygon[Join[d, First[d]]]] /;
                                    First[d] =!= Last[d]

MG0[Polygon[d_List], f_] :=
     Polygon[transformList[f, d]]

MG0[Circle[center_List, r_?NumberQ], f_] :=
     Line[transformCircle[
               f, center, {r, r}, {0, 2Pi}]]
```

```
MG0[Circle[center_List, r_?List], f_] :=
    Line[transformCircle[f, center, r, {0, 2Pi}]]
MG0[Circle[center_List,
                r_?NumberQ, range_List], f_] :=
    Line[transformCircle[f, center, {r, r}, range]]
MG0[Circle[center_List, r_?List, range_List], f_] :=
    Line[transformCircle[f, center, r, range]]
MG0[Disk[center_List, r_?NumberQ], f_] :=
    Polygon[transformCircle[
            f, center, {r, r}, {0, 2Pi}]]
MG0[Disk[center_List, r_?List], f_] :=
    Polygon[transformCircle[f, center, r, {0, 2Pi}]]
MG0[Disk[center_List, r_?NumberQ, range_List], f_] :=
    Polygon[transformCircle[
            f, center, {r, r}, range]]
MG0[Disk[center_List, r_?List, range_List], f_] :=
    Polygon[transformCircle[f, center, r, range]]
MG0[CellArray[array_,
        range_List:{{0,0}, {1,1}}, zrange___], f_] :=
    CellArray[
        array,
        {
            complexToPair[
                f[pairToComplex[First[range]]]],
            complexToPair[
                f[pairToComplex[Last[range]]]]
        },
        zrange
    ]
MG0[Text[expr_, d_List, opts___], f_] :=
    Text[expr,
        complexToPair[f[pairToComplex[d]]], opts]
MG0[expr_, f_] := expr
```

□ RectangularGrid

```
Options[RectangularGrid] =
    {
        PlotPoints -> 14
    };
```

```
RectangularGrid[{{Remin_, Remax_}, {Immin_, Immax_}},
            options___] := Module[
     {plotPoints, n, x},

     (* Extract options *)
     {plotPoints} = {PlotPoints} /.
              {options} /. Options[RectangularGrid];

     If[Head[plotPoints] =!= List,
         plotPoints = {plotPoints, plotPoints}
     ];

     {
     GrayLevel[0.5],
     VerticalLines[{{Remin, Remax}, {Immin, Immax}},
              PlotPoints -> plotPoints[[1]], options],
     GrayLevel[0.0],
     HorizontalLines[{{Remin, Remax}, {Immin, Immax}},
              PlotPoints -> plotPoints[[1]], options]
     }
]
```

□ VerticalLines

```
Options[VerticalLines] =
     {
         PlotPoints -> 14
     };

VerticalLines[{{Remin_, Remax_}, {Immin_, Immax_}},
            options___] := Module[
     {plotPoints, n, x},

     (* Extract options *)
     {plotPoints} = {PlotPoints} /.
              {options} /. Options[VerticalLines];

     Table[
         x = Remin +
                 (Remax - Remin) (n-1)/(plotPoints-1);
         Line[N[{{x, Immin}, {x, Immax}}]],
         {n, 1, plotPoints}
     ]
]
```

□ HorizontalLines

```
Options[HorizontalLines] =
     {
          PlotPoints -> 14
     };

HorizontalLines[{{Remin_, Remax_}, {Immin_, Immax_}},
                    options___] :=
Module[
     {plotPoints, n, x},

     (* Extract options *)
     {plotPoints} = {PlotPoints} /.
               {options} /. Options[HorizontalLines];

     Table[
          x = Immin +
                    (Immax - Immin) (n-1)/(plotPoints-1);
          Line[N[{{Remin, x}, {Remax, x}}]],
          {n, 1, plotPoints}
     ]
]
```

□ PolarGrid

```
Options[PolarGrid] =
     {
          PlotPoints -> 14
     };

PolarGrid[center:{_, _}, radius_, options___] :=
Module[
     {
          plotPoints,
          ncenter, nradius, n, t,
     },

     (* Extract options *)
     {plotPoints} = {PlotPoints}
               /. {options} /. Options[PolarGrid];

     If[Head[plotPoints] =!= List,
          plotPoints = {plotPoints, plotPoints}
     ];

     {
          GrayLevel[0.5],
          PolarLines[center, radius,
               PlotPoints -> plotPoints[[2]], options],
          GrayLevel[0.0],
          PolarCircles[center, radius,
               PlotPoints -> plotPoints[[1]], options]
     }
 ]
```

□ PolarCircles

```
Options[PolarCircles] =
    {
        PlotPoints -> 14
    };

PolarCircles[center:{_, _}, radius_, options___] :=
Module[
    {
        plotPoints,
        nradius, ncenter, n, t,
        minradius, deltaradius
    },

    (* Extract options *)
    {plotPoints} = {PlotPoints}
            /. {options} /. Options[PolarCircles];

    ncenter = N[center];
    nradius = N[radius];
    If[Head[nradius] =!= List, nradius = {0, nradius}];
    minradius = N[nradius[[1]]];
    deltaradius = N[nradius[[2]] - minradius];

    Table[
        Circle[ncenter, minradius +
                deltaradius (n-1)/(plotPoints-1)],
        {n, 1, plotPoints}
    ]
]
```

□ PolarLines

```
Options[PolarLines] =
     {
          PlotPoints -> 14
     };

PolarLines[center:{_, _}, radius_, options___] :=
Module[
     {
          plotPoints,
          nradius, ncenter, n, t,
          minradius, maxradius
     },

     (* Extract options *)
     {plotPoints} = {PlotPoints}
             /. {options} /. Options[PolarLines];

     ncenter = N[center];
     nradius = N[radius];
     If[Head[nradius]=!=List, nradius = {0, nradius}];
     minradius = N[nradius[[1]]];
     maxradius = N[nradius[[2]]];

     Table[
          t = 2Pi n / plotPoints;
          Line[{
               ncenter + minradius N[{Cos[t], Sin[t]}],
               ncenter + maxradius N[{Cos[t], Sin[t]}]
          }],
          {n, 1, plotPoints}
     ]
]
```

□ ComplexMapPlot

```
Clear[ComplexMapPlot];

Options[ComplexMapPlot] =
     {
          AspectRatio -> Automatic,
          Axes -> Automatic,
          IncludePreimage -> True,
          Prolog -> {},
          Epilog -> {}
     };

ComplexMapPlot[f_, var_,
             Graphics[domain_, ___], options___] :=
     ComplexMapPlot[f, var, domain, options];

ComplexMapPlot[f_, var_, domain_, options___] :=
Module[
     {output, mapOptions, showOptions,
      includePreimage},

     {includePreimage} = {IncludePreimage} /.
                 {options} /.
                 Options[ComplexMapPlot];

     mapOptions = Select[
          Join[{options}, Options[ComplexMapPlot]],
          (
                 !FreeQ[#, PlotPoints]
              || !FreeQ[#, PlotDivision]
              || !FreeQ[#, MaxBend]
          )&
     ];

     showOptions = Select[
          Join[{options}, Options[ComplexMapPlot]],
          (
                 FreeQ[#, PlotPoints]
              && FreeQ[#, PlotDivision]
              && FreeQ[#, MaxBend]
              && FreeQ[#, Prolog]
              && FreeQ[#, Epilog]
              && FreeQ[#, IncludePreimage]
          )&
     ];
```

```
        output = Map[First,
            Map[
                Function[ff, MapGraphics[
                    Graphics[{domain}],
                    Function[var2, ff /. var -> var2],
                    mapOptions
                ]],
                Flatten[{f}]
            ]];
        {prolog, epilog} = {Prolog, Epilog} /. {options} /. Options[Cc

        If[TrueQ[includePreimage],
            Show[
                Graphics[{
                    Rectangle[{0,   0}, {1,   1},
                        Graphics[{domain}, showOptions]],
                    Rectangle[{1.1, 0}, {2.1, 1},
                        Graphics[{prolog, output, epilog},
                                showOptions]]
                }],
                Axes -> None,
                PlotRange -> {{0, 2.1}, {0, 1}},
                showOptions
            ],
            Show[
                Graphics[{output, epilog}],
                showOptions
            ]
        ]
    ]
```

■ Finish

End private context:

```
    End[];
```

IteratedMappings

By Theodore W. Gray, Doug Stein, and Jerry Glynn.

This Notebook implements a number of functions for making iterated affine map pictures. It is based on suggestions by Doug Stein, written by Theodore W. Gray, and critiqued by Jerry Glynn.

Implementation

Preparation

Create the context representing the contents of this Notebook:

```
BeginPackage["IteratedMappings`"];
EndPackage[];
```

Load any needed packages:

```
Needs["Graphics`Colors`"];
```

Declare all the globally visible symbols:

```
IteratedMapping;
AffineMap;
```

Begin private context:

```
Begin["IteratedMappings`Private`"];
```

Declare all the private symbols:

```
`MapGraphics;
`transformList;
`transformCircle;
`MG0;
```

■ Definitions

□ MapGraphics

```
`parametricPlotOptions;

transformList[f_, list_] := Map[f, list];

transformCircle[f_, center_, {rx_, ry_},
                {mint_, maxt_}] := Module[
    {t},
    ParametricPlot[
        f[{rx Cos[t], ry Sin[t]} + center],
        {t, N[mint], N[maxt]},
        Release[parametricPlotOptions]
    ][[1,1,1,1]]
]

MapGraphics[Graphics[list_List, opts___],
            f_, options___] :=
    (
    parametricPlotOptions =
        Join[{DisplayFunction -> Identity}, options];
    Graphics[ MG0[list, f], opts ]
    )

MG0[d_List, f_] := Map[ MG0[#, f]& , d ]

MG0[Point[d_List], f_] := Point[f[d]]

MG0[Line[d_List], f_] := Line[transformList[f, d]]

MG0[Rectangle[{xmin_, ymin_}, {xmax_, ymax_}], f_] :=
    MG0[Polygon[{
        {xmin,ymin},
        {xmin,ymax},
        {xmax, ymax},
        {xmax, ymin}}], f]

MG0[Polygon[d_List]] :=
    MG0[Polygon[Join[d, First[d]]]] /;
        First[d] =!= Last[d]

MG0[Polygon[d_List], f_] :=
    Polygon[transformList[f, d]]

MG0[Circle[center_List, r_?NumberQ], f_] :=
    Line[transformCircle[f, center, {r, r}, {0, 2Pi}]]
```

```
MG0[Circle[center_List, r_?List], f_] :=
    Line[transformCircle[f, center, r, {0, 2Pi}]]
MG0[Circle[center_List, r_?NumberQ, range_List], f_] :=
    Line[transformCircle[f, center, {r, r}, range]]
MG0[Circle[center_List, r_?List, range_List], f_] :=
    Line[transformCircle[f, center, r, range]]
MG0[Disk[center_List, r_?NumberQ], f_] :=
    Polygon[
        transformCircle[f, center, {r, r}, {0, 2Pi}]]
MG0[Disk[center_List, r_?List], f_] :=
    Polygon[transformCircle[f, center, r, {0, 2Pi}]]
MG0[Disk[center_List, r_?NumberQ, range_List], f_] :=
    Polygon[transformCircle[f, center, {r, r}, range]]
MG0[Disk[center_List, r_?List, range_List], f_] :=
    Polygon[transformCircle[f, center, r, range]]
MG0[Raster[array_, range_List:{{0,0}, {1,1}},
                                    zrange___], f_] :=
    Raster[array, f /@ range, zrange]
MG0[Text[expr_, d_List, opts___], f_] :=
    Text[expr, f[d], opts]
MG0[expr_, f_] := expr
```

□ IteratedMapping

```
Clear[IteratedMapping];
Options[IteratedMapping] =
    {
        Axes -> Automatic,
        AspectRatio -> Automatic,
        Epilog -> {}
    };
```

```
IteratedMapping[matrices_, iterations_,
                    domain_, options___] :=
Module[
    {output, mapOptions, showOptions, grToGr},

    mapOptions = Select[
        Join[{options}, Options[IteratedMapping]],
        (
                !FreeQ[#, PlotPoints]
            || !FreeQ[#, PlotDivision]
            || !FreeQ[#, MaxBend]
        )&
    ];

    showOptions = Select[
        Join[{options}, Options[IteratedMapping]],
        (
                FreeQ[#, PlotPoints]
            && FreeQ[#, PlotDivision]
            && FreeQ[#, MaxBend]
            && FreeQ[#, Epilog]
        )&
    ];

    grToGr[gr_] := Graphics[Map[First, Map[
        Function[mm, MapGraphics[
                gr,
                Function[p, mm[[1]] . p + mm[[2]]],
                mapOptions
            ]],
            Flatten[{matrices}]
        ]]];

    output = Map[
            First,
            {
                Nest[
                    grToGr,
                    Graphics[{domain}],
                    iterations
                ]
            }
        ];
    epilog = Epilog /. {options} /.
            Options[IteratedMapping];
```

```
        Show[
            Graphics[{output, epilog}],
            showOptions
        ]
    ]
```

■ Finish

End private context:

```
    End[];
```

InequalityPlot

by Theodore W. Gray and Jerry Glynn

This Notebook implements the function **InequalityPlot**, which plots all sorts of inequalities involving one or two curves. See the Examples section for information about how to use **InequalityPlot**.

Implementation

Preparation

Create the context representing the contents of this Notebook:

```
BeginPackage["InequalityPlot`"];
EndPackage[];
```

Declare all the globally visible symbols:

```
InequalityPlot;
```

Begin private context:

```
Begin["InequalityPlot`Private`"];
```

Declare all the private symbols:

```
`iPlot1;
`iPlot2;
`makeAPoly2;
`makePolys2;
```

■ Definitions

□ Usage Message

```
InequalityPlot::usage =
"InequalityPlot[y <= expr, {x, xmin, xmax}] \
plots the area under a curve.  \
InequalityPlot[expr <= y <= expr, {x, xmin, xmax}] \
plots the area between two curves.  All types of \
inequalities (<, <=, >, >=) are supported.  \
Options are passed to Show.";
```

□ Rules for single inequalities

```
InequalityPlot[x_Symbol < y_Symbol,
               range:{x_, __}, options___] :=
    iPlot1[x, range, True, False, options];
InequalityPlot[x_Symbol <= y_Symbol,
               range:{x_, __}, options___] :=
    iPlot1[x, range, True, True, options];
InequalityPlot[x_Symbol > y_Symbol,
               range:{x_, __}, options___] :=
    iPlot1[x, range, False, False, options];
InequalityPlot[x_Symbol >= y_Symbol,
               range:{x_, __}, options___] :=
    iPlot1[x, range, False, True, options];

InequalityPlot[y_Symbol < rhs_,
               range:{__}, options___] :=
    iPlot1[rhs, range, False, False, options];
InequalityPlot[y_Symbol <= rhs_,
               range:{__}, options___] :=
    iPlot1[rhs, range, False, True, options];
InequalityPlot[y_Symbol > rhs_,
               range:{__}, options___] :=
    iPlot1[rhs, range, True, False, options];
InequalityPlot[y_Symbol >= rhs_,
               range:{__}, options___] :=
    iPlot1[rhs, range, True, True, options];
```

```
InequalityPlot[rhs_ < y_Symbol,
                  range:{__}, options___] :=
    iPlot1[rhs, range, True, False, options];
InequalityPlot[rhs_ <= y_Symbol,
                  range:{__}, options___] :=
    iPlot1[rhs, range, True, True, options];
InequalityPlot[rhs_ > y_Symbol,
                  range:{__}, options___] :=
    iPlot1[rhs, range, False, False, options];
InequalityPlot[rhs_ >= y_Symbol,
                  range:{__}, options___] :=
    iPlot1[rhs, range, False, True, options];
```

□ Rules for double inequalities that are between two curves

```
InequalityPlot[lhs_ < y_Symbol < rhs_,
                  range:{__}, options___] :=
    iPlot2[lhs, rhs, range, False, False, options];
InequalityPlot[lhs_ < y_Symbol <= rhs_,
                  range:{__}, options___] :=
    iPlot2[lhs, rhs, range, False, True, options];
InequalityPlot[lhs_ <= y_Symbol < rhs_,
                  range:{__}, options___] :=
    iPlot2[lhs, rhs, range, True, False, options];
InequalityPlot[lhs_ <= y_Symbol <= rhs_,
                  range:{__}, options___] :=
    iPlot2[lhs, rhs, range, True, True, options];

InequalityPlot[lhs_ > y_Symbol > rhs_,
                  range:{__}, options___] :=
    iPlot2[rhs, lhs, range, False, False, options];
InequalityPlot[lhs_ > y_Symbol >= rhs_,
                  range:{__}, options___] :=
    iPlot2[rhs, lhs, range, False, True, options];
InequalityPlot[lhs_ >= y_Symbol > rhs_,
                  range:{__}, options___] :=
    iPlot2[rhs, lhs, range, True, False, options];
InequalityPlot[lhs_ >= y_Symbol >= rhs_,
                  range:{__}, options___] :=
    iPlot2[rhs, lhs, range, True, True, options];
```

□ Rules for double inequalities that are above or below two curves

```
InequalityPlot[lhs_ > y_Symbol < rhs_,
               range:{__}, options___] :=
    iPlot1[Min[lhs, rhs],range,False,False,options];
InequalityPlot[lhs_ > y_Symbol <= rhs_,
               range:{__}, options___] :=
    iPlot1[Min[lhs, rhs],range,False,True,options];
InequalityPlot[lhs_ >= y_Symbol < rhs_,
               range:{__}, options___] :=
    iPlot1[Min[lhs, rhs],range,False,True,options];
InequalityPlot[lhs_ >= y_Symbol <= rhs_,
               range:{__}, options___] :=
    iPlot1[Min[lhs, rhs],range,False,True,options];

InequalityPlot[lhs_ < y_Symbol > rhs_,
               range:{__}, options___] :=
    iPlot1[Max[lhs, rhs],range,True,False,options];
InequalityPlot[lhs_ < y_Symbol >= rhs_,
               range:{__}, options___] :=
    iPlot1[Max[lhs, rhs],range,True,True,options];
InequalityPlot[lhs_ <= y_Symbol > rhs_,
               range:{__}, options___] :=
    iPlot1[Max[lhs, rhs],range,True,True,options];
InequalityPlot[lhs_ <= y_Symbol >= rhs_,
               range:{__}, options___] :=
    iPlot1[Max[lhs, rhs],range,True,True,options];
```

□ iPlot1

```
iPlot1[rhs_, {x_, xmin_, xmax_},
       isGreater_, orEqual_, options___] :=
Module[{lines, points, minY, maxY, extremeY},

    lines = Plot[rhs, {x, xmin, xmax},
                 DisplayFunction->Identity][[1,1]];
    points =
        Partition[Flatten[lines /. Line -> List], 2];

    minY = Min[Transpose[points][[2]]];
    maxY = Max[Transpose[points][[2]]];
    If[TrueQ[isGreater],
        extremeY = maxY + (maxY - minY)/10.,
        extremeY = minY - (maxY - minY)/10.
    ];

    Show[Graphics[Reverse[First[FullGraphics[
        Graphics[{
            GrayLevel[0.5],
            Map[
                Polygon[
                    Join[
                        {{#[[1,1,1]], extremeY}},
                        #[[1]],
                        {{#[[1,-1,1]], extremeY}}
                    ]
                ]&,
                lines
            ],

            GrayLevel[0.0],
            If[TrueQ[orEqual],
                Dashing[{}],
                Dashing[{0.01}]
            ],
            lines
        },
        Axes->Automatic, options]]]]]
    ]
]
```

□ iPlot2

```
`makeAPoly2[expr_] :=
    Polygon[
        Join[expr[[1, 1]], Reverse[expr[[2, 1]]]]]

`makePolys2[expr_] :=
    Map[makeAPoly2, Transpose[Map[First, expr]]]

iPlot2[lhs_, rhs_, {x_, xmin_, xmax_},
        orEqual1_, orEqual2_, options___] :=
Module[{linesLhs, linesRhs, areas},

    linesLhs =
        First[Plot[lhs, {x, xmin, xmax},
            DisplayFunction->Identity
        ]];
    linesRhs =
        First[Plot[rhs, {x, xmin, xmax},
            DisplayFunction->Identity
        ]];

    Off[Plot::notnum];
    Off[Plot::notreal];
    areas =
        First[Plot[
            {
                If[lhs < rhs,
                    Max[lhs, rhs],
                    Undefined
                ],
                If[lhs < rhs,
                    Min[lhs, rhs],
                    Undefined
                ]
            },
            {x, xmin, xmax},
            DisplayFunction->Identity
        ]];
    On[Plot::notnum];
    On[Plot::notreal];
```

```
        Show[Graphics[Reverse[First[FullGraphics[
            Graphics[{
                GrayLevel[0.5],
                makePolys2[areas],

                GrayLevel[0.0],
                If[TrueQ[orEqual1],
                    Dashing[{}],
                    Dashing[{0.01}]
                ],
                linesLhs,

                If[TrueQ[orEqual2],
                    Dashing[{}],
                    Dashing[{0.01}]
                ],
                linesRhs
            },
            Axes->Automatic,
            PlotRange -> {{xmin, xmax}, Automatic},
            options]]]]]
        ]
    ]
```

■ Finish

End private context:

```
    End[];
```

Making Posters

An appendix for the glue-and-paper set, by Theodore W. Gray.

One of the more amusing uses for *Mathematica* is as a poster maker. If you have *Mathematica*, a Macintosh with at least 4MB, and a PostScript-based laser printer, you can print immense graphics and text on hundreds of sheets of paper, which you can then assemble into a poster of almost unlimited size.

Introduction

In this Notebook I will describe in detail how to generate, print, and assemble high quality posters using only standard equipment and low cost materials. Although assembling a large poster can be a lot of work, you can expect to make an 8-foot-square poster for less than $15 in materials, which compares very favorably with the many hundreds of dollars it would cost you otherwise. Besides, haven't you always wanted to make an 8-foot Bessel function in your own backyard?

The techniques discussed in this Notebook are the result of extensive personal experience in the *Mathematica* poster construction business. Although you may find yourself with somewhat different needs or equipment, I think you will find most of the advice practical and useful, not to mention long-winded and pedantic.

Let's start out with one basic reality: You need to have your own personal or company laser printer to do big posters. If you have to pay 25¢ per page for LaserWriter output at a local copy shop, you will quickly panic at the size of your bills. Some of the posters I've done have taken more than 20 hours to print out, although 1 hour is more typical. Dot matrix printers, while they will work in principle, are very slow, and I personally would not have the patience to wait for 100 pages of complex graphics from an ImageWriter.

All the posters I've done have been made on a Mac II, using a LaserWriter IINT. If you have a different brand of laser printer, it shouldn't make much difference, as long as it is PostScript-based and has at least 2MB of memory, like the NT. Non-postscript printers (i.e. the LaserWriter IISC) may have trouble with some graphics, and particularly with very large text. In general, stay away from non-PostScript printers.

The construction techniques I will describe are not for the casual, weekend poster maker. These instructions are for serious people who aren't afraid of an X-Acto knife and a bottle of Elmer's glue.

Making A Poster

Generating the Image

The first step in making a poster is deciding what it should be a poster *of*. Let's assume you want to make a poster of a *Mathematica*- generated PostScript graphic, say a stellated icosahedron. (Later on we'll discuss techniques for doing huge text as well.)

Open a new Notebook window and type in the input for your poster. To make the stellated icosahedron, you might enter:

```
Needs["Graphics`Polyhedra`"];
Show[Graphics3D[Stellate[Icosahedron[]]]];
```

Evaluate the input, and check that what you get is reasonable. *(Please **Save** Your Notebook Now.)*

If you don't have a lot of memory, you might want to start a fresh *Mathematica* session, without starting the local kernel. This will make as much memory as possible available for printing your poster.

Setting Up Your Styles

If you want to include any text in your poster (for example, the input lines that generated your graphic), you will need to make a few changes to the default Styles so the text will print out right. If you don't want to include any text, delete all cells other than the cell containing your graphics and skip this section.

You need to set two properties of the Styles in your Notebook: size (large), and horizontal scrolling (enabled).

1) Open the "Styles" window (using the **Edit Styles** command in the Style menu).
2) Select all the cells from "input" to "name".
3) Choose a size (typically 72 point or larger is good) from the Size submenu

in the Formats menu. (Don't worry if your text looks horrible; it will print fine.)
4) Choose **Scroll Horizontally** from the Formats menu. (This will make the cell contents and names appear only once on the left side of the poster.)
5) Close the Styles window.

If you use any other cell styles (for example, "text"), make sure that they also have the Scroll Horizontally option set. Any cells that don't have this option set will be printed over again in each vertical column of the poster. This is not what you normally want, but it can be useful for printing repeated titles across the whole poster, for example.

Of course, you can also use the Size submenu in the regular Styles menu to set the size of text in your Notebook. It's just easier to set the overall Style initially to a large size so you don't have to change the text in each individual cell.

■ Huge Text

You can also use *Mathematica* to produce very large text-based posters, even though this is not exactly what it is intended for. To make a huge text poster, follow all the same instructions as for a graphics poster, along with the following:

Start by making a cell in a non-wordwrapping Style (for example, in input style). Make sure the Style is set to scroll horizontally (see "Setting Up Your Styles" above).

Type the text you want in a small size. Make sure it's right. Now, select all the text you want huge, and choose the **Custom Size** command in the Size submenu of the Styles menu. Enter the point size. I find that 2000 point (27 inch) type is about the largest I can get on a LaserWriter IINT. You may find a different limit on your printer. Don't be afraid of these huge sizes. They work just fine. Scrolling is a bit slow, and editing is not practical, but printing is no problem.

You may find that huge text sometimes gets clipped on one side or another. This is because there are slight differences between the screen fonts and the printer fonts. If this happens, just add some blank lines (in a much smaller point size) above and below the huge text, and a space at the beginning or end of each line of huge text. This will move the huge text more towards the center of the cell, preventing clipping.

It almost goes without saying that you should use a LaserWriter font! This means Times, Geneva, Courier, and Symbol for almost any printer. The LaserWriter Plus and II have a number of additional fonts built in. You can also use any Downloadable PostScript fonts.

■ Setting the Size of the Graphic

The next step is to drag your graphic out to the size you want your poster to be. If you don't have a lot of memory, or if you intend to make a really large poster, you will want to follow steps 1 and 2; otherwise go directly to step 3.

1) Choose the **Graphics Settings...** command in the Cells menu.
2) Check the "Draw dummy graphics instead of PostScript" option. This prevents *Mathematica* from trying to use ridiculous amounts of memory displaying your graphic. Click the OK button. You will see a little image in the top left of your cell instead of your graphic. (The printed size of the graphic is determined by the bounding box, not by the size of the dummy graphic.)
3) Click on your graphic to select it (you should see a bounding box with a bunch of resizing handles on it).
4) Scroll your Notebook so the bottom right resizing handle of the graphic is visible.
5) Click on the bottom right resizing handle. Hold down the mouse button and drag the cursor outside the window to the right. The Notebook will start to autoscroll rapidly. Watch the numbers in the status field at the bottom of the window. The width and height of the graphic will be displayed in units of points (1/72 of an inch). Release the mouse button when you get to the desired size (for example, 5184 is six feet). *Mathematica* will automatically preserve the aspect ratio of your plot, making it very tall as well as very wide.
6) Release the button, then use the horizontal scroll bar to scroll back a bit. You should see the right edge of your newly huge graphic.

This is a good time to *Save your Notebook* again.

■ Setting the Printing Options

You need to set up a number of the options in the Printing Options dialog box:

1) Choose **Printing Options...** from the File menu.
2) Check the "Print multiple pages for wide cells" and "Print registration marks" options. Make sure the "Print Cell Brackets" option is **not** checked.
3) Set the page margins. I find the following to be the best margins for most cases: Left: 0.5, Right: 0.5, Top: 0.5, Bottom: 0.75.
The larger bottom margin leaves room for the page number. If you make the margins too small, it is harder to glue the pages together.
4) Set the header and footer. I recommend no header, and just the page

number (\p) in the footer. The page number is quite important, since it is about the only way to keep track of which page goes where. It will be covered up when the poster is assembled.

■ Printing Your Poster

Now you are finally ready to start printing your poster. Put plenty of paper in your printer, make sure the boss is not looking for the next few hours, and go for it: Choose **Print...** from the File menu.

Click OK in the dialog box, and start waiting. If you have to interrupt the printing, you can restart it later by specifying a range of pages to print. Just remember that each "Page" is really a whole horizontal row of pages. You can't restart in the middle of a row. Be very very careful that you have not changed the Notebook in any way before restarting the printing. Any change at all may cause a subtle misalignment in the pages, which you will not notice until you have spent hours assembling the first part of the poster. Trust me, I know.

Depending on how large your poster is, and how complex your graphic, it can take quite a while to print the whole thing. Here are the page counts for typical poster sizes:

2x2 foot: 9 pages (3x3 pages)
3x3 foot: 20 pages (4x5 pages)
5x5 foot: 63 pages (7x9 pages)
6x6 foot: 80 pages (8x10 pages)
7x7 foot: 108 pages (9x12 pages)
8x8 foot: 143 pages (11x13 pages)
30x30 foot: 1862 pages (theoretically)

Please check the output from the printer regularly, to make sure you didn't make a silly mistake in setting up some parameter. It is shocking the amount of paper you can waste printing posters incorrectly. Don't be alarmed if the first one or two rows are completely empty. This is normal, since most graphics have a relatively large margin of white space. It is often safe to start printing on page two of a large poster. You can always go back and print the first row if it turns out to be needed.

If you are printing an extremely complex graphic, one which takes a very long time to print each page, it may be worth your while to print two copies of each page (using the Copies: field in the Print dialog box). The two copies can be printed in little more time than a single copy, and if you mess up one copy during assembly, you can use the second as a backup. Since you can print only whole rows at a time, it may take a very long time to print out a replacement page.

■ Trimming the Pages

Once your poster has printed out, collect your stack of pages and take them to a clear, well-lit work area. You will need the following equipment:

1) A piece of thick cardboard or other cutting surface. It will be sliced to bits.
2) A STEEL ruler (no plastic or aluminum, as they would damaged).
3) A good X-Acto razor blade knife and one #19 straight blade for each 30-50 pages in your poster. Don't even think about using scissors.

You want to cut away the top and right edge of each page, leaving the left and bottom for gluing. Do each page individually, laying the steel ruler carefully on the inside of the registration marks, and using the very tip of the X-Acto knife to cut the paper. With a bit of practice you can cut neatly through the paper, leaving a very clean, accurate edge. It is important to use a new blade about every 30-50 pages. Paper is very abrasive stuff, and you will be amazed at how much better the new blade works. Please be careful not to cut the tips of your fingers off. This is what happens if they extend slightly past the edge of the ruler you are holding.

It is of the utmost importance that each page be cut accurately. Any small error will cause a misalignment in the poster, and small misalignments add up to big misalignments as the poster grows.

■ Gluing the Poster Together

Now that you have trimmed all the pages, you can start gluing them together. You will need the following equipment:

1) A work table at least as wide as your poster, with one or two feet of space in back.
2) A sheet of paper to cover the whole table.
3) A bottle of Elmer's white glue.
4) A stack of scrap paper, one sheet for each 4-5 sheets in your poster. You can use all the paper you wasted trying to get your poster to print out correctly.
5) A bottle of Liquid Paper ("Just For Copies" type works best).

You may find the following gluing instructions somewhat ridiculous, but this is what I find works best. Using a Glue-Stick may seem like an attractive alternative, but only for a day or two, until your poster starts falling apart. The advantage of Elmer's is that it can make an absolutely permanent bond. The paper tears before the Elmer's gives way.

Start by laying out the first row of your poster. You may find it helpful to draw a straight line all the way across the piece of paper covering your work table, about in the middle. You can use this line to line up the pages. Make sure everything really does fit together in a reasonable way. (If your first row is blank, consider not assembling it at all, and start with the second row instead.) Once the row is laid out, use the Liquid Paper to cover the page numbers (this keeps them from showing through, especially on sealed posters (see below for information on sealing your poster).

You will be assembling the poster from right to left, with rows going from top to bottom. That is the order in which the pages lay on top of each other. Put the top right page down on the right side of your table. It will not need any glue.

Take the second page from the right in the top row and place it face down on top of your stack of scrap paper. Turn it so the edge that will be glued to the right-most page is away from you, and completely on top of the scrap paper. Run a bead of Elmer's glue all the way across the edge. Take a finger (one without the tip cut off) and spread out the glue, making sure to get it all the way to the edge, all along the edge. The best way to do this is to smear it onto the scrap paper as well, as long as you are careful not to get it on the front surface of the poster paper. You want to use enough glue so the edge is completely soaked with glue, in a strip about 1/2 inch wide. The glue left on the paper should be only a thin film, not a thick coat. The paper will wrinkle slightly. You will now want to wipe your finger on another sheet of your scrap paper.

It is mildly important not to get glue on the front surface of the poster. Although the glue dries clear, it can make black areas look blotchy, and it interacts badly with spray sealers.

Now, turn the page over and CAREFULLY place it down on the right-most page on your table. Line up the edge of the glued sheet with the registration marks on the other sheet. You have a few seconds to move the seam about, but not many. Take a clean sheet of scrap paper (one without any glue on it), cover the seam with the paper, and rub hard with your fist over the whole seam. This helps make a strong bond. If you have the right amount of glue, none will be squeezed out of the bond.

I cannot overstate the importance of getting each and every page lined up perfectly. Any small misalignment will cause headaches in the next row down, when you have to purposely misalign pages around the problem area, just to get things to look reasonable. These sorts of things tend to get worse and worse as you work your way down the poster.

Do the rest of the pages in the first row the same way. I find I can use the same sheet of scrap paper for applying the glue to about 5 sheets of paper before it gets too covered with glue. You will want to keep an eye on the horizontal line drawn on your table, to make sure the first row is straight. If it's a choice between misaligning a registration mark and having an obviously un-straight row, go for a straight row, misaligning the top registration mark, not the bottom one.

You are now ready to start the second row. Start on the right side again. The pages in the second and lower rows each need glue on both the top and right sides (except, of course, the right-most page in each row.) Do everything the same as for the first row, except be sure to line up both the top and side of each page. You may find yourself having to trade inaccuracy in one direction for accuracy in another. Just try to make the best compromise, and hope the misalignment never gets big enough to notice in the end. It is usually better the have the inaccuracy at the top of the page, not the bottom, so that it does not cause more problems in the next row.

Keep sliding the poster back on the table as you add each row. Let it hang off the back edge of the table as it grows. Be patient: assembling a large poster can take days. Well, hours anyway.

■ Retouching the Poster

A bit of creative pen work can improve the final look of your poster. Assuming you have cut and assembled your poster well, the seams between pages should not be very noticeable. One problem is solid black or dark gray areas. Even if the page is cut and pasted perfectly, the edge of the paper itself will always be white, and can be noticeable. The best thing to do is to use a black technical pen to fill in the crack. Black is easy, but gray areas can be done as well. Just try to simulate the gray screen with small dots of ink.

You can use Liquid Paper to cover any registration marks or page numbers that are still visible, as well as any mistakes you make with the pen mentioned above. Liquid Paper interacts well with the spray sealer described below.

If you had serious alignment problems, you can do a lot with a pen and Liquid Paper. Most posters should be repairable with some effort.

■ Finishing the Poster

Once your poster is glued together, it is basically finished. You can now start the next one. If you intend to put your poster to heavy use, you may want to provide it with a strong backing, and a coat of spray sealer on the front.

I think the best sealer to use is Blair Spray-Clear, Matte Finish. This is available in art supply houses, and is rather expensive. It can take about $20 worth of the stuff to do a 6x6 poster. If you want to use a different brand, be sure to test it on a small page first, to make sure it doesn't have some terrible effect on your poster.

You want to give the poster a good soaking. Spray evenly, in slow wide strokes. Watch the paper, and spray at a rate that makes the paper turn clear as it is being sprayed, which indicates that the sealer has penetrated through the paper. Don't worry if it looks blotchy at first. As long as all the paper gets soaked enough, it will look good when it dries. This soaking will make the paper tough and almost plastic-like. Amazingly, it also improves the contrast of the LaserWriter output. As long as you have the paper horizontal when it is sprayed, and don't touch it, there is no danger of smearing with Blair Spray Clear.

It is very important to do all this spraying outdoors. Try to live in California, so you will have good weather for spraying. This stuff is really rather poisonous, and "good ventilation" is usually not good enough when you go through 5 cans in half an hour.

A good backing to use for your poster is ordinary plastic contact "paper", as is commonly used for lining shelves. Get the pure white kind, or the color or pattern will show through to your poster. The trick is to get the stuff onto the back of your poster without totally wrinkling it. This is a dangerous operation. One mistake and the whole thing is ruined.

I do the following (and I've never lost a poster yet): Get a good hard surface at floor level (I use a folding table with the legs folded). It should be at least as long as your poster is, but need not be as wide. Lay the poster face down with one edge on the surface. Cut a length of contact paper as long as the poster is wide. Get another person to hold one end of the contact paper, and remove all the backing. Don't let it touch anything! Grab hold of the other end of the contact paper, and guide it to the edge of the poster. The other person should stand on the other side of the poster, and hold the contact paper high. Let it extend a few inches past the edge of the poster (you can trim it later).

Now, hold your end close to the poster, and have the other person slowly lower the other end. Press your end down, and then start to slowly smooth down the contact paper, having your partner lower her end as you go. Crawl onto the poster, and keep smoothing the contact paper down. Don't ever let it wrinkle. If it does, don't worry, just keep going. You can cure a wrinkle later by very carefully slitting just the plastic with a sharp razor blade knife and smoothing it out.

Once the first piece of contact paper is on, go over the whole surface several times, pressing it down as hard as you can. This will make the contact paper hold.

Add additional pieces of contact paper as needed until the whole poster is covered. Overlap each piece a few inches.

■ Care and Feeding of Your Poster

Unsealed posters should be protected from everything. LaserWriter output tends to smear easily.

Sealed and backed posters are quite tough. They are basically waterproof, though I would worry about outdoor use. One thing to watch out for is that the Spray Clear can take a long time (a week) to dry out totally, even though it is dry to the touch in a few hours. If the poster is rolled tightly, as for shipping, the Spray Clear will stick to the plastic on the back of the poster, causing the poster to rip when it is unrolled. This problem seems to go away once the poster is a few weeks old. If you need to ship a fresh poster, I would suggest rolling in a layer of white paper. The Spray Clear should not stick to the paper.

Visiting Professors and Authors Biographies

Who are these people anyway?

■ Jerry Glynn

Jerry Glynn is a recognized mathematics teacher, and author of the recent book *Exploring Math from Algebra to Calculus with Derive.* He is co-director, with Don Cohen, of The Math Program, a 15 year-old organization for teaching math to students from elementary age through adults. In 1989 Glynn and his wife Joyce started MathWare, a company dedicated to distributing innovative and powerful software and books for the teaching of math.

Glynn received a degree in Engineering from Steven's Institute of Technology in 1960. After teaching high school mathematics on Long Island for 4 years and taking graduate courses in the N.Y. area, Glynn joined The Madison Project, an ambitious project for improving math teaching based at Webster College, St. Louis, in 1965. In 1969 he became founding director of The New City School in St. Louis. From 1972 to 1976 Glynn developed mathematics courseware for the PLATO project at the University of Illinois.

Glynn also gives courses at the University of Illinois for future teachers of elementary and secondary school mathematics. Over the last few years Glynn has been introducing computers into his teaching, with special emphasis on *Mathematica* and Derive, and giving workshops for high school and college teachers. Glynn is the father of four children, David, Brian, Clair, and Erin.

■ Theodore W. Gray

Theodore W. Gray is Director of User Interfaces at Wolfram Research, and the primary author and designer of the *Mathematica* Notebook system.

Gray was a student at Glynn's Math Program for several years as a child. He received a degree in Chemistry from the University of Illinois in 1985, after studying in

Göttingen, Germany. Gray has worked in industrial computational chemistry at the DuPont company, and in theoretical chemistry at the University of California, Berkeley. In 1987 he co-founded Wolfram Research, Inc. together with Stephen Wolfram and others.

■ Daniel R. Grayson

Daniel R. Grayson is one of the original co-authors of *Mathematica*. He helped design the *Mathematica* programming language and designed an extension of the C programming language in which the code for the *Mathematica* kernel is written. He implemented many of the Mathematical parts of *Mathematica*, including arbitrary precision arithmetic, equation solving, matrix manipulations, power series, and computation of various transcendental functions. He also designed and implemented the protocol which is responsible for communication between the front-end and the kernel running on a remote machine.

Grayson is a professor of mathematics at the University of Illinois at Urbana-Champaign. He studied mathematics at the University of Chicago as an undergraduate, and received his PhD from MIT in 1976. He worked at Columbia University and the Institute for Advanced Study, and then moved to Illinois in 1982. His main research interest is in algebraic K-theory and its connections with number theory and geometry.

■ Jerry B. Keiper

Jerry B. Keiper is one of the original co-authors of *Mathematica*, contributing mainly numerical evaluation of special functions and other high-precision numerical evaluation. Keiper was born in Medina, Ohio in 1953 and was raised on a small dairy farm there. He was educated at the Ohio State University, the University of Toledo (Ohio), and received his PhD in Numerical Analysis at the University of Illinois in 1989. His primary Mathematical interests are the evaluation of special functions, especially to very high precision, and the Riemann Hypothesis. His first teaching experience was as a teaching assistant at the Ohio State University in 1974 and he has been involved in education much of the time since then, including a short time in high school and three years at the Federal University of Technology, Bauchi in Nigeria. He has spent a year as an apprentice pipe organ builder and enjoys bicycling and clock repair. He currently resides in Newton, Kansas with his wife and continues to work on *Mathematica*.

Index